Springer Proceedings in Mathematics & Statistics

Volume 198

Springer Proceedings in Mathematics & Statistics

This book series features volumes composed of selected contributions from workshops and conferences in all areas of current research in mathematics and statistics, including operation research and optimization. In addition to an overall evaluation of the interest, scientific quality, and timeliness of each proposal at the hands of the publisher, individual contributions are all refereed to the high quality standards of leading journals in the field. Thus, this series provides the research community with well-edited, authoritative reports on developments in the most exciting areas of mathematical and statistical research today.

More information about this series at http://www.springer.com/series/10533

Ilias S. Kotsireas · Edgar Martínez-Moro
Editors

Applications of Computer Algebra

Kalamata, Greece, July 20–23 2015

Springer

Editors
Ilias S. Kotsireas
Department of Physics and Computer Science
Wilfrid Laurier University
Waterloo, ON
Canada

Edgar Martínez-Moro
Institute of Mathematics
University of Valladolid
Valladolid
Spain

ISSN 2194-1009 ISSN 2194-1017 (electronic)
Springer Proceedings in Mathematics & Statistics
ISBN 978-3-319-56930-7 ISBN 978-3-319-56932-1 (eBook)
DOI 10.1007/978-3-319-56932-1

Library of Congress Control Number: 2017938138

Mathematics Subject Classification (2010): 13P25, 11T71, 65Y04, 68U05, 97N80, 97U50

Printed on acid-free paper

This Springer imprint is published by Springer Nature
The registered company is Springer International Publishing AG
The registered company address is: Gewerbestrasse 11, 6330 Cham, Switzerland

Preface

This Springer Proceedings in Mathematics and Statistics (PROMS) volume is based on a fully refereed selection of full papers submitted after the end of the very successful Applications of Computer Algebra (ACA) conference that took place on July 20–23, 2015 in Kalamata, Greece. This ACA meeting continued a long tradition (the first one started in 1995 and it has continued in yearly based series) and was organized as a series of Special Sessions. There were 16 Special Sessions plus a Poster Session organized at ACA 2015. The ACA Working Group is responsible for approving the Special Sessions via proposals submitted from potential organizers. These 16 sessions covered a wide range of topics within the conference scope: (a) Computer algebra in quantum computing and quantum information theory (b) Human–Computer Algebra Interaction (c) Computer Algebra in Education (d) Computer Algebra in Coding Theory and Cryptography (e) Computational differential and difference algebra (f) Algebraic and Algorithmic Differential and Integral Operator (g) Symbolic summation and integration: algorithms, complexity, and applications (h) Algebraic Graph Theory and its Applications (i) Applied and Computational Algebraic Topology (j) Non-standard Applications of Computer Algebra (k) Polynomial System Solving, Gröbner Basis, and Applications (l) Computational aspects and mathematical methods for finite fields and their applications in information theory (m) Polytopes in Algebra and Computation (n) Gröbner Bases, Resultants and Linear Algebra (o) Computer Algebra Methods for Matrices over Rings and (p) Open Source Software and Computer Algebra.

The papers in this PROMS volume cover all the sessions and they showcase the kind of quality papers presented at the meeting.

The ACA Working Group and the ACA 2015 session organizers did a tremendous work for selecting and scheduling the 162 contributions presented at ACA 2015. Our particular thanks are due to the members of the Local Organizing Committee for handling the local arrangements. The conference's Advisory Committee, Stanly Steinberg, Michael Wester, and Eugenio Roanes-Lozano, and the Scientific Committee (the ACA working group) also deserve special thanks.

The process of deciding the accepted papers to accept was not easy due to the high quality of submissions, thus we are especially grateful to the expert referees. Finally, we would like to express our most sincere thanks to the PROMS staff at Springer for their tireless efforts and continuous support in helping us publish this volume.

January 2017

Ilias S. Kotsireas
General Chair
Waterloo, ON, Canada

Edgar Martínez-Moro
Program Committee Chair
Valladolid, Spain

Contents

Contributors

Anissa Ali Laboratoire QUARTZ EA 7393, Saint-Ouen, France

Vesna Berec University of Belgrade, Belgrade, Serbia; Institute of Nuclear Sciences Vinca, Belgrade, Serbia

Mijail Borges-Quintana Department of Mathematics, Faculty of Natural and Exact Sciences, University of Oriente, Santiago de Cuba, Cuba

Miguel A. Borges-Trenard Department of Mathematics, Faculty of Natural and Exact Sciences, University of Oriente, Santiago de Cuba, Cuba

Martin Bossert Institute of Communications Engineering, Ulm University, Ulm, Germany

M.E. Canut Díaz Velarde FES-Acatlán, UNAM, Naucalpan, Estado de México, Mexico

Thierry Dana-Picard Jerusalem College of Technology, Jerusalem, Israel

Petroula Dospra Department of Mathematics, Aristotle University of Thessaloniki, Thessaloniki, Greece

G.H.E. Duchamp LIPN - UMR 7030, CNRS, Villetaneuse, France

Juan García Escudero Facultad de Ciencias, Universidad de Oviedo, Oviedo, Spain

Marc Ethier Division of Computational Mathematics, Faculty of Mathematics and Computer Science, Jagiellonian University, Kraków, Poland; University of Saint-Boniface, Winnipeg, MB, Canada

Tetsuo Fukui Mukogawa Women's University, Nishinomiya, Japan

Bernhard Garn SBA Research, Vienna, Austria

Smaranda Laura Goţia Department of Physiology, Victor Babes University of Medicine and Pharmacy, Timisoara, Romania

Anna Grim University of St. Thomas, St. Paul, MN, USA

Roman Hašek University of South Bohemia, České Budějovice, Czech Republic

Martin Helmer Department of Mathematics, University of California Berkeley, Berkeley, CA, USA

V. Hoang Ngoc Minh Université Lille II, Lille, France

Karim Ishak Institute of Communications Engineering, Ulm University, Ulm, Germany

Grzegorz Jabłoński Division of Computational Mathematics, Faculty of Mathematics and Computer Science, Jagiellonian University, Kraków, Poland; Institute of Science and Technology Austria, Klosterneuburg, Austria

J.J. Jiménez Zamudio FES-Acatlán, UNAM, Naucalpan, Estado de México, Mexico

Deepak Kapur Department of Computer Science, University of New Mexico, Albuquerque, NM, USA

O.G. Karamitrou University of Patras, Patras, Greece

Zoltán Kovács The Private University College of Education of the Diocese of Linz, Linz, Austria

Jan Krupa Department of Applied Mathematics, Warsaw University of Life Sciences (SGGW), Warsaw, Poland

E.D. Kuznetsov Ural Federal University, Yekaterinburg, Russia

Grégoire Lecerf Laboratoire d'informatique, UMR 7161 CNRS, Campus de l'École polytechnique, Palaiseau, France

Robert H. Lewis Fordham University, New York, NY, USA

J. López-García FES-Acatlán, UNAM, Naucalpan, Estado de México, Mexico

Edgar Martínez-Moro Institute of Mathematics IMUVa, University of Valladolid, Valladolid, Castilla, Spain

Ryutaroh Matsumoto Department of Communications and Computer Engineering, Tokyo Institute of Technology, Tokyo, Japan

P. Mavridi University of Patras, Patras, Greece

Manfred Minimair Department of Mathematics and Computer Science, Seton Hall University, South Orange, NJ, USA

J.A. Miszczak Institute of Theoretical and Applied Informatics, Polish Academy of Sciences, Gliwice, Poland; Applied Logic, Philosophy and History of Science Group, University of Cagliari, Cagliari, Italy

Mireille Moinet Laboratoire QUARTZ EA 7393, Saint-Ouen, France

Marian Mrozek Division of Computational Mathematics, Faculty of Mathematics and Computer Science, Jagiellonian University, Kraków, Poland

Sven Müelich Institute of Communications Engineering, Ulm University, Ulm, Germany

Ngo Quoc Hoan Université Paris XIII, Villetaneuse, France

Darian Onchis-Moaca Faculty of Mathematics, University of Vienna, Vienna, Austria; Faculty of Mathematics and Computer Science, West University of Timisoara, Timişoara, Romania

Vadim Olshevsky University of Connecticut, Storrs, USA

Victor Y. Pan Departments of Mathematics and Computer Science, Lehman College and the Graduate Center of the City University of New York, Bronx, NY, USA; Ph.D. Programs in Mathematics and Computer Science, The Graduate Center of the City University of New York, New York, NY, USA

Sirani M. Perera Embry-Riddle Aeronautical University, Daytona Beach, USA

K. Penson Université Paris VI - LPTMC, Paris Cedex 05, France

A.S. Perminov Ural Federal University, Yekaterinburg, Russia

Dimitrios Poulakis Department of Mathematics, Aristotle University of Thessaloniki, Thessaloniki, Greece

Sven Puchinger Institute of Communications Engineering, Ulm University, Ulm, Germany

Denis Raux Laboratoire d'informatique, UMR 7161 CNRS, Campus de l'École polytechnique, Palaiseau, France

Pedro Real Department of Applied Mathematics I, University of Seville, Seville, Spain

Diego Ruano Department of Mathematical Sciences, Aalborg University, Aalborg, Denmark

Philippe Serre Laboratoire QUARTZ EA 7393, Saint-Ouen, France

K.N. Sgarbas University of Patras, Patras, Greece

Chehrzad Shakiban University of St. Thomas, St. Paul, MN, USA

P. Simonnet Université de Corse, Corte, France

Dimitris E. Simos SBA Research, Vienna, Austria

Adam Strzeboński Wolfram Research Inc., Champaign, IL, USA

Yao Sun SKLOIS, Institute of Information Engineering, CAS, Beijing, China

Elias P. Tsigaridas POLSYS Project, INRIA Paris-Rocquencourt UPMC, Univ Paris 06, LIP6, Paris, France

C. Tsimpouris University of Patras, Patras, Greece

Joris van der Hoeven LIX, CNRS, École polytechnique, Palaiseau Cedex, France; Laboratoire d'informatique, UMR 7161 CNRS, Campus de l'École polytechnique, Palaiseau, France

Elena Varbanova Faculty of Applied Mathematics and Informatics, Technical University of Sofia, Sofia, Bulgaria

Dingkang Wang KLMM, Academy of Mathematics and Systems Science, CAS, Beijing, China

Włodzimierz Wojas Department of Applied Mathematics, Warsaw University of Life Sciences (SGGW), Warsaw, Poland

Jan Zahradník University of South Bohemia, České Budějovice, Czech Republic

Simone Zappalá Faculty of Mathematics, University of Vienna, Vienna, Austria

Nurit Zehavi Weizmann Institute of Science, Rehovot, Israel

David G. Zeitoun Orot College of Education, Rehovot, Israel

Jie Zhou KLMM, Academy of Mathematics and Systems Science, CAS, Beijing, China

An Algebraic Method to Compute the Mobility of Closed-Loop Overconstrained Mechanisms

Anissa Ali, Mireille Moinet and Philippe Serre

Abstract In mechanical engineering, the degree of freedom or mobility is a fundamental property of solid assemblies. To compute it, classical formulas fail when closed-loop overconstrained mechanisms are concerned. Another way to define the mobility is to consider the dimension of the algebraic variety representing the closure of the loop. The approach described here, consists in computing the conditions that ensure that an overconstrained mechanism is mobile.

Keywords Computer-aided design · Mobility · Groebner basis

1 Introduction

The mobility or degree of freedom (DOF) is a fundamental property in mechanical engineering. In mechanics, the degree of freedom of a mechanism is the number of independent parameters that define its spatial configuration. There exists a lot of formulas for the computation of the DOF. However, these formulas only permit to know if a mechanism is mobile or not and are not accurate for some mechanisms. As a result, during the process of design or redesign of a mechanism, the modification of parameters will lead to the lost of the mobility. To our knowledge, no computer-aided design softwares are able to help the designer in this specific situation.

To overcome these problems, our approach consists in finding relationships between parameters that will ensure that the mechanism is mobile. These relationships will be called "mobility condition" and will be obtained using the theory of Groebner basis. In the following, we will describe briefly this approach.

A. Ali (✉) · M. Moinet · P. Serre
Laboratoire QUARTZ EA 7393, 3 rue Fernand Hainaut, 93407 Saint-Ouen, France
e-mail: anissa.ali@supmeca.fr

M. Moinet
e-mail: mireille.moinet@supmeca.fr

P. Serre
e-mail: philippe.serre@supmeca.fr

I.S. Kotsireas and E. Martínez-Moro (eds.), *Applications of Computer Algebra*,
Springer Proceedings in Mathematics & Statistics 198,
DOI 10.1007/978-3-319-56932-1_1

First of all, Sect. 2 gives a definition and an algebraic representation of a closed-loop mechanism. Section 3, shows how the theory of Groebner basis is helpful to compute the mobility condition. Then, Sect. 4 describes a case study presented in [5]. To conclude, some challenges that our approach raises are outlined.

2 Algebraic Representation of a Closed-Loop Overconstrained Mechanism

A closed-loop mechanism is a set of rigid bodies connected by mechanical joints that forms a loop. Here, joints are supposed to be ideal. The most familiar joints are the revolute joint (or hinged joint) and the prismatic joint (or sliding joint). The other joints are modelled as combinations of revolute and prismatic joints. A classic way to represent a mechanism is to use a graph.

Figure 1 shows a closed-loop mechanism with n rigid bodies. The vertices B_i represent the rigid body i and the edges represent the joint i.

To model the closure of the mechanism, a set of polynomial equations with rational coefficients is generated. First, two types of parameters are defined: dimensional parameters (D) and positional ones (P). Dimensional parameters give a geometric description of the rigid bodies. Positional parameters represent the relative position between two rigid bodies.

Frames are then constructed in each side of the rigid bodies. In Fig. 2, each R_i denotes a frame.

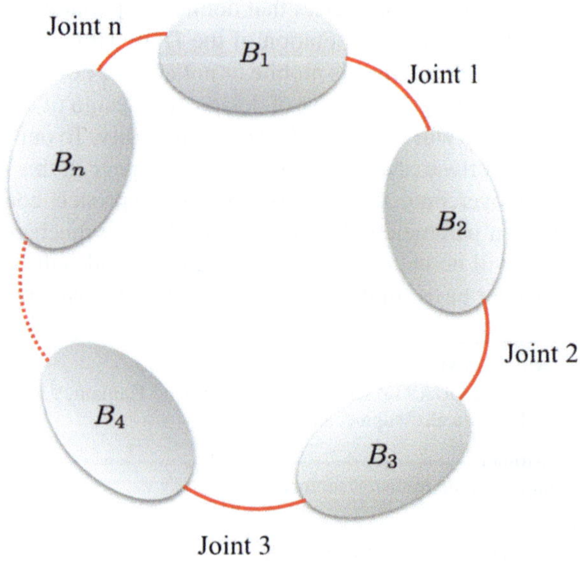

Fig. 1 A closed-loop mechanism with n rigid bodies

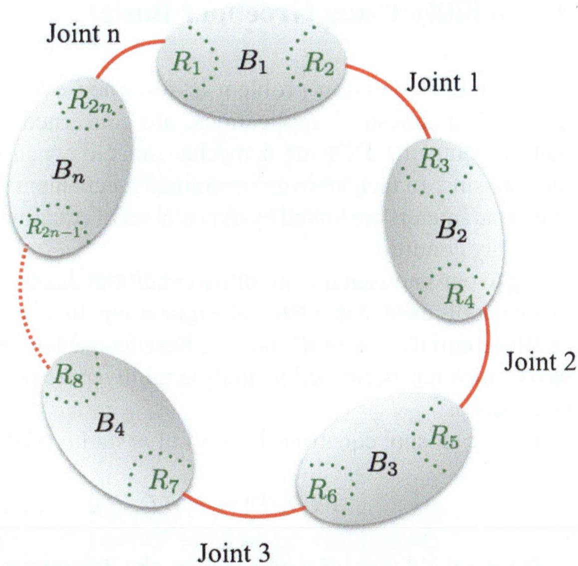

Fig. 2 Frames construction in a closed-loop mechanism with n rigid bodies

Then two types of displacements are defined : dimensional displacements (D_{disp}) and positional ones (P_{disp}). Dimensional displacements represent the displacements in the rigid body and are function of D. Positional displacements are the displacements in the joints and are function of P. The displacements are expressed using homogeneous matrices or dual quaternions.

Assuming that n is the number of rigid bodies in the loop, the closure equations are given by

$$\prod_{i=1}^{n} D_{\text{disp}_i} P_{\text{disp}_i} = I \tag{1}$$

where I denotes the identity displacement.

A polynomial transformation is then done (and a rational conversion of the coefficient if necessary). The polynomial closure equations will be noted

$$F(D, P) = 0 \tag{2}$$

When dealing with overconstrained mechanisms, $F(D, P)$ represents an overconstrained algebraic system with respect to D and an underconstrained one with respect to P.

3 Mobility Using Groebner Basis

The mobility condition problem arises when overconstrained mechanisms, often designed in industrial applications, are concerned. Indeed, well-known formulas that compute the DOF of a mechanism fail when dealing with overconstrained mechanisms. In fact, an overconstrained mechanism is mobile only when its dimensional parameters are linked by a special set of equations. We will call these equations: "mobility condition".

Up to our knowledge, "mobility condition" is out of reach of the classical mathematical tools used in mechanical engineering, that is to say, rigid body motion theory and differential calculus. Hence, we have focused our mind on the theory of Groebner basis which has been used to analyse multivariate polynomial mechanical problems (see [3, 4]).

Let the closure equations be a set of m multivariable polynomials.

$$F(D, P) = (f_1(D, P), \dots, f_m(D, P)) \tag{3}$$

During the design of a mechanism, the dimensional parameters are given by the designer and the positional parameters are then computed so that the mechanism can be displayed. As a result, the unknowns of the closure equations are the positional parameters $P = (p_1, \dots, p_r)$. The coefficients are polynomials which unknowns are the dimensional parameters $D = (d_1, \dots, d_s)$. Hence, each component of the closure equations is written as

$$f_i(D, p_1, ..., p_r) = \sum_{(j_1,...,j_n)} c_{i,(j_1,...,j_n)}(D) p_1^{j_1} ... p_r^{j_n} \tag{4}$$

Looking for the mobility condition is tantamount to solving the following problem:

$$(Pb) \begin{cases} \textit{Let } F(D, P) \textit{ be a set of polynomial equations.} \\ \textit{Find the relationships such that the system has infinitely many solutions.} \end{cases}$$

In an algebraic point of view, a definition of the mobility of a mechanism is then

Definition 1 Let I be the ideal generated by $< f_1(D, P), .., f_m(D, P) >$.
A mechanism is mobile when the algebraic variety $V_{\mathbb{Q}[d_1,...,d_s]}(I)$ is infinite.

In the theory of Groebner basis, there exists a criterion (see [1, 2]) that determines whether a system of polynomial equations has only finitely many solutions. This criterion will help us to solve the mobility condition problem.

Let f be a polynomial and $<$ be an admissible ordering. $LM(f)$ and $LC(f)$ will represent the leading monomial and leading coefficient of f with respect to (wrt) $<$. Then the leading term of f will be noted $\mathrm{LT}(f) = \mathrm{LM}(f) \times \mathrm{LC}(f)$.

Theorem 1 *Let I be an ideal generated by $< f_1, ..f_m >$ in $\mathbb{K}[x_1, \dots, x_n]$, G a Groebner basis of I wrt $<$. The algebraic variety $V(I)$ is finite if and only if*

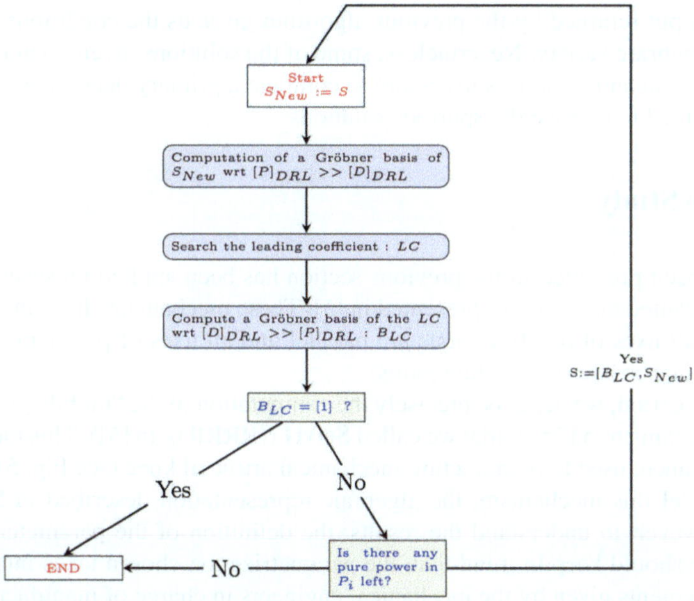

Fig. 3 Mobility computation algorithm

$$\forall i \in \{1, \ldots, n\},\ \exists\, g_i \in G \textit{ such that } LT(g_i) = x_i^{k_i}$$

According to this theorem, one can easily understand that to make an algebraic variety infinite, it is sufficient to eliminate a pure power in an unknown. As a result, here is another definition of the mobility that derives directly from Theorem 1.

Definition 2 Let I be the ideal generated by $< f_1, ..f_m >$, G a Groebner basis of I wrt $<$. The mechanism modelled by I is mobile if and only if

$$\exists\, i \in \{1, \ldots, n\},\ \forall\, g_i \in G \textit{ such that } LC(g_i) = C(d_1, ...d_s) = 0$$

where $C(d_1, \ldots, d_s)$ is a polynomial which unknowns are the dimensional parameters.

In our analysis, a block ordering is used and the Degree Reverse Lexicographical (DRL) monomial ordering is applied in each block (Fig. 3 details the mobility computation algorithm). Not only does this ordering allow to make the analysis of polynomials that have polynomials as coefficients but it is also known to be the most efficient monomial ordering.

We assume that $[p_1, \ldots, p_r]_{\mathrm{DRL}} >> [d_1, \ldots, d_s]_{\mathrm{DRL}}$. The algorithm that computes the mobility condition is as follows:

Our approach is implemented in Maple 18 and Groebner bases computation are done using the FGb package (version 1.61) written by Jean-Charles FAUGERE.

The output returned by the previous algorithm gives us the conditions to have a infinite algebraic variety. Nevertheless, some of the solutions given do not have any mechanical meaning. So it is necessary to compute a primary decomposition of the ideal obtained to remove the spurious solutions.

4 Case Study

The approach presented in the previous section has been applied on several simple overconstrained mechanisms presented in [5]. These mechanisms have those particularities: all axes of revolute joints are parallel and all axes of prismatic joints are perpendicular to axes of revolute joints.

In this section, we describe precisely the computation of the "mobility condition" of the mechanism in Fig. 4, that we called **Selvi1** ($(RRRR)_E$ in [5]). This mechanism is, for instance, used to manufacture mechanical artificial knee (see Fig. 5).

To model this mechanism, the algebraic representation described in Sect. 1 is used. However, to understand the results, the definition of the parameters will be given. We should keep in mind that, the parametrization chosen takes into account the requirements given by the mechanical engineers in charge of manufacturing the mechanism. Indeed, one of the main requirement is to ensure that the rigid bodies do not overlap.

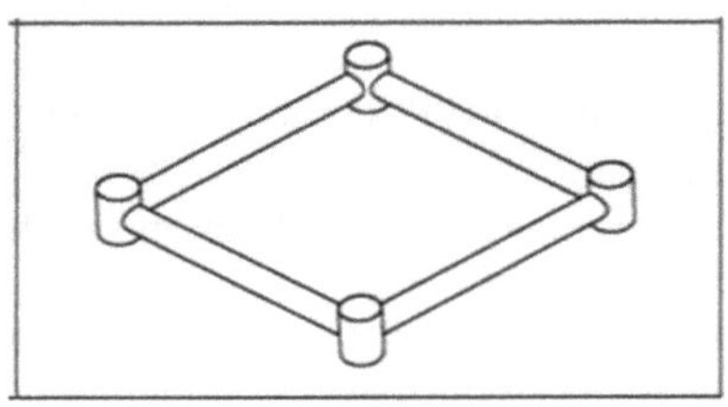

Fig. 4 Selvi1 (figure extracted from [5])

Fig. 5 Mechanical knee joint

Fig. 6 Parametrization of rigid body i

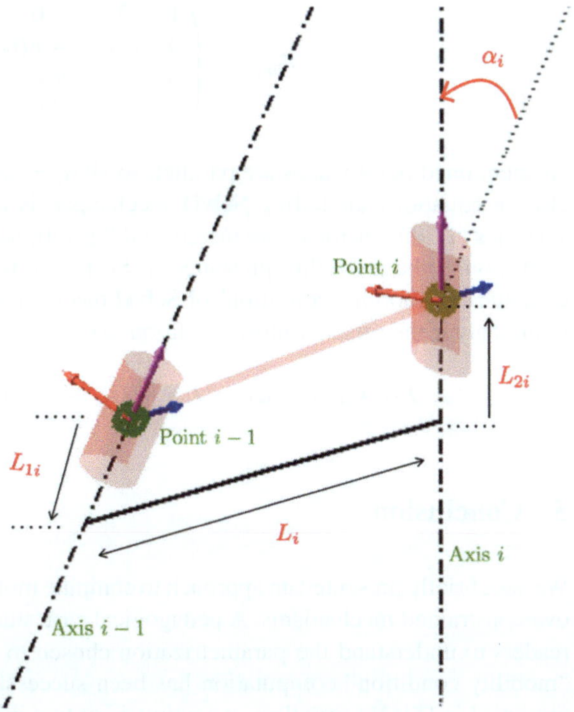

Selvi1 mechanism is made of four rigid bodies. Each of them are connected together by a revolute joint. As a result, dimensional parameters are the same for all rigid bodies. Figure 6 gives an illustration of the dimensional parameters chosen.

Assuming that a revolute joint is modelled by a line and a point and that the number of the rigid body is i, we have

1. α_i is the angle between Axis $i-1$ and Axis i.
2. L_i is the length of the common perpendicular between Axis $i-1$ and Axis i.
3. L_{1i} (resp. L_{2i}) is the signed distance between point $i-1$ (resp. i) and the foot of the common perpendicular that belongs to Axis $i-1$ (resp. Axis i). These distances will be called offsets.

Hence the displacement in a rigid body i is given by this following homogeneous matrix:

$$D_{\text{disp}_i} = \begin{pmatrix} 1 & 0 & 0 & 0 \\ L_i & 1 & 0 & 0 \\ -\sin(\alpha_i)L_{2i} & 0 & \cos(\alpha_i) & -\sin(\alpha_i) \\ L_{1i} + \cos(\alpha_i)L_{2i} & 0 & \sin(\alpha_i) & \cos(\alpha_i) \end{pmatrix}$$

Positional parameters correspond to the angle between the common perpendicular i and $i+1$. This angle is noted t_i. The displacement in a joint i is then given by this following homogeneous matrix:

$$P_{\mathrm{disp}_i} = \begin{pmatrix} 1 & 0 & 0 & 0 \\ 0 & \cos(t_i) & -\sin(t_i) & 0 \\ 0 & \sin(t_i) & \cos(t_i) & 0 \\ 0 & 0 & 0 & 1 \end{pmatrix}$$

As mentioned before, axes are parallel, so all α_i equal zero. Hence, the polynomial closure equations modelling **Selvi1** mechanism is a set of 10 equations with 20 variables (12 dimensional parameters and 8 positional parameters).

Applying the algorithm presented in Sect. 4, it takes approximatively 9.5 s[1] to compute the "mobility condition" of **Selvi1** mechanism. **Selvi1** mechanism is mobile if and only if the offsets follow condition (5).

$$L_{11} + L_{21} + L_{12} + L_{22} + L_{13} + L_{14} + L_{23} + L_{24} = 0 \quad (5)$$

5 Conclusion

We have briefly presented an approach to compute mobility condition for the study of overconstrained mechanisms. A pedagogical case study has been used to enable the readers to understand the parametrization chosen to describe the mechanism. The "mobility condition" computation has been successful for two other mechanisms presented in [5]. Nevertheless, we pointed out two major problems.

First, it is necessary to compute a primary decomposition of the result obtained to select the components with mechanical meaning. However the algorithm provided in Maple 18, which uses a former version of FGb, fails for some examples. For the future, it would be interesting to find a more efficient primary decomposition algorithm. Then, the computation of a Groebner basis is necessary to apply our method. Indeed, it may fail because of unreasonable computing time for industrial applications. To overcome these difficulties, a lot of promising methods are proposed for the future.

A first try will be to do a semi-numerical study. Indeed, in mechanical engineering it is usual to study a mechanism with some dimensional parameters that are fixed. By doing this, the closed-loop equations will be simpler and it will speed up the computation of the Groebner basis. However, we should keep in mind that the main difficulty is the choice of the dimensions that become numeric. Indeed, if the dimensional parameters chosen are linked, there is a high chance that the study will lead to no solution.

---

[1]CPU computational time given by Maple 18

References

1. COX, D.A., Little, J., Oshea, D.: Ideals, Varieties, and Algorithms: An Introduction to Computational Algebraic Geometry and Commutative Algebra. Springer (2007)
2. Faugère, J.C.: Polynomial systems, computer algebra and applications (Lecture 2-13-1). http://www-polsys.lip6.fr/~jcf/Teaching/index.html
3. Husty, M.L., Schröcker, H.P.: A proposal for a new definition of the degree of freedom of a mechanism. In: Conference on Interdisciplinary Applications of Kinematics (IAK) (2008)
4. Rameau, J.F., Serre, P.: Computing mobility condition using Groebner basis. Mech. Mach. Theory, **91**, 21–38 (2015). doi:10.1016/j.mechmachtheory.2015.04.003
5. Selvi, O.Structural and kinematic synthesis of overconstrained mechanisms. http://library.iyte.edu.tr/tezler/doktora/makinamuh/T000992.pdf (2012)

Simplicial Topological Coding and Homology of Spin Networks

Vesna Berec

Abstract We study the commutation of the stabilizer generators embedded in the q-representation of higher dimensional simplicial complex. The specific geometric structure and topological characteristics of 1-simplex connectivity are generalized to higher dimensional structure of spin networks encoded in ordered complex via combinatorial optimization of a closed compact space. Obtained results of a consistent homology-chain basis are used to define connectivity and dynamical self organization of spin network system via continuous sequences of simplicial maps.

Keywords Spin network · Simplicial complex · Graph state · Combinatorial optimization · Quantum code

1 Introduction

Spin networks [1–6] can be presented by purely combinatorial structures: one-dimensional simplicial complexes with edges labeled by numbers $j = 0, 1/2, 1, 3/2$, etc. These numbers stand for total angular momentum or "spin". The imposed condition is that three edges meet at each vertex, with the corresponding spins: j_1, j_2, j_3, adding up to an even integer and satisfying the triangle inequality. These rules are motivated by the quantum properties of angular momentum: if we combine a component with spin j_1 and a component with spin j_2, the spin j_3 of the unit system satisfies exactly latter constraint. In such setting, given that $\mathbb{F}$ is a general field, a spin network represents quantum states of $\mathbb{F}$-geometry on $d = 3 + 1$ dimensional space defined by tensor product states

$$H_{j_1 j_2 j_3 j_4} \equiv \bigotimes_{i=1}^{4} H_{ji}, \quad H_{\perp} \equiv \bigoplus_{\{J\}} H^0_{\{J\}}. \tag{1}$$

V. Berec (✉)
University of Belgrade, Studentski Trg 1, Belgrade, Serbia
e-mail: bervesn@gmail.com

V. Berec
Institute of Nuclear Sciences Vinca, P.O. Box 522, Belgrade, Serbia

I.S. Kotsireas and E. Martínez-Moro (eds.), *Applications of Computer Algebra*, Springer Proceedings in Mathematics & Statistics 198,
DOI 10.1007/978-3-319-56932-1_2

where $\{J\}$ runs over the set of ordered 4-tuples of integers or half-integers such that $H^0_{\{J\}}$ is nonempty complex obtained from the n-skeleton H^n, constructed from H^{n-1} by attaching n-simplexes via maps $\phi : K^{n-1} \to H^{n-1}$. In the PR model [1] a partition function is defined for a given three-dimensional simplicial complex [by deforming SU(2) to a quantum group [4], where the partition function depends only on the topology of the manifold which is triangulated by the simplicial complex] by means of the following: to each edge of the complex is associated a spin [i.e., an irreducible unitary SU(2) representation, determined only by its dimension $d \equiv 2j + 1$].

In particular case, we are interested in the homomorphisms of the simplicial q-th homology group which represents the free abelian group generated by the q-cycles, and their induced mapping on the stabilizer group (S_G) basis. Assuming that Γ and S_G are free abelian groups with bases $g_1, \ldots, g_n$ and $g'_1, \ldots, g'_m$, respectively, if $f : \Gamma \to S_G$ is a homomorphism, then $f(g_j) = \sum_{i=1}^{m} (-1)^i \lambda_{ij} g'_i$ for unique integers λ_{ij}, where the parity of any transposition is -1. More general, giving that K is a simplicial complex, and S_G is an abelian group, then for non-negative integer q, to each $(q+1)$-tuple $(x_0, x_1, \ldots, x_q)$ of vertices spanning a simplex $\sigma_q(K)$, there corresponds an element $\alpha(x_0, x_1, \ldots, x_q)$ of S_G defining a homomorphism $\alpha : C_q(K) \to S_G$, where $C_q(K)$ denotes the corresponding chain group, i.e., finitely generated abelian group on the oriented simplices.

This paper is organized as follows. After introducing basic concepts, in Sect. 3 we present a realization of the spin networks in terms of simplicial manifolds, associated with the properties of the fundamental groups. A distinctive feature of these groups is that they are topological invariant, i.e., topological spaces of the same homotopy description have the same fundamental group, and a loop differentiable property [7]. Details of the stabilizer formalism with the implementation to spin network unit on graph state are discussed in Sect. 4.

2 Preliminaries

2.1 Simplicial Complexes

Let $x_0, \ldots, x_q$ be points geometrically independent in $\mathbb{R}^m$ where $m \geqslant q$. The q-simplex $\sigma_q = \langle x_0, \ldots, x_q \rangle$ is a compact (bounded and closed) subset of $\mathbb{R}^m$, given by

$$\sigma_q = \left\{ v \in \mathbb{R}^m \,|\, v = \sum_{i=0}^{q} c_i x_i, c_i \geqslant 0, \sum_{i=0}^{q} c_i = 1 \right\}. \tag{2}$$

For an integer n such that $0 \leqslant n \leqslant q$, $n+1$ points define a n-simplex $\sigma_n = \langle x_{i_0}, \ldots, x_{i_n} \rangle$ denoted as n-face of σ_q. In particular, $\mathcal{K}$ represent a set of finite number of simplexes in $\mathbb{R}^m$ called simplicial complex [8, 9] if

- $\sigma \in \mathcal{K}$ and $\sigma' \leqslant \sigma$, then $\sigma' \in \mathcal{K}$.

- $\sigma, \sigma' \in \mathcal{K}$, then the intersection $\sigma \cap \sigma'$ is either empty set or a common face of σ and σ', i.e., either $\sigma \cap \sigma' = \emptyset$ or $\sigma \cap \sigma' \leqslant \sigma$, and $\sigma \cap \sigma' \leqslant \sigma'$.

Let $\sigma_q = \left[x_0, \ldots, x_q\right] (q > 0)$ denote an oriented q-simplex, then the boundary $\partial_q \sigma_q$ of σ_q is an $(q-1)$-chain where ∂_q, called boundary operator, defines a homomorphism map $\partial_q : C_q(K) \to C_{q-1}(K)$. For K representing the n-dimensional simplicial complex, there exists a sequence of free Abelian groups and homomorphisms, called chain complex [10, 11]:

$0 \xrightarrow{i} C_n(K) \xrightarrow{\partial_n} C_{n-1}(K) \xrightarrow{\partial_{n-1}} \cdots \xrightarrow{\partial_2} C_1(K) \xrightarrow{\partial_1} C_0(K) \xrightarrow{\partial_0} 0,$ where $i : 0 \to C_n(K)$.

Let $\mathcal{K}$ be a finite simplicial complex, where $|\mathcal{K}|$ represents the union of all the simplices $\sigma \in \mathcal{K}$. A topological space X which is homemeomorphic to $|\mathcal{K}|$ represents a polyhedron where $\mathcal{K}$ is a triangulation of X. If X and $K_q \subseteq \mathcal{K}$ are simplicial complexes, a morphism $\phi : \mathcal{K} \to K_q$ is a function $\phi : \mathcal{K}(\sigma_0) \to K_q(\sigma_0)$, where σ_0 denotes 0-simplexes or vertices, such that if $\sigma_q \in \mathcal{K}$ is a q-simplex spanned by the affinely independent set $x_0, \ldots, x_q$ of $(q+1)$ points, then the elements of the set $\phi(x_0), \ldots, \phi(x_q)$ form an affinely independent set of points spanning a simplex $\phi_\sigma \in K_q$, where $\dim \phi_\sigma \leq \dim \sigma$. In particular, a morphism $\phi : \mathcal{K} \to K_q$ of simplicial complexes for distinct elements $x_i \to \phi(x_i)$ determines a unique map of the simplex σ to ϕ_σ, by generating a piecewise-affine map of spaces $|\phi| : |\mathcal{K}| \to \left|K_q\right|$, where $|\cdot|$ is a functor from the category K of simplicial complexes to the category TOP of topological spaces. Considering X as a topological space and assigning a base point [12] $*p \in X$, a loop established at p is a path $\alpha : [0, 1] \to X$ with $\alpha(0) = \alpha(1) = p$. Then a map: $P : [0, 1]^2 \to X$ with $P(t, 0) = \alpha(t)$, $P(t, 1) = \beta(t)$ and $P(0, \tau) = P(1, \tau) = p$, $\forall (t, \tau) \in [0, 1]$ determines two homotopic loops α, β which can be deformed one from other via other loops on the set of common paths, defining an equivalence relation. The homotopy class of α loop is denoted as $[\alpha]$. In particular, two loops α, β are denoted with the homotopy classes $[\alpha][\beta] = \alpha * \beta$ for the path

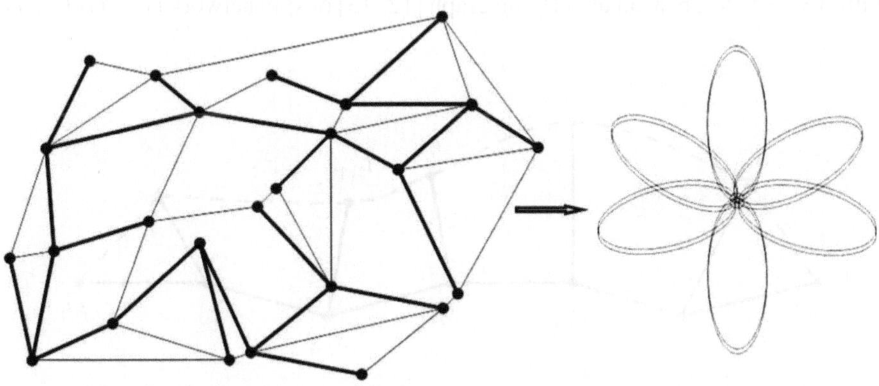

Fig. 1 Spin network represented via subgraph $X \in G$, is a maximal tree which is homotopy equivalent to a wedge of *circles*

that goes twice around α, then around β, such that $\alpha * \beta(t) = \alpha(2t), t \leqslant 1/2$ and $\beta(2t-1), t \leqslant 1/2$, see Fig. 1. right, which illustrates the wedge of six circles generated by gluing together a collection of spaces at a base point twice around loops α, β.

3 Homotopy of Spin Networks Embeded in Simplicial Complex

Let V be the vertex set and $e_1, e_2, \ldots, e_n$ be the sequence of edges on $V \times V$, connected along a path from a point a to a point b on the surface S, given by: $e_i = P_i P_{i+1}$, $P_1 = a$, $P_{n+1} = b$, where distinct edges possess orientation which coincides with the path direction. Then the path can be associated to the 1-chain: $e_1 + e_1 \cdots + e_n$. A linear transformation of the 1-chain module is associated by each group element action $g \in \Gamma$ which permutes the edges in either the successive mirror or the dual tiling, defining: $\alpha_1 e_1 + \alpha_2 e_2 + \cdots \alpha_n e_n \rightarrow \alpha_1 g e_1 + \alpha_2 g e_2 + \cdots \alpha_n g e_n$. In general, the Γ-action commutes with the boundary operator, i.e., $\partial g n = g \partial n$ for every chain, where

$$gZ_n(S;\mathbb{R}) = Z_n(S;\mathbb{R}) = \ker\partial_n : C_n(S;\mathbb{R}) \rightarrow C_{n-1}(S;\mathbb{R}),$$

$$gB_n(S;\mathbb{R}) = B_n(S;\mathbb{R}) = \mathrm{im}\partial_n : C_{n+1}(S;\mathbb{R}) \rightarrow C_n(S;\mathbb{R}),$$

resulting that distinct elements of Γ map homology classes to homology classes, yielding a linear action of Γ on $H_n(S;\mathbb{R}) = \frac{Z_n(S;\mathbb{R})}{B_n(S;\mathbb{R})}$. Then, a corresponding vertex set V represents a submodule for $V \subseteq H_n(S;\mathbb{R})$ which is Γ-invariant or a Γ-submodule if $gV = V, \forall g \in \Gamma$. Such action of group Γ on the homology chain is known as the homology representation.

Let Γ be a group, where $S \subseteq \Gamma$ is a generator subset. Let $\bar{S}$ be a set of inverses of S with $A = S \sqcup \bar{S}$. Then, an underlying graph [12, 13] of spin network $G = G(\Gamma, S)$ is

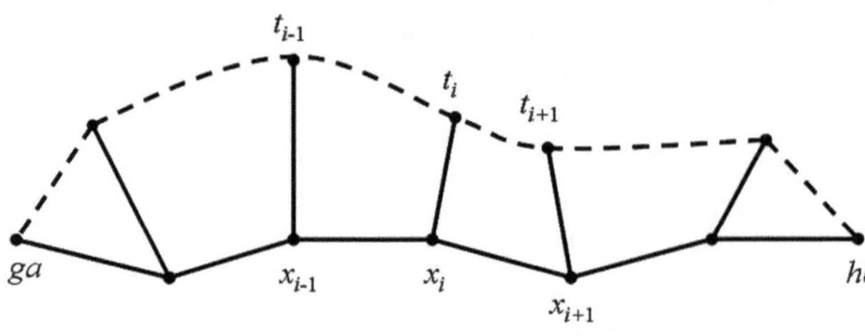

Fig. 2 Construction of a spin network by the union of the set of flat connections which can be defined over the multiply connected manifolds [14], given by unit intervals of a finite set of *curves* crosshatching only at their endpoints of the metric space [15]

established by connecting vertices $g, h \in V_G$, where the set $V_G \subset \Gamma$, by establishing edge in A under condition

$$(g, h \in V_G) = \begin{cases} 1, & if\ g^{-1}h \in A \\ 0, & \text{otherwise.} \end{cases} \tag{3}$$

That is, for distinct $g \in \Gamma$ and $a \in A$ there is an edge relating g to ga. In particular, the directed edge from g to ga is defined as the element a.

Given any $g, h \subseteq \Gamma$, let $\alpha \subseteq V_G$ be a geodesic connecting ga to a point hb, by selecting a sequence of points, $ga = x_0, x_1, \ldots, x_n = hb$, see Fig. 2, along α, such that $d(x_i, x_{i+1}) \leqslant 1, \forall i$. For each i, $g_i \in V_G$ are selected so that $\alpha : [a, b] \to [0, 1]$.

Definition 1 Given a metric space (M, d), let $I \subseteq R$ be an interval. A path which denotes unit interval (geodesic) is
$\gamma : I \to \{M \mid d(\gamma(t), \gamma(\tau)) = |t - \tau|, \forall (t, \tau) \in I\}$.

Assuming $\gamma : [a, b] \to M$ is an arbitrary path, its length is represented as

$$\sup \left\{ \sum_{i=1}^{n} d(\gamma(t_{i-1}), \gamma(t_i)) | \, a = t_0 < t_1 < \cdots < t_n = b \right\}. \tag{4}$$

Theorem 1 *Let X_σ and $X_{\sigma'}$ be subspaces of X such that the covering dimension of simplexes σ, σ' is maximal covering of X. Let $\gamma_i : X_{\sigma\sigma'} \to X_\gamma$ and $\gamma_j : X_\gamma \to X$ be the inclusions, resulting that $h_\gamma : \Pi(X_\gamma) \to \Lambda$ are functors into a groupoid defining commutativity relation $h_{\sigma'}\Pi(i_{\sigma'}) = h_\sigma \Pi(i_\sigma)$, i.e., a different path γ gives the same result, where a unique functor $\lambda : \Pi(X) \to \Lambda$ is defined such that $h_{\sigma'} = \lambda\Pi(j_{\sigma'})$, $h_\sigma = \lambda\Pi(j_\sigma)$ as*

$$\begin{array}{ccc} \Pi(X_{\sigma\sigma'}) & \xrightarrow{\Pi(i_\sigma)} & \Pi(X_\sigma) \\ \Pi(i_{\sigma'}) \downarrow & & \downarrow \Pi(j_\sigma) \\ \Pi(X_{\sigma'}) & \xrightarrow{\Pi(j_{\sigma'})} & \Pi(X) \end{array} \tag{5}$$

is a pushout in groupoid of the inclusions $X_\sigma \supset X_{\sigma\sigma'} \subset X_{\sigma'}$.

Proof In particular, a path $\gamma : [a, b] \to X$ represents a morphism $[\gamma]$ in $\Pi(X_\gamma)$ from $\gamma(a)$ to $\gamma(b)$ if we arrange it with an increasing homeomorphism $\alpha : [a, b] \to [0, 1]$. If $a = t_0 < t_1 < \cdots < t_n = b$ then γ establishes the composition of the morphisms $[\gamma |[t_i, t_{i+1}]]$. Let $\gamma : I \to X$ be a path and let $\sigma : \{0, \ldots, n\} \to \{0, 1\}$ $|\gamma([t_i, t_{i+1}]) \subset X(\sigma_i)$, then there exists a decomposition in affine space: $0 = t_0 < t_1 < \cdots < t_{n+1} = 1$ such that: $[\gamma |[t_i, t_{i+1}]] \subset X(\sigma_i)$, $i = 0, \ldots, n$. The construction $[\gamma |[t_i, t_{i+1}]]$ as path γ_i in X_{γ_i}, produces composition

$$[\gamma] = \Pi\left(\sigma_{\gamma(n)}\right)[\gamma_n] \circ \cdots \circ \Pi\left(\sigma_{\gamma(0)}\right)[\gamma_0]\,. \tag{6}$$

If subdivision λ exists, then $\lambda[\gamma] = h_{\gamma(n)}[\gamma_n] \circ \cdots \circ h_{\gamma(0)}[\gamma_0]$ is inclined by the homotopy composition of the path.
Let $h : \mathcal{K} \to X$ be a homotopy of paths from a to b. We consider edge-paths in the subsets of 3-simplex ($\mathcal{K}_3$) which path-connect coordinates $a = (000)$ and $b = (111)$, see Fig. 3. These paths differ from $h_{\gamma(0)}$ and $h_{\gamma(1)}$ by composition with a constant interval. h generates two paths in $\mathcal{K}$, which give the same result since they differ by a homotopy on subinterval which belongs to the subsets $\sigma_i \in \mathcal{K}_3$, $i = 1, \ldots, 4$. □

Given a topological space X representing the union of subsets $X_\sigma, X_{\sigma'}$, general properties of X encompassed from those of $X_\sigma, X_{\sigma'}$, and $X_{\sigma\sigma'} = X_\sigma \cap X_{\sigma'}$ can be inferred from the Theorem 2.

Theorem 2 [15, 16]. *Let K_0 and K_1 be subspaces of simplicial complex $\mathcal{K}$ such that the maximal dimension simplexes $\sigma_0 \in K_0$, $\sigma_1 \in K_1$, represent covering of X. Considering $\gamma_i : K_{01} = K_0 \cap K_1 \to K_\gamma$ and $\gamma_j : K_\gamma \to X$ as inclusions, in particular, let K_0, K_1, K_{01} be path connected with base $* \in K_{01}$. Then Eq. (7)*

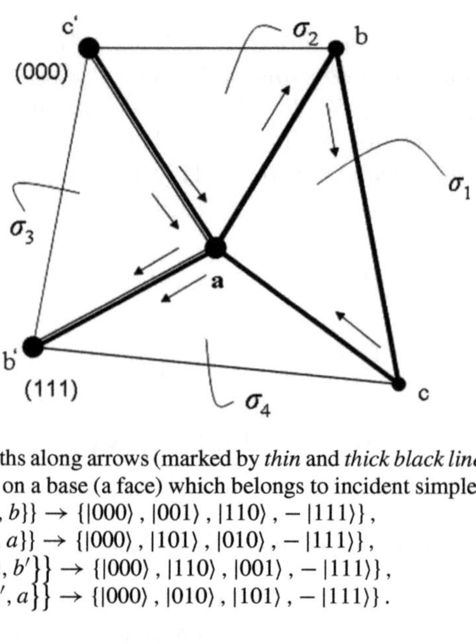

Fig. 3 Two different paths along arrows (marked by *thin* and *thick black lines*) induce the following stabilizer generator sets on a base (a face) which belongs to incident simplexes (see Theorem 2 and Sect. 4):

$$\begin{aligned}
\sigma_1 \cap \sigma_2 &= \{\{a, b\}\} \to \{|000\rangle, |001\rangle, |110\rangle, -|111\rangle\},\\
\sigma_4 \cap \sigma_1 &= \{\{c, a\}\} \to \{|000\rangle, |101\rangle, |010\rangle, -|111\rangle\},\\
\sigma_3 \cap \sigma_4 &= \left\{\{a, b'\}\right\} \to \{|000\rangle, |110\rangle, |001\rangle, -|111\rangle\},\\
\sigma_2 \cap \sigma_3 &= \left\{\{c', a\}\right\} \to \{|000\rangle, |010\rangle, |101\rangle, -|111\rangle\}.
\end{aligned}$$

$$\begin{array}{ccc} \pi_1(K_{01},*) & \xrightarrow{\pi(i_{1*})} & \pi_1(K_1,*) \\ \pi_1(i_{0*}) \downarrow & & \downarrow \pi_1(j_{1*}) \\ \pi_1(K_0,*) & \xrightarrow{\pi_1(j_{0*})} & \pi_1(X,*) \end{array} \tag{7}$$

is a pushout in topological space of the inclusions $K_0 \supset K_{01} \subset K_1$*, representing a fundamental group.*

Proof Assuming that simplicial complex $\mathcal{K}$ is path connected and $z \in \mathcal{K}$, where $z = *$, then $r : \Pi(\mathcal{K}) \to \pi_1(\mathcal{K}, z)$ induces morphism compositions over the full subset z. For each $z \in \mathcal{K}$ exists a morphism such that $u_z = \text{id}$, $u_y \alpha u_x^{-1}$ where α: $x \to y$, represented by:

$$\begin{array}{ccccc} \Pi(K_0) & \longleftarrow & \Pi(K_{01}) & \longrightarrow & \Pi(K_1) \\ \downarrow r_1 & & \downarrow r_{01} & & \downarrow r_0 \\ \pi_1(K_0,*) & \leftarrow & \pi_1(K_{01},*) & \rightarrow & \pi_1(K_1,*). \end{array} \tag{8}$$

Precisely, restriction of $\mathcal{K}$ to subcomplexes: K_{01}, K_0, K_1, and X with a base point $z = *$, yields a commutative relation where morphisms in $\Pi(X)$ are respectively assigned by the composition of morphisms in $\Pi(K_0)$ and $\Pi(K_1)$, likewise, the group $\pi_1(X, *)$ is formed by the images of j_0* and j_1*. □

4 Application to Graph State and Spin Network

Graph state is represented in scope of the stabilizer formalism [17, 18] via tensor products of Pauli operators σ_X and σ_Z, whose composition and structure are based on the complexity of the underlying graph which can be seen as one-dimensional simplicial complex. The stabilizers establish a group (S_G) under multiplication, formed from n generators g_i, associated to a number of vertices x_i of the graph [19]. In particular, stabilizer generators are induced on the vertex set V_G of a graph G by the bijective mapping $(\Gamma(V_G), A) \to (S_G, \cdot)$, see Sect. 3, Eq. (3). Graph state is obtained by relating each vertex $x_i \in V_G$ with a stabilizer generator $g_i = \sigma_X^i \sigma_Z^{ij}$, where $g_i |G\rangle = |G\rangle$, $\forall i = 1, \ldots, n$. The stabilizer generators [20–22] g_i for n graph state generate the complete Abelian stabilizer group S_G of $|G\rangle$ with multiplication. The group S_G consists of 2^n elements which uniquely represent a graph state

$$|G\rangle = \left\{ \sum_{i=1}^{2^n} \alpha_i |x_i\rangle = \sum_i \alpha_i S_G^i |x_i\rangle, \sum_i |\alpha_i|^2 = 1 \right\}. \tag{9}$$

The stabilizer group S_G is formed from a set of $n-k$ generators $g_1, \ldots, g_{n-k}$, which: (a) commute; (b) are unitary and Hermitian; and (c) $g_i^2 = I$. Each element of the stabilizer group S_G can be expressed as a product of the generators as $S_k = g_1^{\alpha_1} \cdots g_{n-k}^{\alpha_{n-k}}$, $S_k \in S_G$, $\alpha_i \in \{0, 1\}, i = 1, \ldots, n-k$, where $S \subseteq G_n$ with G_n denoting a corresponding Pauli group for n qubit state.

Definition 2 Stabilizer code of length n is represented by the fixed point set [23] $S_k = \{I, X, Y, Z\}$ of Pauli operators:

$$I = \begin{pmatrix} 0 & 1 \\ 1 & 0 \end{pmatrix}, \quad X \equiv \sigma_X = \begin{pmatrix} 0 & 1 \\ 1 & 0 \end{pmatrix}, \quad Y \equiv \sigma_Y = \begin{pmatrix} 0 & -i \\ i & 0 \end{pmatrix}, \quad Z \equiv \sigma_Z = \begin{pmatrix} 1 & 0 \\ 0 & -1 \end{pmatrix},$$

$k = 1 \ldots, n$ such that $S_1, S_2, \ldots, S_k$ are acting over n qubits (i.e., over $\left(\mathbb{C}^2\right)^{\otimes n}$).

When stabilizers S are composed of elements $\{\sigma_i\}_{i=X,Y,Z}$ of $\{I, X\}^{\otimes n}$ and $\{\sigma_i\}_{i=X,Y,Z}$ of $\{I, Z\}^{\otimes n}$, it can be seen that $\left[\sigma_i \sigma_j\right] = 2i\varepsilon_{ijk}\sigma_k$ and $\left\{\sigma_i \sigma_j\right\} = 2\delta_{ij}$. Precisely, I represents the identity matrix of size 2, X denotes the Pauli matrix encoding the bit flip error and Z denotes the Pauli matrix describing the phase error. The isomorphisms between $\{I, X\}, \{I, Z\}$ and the vector space $\mathbb{F}_2$ makes possible establishing a connection between classical and quantum codes. On the basis of these isomorphisms, the stabilizers relate to binary vectors and the commutation relation corresponds to the orthogonality relation in $\mathbb{F}_2^n$.

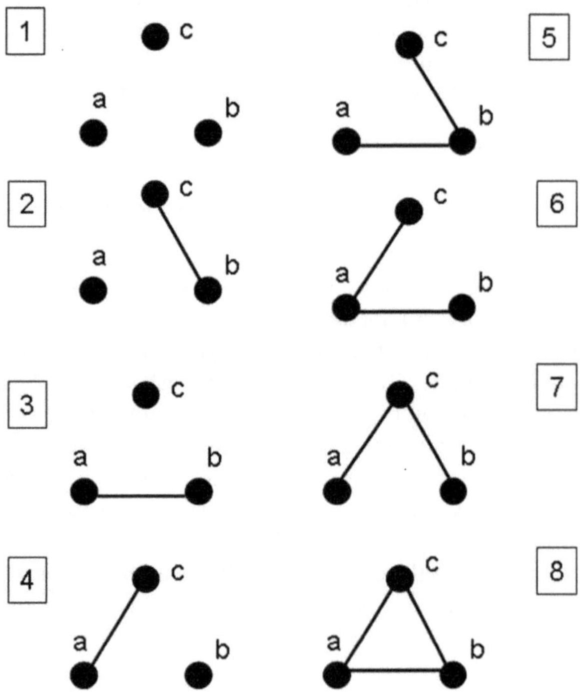

Fig. 4 Stabilizer generators for three-partite graph states representing elementary segment of spin network, see Eqs. (10, 11)

In particular, stabilizers of the graph state, given in Fig. 4, are represented by each row of the binary matrix [24, 25]

$$\begin{pmatrix} 0\,0\,0 & 0\,0\,0 \\ 1\,0\,0 & 0\,0\,1 \\ 0\,1\,0 & 0\,0\,1 \\ 0\,0\,1 & 1\,1\,0 \\ 1\,1\,0 & 0\,0\,0 \\ 1\,0\,1 & 1\,1\,1 \\ 0\,1\,1 & 1\,1\,1 \\ 1\,1\,1 & 1\,1\,0 \end{pmatrix} \tag{10}$$

where nodes of element of the spin network $\{a, b, c\}$, where: $a + b \geqslant c = 2k,\ a + c \geqslant b = 2k,\ b + c \geqslant a = 2k$ are encoded in the graph state ($n = 3$) establishing incidence relations via following generators:

$$\begin{aligned}
&(1)\,\{\{a\}, \{b\}, \{c\}\} \rightarrow \{|000\rangle\}, \\
&(2)\,\{\{a\}\} \rightarrow \{|000\rangle, |001\rangle, |010\rangle, -|011\rangle\}, \\
&(3)\,\{\{b\}\} \rightarrow \{|000\rangle, |100\rangle, |001\rangle, -|101\rangle\}, \\
&(4)\,\{\{c\}\} \rightarrow \{|000\rangle, |010\rangle, |100\rangle, -|110\rangle\}, \\
&(5)\,\{\{a, c\}\} \rightarrow \{|000\rangle, |010\rangle, |101\rangle, |111\rangle\}, \\
&(6)\,\{\{b, c\}\} \rightarrow \{|000\rangle, |100\rangle, |011\rangle, |111\rangle\}, \\
&(7)\,\{\{a, b\}\} \rightarrow \{|000\rangle, |001\rangle, |110\rangle, |111\rangle\}, \\
&(8)\,\{\{a, b, c\}\} \rightarrow \{|100\rangle, |010\rangle, |001\rangle, -|111\rangle\},
\end{aligned} \tag{11}$$

where (5–7) represent standard three-qubit flip code on the code subspace: $V_S = \{|000\rangle, |111\rangle\}$ for stabilizer set $S = \{I, Z_1Z_2, Z_2Z_3, Z_1Z_3\}, I = (Z_1Z_2)^2$.

5 Conclusion

We have analyzed and demonstrated implementation of graph states in composing the spin networks architectures. The characterization of graph states is utilized via the underlying graph construction defined in terms of affine simplexes with respect to path-connection induced homeomorphisms and polytope construction herein. Future outlook is implementation of higher dimensional homologies in order to establish a self-correcting memory which allows secure data processing without continual active error correction via stabilizer measurement.

Acknowledgements Author acknowledges the sponsorship provided through FP7 EU Commission framework.

References

1. Penrose, R.: Applications of negative dimensional tensors. In: Welsh, D. (ed.) Combinatorial Mathematics and its Applications, pp. 221–244. Academic Press, New York (1971)
2. Rovelli, C., Smolin, L.: Loop space representation of quantum general relativity. Nucl. Phys. B **331**, 80–152 (1990)
3. Rovelli, C., Smolin, L.: Discreteness of area and volume in quantum gravity. Nucl. Phys. B **442**, 593–619 (1995)
4. Seth, M.A.: A spin network primer. Am. J. Phys. **67**(11), 972 (1999)
5. Baez, J.C.: Spin networks in gauge theory. Adv. Math. **117**(2), 253 (1996)
6. Baez, J.C.: Diffeomorphism-invariant spin network states. J. Funct. Anal. **158**, 253–266 (1998)
7. Bartolo, C., Di Gambini, R., Griego, J., Pullin, J.: Consistent canonical quantization of general relativity in the space of Vassiliev invariants. Phys. Rev. Lett. **84**(11), 2314–2317 (2000)
8. Grünbaum, B.: Convex Polytopes, 2nd edn. Springer, New York (2003)
9. Ziegler, G.M.: Lectures on Polytopes. Springer, Berlin (1995)
10. Whitehead, G.W.: Elements of Homotopy Theory. Springer, New York (1978)
11. Gray, B.: Homotopy Theory. Pure and Appl. Math. 64, Academic Press, New York (1975)
12. Griffiths, H.B.: The fundamental group of two spaces with a common point. Quart. J. Math. **5**, 175–190 (1954)
13. Diestel, R.: Graph Theory, Graduate Texts in Mathematics, vol. 173. Springer, Heidelberg (2005)
14. Rosen, K.H.: Handbook of Discrete and Combinatorial Mathematics. CRC, Boca Raton (1999)
15. Seifert, H.: Konstruktion dreidimensionaler geschlossener Räume. Ber. Sächs. Akad. Wiss. **83**, 26–66 (1931)
16. van Kampen, E.H.: On the connection between the fundamental group of some related spaces. Am. J. Math. **55**, 261–267 (1933)
17. Gottesman, D.: Stabilizer codes and quantum error correction. Ph.D. thesis, California Institute of Technology (1997)
18. Nielsen, M.A., Chuang, I.L.: Quantum Computation and Quantum Information, Cambridge Series on Information and the Natural Sciences, 1st edn. Cambridge University Press, Cambridge (2004)
19. Hatcher, A.: Algebraic Topology. Cambridge University Press, Cambridge (2002)
20. Berec, V.: Phase space dynamics and control of the quantum particles associated to hypergraph states. EPJ Web Conf. **95**, 04007 (2015)
21. Berec, V.: Non-Abelian topological approach to non-locality of a hypergraph state. Entropy **17**(5), 3376–3399 (2015)
22. Goebel, K., Kirk, W.A.: Topics in metric fixed point theory. In: Cambridge Studies in Advanced Mathematics, vol. 28. Cambridge University Press, Cambridge, New York (1990)
23. Gottesman, D.: Class of quantum error-correcting codes saturating the quantum Hamming bound. Phys. Rev. A **54**, 1862 (1996)
24. Calderbank, A., Rains, E., Shor, P., Sloane, N.: Quantum error correction and orthogonal geometry. Phys. Rev. Lett. **78**, 405 (1997)
25. Pemberton-Ross, P.J., Kay, A.: Perfect quantum routing in regular spin networks. Phys. Rev. Lett. **106**(2), 020503 (2011)

Trial Set and Gröbner Bases for Binary Codes

Mijail Borges-Quintana, Miguel A. Borges-Trenard and Edgar Martínez-Moro

Abstract In this work, we show the connections between trial sets and Gröbner bases for binary codes, which give characterizations of trial sets in the context of Gröbner bases and algorithmic ways for computing them. In this sense, minimal trial sets will be characterized as trial sets associated with minimal Gröbner bases of the ideal associated to a code.

Keywords Binary linear codes · Test set · Groebner basis

1 Introduction

The concept of trial set for linear codes was introduced in [6]. This set of codewords can be used to derive and algorithm for complete decoding in a similar way that a gradient decoding algorithm uses a test set (see [2]). A trial set allows to characterize the so-called *correctable errors* and to investigate the monotone structure of correctable and uncorrectable errors. Also important bounds on the error–correction capability of binary codes beyond half of minimum distance using trial sets are presented in [6]. One problem posted in the conclusion of [6] was the importance of characterizing minimal trial sets for families of binary codes.

Supported by a Postdoctorate scholarship at the University of Valladolid (09-2014 to 02-2015) by Erasmus Mundus Program, Mundus Lindo Project.

M. Borges-Quintana · M.A. Borges-Trenard
Department of Mathematics, Faculty of Natural and Exact Sciences,
University of Oriente, Santiago de Cuba, Cuba
e-mail: mijail@uo.edu.cu

M.A. Borges-Trenard
e-mail: mborges@uo.edu.cu

E. Martínez-Moro (✉)
Institute of Mathematics IMUVa, University of Valladolid, Valladolid,
Castilla, Spain
e-mail: edgar.martinez@uva.es

I.S. Kotsireas and E. Martínez-Moro (eds.), *Applications of Computer Algebra*,
Springer Proceedings in Mathematics & Statistics 198,
DOI 10.1007/978-3-319-56932-1_3

The ideal associated with any linear code (code ideal for simplicity) was introduced in [3] together with some applications of Gröbner bases theory in this context, such as the reduction process by Gröbner bases of code ideals with respect to (w.r.t.) specific orders that corresponds to the decoding process of the code.

The outline of this contribution is as follows, in Sect. 2 we state the main concepts and results related with binary codes, trial sets, the code ideals, and Gröbner bases which are needed for an understanding of this work. The connection between trial sets for binary codes and Gröbner bases for the corresponding code ideal is presented in Sect. 3.

2 Preliminaries

2.1 Binary Codes

By $\mathbb{Z}$, $\mathbb{K}$, $\mathbf{X}$, $\mathbb{K}[\mathbf{X}]$, and $\mathbb{F}_2$, we denote the ring of integers, an arbitrary field, the set of n variables $\{x_1, \ldots, x_n\}$, the polynomial ring in the n variables of $\mathbf{X}$ over the field $\mathbb{K}$ and the finite field with 2 elements.

A *binary linear code* $\mathscr{C}$ over $\mathbb{F}_2$ of length n and dimension k, or an $[n, k]$ binary code for short, is a k-dimensional subspace of $\mathbb{F}_2^n$. We will call the vectors $\mathbf{v}$ in $\mathbb{F}_2^n$ words and in the particular case where $\mathbf{v} \in \mathscr{C}$, codewords. For every word $\mathbf{v} \in \mathbb{F}_2^n$ its *support* is defined as $\mathrm{supp}(\mathbf{v}) = \{i \mid v_i \neq 0\}$ and its *Hamming weight*, $\mathrm{w}_H(\mathbf{v})$ is the cardinality of $\mathrm{supp}(\mathbf{v})$.

The *Hamming distance*, between two vectors $\mathbf{x}, \mathbf{y} \in \mathbb{F}_2^n$ is $d_H(\mathbf{x}, \mathbf{y}) = \mathrm{w}_H(\mathbf{x} - \mathbf{y})$. The *minimum distance* $d(\mathscr{C})$ of a linear code $\mathscr{C}$ is defined as the minimum weight among all nonzero codewords. In addition, we have $\mathbf{x} \subset \mathbf{y}$ provided that $\mathrm{supp}(\mathbf{x}) \subset \mathrm{supp}(\mathbf{y})$.

For the rest of this section, we will follow [6]. We will consider $\prec$ a so-called α-ordering on $\mathbb{F}_2^n$, a weight compatible total ordering on $\mathbb{F}_2^n$ which is monotone, that is:

$$\left.\begin{array}{l}\text{for any } \mathbf{y_1},\ \mathbf{y_2}\ s.t.\ 2 \leq \mathrm{w}_H(\mathbf{y_1}) = \mathrm{w}_H(\mathbf{y_2}) < n \text{ and } \mathrm{supp}(\mathbf{y_1}) \cap \mathrm{supp}(\mathbf{y_2}) \neq \emptyset \\ \text{and for any } i \in \mathrm{supp}(\mathbf{y_1}) \cap \mathrm{supp}(\mathbf{y_2}) \text{ and vectors } \mathbf{x_1} \text{ and } \mathbf{x_2} \text{ defined by} \\ \mathrm{supp}(\mathbf{x_1}) = \mathrm{supp}(\mathbf{y_1}) \setminus \{i\} \text{ and } \mathrm{supp}(\mathbf{x_2}) = \mathrm{supp}(\mathbf{y_2}) \setminus \{i\} \text{ then } \mathbf{y_1} \prec \mathbf{y_2} \text{ if} \\ \mathbf{x}_1 \prec \mathbf{x}_2. \end{array}\right\} \quad (1)$$

$E^0(\mathscr{C})$ will denote the set of correctable errors of a binary code $\mathscr{C}$ which is the set of the minimal elements w.r.t. $\prec$ in each coset of $\mathbb{F}_2^n/\mathscr{C}$, and the elements of $E^1(\mathscr{C}) = \mathbb{F}_2^n \setminus E^0(\mathscr{C})$ are called uncorrectable errors. A *trial set* $T \subset \mathscr{C} \setminus \mathbf{0}$ of the code $\mathscr{C}$ is a set which has the property $\mathbf{y} \in E^0(\mathscr{C})$ if and only if $\mathbf{y} \preceq \mathbf{y} + \mathbf{c}$, for all $\mathbf{c} \in T$. A trial set provides an algorithmic way of computing a correctable error nearest to a given vector $\mathbf{y}$.

Since we choose a monotone α-ordering on $\mathbb{F}_2^n$, the set of correctable and uncorrectable errors form a monotone structure, namely, that if $\mathbf{x} \subset \mathbf{y}$, then $\mathbf{x} \in E^1(\mathscr{C})$ implies $\mathbf{y} \in E^1(\mathscr{C})$ and $\mathbf{y} \in E^0(\mathscr{C})$ implies $\mathbf{x} \in E^0(\mathscr{C})$.

By $M^1(\mathscr{C})$ we will denote the set of minimal uncorrectable errors, i.e., the set of $\mathbf{y} \in E^1(\mathscr{C})$ such that, if $\mathbf{x} \subseteq \mathbf{y}$ and $\mathbf{x} \in E^1(\mathscr{C})$, then $\mathbf{x} = \mathbf{y}$. In a similar way, the set of maximal correctable errors is the set $M^0(\mathscr{C})$ of elements $\mathbf{x} \in E^0(\mathscr{C})$ such that, if $\mathbf{x} \subseteq \mathbf{y}$ and $\mathbf{y} \in E^0(\mathscr{C})$, then $\mathbf{x} = \mathbf{y}$.

For $\mathbf{c} \in \mathscr{C} \setminus \{\mathbf{0}\}$, a *larger half* is defined as a minimal word $\mathbf{u}$ in the ordering $\preceq$ such that $\mathbf{u} + \mathbf{c} \prec \mathbf{u}$. The set of larger halves for a codeword $\mathbf{c}$ is denoted by $\mathrm{L}(\mathbf{c})$, and for $U \subseteq \mathscr{C} \setminus \{\mathbf{0}\}$ the set of larger halves for elements of U is denoted by $\mathrm{L}(U)$. Note that $\mathrm{L}(\mathscr{C}) \subseteq E^1(\mathscr{C})$.

For any $\mathbf{y} \in \mathbb{F}_2^n$, let the set $H(\mathbf{y}) = \{\mathbf{c} \in \mathscr{C} : \mathbf{y} + \mathbf{c} \prec \mathbf{y}\}$. Then $\mathbf{y} \in E^0(\mathscr{C})$ if and only if $H(\mathbf{y}) = \emptyset$, and that $\mathbf{y} \in E^1(\mathscr{C})$ if and only if $H(\mathbf{y}) \neq \emptyset$. Theorem 1 in [6] provides a characterization of the set $M^1(\mathscr{C})$ in terms of $H(\cdot)$ and larger halves of the set of minimal codewords $M(\mathscr{C})$.

Proposition 1 (Corollary 3, [6]) *Let $\mathscr{C}$ be a binary code and $T \subseteq \mathscr{C} \setminus \{\mathbf{0}\}$. The following statements are equivalent*

1. *T is a trial set for $\mathscr{C}$.*
2. *If $\mathbf{y} \in M^1(\mathscr{C})$, then $T \cap H(\mathbf{y}) \neq \emptyset$.*
3. *$M^1(\mathscr{C}) \subseteq \mathrm{L}(T)$.*

Gröbner Bases and Binary Codes

We define the following characteristic crossing function: $\Delta : \mathbb{F}_2 \to \mathbb{Z}$ which replace the class of 0, 1 by the same symbols regarded as integers. This map will be used with matrices and vectors acting coordinate wise. Also, for the reciprocal case, we defined $\nabla : \mathbb{Z} \to \mathbb{F}_2$. Let $\mathbf{a} = (a_1, \ldots, a_n)$ be an n-tuple of elements of the field $\mathbb{F}_2$. We will adopt the following notation:

$$\mathbf{x}^{\mathbf{a}} = x_1^{\Delta a_1} \cdots x_n^{\Delta a_n} \in [\mathbf{X}]. \tag{2}$$

The code ideal can be given by the two equivalent formulas in (3) and (4) below, the equivalence between (3) and (4) was proved in [5]. Let W be a generator matrix of an $[n, k]$ binary code $\mathscr{C}$ (the row space of the matrix generates $\mathscr{C}$) and $\mathbf{w}_i$ denotes its rows for $i = 1, \ldots k$.

$$I(\mathscr{C}) = \langle \mathbf{x}^{\mathbf{a}} - \mathbf{x}^{\mathbf{b}} \mid \mathbf{a} - \mathbf{b} \in \mathscr{C} \rangle \subseteq \mathbb{K}[\mathbf{X}]. \tag{3}$$

$$I(\mathscr{C}) = \langle \{\mathbf{x}^{\mathbf{w}_i} - 1 : i = 1, \ldots k\} \cup \{x_i^2 - 1 : i = 1, \ldots, n\} \rangle \subseteq \mathbb{K}[\mathbf{X}]. \tag{4}$$

Note that $I(\mathscr{C})$ is a zero-dimensional ideal since the quotient ring $R = \mathbb{K}[\mathbf{X}]/I(\mathscr{C})$ is a finite dimensional vector space and its dimension is equal to the number of cosets in $\mathbb{F}_2^n/\mathscr{C}$.

For every element $\mathbf{x}^a$ in the monoid $[\mathbf{X}]$, with $a \in \mathbb{N}^n$, we have a corresponding vector $\nabla(a) \in \mathbb{F}_2^n$, and viceversa, any vector $\mathbf{w} \in \mathbb{F}_2^n$ has a unique standard representation $\mathbf{x}^{\mathbf{w}}$ (the exponents of the variables are 0 or 1) as an element of [**X**]. A term order on [**X**] (see [1]) is a total order $<$ on [**X**] satisfying the following two conditions:

1. $1 < \mathbf{x}^a$ for all $\mathbf{x}^a \in [\mathbf{X}]$, $\mathbf{x}^a \neq 1$.
2. If $\mathbf{x}^a < \mathbf{x}^b$, then $\mathbf{x}^a\mathbf{x}^\gamma < \mathbf{x}^b\mathbf{x}^\gamma$, for all $\mathbf{x}^\gamma \in [\mathbf{X}]$.

A total degree term order is a term order such that $\mathbf{x}^a < \mathbf{x}^b$ provided that $\sum_{i=1}^n a_i < \sum_{i=1}^n b_i$. Examples of such orders are the Degree and Degree Reverse Lexicographical orders (see [1]).

Let $<$ be a term order, let us $\mathrm{T}(f)$ denotes the maximal term of a polynomial f with respect to the order $<$. The set of maximal terms of the set $F \subseteq \mathbb{K}[X]$ is denoted $\mathrm{T}\{F\}$ and $\mathrm{T}(F)$ denotes the semigroup ideal generated by $\mathrm{T}\{F\}$. Finally, $\langle F\rangle$ is the polynomial ideal in $\mathbb{K}[\mathbf{X}]$ generated by F. In particular, for the code ideal $I(\mathscr{C})$, $\mathrm{T}(I(\mathscr{C}))$ is the set of maximal terms and $N(I(\mathscr{C})) = [\mathbf{X}] \setminus \mathrm{T}(I(\mathscr{C}))$ the set of words in canonical forms. We emphasize that there is a one to one correspondence between $N(I(\mathscr{C}))$ and the cosets in $\mathbb{F}_2^n/\mathscr{C}$. One characterization of Gröbner bases is that G is a *Gröbner basis* of the ideal $\langle \mathrm{G}\rangle$ if and only if $\mathrm{T}(\langle \mathrm{G}\rangle) = \mathrm{T}(\mathrm{G})$ (see [1]).

3 Gröbner Bases and Trial Set for Binary Codes

Any total degree compatible order induces an α-ordering monotone $\prec$ on $\mathbb{F}_2^n$ such that $\mathbf{v} \prec \mathbf{w}$ if $\mathbf{x}^{\mathbf{v}} < \mathbf{x}^{\mathbf{w}}$ for any $\mathbf{v}, \mathbf{w} \in \mathbb{F}_2^n$. On the other hand, given an α-ordering monotone on $\mathbb{F}_2^n$ we could define a total order on [**X**] which may not be a term order, a class of these orders on [**X**] were called in [3] error-vector orderings.

In this work, we will focus in the first situation, the α-ordering monotone which is defined in [6] it is derived from the Degree Lexicographical order. In general, let $<$ be a total degree term order on [**X**], and let $\prec$ be the corresponding α-ordering monotone on $\mathbb{F}_2^n$.

Proposition 2 (Correctable and uncorrectable errors and canonical forms and maximal terms) *Let* $\mathbf{x}^w \in [\mathbf{X}]$, $w \in \mathbb{N}^n$ *then*

1. *If* $\mathbf{x}^w$ *is not the standard representation of the word* $\nabla(w)$ *in* $\mathbb{F}_2^n$, *then it is a maximal term, i.e.,* $\mathbf{x}^w \in \mathrm{T}(I(\mathscr{C}))$.
2. *If* $\nabla(w) \in E^1(\mathscr{C})$, *then* $\mathbf{x}^w \in \mathrm{T}(I(\mathscr{C}))$.
3. *If* $\mathbf{x}^w$ *is the standard representation of the word* $\nabla(w)$ *and* $\nabla(w) \in E^0(\mathscr{C})$, *then* $\mathbf{x}^w$ *is a canonical form, i.e.,* $\mathbf{x}^w \in N(I(\mathscr{C}))$.
4. *If* $\mathbf{x}^w$ *is the standard representation of the word* $\nabla(w)$ *and* $\nabla(w) \in M^1(\mathscr{C})$, *then* $\mathbf{x}^w$ *is an irredundant maximal term, i.e.,* $\mathbf{x}^w \notin \mathrm{T}(I(\mathscr{C})) \setminus \{\mathbf{x}^w\}$ *and is a maximal term of any Gröbner basis of* $I(\mathscr{C})$ *w.r.t.* $<$.

The set of irredundant maximal terms are the maximal terms of any minimal Gröbner basis, for example, of the reduced Gröbner basis. For simplicity, we will assume that the coefficients of the maximal terms in a Gröbner basis are positive.

Proof (1): If $\mathbf{x}^w$ is not the standard representation of the word $\nabla(w)$ in $\mathbb{F}_2^n$, then $\mathbf{x}^w$ is a multiple of some x_i^2, for $i = 1, \ldots, n$, and $x_i^2 \in \mathrm{T}(I(\mathscr{C}))$ (see (4)).
(2): Let $w \in \mathbb{N}^n$ $s.t.$ $\nabla(w) \in E^1(\mathscr{C})$ then, $H(\nabla(w)) \neq \emptyset$, so there exists $\mathbf{c} \in \mathscr{C}$ such that $\mathbf{a} = \nabla(w) + \mathbf{c} \prec \nabla(w)$, which means also $\mathbf{x}^\mathbf{a} < \mathbf{x}^w$. By (3), $\mathbf{x}^w - \mathbf{x}^\mathbf{a} \in I(\mathscr{C})$; therefore, $\mathbf{x}^w \in \mathrm{T}(I(\mathscr{C}))$.
(3): Since $\nabla(w) \in E^0(\mathscr{C})$ it is clear that $\mathbf{x}^w$, the standard representation of the word $\nabla(w)$, is the minimal element on $[\mathbf{X}]$ according to $<$ among the elements $\mathbf{x}^v \in [\mathbf{X}]$ such that $\nabla(v) - \nabla(w) \in \mathscr{C}$. Then, $\mathbf{x}^w \in N(I(\mathscr{C}))$.
(4): $\nabla(w) \in M^1(\mathscr{C}) \subset E^1(\mathscr{C})$ implies by Proposition 2 that $\mathbf{x}^w$ is a maximal term and $\mathbf{x}^w \notin \mathrm{T}(I(\mathscr{C})) \setminus \{\mathbf{x}^w\}$, by definition of $M^1(\mathscr{C})$.

Definition 1 (*Gröbner codewords* [4]) Let G be a Gröbner basis for $I(\mathscr{C})$ w.r.t. $<$, the set of Gröbner codewords $\mathscr{C}_\mathrm{G}$ corresponding to G are the codewords associated with G by $\mathscr{C}_\mathrm{G} = \{\mathbf{c} \in \mathscr{C} : \mathbf{c} = \mathbf{w} + \mathbf{v},\ \text{s.t. } \mathbf{x}^\mathbf{w} - \mathbf{x}^\mathbf{v} \in \mathrm{G},\ \mathbf{w}, \mathbf{v} \in \mathbb{F}_2^n,\ \mathbf{v} \prec \mathbf{w}\}$.

Theorem 1 *Let* G *be a Gröbner basis for* $I(\mathscr{C})$ *w.r.t.* $<$, *then* $\mathscr{C}_\mathrm{G}$ *is a trial set.*

Proof We will prove the statement 2 of Proposition 1. Let $\mathbf{w} \in M^1(\mathscr{C})$, then $\mathbf{x}^\mathbf{w} \in \mathrm{T}\{\mathrm{G}\}$ (see Proposition 2) and by Definition 1 there exists $\mathbf{c} \in \mathscr{C}_\mathrm{G}$ s.t. $\mathbf{c} = \mathbf{w} + \mathbf{v}$ s.t. $\mathbf{v} \prec \mathbf{w}$. Thus $\mathbf{c} + \mathbf{w} = \mathbf{v} \prec \mathbf{w}$ and $\mathbf{c} \in H(\mathbf{w})$.

Theorem 2 *Let* T *be a trial set, the set* $\mathrm{G}_T = \{\mathbf{x}^\mathbf{w} - \mathbf{x}^\mathbf{v} : \mathbf{w} \in \mathrm{L}(\mathbf{c})$ *for some* $\mathbf{c} \in T$ *and* $\mathbf{v} = \mathbf{c} - \mathbf{w}\} \cup \{x_i^2 - 1 : i = 1, \ldots, n\}$ *is a Gröbner basis for* $I(\mathscr{C})$ *w.r.t.* $<$.

Proof If $\mathbf{x}^u$ is a maximal term which is not the standard representation of $\nabla(u)$, then it can be reduced to the standard representation of $\nabla(u)$ by means of the set $\{x_i^2 - 1 : i = 1, \ldots, n\}$. Thus, let us assume that $\mathbf{x}^\mathbf{u} \in \mathrm{T}(I(\mathscr{C}))$ and $\mathbf{u} \in E^1(\mathscr{C})$. It is clear that there exists $\mathbf{w} \subseteq \mathbf{u}$ s.t. $\mathbf{w} \in M^1(\mathscr{C})$, $\mathbf{w} \in M^1(\mathscr{C})$ implies there exists $\mathbf{c} \in T$ s.t. $\mathbf{w} \in \mathrm{L}(\mathbf{c})$ (by Proposition 1.3). Let $\mathbf{v} = \mathbf{c} - \mathbf{w}$, then we have $\mathbf{x}^\mathbf{w} - \mathbf{x}^\mathbf{v} \in \mathrm{G}_T$ and $\mathbf{x}^\mathbf{w} \mid \mathbf{x}^\mathbf{u}$ (remember $\mathbf{w} \subseteq \mathbf{u}$). Consequently, G_T is a Gröbner basis for $I(\mathscr{C})$.

3.1 An Example

Let G be a generator matrix of the $[7, 3]$ binary code $\mathscr{C}$ over $\mathbb{F}_2^6$ given by

$$G = \begin{pmatrix} 1\,0\,0\,1\,1\,0\,0 \\ 0\,1\,0\,0\,1\,1\,0 \\ 0\,0\,1\,0\,0\,1\,1 \end{pmatrix}.$$

From this matrix we obtain a set of generating polynomials for $I(\mathscr{C})$ as in Eq. (4) and then we compute the reduced Gröbner basis of $I(\mathscr{C})$ with respect to the Degree Reverse Lexicographical order (see [5]), that is

$$\begin{aligned}\text{G}b = \{&x_1^2 - 1,\ x_2^2 - 1,\ x_3^2 - 1,\ x_4^2 - 1,\ x_5^2 - 1,\ x_6^2 - 1,\ x_7^2 - 1,\\ &x_6x_7 - x_3,\ x_3x_7 - x_6,\ x_5x_6 - x_2,\ x_3x_6 - x_7,\ x_2x_6 - x_5,\ x_4x_5 - x_1,\\ &x_3x_5 - x_2x_7,\ x_2x_5 - x_6,\ x_1x_5 - x_4,\ x_2x_4 - x_1x_6,\ x_1x_4 - x_5,\ x_2x_3 - x_5x_7,\\ &x_1x_2 - x_4x_6\}.\end{aligned}$$

Applying Definition 1 we have

$$\begin{aligned}\mathscr{C}_{\text{G}b} = \{&(0, 0, 1, 0, 0, 1, 1),\ (0, 1, 0, 0, 1, 1, 0), (1, 0, 0, 1, 1, 0, 0),\\ &(0, 1, 1, 0, 1, 0, 1), (1, 1, 0, 1, 0, 1, 0)\}\end{aligned}$$

which is a trial set (see Theorem 1). Now starting from $\mathscr{C}_{\text{G}b}$, if we compute the set of larger halves of these codewords and apply Theorem 2, we get a set of binomials which contains Gb; therefore, a Gröbner basis is obtained and it may be reduced to the reduced Gröbner basis Gb. In more details, let Gt be the set of binomials obtained from $\mathscr{C}_{\text{G}b}$ by applying Theorem 2. Note that $(1, 1, 0, 0, 0, 0, 0)$ and $(0, 1, 0, 1, 0, 0, 0)$ are larger halves of $(1, 1, 0, 1, 0, 1, 0)$, then $\{x_1x_2 - x_4x_6,\ x_2x_4 - x_1x_6\} \in \text{G}t$. In the same way, it can be seen that $\text{G}b \subset \text{G}t$. On the other hand, $(0, 1, 0, 0, 1, 0, 0)$ is a larger half of $(0, 1, 1, 0, 1, 0, 1)$, then $x_2x_5 - x_3x_7 \in \text{G}t$, but $x_2x_5 - x_3x_7 \notin \text{G}b$. Note that this binomial can be reduced to zero w.r.t. Gb. Thus applying Theorem 2 we get a Gröbner basis Gt which contains the reduced Gröbner basis Gb.

3.1.1 Decoding

Let $\mathbf{y} = (1, 0, 0, 1, 1, 1, 0)$, the corresponding word in $[\mathbf{X}]$ is $\mathbf{w} = x_1x_4x_5x_6$. Now we will use the Gröbner basis Gb to reduce $\mathbf{w}$ to its canonical form.

$\mathbf{w} - x_1x_4(x_5x_6 - x_2) = x_1x_2x_4 = \mathbf{w}_1$, $\mathbf{w}_1 - x_4(x_1x_2 - x_4x_6) = x_4^2x_6 = \mathbf{w}_2$, $\mathbf{w}_2 - x_6(x_4^2 - 1) = x_6$.

Therefore, the error vector corresponding to $\mathbf{y}$ is $(0, 0, 0, 0, 0, 1, 0)$ and the codeword is $(1, 0, 0, 1, 1, 0, 0)$ (see [3] for more details).

Now, we will show the decoding process with the trial set $\mathscr{C}_{\text{G}b}$ and, at the same time, we will show the analogy with the previous use of the Gröbner basis Gb.

$\mathbf{y} - (0, 1, 0, 0, 1, 1, 0) = (1, 1, 0, 1, 0, 0, 0) = \mathbf{y}_1$,

$\mathbf{y}_1 - (1, 1, 0, 1, 0, 1, 0) = (0, 0, 0, 0, 0, 1, 0) = \mathbf{y}_2$.

Note how in the previous sequence of steps the weight is decreasing from $\mathbf{y}$ to $\mathbf{y}_2$. Also $\mathbf{y}_2$ cannot be reduced by the trial set, which means that $\mathbf{y}_2$ is a correctable error and the corresponding codeword to $\mathbf{y}$ is $\mathbf{y} - \mathbf{y}_2 = (1, 0, 0, 1, 1, 0, 0)$, as it is expected.

Remark 1 It is clear that the trial set associated to a Gröbner basis is a smaller structure and, as it is showed above, the reduction process for decoding can be done in a similar way as the Gröbner basis does. This could be interesting to take it into account while studying some properties of the code ideal; i.e, to analyze the use of the trial set instead of the Gröbner basis.

3.2 Minimal Trial Sets and Minimal Gröbner Bases

A minimal trial set is a trial set such that any proper subset is not a trial set. Having an smaller trial set would be an smaller set that could be used for gradient decoding, although smaller trial sets do not necessarily ensure more efficiency. In [6] the authors shows an advantage of having a minimal trial set, since the size of trial sets are used to derive some important bounds on the error correction beyond half the minimum distance.

By Proposition 1.3, the set of larger halves of a trial set T should contain at least the set $M^1(\mathscr{C})$, by Theorem 2 and Proposition 2 this means that the corresponding Gröbner basis G_T should contain at least the irredundant maximal terms (this is the case for any Gröbner basis); therefore, there is a direct connection between minimal trial sets and minimal Gröbner bases. In particular, a distinguished minimal trial set would be the set of Gröbner codewords corresponding to the reduced Gröbner basis.

References

1. Adams,W.W., Loustaunau, Ph.: An introduction to Gröobner bases. In: Graduate Studies in Mathematics, vol. 3. American Mathematical Society, Providence (1994)
2. Barg, A.: Complexity issues in coding theory. In: Handbook of Coding Theory, vol. I, pp. 649–754. North-Holland, Amsterdam (1998)
3. Borges-Quintana, M., Borges-Trenard, M.A., Martínez-Moro, E.: On a Gröbner bases structure associated to linear codes. J. Discret. Math. Sci. Cryptogr. **10**(2), 151–191 (2007)
4. Borges-Quintana, M., Borges-Trenard, M.A., Martínez-Moro, E.: A Gröbner representation for linear codes. In: Advances in Coding Theory and Cryptography, Ser. Coding Theory Cryptololgy, vol. 3, pp. 17–32. World Sci. Publ., Hackensack (2007)
5. Borges-Quintana, M., Borges-Trenard, M.A., Fitzpatrick, P., Martínez-Moro, E.: Gröbner bases and combinatorics for binary codes. Appl. Algebra Eng. Comm. Comput. **19**(5), 393–411 (2008)
6. Helleseth, T., Kløve, T., Vladimir, I.L.: Error–correction capability of binary linear codes. IEEE Trans. Inf. Theory **51**(4), 1408–1423 (2005)
7. Huffman, W.C., Pless, V.: Fundamentals of Error-Correcting Codes. Cambridge University Press, Cambridge (2003)

Automated Study of Envelopes of One-Parameter Families of Surfaces

Thierry Dana-Picard and Nurit Zehavi

Abstract Learning mathematics in a technology-rich environment enables to revive classical topics which have been removed from the curriculum a long time ago. Theoretical issues and their applications can be studied within an experimental process, using automated proofs. We present how envelopes of one-parameter families of surfaces in 3D space and some of their properties can be presented using technology. This approach may be useful for students in an engineering curriculum and for in-service/pre-service teachers. Working with technology and taking advantage of both algebraic symbolic features, such as algorithms computing Gröbner bases, and visualization tools, educational and professional profit is obtained such as reviving classical topics from differential geometry, broadening horizons, introducing new topics. The purpose is also to enhance the learners experimental skills. In such a framework, conversion between various registers of representation is an important issue.

Keywords Parametric families of surfaces · Envelopes · Gröbner bases · Implicitization

1 Introduction

Computer algebra systems (CAS) are all purpose software packages that facilitate speedy symbol manipulations and symbolic computations. They provide also rich graphical tools. As such they can be used to solve problems and prove theorems. As educators, we must identify pedagogical advantages of the technology for broadening horizons and introducing topics from the classical mathematics as recommended in [6]. Topics that students considered either too technical or too theoretical become reachable by using the technology for visualization, experimentation, and

T. Dana-Picard (✉)
Jerusalem College of Technology, Havaad Haleumi Street 21, Jerusalem, Israel
e-mail: ndp@jct.ac.il

N. Zehavi
Weizmann Institute of Science, Rehovot, Israel
e-mail: nurit.zehavi10@gmail.com

I.S. Kotsireas and E. Martínez-Moro (eds.), *Applications of Computer Algebra*, Springer Proceedings in Mathematics & Statistics 198,
DOI 10.1007/978-3-319-56932-1_4

automatic proving. The technology includes CAS, but also dynamical geometry software (DGS). The illustration of classical concepts and methods in differential geometry using technology provides examples of the new educational possibilities (see [7, 15]).

In this chapter, we present a study of envelopes of families of surfaces in 3D space, relying strongly on automated methods. The reason for this was primarily that envelopes are a beautiful classical topic. We quote Hilbert in his 1900 address (available in [13]) drawing students and teachers attention to the importance of this topic

> who would give up the picture of a family of curves or surfaces with its envelope which plays so important a part in differential geometry, in the theory of differential equations, in the foundation of the calculus of variations and in other purely mathematical sciences?

Envelopes have numerous concrete applications

- They are related to the theory of caustics and wave fronts, i.e., to geometrical optics and to the theory of singularities ([2, 19], quoted in [3].
- Envelopes have applications in robotics and kinematics: rigid body motion in the plane, in three-space, collision avoidance of robot motion, construction of gears, etc. (see [17]).
- A connection can be made to realistic mathematics for mechanical design, where envelopes are used extensively for designing complex objects as in [4].
- Cavities concealed in the soil are explored by sending waves and modeling the shape of closed surfaces defined by the reflection points of the wave on the cavity walls, then looking for an envelope of the modelized surfaces.
- Envelopes have frequent applications in advanced domains of science and technology, such as cosmology.

More than 50 years ago, Thom claimed in [19] that a reason for the disappearance of envelopes from the syllabus is that the theory is not enough developed, with not so many theorems, too many special cases, etc. We may add the visualization problems encountered by students.

Topics in differential geometry integrate methods from algebra, geometry, and calculus, and thus invite making mathematical connections, which is a key goal for learning and teaching. When one method provides better insight (to certain learners) it could help the study of a more challenging method or of a more challenging topic. In Sect. 3.2 we show that sometimes an algebraic method is more efficient than a purely analytic one. The converse may occur also. Identifying such processes and analyzing them is a key goal for learning and teaching. Every CAS works in different registers, algebraic, numerical and graphical, two different CAS presenting different levels of implementation for each register. In each register, the implemented algorithms are a consequence of theoretical developments in Mathematics.

A central tool we use in this chapter is the automatic solution of nonlinear systems of polynomial equations, based on algorithms for the computation of Gröbner bases. The algorithms are described in [1, 5]. This method helps transforming classical topics into more modern ones. For example, Pech proves in [16] a large set of results in Geometry using both classical methods and Gröbner bases computations. These methods are used in [8] for the study of closed paths of light in various billiards (Fermat curves, astroids). In [11], Gröbner bases computations have been used for the study of bisoptic curves. In every occurrence, the first step consists in translating the given data into polynomial equations. The results are obtained in polynomial form. Sometimes a mixed treatment is useful, algebraic, and analytic. The CAS dynamical features (slider bar, move a point, a figure, etc.) have a central role in the experimental work.

Different approaches to envelopes appear in the literature. We recall here the three approaches described by Kock in [14]; we refer also to his discussion of the problems inherent to each approach.

Let $\{S_t\}_{t\in\mathbb{R}}$ be a one-parameter family of surfaces in 3D space.

- *Synthetic*: The characteristic C_t is the limit surface of the family of intersections $S_t \cap S_{t+h}$ as $h \to 0$. If it exists, the envelope E is the union of the characteristics; This point of view has been exploited in [9], for families of plane curves.
- *Impredicative*: If it exists, the envelope E is a surface with the following property: for each point $M \in E$, there exists a unique value $t_M \in \mathbb{R}$ of the parameter such that E is tangent to S_{t_M} at M. The locus of points where E touches a surface S_t is called the E-characteristic C_t.
- *Analytic*: There are two descriptions, according to whether the family is given by an implicit or a parametric presentation.
  - Implicit: The envelope of a one-parameter family of surfaces given by an implicit equation $F(x, y, z, t) = 0$ is determined by the solution set of the following system of equations
$$\begin{cases} F(x, y, z, t) & = 0 \\ \frac{\partial F}{\partial t} F(x, y, z, t) & = 0 \end{cases}$$
  - Parametric:The envelope of a one-parameter family of surfaces given by a parametrization $(M(u, v, t))_{u,v}$ is determined by the solution of the following equation:
$$\det\left(\frac{\partial M}{\partial u}, \frac{\partial M}{\partial v}, \frac{\partial M}{\partial t}, \right) = 0.$$

In this chapter, we focus on families of surfaces with implicit presentation. Among the problems mentioned in [14], saying *the* envelope is mostly improper in many cases, and should be replaced by *an* envelope: when *the* envelope found is the union of disjoint components, each component alone may be an envelope according to the "impredicative" definition. Another problem is the imprecision of a definition of the "limit curve" mentioned above (which space? how is the topology defined?). As CAS-based methods provide both graphical experimentation and automatic solution of equations, all the components of a possible envelope are discovered together.

In the next section, we recall briefly two important examples of envelopes of families of curves in the plane. Noting their specific features has an importance when performing the transition towards families of surfaces in the 3D space.

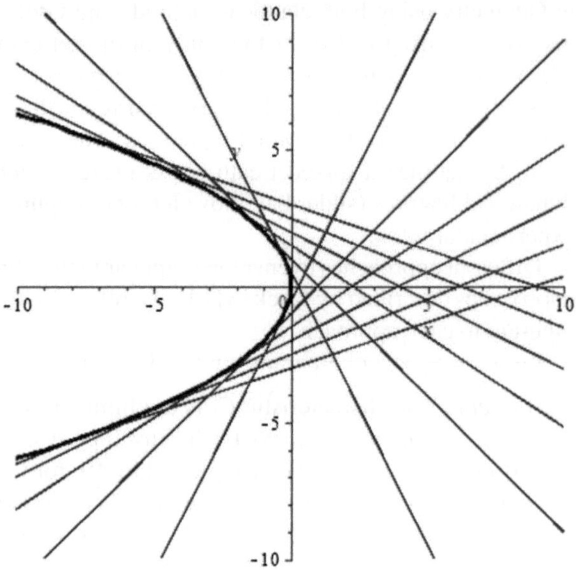

Fig. 1 The envelope of a family of *lines* in the plane

2 Envelopes of One-Parameter Families of Plane Curves

A family of plane curves is given by an equation of the form $F(x, y, t) = 0$, with parameter $t \in \mathbb{R}$. An envelope of the family, if it exists, is a curve tangent to every curve in the family. It can be shown that this envelope is the solution set of the system of equations

$$\begin{cases} F(x, y, t) &= 0 \\ \frac{\partial F}{\partial t}(x, y, t) &= 0 \end{cases} \tag{1}$$

Figure 1 shows the envelope of the family of lines given by the equation $x + ty = t^2$, where $t \in \mathbb{R}$; this envelope is the parabola with equation $y = -x^2/4$, and Fig. 2 shows the envelope of the family of circles with radius 1 centered on the ellipse whose equation is $x^2/4 + y^2 = 1$. Here *the* envelope has two components, each component is *an* envelope of the family of circles.

Figures 1 and 2 have been obtained with GeoGebra[1], in which a slider bar is implemented. The usage of the slider bar provides a dynamical environment and enables to build envelopes experimentally. Figure 2 shows how this is performed. On the left side, a partial construction is shown, and a global construction is displayed on the right side. This one has been obtained using the slider bar, which yields a uniform spacing between circles. Another possibility exists, less clear, using the **Move** command. In this case, spacing between neighboring circles is not uniform, the appearance of the envelopes being thus slightly different.

[1]Freely downloadable for www.geogebra.org.

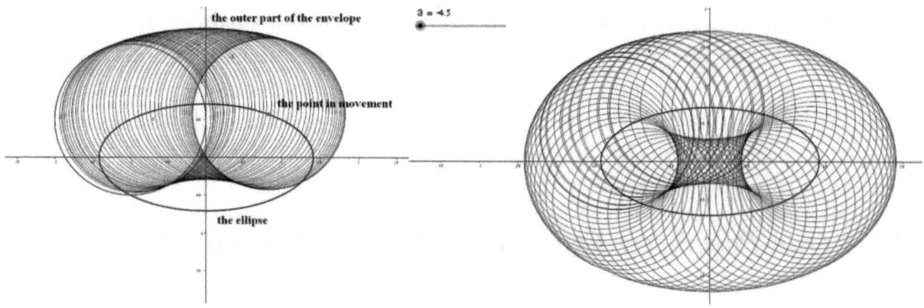

Fig. 2 Dynamical exploration of the envelope of a family of circles

3 Envelopes of One-Parameter Families of Surfaces in 3D Space

The transition to parameterized families of surfaces in 3D space rely on the same analytic and algebraic techniques, but with different graphical features of the CAS. GeoGebra has a slider bar for direct manipulation of plots in 2D, but for such dynamical features in 3D, we had to use another software.[2] This makes transition from 2D to 3D somehow critical (see [20]).

A one-parameter family of surfaces is defined by an equation of the form $F(x, y, z, t) = 0$, where t is real parameter. We recall the fundamental theorem for an analytic determination of an envelope, with a short proof.

Theorem 1 *If it exists, the envelope of the given family of surfaces is determined by the system of equations*

$$\begin{cases} F(x, y, z, t) &= 0 \\ \frac{\partial F}{\partial t}(x, y, z, t) &= 0 \end{cases}. \tag{2}$$

Proof We increase the parameter by a small h and apply Taylor's theorem; we obtain

$$\underbrace{F(x, y, z, t+h)}_{=0} = \underbrace{F(x, y, z, t)}_{=0} + h\frac{\partial F}{\partial t}(x, y, z, t+\theta h),$$

whence:

$$\frac{\partial F}{\partial t}(x, y, z, t+\theta h) = 0.$$

If $h \to 0$, then this equation reduces to $\frac{\partial F}{\partial t}(x, y, z, t) = 0$. Thus, the desired envelope is the solution set of System (2).

[2]It is called MathStudio; a web version is at http://mathstud.io/welcome/. The original software runs under a different operating system. Here it may provide a dynamical display, but without a **Trace** option. Another problem relies in that it is disconnected from the symbolic computations we need. Therefore we do not comment its usage here.

In the two next subsections, we show examples of *canal surfaces*, i.e., envelopes of families of spheres with constant radius, already studied by Monge; see [18], p. 166. Section 3.3 is devoted to envelopes of a family of planes. Singular points appear, whose set is a space curve called the *edge of regression* of the envelope.

3.1 A One-Parameter Family of Spheres with Aligned Centers

We consider the family of spheres of radius 2 centered on the x-axis. The following rows of Maple code provide a plot of a couple of spheres of this family, displayed on Fig. 3(a). The coordinate axes are emphasized and labeled. Note the usage of the specific command **sphere** from the package **plottools**, avoiding the usage of the **plot3d** and **implicitplot3d** commands, which require as input a analytic description of the one-parameter family of spheres. Another advantage is the high accuracy of the plot. The **plots** package is used for the sake of the **display** command.

```
> restart; with(plottools); with(plots):
> n := 4:
> coordaxes := plot3d({[t, 0, 0],[0,t,0],[0,0,t]},
   t = -10 .. 10, s = -8 .. 8, scaling = constrained):
> p := proc (k) -> sphere([k, 0, 0], 2) end proc:
> familyspheres:=seq([p(1.5*k)], k = -n .. n):
> display(familyspheres, coordaxes, scaling = constrained,
  transparency = .5, axes = normal);
```

Figure 3a leads to conjecture that an envelope exists and it is a cylinder wrapping the spheres; see Fig. 3b. In order to prove this, we need to solve the system of Eq. (2).

The centers of the given spheres are the points of the x-axis, thus the general equation for the spheres is

$$(x - t)^2 + y^2 + z^2 = 4. \tag{3}$$

Let $F(x, y, z, t) = (x - t)^2 + y^2 + z^2 - 4 = x^2 + y^2 + z^2 - 2tx + t^2 - 4$. Now, System (2) reads

$$\begin{cases} x^2 + y^2 + z^2 - 2tx - 4 & = 0 \\ 2x & = 0 \end{cases}$$

By substitution, we obtain the following equation:

$$y^2 + z^2 = 4, \tag{4}$$

which defines a cylinder whose axis is the $x - axis$ and whose base has radius 2; see Fig. 3(b).

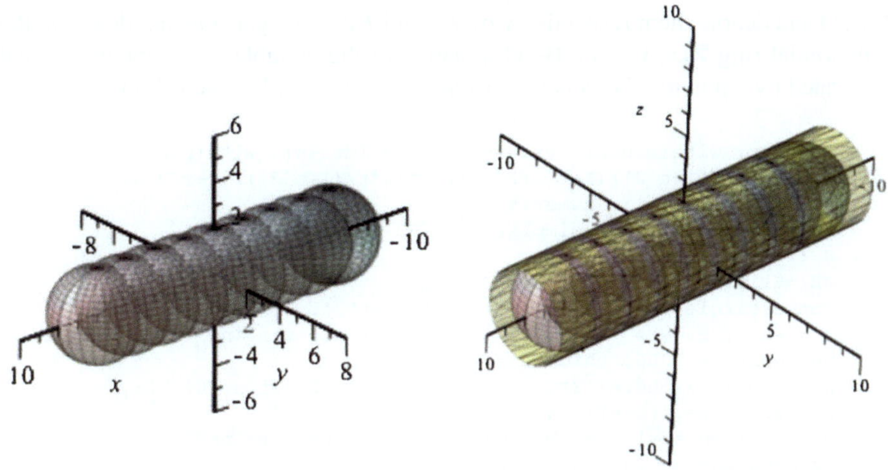

(a) visualization of the family of spheres (b) visualization of the envelope

Fig. 3 Envelope of a family of spheres

Note that the commands used for creating Fig. 3 induce the choice of a mesh well-suited for the spheres: the lines defining the mesh are parallels and meridians of the spheres. For the cylinder, the defining lines of the mesh are generating lines of the cylinder and circles. Therefore the plots are very accurate.The choice of the **implicitplot3d** command would have implied an ordinary triangular mesh, whence a less accurate plot. In [14], Kock calls this a *coarse* mesh. This issue is addressed in [21]. We recall only the fact that the mesh divides the domain into cells. The values of the given function are computed for interior points of a cell by interpolation of values of the functions on the border of the cell. Different choices of the mesh may lead to more or less accurate plots, sometimes the plot may look very strange; see [10].

3.2 Unit Spheres Centered on a Circle

Denote by $\mathscr{C}$ the circle in the xy-plane whose center is the origin and radius equal to 2. We consider the one-parameter family of unit spheres S_t centered on $\mathscr{C}$; see Fig. 4. A general equation for the spheres is

$$\left(x - 2\frac{1-t^2}{1+t^2}\right)^2 + \left(y - 2\frac{2t}{1+t^2}\right)^2 + z^2 - 1 = 0. \tag{5}$$

We denote by $F(x, y, z, t)$ the left-hand side in Eq. 5. The goal of first rows is to perform an algebraic process. We consider the numerators of left-hand sides of

Eq. (1) and denote them respectively by $F1$ and $F2$. They generate an ideal J in the polynomial ring $\mathbb{R}[x, y, z, t]$. By elimination of the variable t, we obtain an ideal generated by a polynomial denoted by $envpoly$. The Maple code follows:

```
> restart; with(plots): with(PolynomialIdeals): with(plottools):
> F := (x-2*(1-t^2)/(1+t^2))^2+(y-2*(2*t/(1+t^2)))^2+z^2-1;
> simplify(%); F1 := numer(%):
> derF := diff(F, t); simplify(%); F2 := numer(derF):
> J := <F1, F2>:
> JE:=EliminationIdeal(J, {x, y, z}):
> centercircle := plot3d([2*(1-t^2)/(1+t^2), 2*(2*t/(1+t^2)), 0],
  t = -5 .. 5, s = -10 .. 10, axes = normal, scaling = constrained,
  numpoints = 3000, thickness = 3):
> n:=15:p:=k->sphere([2*cos(2*k*Pi/n),2sin(2k*Pi/n),0],1):
> boules := seq([p(k)], k = 0 .. n-1);
> display(centercircle, boules, scaling = constrained,
  transparency = .4, axes = boxed);
```

The first **display** command produces a plot for some spheres and the circle $\mathscr{C}$, visible because of the transparency option. The circle and the sphere have been plotted using a well-fitted command for this purpose, which ensures a high accuracy of the plot.

We assigned the name *envpoly* to the generator of the ideal JE which has been computed with the command **EliminationIdeal**; we have

$$\begin{aligned} envpoly =& 9x^4+18y^2x^2+9y^4-10x^6-30y^2x^4+6z^2x^4-30y^4x^2+12z^2y^2x^2-10y^6 \\ &+6z^2y^4+x^8+4y^2x^6+2z^2x^6+6y^4x^4+6z^2y^2x^4+z^4x^4+4y^6x^2+6z^2y^4x^2 \\ &+2z^4y^2x^2+y^8+2z^2y^6+z^4y^4. \end{aligned}$$

The equation $envpoly = 0$ is an implicit equation for the desired envelope. Now we can plot the surfaces; the **implicitplot3d** command is the standard one for that.

```
> envlp := implicitplot3d(envpoly = 0, x = -3 .. 3, y = -3 .. 3,
   z = -2 .. 2,transparency = .5, axes = normal, numpoints = 3000,
   axes = boxed, scaling = constrained, color = yellow);
> display(boules, envlp, centercircle, scaling = constrained,
  axes = boxed);
```

This **display** command produces the same plot as before with the envelope added. Regarding the envelope, it was necessary to use **implicitplot3d** which produces a *coarse* plot, with a triangular mesh. Nevertheless the plot is good enough to show the envelope and the family of spheres in Fig. 4(b).

Actually a parametric representation of the envelope is available, using the **solve** command. A couple of problems arise to produce a plot which could be good enough. We discuss this point in Sect. 4.

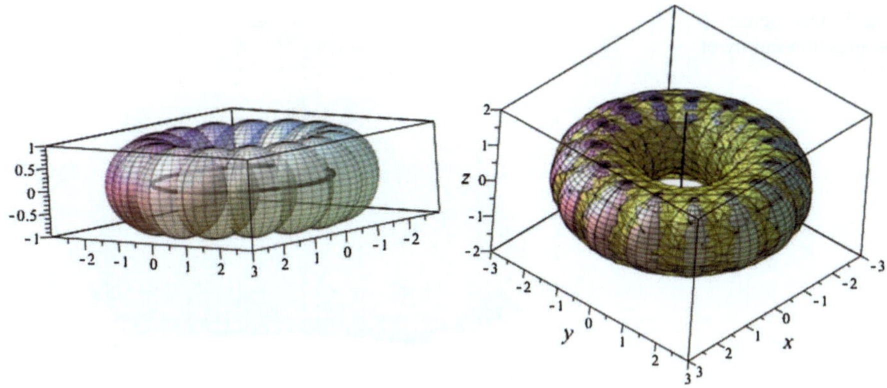

(a) The family and the circle (b) The family within the envelope

Fig. 4 Envelope of a family of unit spheres centered on a circle

3.3 A One-Parameter Family of Planes

In [9], we studied a geometric-progression family of lines in the plane, namely the family given by the equation $x + ty = t^2$, where t is a real parameter. By different ways, either synthetic or analytic, we discovered that this family has an envelope, namely the parabola whose equation is $x = -y^2/4$. This is displayed in Fig. 1. The example of a geometric-progression family of planes will show that the transition from 2D to 3D introduces new properties and a need for other tools, both theoretical and technological.

We consider now the family of planes in 3D space given by the following equation:

$$x + ty + t^2 z = t^3, \; t \in \mathbb{R}. \tag{6}$$

Displaying a couple of planes of the family is unilluminating and does not contribute to intuition whether an envelope exist or not (see Fig. 5, obtained with the DPGraph software[3]).

For every one-parameter family of planes, the following holds (see [12]):

Theorem 2 *Let* $\{S_t\}$ *be a* 1*-parameter family of planes in the* 3*-dimensional space. If the family has an envelope, then*

1. *The envelope has an edge of regression (cuspidal edge).*
2. *The envelope is a ruled surface, whose generators are the tangents to the edge of regression.*

The edge of regression is a space curve on the surface whose points are the singular points of the envelope.

---

[3] www.dpgraph.org.

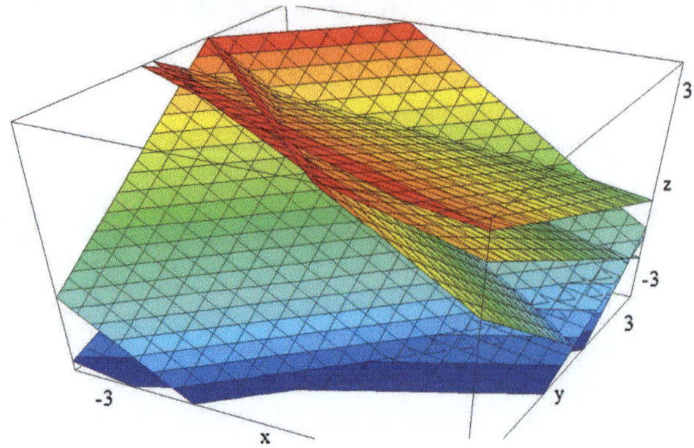

Fig. 5 Geometric-progression family of planes

Proof Suppose that the planes S_t are given by the equation $u_1(t)x + u_2(t)y + u_3(t)z + u_4(t) = 0$, where u_1, u_2, u_3 and u_4 are real functions of the real parameter t. Then an envelope is given by intersecting the planes S_t with the first derivative planes, whose equations are $u_1'(t)x + u_2(t)y + u_3'(t)z + u_4'(t) = 0$. If the following condition holds:

$$\forall i, j, \in \{1, 2, 3\}, \begin{vmatrix} u_i(t) & u_j(t) \\ u_i'(t) & u_j'(t) \end{vmatrix} = 0, \tag{7}$$

there is no envelope. Otherwise, the characteristic curve exists and for each value of t, it is a line. Therefore the envelope exists and is a ruled surface.

As an example, consider the geometric-progression family of planes, given by the following equation:

$$x + ty + t^2 z = t^3, \ t \in (R). \tag{8}$$

If it exists, an envelope of this family is determined by the solutions of the systems of Eq. (2), namely here

$$\begin{cases} x + ty + t^2 z - t^3 & = 0 \\ y + 2tz - 3t^2 & = 0 \end{cases}. \tag{9}$$

The solutions of this system of equations are given by

$$\begin{cases} x = t^2(s - 2t) \\ y = t(3t - 2s) \\ z = s \end{cases}, \ s, t, \in (R). \tag{10}$$

Equations (10) determine a surface in the 3D space (displayed in Fig. 6). Here an implicit equation may be obtained, either by hand or (as equations are polynomial) using the elimination implemented in the Gröbner bases package of the software. Both processes yield the following implicit equation:

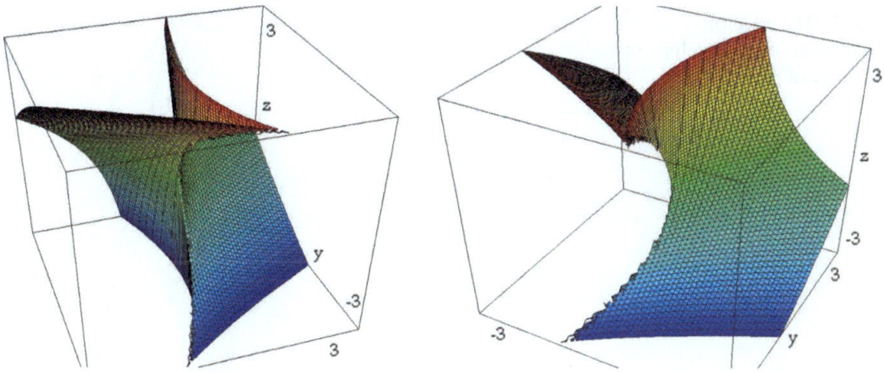

Fig. 6 Exploration of the envelope of a family of planes

$$27x^2 + 18xyz + 4xz^3 - 4y^3 - y^2z^2 = 0. \tag{11}$$

Figure 6 shows two views of the envelope, plotted using DPGraph. In order to determine the mesh, the software chose two kinds of lines, one of them is precisely the tangents to the edge of regression. Note that the edge of regression does not look so smooth. This is common situation close to singular points. We could improve this by choosing a finer mesh (DPGraph allows that), the cost of this change being that the surface could be covered with black lines. Figure 8 shows a better plot, obtained with Maple's parametric plot.

Using standard methods from Calculus (here it can be hand-made, in general the CAS has the necessary commands), we find the following parametric presentation for the edge of regression:

$$\begin{cases} x = \frac{r^3}{27} \\ y = -\frac{r^2}{3} \\ z = r \end{cases}, \ r \in \mathbb{R}, \tag{12}$$

and a general parametric presentation for characteristic lines is as follows:

$$\begin{cases} x = t^2 s - 2t^3 \\ y = -2ts + 3t^2 = 0 \\ z = s \end{cases}, \ s, t \in \mathbb{R}. \tag{13}$$

The curve alone is displayed in Fig. 7, from a point of view enabling to see that it is a nonplanar curve.

It can be proven that characteristic lines are tangent to the edge of regression of the envelope of the family of planes. This is clear on Fig. 8. In fact, the edge of regression is an envelope of the family of generators of the envelope of the family of planes (see [12]).

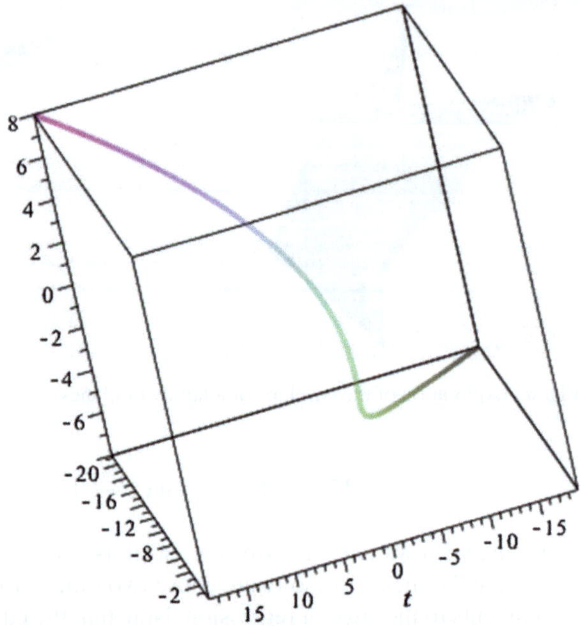

Fig. 7 The edge of regression of the envelope of a family of planes

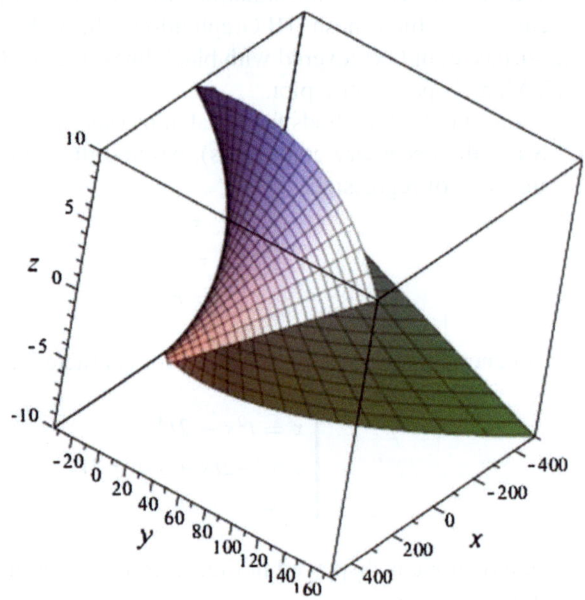

Fig. 8 The envelope as a rules surface

4 Discussion

As other classical topics in differential geometry, the study of envelopes of surfaces in 3D space provides an opportunity to discover new topics beyond the scope of the regular curriculum, sometimes together with applications to practical situations. For example, not every student knows what a ruled surface is: the problem studied in Sect. 3.3 is a good motivation for the student to acquire this new mathematical knowledge. Moreover new computation skills with technology may be developed, in particular for the experimental aspect of the work (e.g., exploring the existence of cusps, as in Fig. 1b). For this, the availability in the software of a slider bar is a central issue. Among these skills, ability to switch between different registers of representation may be improved, within mathematics themselves (parametric vs implicit) and with the computer (algebraic, graphical, etc.).

The algebraic engine we used in different CAS was computations of Gröbner bases for two purposes (see [1]).

1. to solve the given system of equations, which yields a parametric representation of the envelopes; the choice of a suitable ordering on the variables is generally made by the CAS itself, but sometimes the freedom of choice may lead to shorter solution process.
2. to look for an implicitization of this parametric representation. Here an elimination order is the core of the process.

These algorithms come from abstract algebra, a domain that not every student learning differential geometry masters. This can be a motivation to acquire new knowledge. Finally this new knowledge will be composed of both mathematical theory and computational skills.

Both for the parametric presentation and for the implicit equation, plotting may be not so easy, despite the availability of specific commands of the software with a lot of options. An **implicitplot** command exists in every software among those mentioned in Sect. 1. As it uses standard choices for the mesh (see [21]), the plot may be quite coarse, as mentioned in [14]. Parametric plotting should be better, as we can see in previous sections. Nevertheless, here too problems may appear.

Consider the example in Sect. 3.2 and add the following command row into the code:

```
> solve(F1 = 0 and F2 = 0, {x, y, z});
```

The output reads as follows:

$$x = -\frac{1}{2}\frac{y(-1+t^2)}{t}, y = y, z = \frac{1}{2}\frac{\sqrt{-y^2t^4+8yt^3-2y^2t^2-12t^2+8yt-y^2}}{t},$$

$$x = -\frac{1}{2}\frac{y(-1+t^2)}{t}, y = y, z = \frac{1}{2}\frac{-\sqrt{-y^2t^4+8yt^3-2y^2t^2-12t^2+8yt-y^2}}{t}.$$

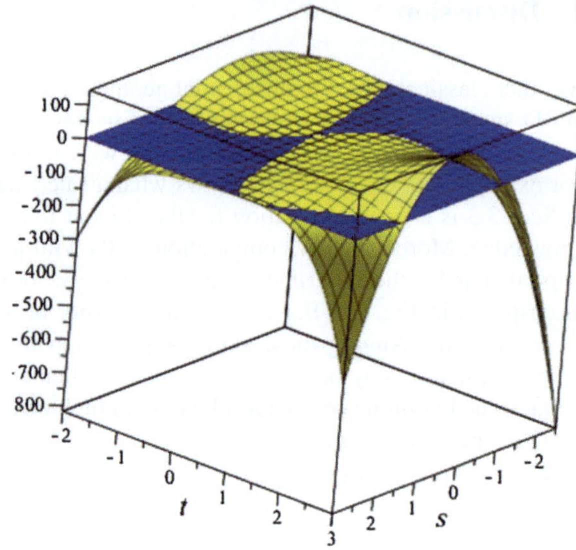

Fig. 9 Looking for suitable parameter range

The envelope is symmetric about the xy-plane and is given here decomposed into two parts: one for positive z (the first formulas) and one for negative z (the second formulas). Plotting will require substitution of a parameter s instead of y, then will be performed using the following structure of commands:

```
> p1:=P( "first parametrization", s= ..., t= ..., options):
> p2:=P( "second parametrization", s= ..., t= ..., options):
> display(p1,p2,scaling= ..., other options);
```

Finding a suitable range for the parameters s, t for the sake of plotting may be nontrivial. In our example, this requires the solution of a two-variable inequality, namely $-s^2t^4 + 8st^3 - 2s^2t^2 - 12t^2 + 8st - s^2 \geq 0$. A graphical solution of the inequality is provided by the CAS. Figure 9 shows the surface defined by the polynomial in variables s, t which must be nonnegative for the parametrization to be well defined. The surface is intersected by the zero plane, so the areas above correspond to. Or maybe not. In this second case, we are left with only the implicit plot possibility for the envelope, but even in the first case, the plot may be inaccurate because of the constraints of the software. When such an implicitization is not to be found, algorithms exist for an approximate implicitization (see [17]).

The transition from 2D to 3D is nontrivial. Automated deduction of the existence of an envelope uses the same methods in both cases, but visualization in 3D may require more abstraction skills. The animation tools implemented in many CAS for working with planar objects may exist also for a 3D setting, but they may not be at the same development level. Another feature is the **Trace** option; we saw in

Sect. 2 the efficiency of working with the slider bar together with **Trace on** (Fig. 2). This is a motivating constraint for more profound work with new commands in order to develop new mathematical knowledge, new mathematical skills and new computational-graphical tools.

References

1. Adams, W., Loustaunau, P.: An Introduction to Gröbner Bases. Graduate Studies in Mathematics, vol. 3. American Mathematical Society, Providence (1994)
2. Arnold, V.I.: On the envelope theory. Uspecki Math. Nauk. **3**(31), 248–249 (1976) (in Russian)
3. Capitanio, G.: On the envelope of 1-parameter families of curves tangent to a semicubic cusp. Comptes Rendus lAcad. Sci. **3**(335), 249–254 (2002)
4. Conkey, J., Joy, K.I.: Using isosurface methods for visualizing the envelope of a swept trivariate solid. In: Proceedings of Pacific Graphics 2000, Hong Kong, pp. 272–280
5. Cox, D., Little, J., OShea, D.: Ideals, Varieties, and Algorithms: An Introduction to Computational Algebraic Geometry and Commutative Algebra. Undergraduate Texts in Mathematics. Springer, New York (1992)
6. Cuoco, A., Levasseur, K.: Classical mathematics in the age of CAS. In: Fey, J., et al. (eds.) Computer Algebra System in Secondary School Mathematics Education, pp. 97–116. NCTM, Reston (2003)
7. Dana-Picard, T.: Technology as a bypass for a lack of theoretical knowledge. Int. J. Technol. Math. Educ. **11**(3), 101–109 (2005)
8. Dana-Picard, T., Naiman, A.: Closed paths of light trapped in a closed Fermat curves. Int. J. Math. Educ. Sci. Technol. **33**(6), 865–877 (2002)
9. Dana-Picard, T., Zehavi, N.: Revival of a classical topic in Differential geometry: the exploration of envelopes in a computerized environment. Preprint (2015)
10. Dana-Picard, T., Kidron, I., Zeitoun, D.: Strange 3D plots. In: Pitta-Pantazi, D., Philipou, G., (eds.) Proceedings of the Fifth Congress of the European Society for Research in Mathematics Education, Larnaca, Cyprus, pp. 1379–1388 (2007)
11. Dana-Picard, T., Mann, G., Zehavi, N.: Bisoptic curves of a hyperbola. Int. J. Math. Educ. Sci. Technol. **45**(5), 762–781 (2014)
12. Ferreol, R.: Surface enveloppe d'une famille de surfaces. Mathcurve (2009). http://www.mathcurve.com/surfaces/enveloppe/enveloppe.shtml
13. Hilbert, D.: Talk at International Congress of Mathematicians, Paris (1900). http://www-history.mcs.st-and.ac.uk/Extras/Hilbert_Problems.html
14. Kock, A.: Envelopes—notion and definiteness. Beiträge zur Algebra und Geometrie (Contributions to Algebra and Geometry) **48**, 345–350 (2007). www.emis.de/journals/BAG/vol.48/no.2/b48h2koc.pdf
15. Migliozzi, N.: Students stuff envelopes. Math. Educ. Res. **4**(2), 46–50 (1995)
16. Pech, P.: Selected Topics in Geometry with Classical vs. Computer Proving. World Scientific Publishing, Singapore (2007)
17. Pottman, H., Peternell, M.: Envelopes—computational theory and applications. In: Proceedings of Spring Conference in Computer Graphic, Budmerice, Slovakia, pp. 3–23 (2000)
18. Struik, D.: Lectures on classical differential geometry. Addison-Wesley, MA: Cambridge (1950) [2nd ed. (1961), Republished by Dover (1988)]
19. Thom, R.: Sur la théorie des enveloppes. J. Math. Pures Appl. **XLI**(2), 177–192 (1962)

20. Yerusalmy, M.: Does Technology Transform the Content of Algebra Curricula? An Analysis of Critical Transitions for Learning and Teaching. International Congress on Mathemaical Education, Copenhagen (2004)
21. Zeitoun, D., Dana-Picard, Th.: Accurate visualization of graphs of functions of two real variables. Int. J. Comput. Math. Sci. **4**(1), 1–11 (2010). http://www.waset.org/journals/ijcms/v4/v4-1-1.pdf

Complex Roots of Quaternion Polynomials

Petroula Dospra and Dimitrios Poulakis

Abstract In this paper, using hybrid Bézout matrices, we give necessary and sufficient conditions, for a quaternion polynomial to have a complex root, a spherical root, and a complex isolated root. These conditions can be easily checked since these matrices are implemented in the computational system MAPLE. Moreover, we compute upper bounds for the norm of the roots of a quaternion polynomial.

Keywords Quaternion polynomial · Bézout Matrices · Spherical Root · Isolated Root

1 Introduction

Let $\mathbb{R}$ and $\mathbb{C}$ be the fields of real and complex numbers, respectively. We consider the division ring of real quaternions,

$$\mathbb{H} = \{x_0 + x_1\mathbf{i} + x_2\mathbf{j} + x_3\mathbf{k} : \ x_0, x_1, x_2, x_3 \in \mathbb{R}\},$$

where the elements $\mathbf{i}, \mathbf{j}, \mathbf{k}$ satisfy the following multiplication rules:

$$\mathbf{i}^2 = \mathbf{j}^2 = \mathbf{k}^2 = -1, \quad \mathbf{ij} = -\mathbf{ji} = \mathbf{k}, \quad \mathbf{jk} = -\mathbf{kj} = \mathbf{i}, \quad \mathbf{ki} = -\mathbf{ik} = \mathbf{j}.$$

Let $q = x_0 + x_1\mathbf{i} + x_2\mathbf{j} + x_3\mathbf{k}$ be an element of $\mathbb{H}$. The *real* and the *imaginary part* of q are the elements $\operatorname{Re} q = x_0$ and $\operatorname{Im} q = x_1\mathbf{i} + x_2\mathbf{j} + x_3\mathbf{k}$, respectively. The *conjugate* of q is defined to be the quaternion $\bar{q} = x_0 - x_1\mathbf{i} - x_2\mathbf{j} - x_3\mathbf{k}$ and its *norm* the quantity

$$|q| = \sqrt{q\bar{q}} = \sqrt{x_0^2 + x_1^2 + x_2^2 + x_3^2}.$$

P. Dospra · D. Poulakis (✉)
Department of Mathematics, Aristotle University of Thessaloniki,
54124 Thessaloniki, Greece
e-mail: poulakis@math.auth.gr

P. Dospra
e-mail: petroula.dospra@gmail.com

I.S. Kotsireas and E. Martínez-Moro (eds.), *Applications of Computer Algebra*,
Springer Proceedings in Mathematics & Statistics 198,
DOI 10.1007/978-3-319-56932-1_5

Let $q, q' \in \mathbb{H}$. We say that the quaternions q and q' are *congruent* or *equivalent*, and we write $q \sim q'$, if there is $w \in \mathbb{H} \setminus \{0\}$ such that $q' = wqw^{-1}$. The congruence relation is an equivalence relation in $\mathbb{H}$. By [27], we have $q \sim q'$ if and only if $\mathrm{Re}\, q = \mathrm{Re}\, q'$ and $|q| = |q'|$. The *congruence class* of q is the set

$$[q] = \{q' \in \mathbb{H}/\ q' \sim q\} = \{q' \in \mathbb{H}/\ \mathrm{Re}\, q = \mathrm{Re}\, q',\ |q| = |q'|\}.$$

In every class $[q]$ there is exactly one complex number z and its conjugate $\bar{z}$, which are $x_0 \pm \mathbf{i}\sqrt{x_1^2 + x_2^2 + x_3^2}$. The quaternions are implemented in the computational package MAGMA and so, we can easily perform basic operations with them.

A polynomial with coefficients in $\mathbb{H}$ is an expression of the form

$$f(t) = a_0 t^n + a_1 t^{n-l} + \cdots + a_n,$$

where $a_0, \ldots, a_n \in \mathbb{H}$. Equality of two such polynomials is defined in the usual way. If $a_0 \neq 0$, then n is called the degree of $f(x)$. The addition and the multiplication of quaternion polynomials are defined in the same way as the commutative case, where the variable t is assumed to commute with quaternion coefficients [17, Chap. 5, Sect. 16]. The evaluation of $f(t)$ at $q \in \mathbb{H}$ is the element

$$f(q) = a_0 q^n + a_1 q^{n-l} + \cdots + a_n.$$

Note that the evaluation at q is not in general a ring homomorphism from $\mathbb{H}[t]$ to $\mathbb{H}$.

A quaternion q is said to be a *zero* or a *root* of $f(t)$ if $f(q) = 0$. The polynomial $B(t) \in \mathbb{H}[t]$ is called a *right factor* of $Q(t)$ if there exists $C(t) \in \mathbb{H}[t]$ such that $Q(t) = C(t)B(t)$. By [17, Proposition 16.2], q is a root of $f(t)$ if and only if $t - q$ is a right factor of $Q(t)$.

Let q be a root of $f(t)$. If q is not real and has the property that $f(z) = 0$ for all $z \in [q]$, then we will say that q generates a *spherical root*. For short, we will also say that q is, rather than generates, a spherical root. If q is real or does not generate a spherical zero, it is called an *isolated root*. By [11, Theorem 4], we have that if two elements of a class are zeros of $f(t)$, then all elements of this class are also zeros of $f(t)$. On the other hand, every congruence class contains exactly one complex number z and its conjugate $\bar{z}$. It follows that the pairs of complex numbers $\{z, \bar{z}\}$ which are roots of $f(t)$ determine all spherical roots of $f(t)$.

The roots of a quaternion polynomial and its expression as a product of linear factors have been investigated in several papers [5, 8–11, 13, 16, 19, 21–24, 26]. Furthermore, quaternion polynomials are used for the presentation of a particular class of space curves, namely the Pythagorean hodograph curves [4]. Such a curve can be generated by another of lower degree if and only if its associated quaternion polynomial has a complex root [4, Chap. 6].

In this paper, we deal with the conditions under which a quaternion polynomial has a complex root. Our study contains the case of spherical roots since such roots correspond to pairs of complex numbers $\{z, \bar{z}\}$ which are roots of $f(t)$. We use hybrid

Bézout matrices and we give necessary and sufficient conditions for a quaternion polynomial to have a complex root. Furthermore, we present necessary and sufficient conditions for a such polynomial to have a spherical root and a complex isolated root. We have chosen the approach of hybrid Bézout matrices since these matrices have better computational behavior than others [3]. Thus, we provide a practical way to check easily the existence of this kind of roots by using a tool which is already implemented in the computational package MAPLE. Moreover, using some results from complex polynomials, we give upper bounds for the norm of the roots of a quaternion polynomial which are comparable with previous ones [20].

The paper is organized as follows. Section 2 is devoted to the presentation of Barnett's theorem with hybrid Bézout matrices. In Sect. 3, necessary and sufficient conditions are given for a quaternion polynomial to have a complex root. Further, we apply this condition in quadratic polynomials and we compute the roots in the case where one of them is complex. In Sect. 4, we give two necessary and sufficient conditions for a quaternion polynomial to have a spherical and a complex isolated root, respectively. Using some results for the roots of complex polynomials, we compute, in Sect. 5, upper bounds for the norm of spherical roots and complex isolated roots of a quaternion polynomial. Furthermore, we give an upper bound for the norm of an arbitrary root. The last section concludes the paper.

2 Barnett's Theorems

In 1971, Barnett computed the degree (resp. coefficients) of the greatest common divisor of several univariate polynomials with coefficients in an integral domain by means of the rank (resp. linear dependencies of the columns) of several matrices involving theirs coefficients [1, 2]. In [3], a formulation of Barnett's results is given using Bézout matrices, Henkel matrices and hybrid Bézout matrices. As it is mentioned in [3], the hybrid Bézout matrices have the best computational behavior.

In this section, we recall the formulation of Barnett's theorem for hybrid Bézout matrices [3] which we shall use for the presentation of our results. We could equally use another formulation of Barnett's results given in [3] or to use another approach [7, 15, 25], but we have chosen the above one as more efficient in computations [3, Sect. 6.3]. Note also that these matrices are implemented in the Computer Algebra System MAPLE (see LinearAlgebra[BezoutMatrix] or linalg[bezout]).

Let $\mathbb{F}$ be a field of characteristic zero and $\mathbb{F}[t]$ the ring of polynomials with coefficients in $\mathbb{F}$. Consider polynomials

$$P(t) = p_0 t^n + p_1 t^{n-1} + \cdots + p_n \quad \text{and} \quad Q(t) = q_0 t^m + q_1 t^{m-1} + \cdots + q_m,$$

in $\mathbb{F}[t]$ with $n \geq m$. The hybrid Bézout matrix, denoted by $\mathrm{Hbez}(P, Q)$, is a square matrix of size n whose entries are defined by

- for $1 \leq i \leq m$, $1 \leq j \leq n$, the (i, j)-entry is the coefficient of t^{n-j} in the polynomial

$$K_{m-i+1} = (p_0 t^{m-i} + \cdots + p_{m-i})(q_{m-i+1} t^{n-m+i-1} + \cdots + q_m t^{n-m}) - (p_{m-i+1} t^{n-m+i-1} + \cdots + p_n)(q_0 t^{m-i} + \cdots + q_{m-i})$$

- for $m+1 \leq i \leq n$, $1 \leq j \leq n$, the (i, j)-entry is the coefficient of t^{n-j} in the polynomial $t^{n-i} Q(t)$.

Let $R(P, Q)$ be the well-known Sylvester resultant of $P(t)$ and $Q(t)$. By [3, Corollary 5.2], we have

$$\det(\mathrm{Hbez}(P, Q)) = R(P, Q).$$

It follows that $\det(\mathrm{Hbez}(P, Q)) = 0$ if and only if $\deg(\gcd(P, Q)) \geq 1$.

Now, let $P(t)$, $Q_1(t), \ldots, Q_k(t)$ be a family of polynomials in $\mathbb{F}[t]$ with $n = \deg P$ and $\deg Q_j \leq n-1$ for every $j \in \{1, \ldots, k\}$. Set

$$\mathscr{BH}_P(Q_1, \ldots, Q_k) = \begin{pmatrix} \mathrm{Hbez}(P, Q_1) \\ \vdots \\ \mathrm{Hbez}(P, Q_k) \end{pmatrix}.$$

Let $D(t)$ be the greatest common divisor of $P(t)$, $Q_1(t), \ldots, Q_k(t)$ over $\mathbb{F}$. Barnett's result that we shall use is given in the following lemma:

Lemma 1 *Let $C_1, \ldots, C_n$ be the columns and r the rank of $\mathscr{BH}_P(Q_1, \ldots, Q_k)$. Then, we have*

$$\deg D = n - r.$$

Proof See [3, Theorem 5.1]. □

By [3, Sect. 6], the computation of hybrid Bézout matrices and their rank require $O(n^2)$ arithmetic operations.

Remark 1 Let $A_1(t), \ldots, A_k(t)$ be polynomials of $\mathbb{F}[t] \setminus \mathbb{F}$. We compute the remainder $\tilde{A}_j(t)$ of the division of $A_j(t)$ by $A_1(t)$ $(j = 2, \ldots, k)$ and we have that $\deg A_1 > \deg A_i$ $(i = 2, \ldots, k)$ and

$$\gcd(A_1, \ldots, A_k) = \gcd(A_1, \tilde{A}_2, \ldots, \tilde{A}_k).$$

Thus, we can also apply the above lemma even if the condition on the degrees of polynomials is not fulfilled.

3 Complex Roots

In this section, we give necessary and sufficient conditions for a quaternion polynomial to have a complex root and we apply this result for computing the roots of a quadratic polynomial in this case.

Theorem 1 *Let $Q(t) \in \mathbb{H}[t] \setminus \mathbb{C}[t]$ be a monic polynomial with* $\deg Q = n \geq 1$ *and* $f(t), g(t) \in \mathbb{C}[t]$ *with* $f(t)g(t) \neq 0$ *such that* $Q(t) = f(t) + \mathbf{k}g(t)$. *Set* $E(t) = \gcd(f(t), g(t))$. *Then the roots of $E(t)$ are precisely the complex roots of $Q(t)$. Furthermore, the following are equivalent:*
(a) *$Q(t)$ has a complex root.*
(b) $\deg E(t) > 0$.
(c) $\det(\mathrm{Hbez}(f, g)) = 0$.
(d) $R(f, g) = 0$.

Proof Let $z \in \mathbb{C}$. Then z is a root of $Q(t)$ if and only if there exists $A(t) \in \mathbb{H}[t] \setminus \mathbb{C}[t]$ such that $Q(t) = A(t)(t - z)$. Write $A(t) = a(t) + \mathbf{k}b(t)$, where $a(t), b(t) \in \mathbb{C}[t]$. Thus, z is a root of $Q(t)$ if and only if we have

$$f(t) = a(t)(t - z) \quad \text{and} \quad g(t) = b(t)(t - z).$$

Therefore, the complex roots of $Q(t)$ are exactly the roots of $E(t)$. Hence $Q(t)$ has a complex root if and only if the greatest common divisor $E(t)$ is not 1. Thus, we get the equivalence of (a) and (b).

By Lemma 1, we have

$$\deg E = n - \mathrm{rankHbez}(f, g).$$

It follows that $\deg E > 0$ if and only if we have

$$n > \mathrm{rankHbez}(f, g)$$

which is equivalent to $\det(\mathrm{Hbez}(f, g)) = 0$. Thus, we obtain the equivalence of (b) and (c). Finally, since $\det(\mathrm{Hbez}(f, g)) = R(f, g)$, the equivalence of (c) and (d) follows. □

Example 1 Consider the polynomial

$$Q(t) = t^4 - (2 + \mathbf{k})t^3 + (3 + \mathbf{j} + 2\mathbf{k})t^2 - 2(1 + \mathbf{j} + \mathbf{k})t + 2(1 + \mathbf{j}).$$

We shall examine, using Theorem 1, whether or not $Q(t)$ has a complex root. We have $Q(t) = f(t) + \mathbf{k}g(t)$, where

$$f(t) = t^4 - 2t^3 + 3t^2 - 2t + 2, \quad g(t) = -t^3 + (2 + \mathbf{i})t^2 - 2(1 + \mathbf{i})t + 2\mathbf{i}.$$

Using MAPLE we obtain the hybrid Bézout matrix of $f(t)$ and $g(t)$

$$\text{Hbez}(f, g) = \begin{pmatrix} -2+2\mathbf{i} & 6-2\mathbf{i} & -8 & 4+4\mathbf{i} \\ 1-2\mathbf{i} & -4+3\mathbf{i} & 6-2\mathbf{i} & -4-2\mathbf{i} \\ \mathbf{i} & 1-2\mathbf{i} & -2+2\mathbf{i} & 2 \\ -1 & 2+\mathbf{i} & -2-2\mathbf{i} & 2\mathbf{i} \end{pmatrix}.$$

Furthermore, we get $\text{rank}\,\text{Hbez}(f, g) = 1$ and hence we have $\det \text{Hbez}(f, g) = 0$. By Theorem 1, the polynomial $Q(t)$ has a complex root.

The case of quadratic quaternion equations has been studied in [6, 12, 14, 19]. The next theorem uses the previous result and provides their solutions in the special case where one of them is complex.

Theorem 2 *Let $Q(t) = t^2 + q_1 t + q_0$ be a quadratic polynomial of $\mathbb{H}[t] \setminus \mathbb{C}[t]$ with no real factor. Set $q_1 = b_1 + \mathbf{k}c_1$ and $q_0 = b_0 + \mathbf{k}c_0$, where $b_0, b_1, c_0, c_1 \in \mathbb{C}$. Then $Q(t)$ has a complex root if and only if*

$$c_0^2 - c_0 b_1 c_1 + b_0 c_1^2 = 0.$$

In this case, we have $c_0 c_1 \neq 0$, and the roots of $Q(t)$ are

$$q = -\frac{c_0}{c_1}, \qquad \sigma = (q - \bar{p})^{-1} p (q - \bar{p}),$$

where $p = -(b_0 c_1 / c_0 + \mathbf{k}c_1)$.

Proof Set $Q(t) = f(t) + \mathbf{k}g(t)$, where $f(t) = t^2 + b_1 t + b_0$ and $g(t) = c_1 t + c_0$. We have

$$\det(\text{Hbez}(f, g)) = c_0^2 - c_0 b_1 c_1 + b_0 c_1^2$$

and by Theorem 1, $Q(t)$ has a complex root if and only if the above quantity is zero.

Suppose now that $Q(t)$ has a complex root q. If $c_1 = 0$, then the above equality implies $c_0 = 0$ and hence $Q(t) \in \mathbb{C}[t]$ which is a contradiction. Thus $c_1 \neq 0$. If $c_0 = 0$, then we deduce $b_0 = 0$, and so t is a factor of $Q(t)$ which is a contradiction. Therefore $c_0 \neq 0$.

Since $Q(q) = 0$, we have $g(q) = 0$ and $f(q) = 0$. It follows that $q = -c_0/c_1$ and $f(t) = (t - b_0/q)(t - q)$. Thus, we have the factorization

$$Q(t) = (t - p)(t - q),$$

where $p = -(b_0 c_1 / c_0 + \mathbf{k}c_1)$. If $p = \bar{q}$, then we have $b_0 c_1 / c_0 + \mathbf{k}c_1 = \bar{c}_0 / \bar{c}_1$. It follows that $c_1 = 0$ which is a contradiction. Thus, we have $p \neq \bar{q}$, and so, [22, Lemma 1] yields

$$Q(t) = (t - (p - \bar{q}) q (p - \bar{q})^{-1})(t - (q - \bar{p})^{-1} p (q - \bar{p})).$$

Hence, the other root of $Q(t)$ is $\sigma = (q - \bar{p})^{-1} p (q - \bar{p})$. □

Example 2 We consider the polynomial

$$Q(t) = t^2 + (1 - \mathbf{i} + \mathbf{j} + \mathbf{k})t + 1 - (1/2)\mathbf{i} + \mathbf{j}.$$

First, we shall examine whether or not this polynomial has a complex root. We write

$$1 - \mathbf{i} + \mathbf{j} + \mathbf{k} = 1 - \mathbf{i} + \mathbf{k}(1 + \mathbf{i}), \quad 1 - (1/2)\mathbf{i} + \mathbf{j} = 1 - (1/2)\mathbf{i} + \mathbf{k}\mathbf{i}.$$

Following the notations of Theorem 2, we set

$$b_1 = 1 - \mathbf{i}, \quad c_1 = 1 + \mathbf{i}, \quad b_0 = 1 - (1/2)\mathbf{i}, \quad c_0 = \mathbf{i}$$

and we get:

$$c_0^2 - c_0 b_1 c_1 + b_0 c_1^2 = 0.$$

Then, Theorem 2 implies that $Q(t)$ has a complex root. Furthermore, this root is

$$q = -\frac{c_0}{c_1} = -\frac{\mathbf{i}}{1 + \mathbf{i}} = -\frac{1 + \mathbf{i}}{2}.$$

Next, we compute

$$p = -(b_0 c_1 / c_0 + \mathbf{k} c_1) = \frac{-1 + 3\mathbf{i}}{2} - \mathbf{j} - \mathbf{k}.$$

Thus, the other root is

$$\sigma = (q - \bar{p})^{-1} p (q - \bar{p}) = -\frac{1}{2} + \frac{5}{6}\mathbf{i} - \frac{4}{3}\mathbf{j} - \frac{4}{3}\mathbf{k}.$$

4 Spherical and Complex Isolated Roots

In this section, we give two necessary and sufficient conditions for a quaternion polynomial to have a spherical root and an isolated complex root, respectively.

Let $Q(t)$ be a monic polynomial in $\mathbb{H}[t] \setminus \mathbb{C}[t]$. Suppose that $f_1(t)$, $f_2(t)$, $f_3(t)$, $f_4(t)$ are polynomials of $\mathbb{R}[t]$ such that

$$Q(t) = f_1(t) + f_2(t)\mathbf{i} + f_3(t)\mathbf{j} + f_4(t)\mathbf{k}.$$

Denote by $D(t)$ its greatest common divisor. The polynomial $Q(t)$ has a real root if and only if the polynomials $f_1(t)$, $f_2(t)$, $f_3(t)$, $f_4(t)$ have a common real root. Thus, we have that $Q(t)$ has a real root if and only if $D(t)$ has a real root.

Write $D(t) = D_1(t)D_2(t)$, where $D_1(t)$ and $D_2(t)$ are polynomials of $\mathbb{R}[t]$. If $D_1(t) \notin \mathbb{R}$, then it has only real roots, and if $D_2(t) \notin \mathbb{R}$, then it has only non real roots. Then, the polynomial $Q(t)$ has a spherical (respectively complex isolated) root if and only if the polynomial $Q(t)/D_1(t)$ does. Thus, in order to study the existence of spherical roots and complex isolated roots of $Q(t)$ is enough to study the case where $Q(t)$ has no real root.

Theorem 3 *Suppose that the quaternion polynomial $Q(t)$ has no real root. Then the pairs of complex conjugate roots of $D(t)$ define all the spherical roots of $Q(t)$. Furthermore, $Q(t)$ has a spherical root if and only if the following inequality holds:*

$$n > \operatorname{rank} \mathscr{B}\mathscr{H}_{f_1}(f_2, f_3, f_4).$$

Proof If $D(t)$ has a real root, then $Q(t)$ has a real root which is a contradiction. Hence, $D(t)$ has no real root. Suppose now that $z \in \mathbb{C} \setminus \mathbb{R}$ is a root of $D(t)$. Thus, its conjugate $\bar{z}$ is also a root of $D(t)$. It follows that z and $\bar{z}$ are roots of $Q(t)$. Therefore, z is a spherical root of $Q(t)$.

Setting $f(t) = f_1(t) + f_2(t)\mathbf{i}$ and $g(t) = f_4(t) + f_3(t)\mathbf{i}$, we have:

$$Q(t) = f(t) + \mathbf{k}g(t).$$

Suppose that $Q(t)$ has a spherical root q. Let z and $\bar{z}$ be the only complex numbers of the class of q. Then we have that $Q(z) = Q(\bar{z}) = 0$, whence we get

$$f(z) = f(\bar{z}) = 0 \quad \text{and} \quad g(z) = g(\bar{z}) = 0.$$

It follows that the real polynomial $(t - z)(t - \bar{z})$ divides $f(z)$ and $g(z)$ and hence $f_1(t), f_2(t), f_3(t), f_4(t)$. Thus, $(t - z)(t - \bar{z})$ divides $D(t)$. Therefore, the pairs of complex conjugate roots of $D(t)$ define all the spherical roots of $Q(t)$. It follows that $Q(t)$ has a spherical root if and only if $\deg D(t) > 0$. On the other hand side, Lemma 1 yields

$$\deg D(t) = n - \operatorname{rank} \mathscr{B}\mathscr{H}_{f_1}(f_2, f_3, f_4).$$

Thus, $Q(t)$ has a spherical root if and only if we have

$$n > \operatorname{rank} \mathscr{B}\mathscr{H}_{f_1}(f_2, f_3, f_4).$$

□

Corollary 1 *If $D(t) = 1$, then the polynomial $Q(t)$ has no spherical or real root.*

Example 3 In [23, Example 2], the roots of the following polynomial have been computed:

$$P(t) = t^6 + (\mathbf{i} + 3\mathbf{k})t^5 + (3 + \mathbf{j})t^4 + (5\mathbf{i} + 15\mathbf{k})t^3 + (-4 + 5\mathbf{j})t^2 + (6\mathbf{i} + 18\mathbf{k})t - 12 + 6\mathbf{j}.$$

We shall see if $P(t)$ has a spherical root. We have

$$\begin{aligned}
f_1(t) &= t^6 + 3t^4 - 4t^2 - 12,\\
f_2(t) &= t^5 + 5t^3 + 6t,\\
f_3(t) &= t^4 + 5t^2 + 6,\\
f_4(t) &= 3t^5 + 15t^3 + 18t.
\end{aligned}$$

We easily see that $f_3(t)$ has no real root. It follows that $P(t)$ has no real root. Using MAPLE, we get

$$\mathrm{Hbez}(f_1, f_2) = \begin{pmatrix} 0 & 12 & 0 & 60 & 0 & 72 \\ 10 & 0 & 50 & 0 & 60 & 0 \\ 0 & 10 & 0 & 50 & 0 & 60 \\ 2 & 0 & 10 & 0 & 12 & 0 \\ 0 & 2 & 0 & 10 & 0 & 12 \\ 1 & 0 & 5 & 0 & 6 & 0 \end{pmatrix},$$

$$\mathrm{Hbez}(f_1, f_3) = \begin{pmatrix} 10 & 0 & 50 & 0 & 60 & 0 \\ 0 & 10 & 0 & 50 & 0 & 60 \\ 2 & 0 & 10 & 0 & 12 & 0 \\ 0 & 2 & 0 & 10 & 0 & 12 \\ 1 & 0 & 5 & 0 & 6 & 0 \\ 0 & 1 & 0 & 5 & 0 & 6 \end{pmatrix},$$

and

$$\mathrm{Hbez}(f_1, f_4) = \begin{pmatrix} 0 & 36 & 0 & 180 & 0 & 216 \\ 30 & 0 & 150 & 0 & 180 & 0 \\ 0 & 30 & 0 & 150 & 0 & 180 \\ 6 & 0 & 30 & 0 & 36 & 0 \\ 0 & 6 & 0 & 30 & 0 & 36 \\ 3 & 0 & 15 & 0 & 18 & 0 \end{pmatrix}.$$

Next, we consider the matrix

$$\mathscr{BH}_{f_1}(f_2, f_3, f_4) = \begin{pmatrix} \mathrm{Hbez}(f_1, f_2) \\ \mathrm{Hbez}(f_1, f_3) \\ \mathrm{Hbez}(f_1, f_4) \end{pmatrix}.$$

Using MAPLE, we get rank $\mathscr{BH}_{f_1}(f_2, f_3, f_4) = 2$. Since we have $\deg P(t) = 6$ and $P(t)$ has no real root, Theorem 3 implies that $P(t)$ has a spherical root.

Let $f(t) = f_1(t) + f_2(t)\mathbf{i}$ and $g(t) = f_4(t) + f_3(t)\mathbf{i}$. Then $Q(t) = f(t) + \mathbf{k}g(t)$. We denote by $E(t)$ the greatest common divisor of $f(t)$ and $g(t)$. Since $D(t)$ divides $f_1(t), f_2(t), f_3(t), f_4(t)$, we deduce that $D(t)$ divides $f(t)$ and $g(t)$. It follows that $D(t)$ divides $E(t)$, and so we have $E(t) = D(t)\tilde{E}(t)$, where $\tilde{E}(t) \in \mathbb{C}[t]$.

Theorem 4 *Suppose that the quaternion polynomial $Q(t)$ has no real root. Then the roots of $\tilde{E}(t)$ are exactly the complex isolated roots of $Q(t)$. Furthermore, $Q(t)$*

has an isolated complex root if and only if the following inequality holds:

$$\operatorname{rank} \operatorname{Hbez}(f, g) < \operatorname{rank} \mathscr{BH}_{f_1}(f_2, f_3, f_4).$$

Proof By Theorem 3, we have $Q(t) = D(t)P(t)$, and $P(t)$ is a quaternion polynomial with only isolated non real roots. Then $Q(t)$ has an isolated complex root if and only if $P(t)$ has a complex root. Setting

$$\alpha(t) = f(t)/D(t) \quad \text{and} \quad \beta(t) = g(t)/D(t),$$

we get:

$$P(t) = \alpha(t) + \mathbf{k}\beta(t).$$

Furthermore, we have

$$\gcd(\alpha(t), \beta(t)) = \tilde{E}(t).$$

Theorem 1 implies that $P(t)$ has a complex root if and only if $\deg \tilde{E}(t) > 0$. Lemma 1 yields:

$$\deg E(t) = n - \operatorname{rank} \operatorname{Hbez}(f, g) \quad \text{and} \quad \deg D(t) = n - \operatorname{rank} \mathscr{BH}_{f_1}(f_2, f_3, f_4).$$

Thus, since $\deg \tilde{E} = \deg E - \deg D$, we deduce that $Q(t)$ has a complex root if and only if we have

$$\operatorname{rank} \operatorname{Hbez}(f, g) < \operatorname{rank} \mathscr{BH}_{f_1}(f_2, f_3, f_4).$$

□

Corollary 2 *If $E(t) = D(t)$, then the polynomial $Q(t)$ has no complex isolated root.*

Example 4 Consider the polynomial

$$R(t) = t^4 - (2 + \mathbf{k})t^3 + (3 + \mathbf{j} + 2\mathbf{k})t^2 - 2(1 + \mathbf{j} + \mathbf{k})t + 2(1 + \mathbf{j}).$$

We shall apply Theorem 4 to check if $R(t)$ possess an isolated complex root. We write

$$\begin{aligned} f_1(t) &= t^4 - 2t^3 + 3t^2 - 2t + 2, \\ f_2(t) &= 0, \\ f_3(t) &= t^2 - 2t + 2, \\ f_4(t) &= -t^3 + 2t^2 - 2t. \end{aligned}$$

The polynomial $f_3(t)$ has no real root, and so, $R(t)$ has no real root. Further, we have

$$f(t) = t^4 - 2t^3 + 3t^2 - 2t + 2, \quad g(t) = -t^3 + 2t^2 - 2t + (t^2 - 2t + 2)\mathbf{i}.$$

Using MAPLE, we get

$$\mathrm{Hbez}(f, g) = \begin{pmatrix} -2+2\mathbf{i} & 6-2\mathbf{i} & -8 & 4+4\mathbf{i} \\ 1-2\mathbf{i} & -4+3\mathbf{i} & 6-2\mathbf{i} & -4-2\mathbf{i} \\ \mathbf{i} & 1-2\mathbf{i} & -2+2\mathbf{i} & 2 \\ -1 & 2+\mathbf{i} & -2-2\mathbf{i} & 2\mathbf{i} \end{pmatrix}.$$

Furthermore, we compute rank $\mathrm{Hbez}(f, g) = 1$. On the other hand, we have

$$\mathrm{Hbez}(f_1, f_2) = 0,$$

$$\mathrm{Hbez}(f_1, f_3) = \begin{pmatrix} -1 & 4 & -6 & 4 \\ 0 & -1 & 2 & -2 \\ 1 & -2 & 2 & 0 \\ 0 & 1 & -2 & 2 \end{pmatrix}$$

and

$$\mathrm{Hbez}(f_1, f_4) = \begin{pmatrix} -2 & 6 & -8 & 4 \\ 1 & -4 & 6 & -4 \\ 0 & 1 & -2 & 2 \\ -1 & 2 & -2 & 0 \end{pmatrix}.$$

Next, we built the matrix

$$\mathscr{BH}_{f_1}(f_2, f_3, f_4) = \begin{pmatrix} \mathrm{Hbez}(f_1, f_2) \\ \mathrm{Hbez}(f_1, f_3) \\ \mathrm{Hbez}(f_1, f_4) \end{pmatrix}$$

and using MAPLE we get rank $\mathscr{BH}_{f_1}(f_2, f_3, f_4) = 2$. Thus, we have

$$\mathrm{rank}\, \mathrm{Hbez}(f, g) = 1 < 2 = \mathrm{rank}\, \mathscr{BH}_{f_1}(f_2, f_3, f_4).$$

Hence, Theorem 4 implies that $R(t)$ has a complex isolated root.

5 Bounds for the Size of the Roots

In [20, Sect. 4] some upper bounds for the size of roots of quaternion polynomials are given. In this section we compute new bounds which are comparable with the bound given in [20, Theorem 4.2], and better in some cases (see Remark 2 below). Furthermore, we give upper bounds for the size of spherical and complex isolated roots.

Let

$$Q(t) = a_0 t^n + a_1 t^{n-1} + \cdots + a_n.$$

be a quaternion polynomial. We define the *height* of $Q(t)$ to be the quantity

$$H(Q) = \max\{1, |a_1/a_0|, \ldots, |a_n/a_0|\}.$$

We write $Q(t) = f(t) + \mathbf{k}g(t)$, where $f(t), g(t) \in \mathbb{C}[t]$, and

$$f(t) = f_1(t) + f_2(t)\mathbf{i}, \quad g(t) = g_1(t) + g_2(t)\mathbf{i},$$

where $f_1(t), f_2(t), g_1(t), g_2(t) \in \mathbb{R}[t]$. We set

$$\mathcal{H}_1 = \min\{H(f), H(g)\} \quad \text{and} \quad \mathcal{H}_2 = \min\{H(f_1), H(f_2), H(g_1), H(g_2)\}.$$

Theorem 5 *Suppose that the polynomial $Q(t)$ is monic and ρ is a root of $Q(t)$. If ρ is a spherical root, then we have*

$$|\rho| < 1 + \mathcal{H}_2^{1/2}.$$

If ρ is an isolated complex root, then we have

$$|\rho| < 1 + \mathcal{H}_1.$$

In the general case, the following inequality holds:

$$|\rho| < 1 + H(Q).$$

Proof Suppose that ρ is a spherical root of $Q(t)$. Then there is $z \in \mathbb{C} \setminus \mathbb{R}$ in the class of ρ which is also a root of $Q(t)$. By Theorem 2, z is a common complex root of $f_1(t), f_2(t), g_1(t), g_2(t)$. Thus, [18, Corollary 3] implies that $|z| < 1 + \mathcal{H}_2^{1/2}$. Since $|\rho| = |z|$, we obtain $|\rho| < 1 + \mathcal{H}_2^{1/2}$.

Suppose that ρ is an isolated root. If $\rho \in \mathbb{C}$, then Theorem 1 implies that ρ is a common root of $f(t)$ and $g(t)$. Hence [18, Corollary 2] yields $|\rho| < 1 + \mathcal{H}_1$.

Suppose next that ρ is an isolated noncomplex root. If $|\rho| \leq 1$, then the result is true. Suppose that $|\rho| > 1$. Since ρ is a root of $Q(t)$, there is $G(t) \in \mathbb{H}[t]$ such that $Q(t) = G(t)(t - \rho)$. Write

$$G(t) = t^{n-1} + b_1 t^{n-2} + \cdots + b_{n-1}.$$

Then

$$Q(t) = G(t)(t - \rho) = t^n + (b_1 - \rho)t^{n-1} + (b_2 - b_1\rho)t^{n-2} + \cdots + b_{n-1}\rho.$$

It follows that

$$a_1 = b_1 - \rho, \quad a_2 = b_2 - b_1\rho, \quad a_3 = b_3 - b_2\rho, \ldots, \quad a_n = b_{n-1}\rho.$$

Let i be the smallest index such that $H(G) = |b_i|$. Then we have

$$H(Q) \geq |b_i - b_{i-1}\rho| \geq ||b_i| - |b_{i-1}\rho|| \geq |H(G) - |b_{i-1}\rho|| > H(G)|1 - |\rho||,$$

whence we deduce the result. □

Remark 2 In case where $a_0 = 1$, [20, Theorem 4.2] yields that the roots ρ of $Q(t)$ satisfy

$$|\rho| \leq \max\{1, \sum_{i=1}^{n} |a_i|\}.$$

If we have

$$\sum_{i=1}^{n} |a_i| > 1 + H(Q),$$

then Theorem 5 gives a better upper bound.

Corollary 3 *Let $Q(t) \in \mathbb{H}[t] \setminus \mathbb{H}$ be a monic polynomial. Then $Q(t)$ has at most a finite number of roots $\mathbf{x}$ of the form $\mathbf{x} = x_1 + x_2\mathbf{i} + x_3\mathbf{j} + x_4\mathbf{k}$, where x_1, x_2, x_3, x_4 are integers.*

6 Conclusion

In this paper, we have used hybrid Bézout matrices to formulate necessary and sufficient conditions for a quaternion polynomial to have a complex root, a spherical root and an isolated complex root. The Bézout matrices are implemented in the computational system MAPLE and so, they give us an efficient practical tool for checking the existence of the above kind of roots. We have also computed upper bounds for the norm of spherical and complex isolated roots of a quaternion polynomial. Finally, such a bound is given for an arbitrary root.

References

1. Barnett, S.: Greatest common divisor of several polynomials. Proc. Camb. Philos. Soc. **70**, 263–268 (1971)
2. Barnett, S.: Polynomials and Linear Control Systems. Marcel Dekker, New York (1983)

3. Diaz-Toca, G.M., Gonzalez-Vega, L.: Barnett's theorems about the greatest common divisor of several univariate polynomials through Bezout-like matrices. J. Symb. Comput. **34**, 59–81 (2002)
4. Dospra, P.: Quaternion polynomials and rational rotation minimizing frame curves, PhD Thesis, Agricultural University of Athens (2015)
5. Eilenberg, S., Niven, I.: The "fundamental theorem of algebra" for quaternions. Bull. Am. Math. Soc. **50**, 246–248 (1944)
6. Farouki, R., Dospra, P., Sakkalis, T.: Scalar-vector algorithm for the roots of quadratic quaternion polynomials, and the characterization of quintic rational rotation-minimizing frame curves. J. Symb. Comput. **58**, 1–17 (2013)
7. Fatouros, S., Karcanias, N.: Resultant properies of the GCD of many polynomials and a factorization representation of GCD. Int. J. Control. **76**(16), 1666–1683 (2003)
8. Gentili, G., Struppa, D.C., Vlacc, F.: The fundamental theorem of algebra for Hamilton and Cayley numbers. Math. Z. **259**, 895–902 (2008)
9. Gentili, G., Stoppato, C.: Zeros of regular functions and polynomials of a quaternionic variable. Mich. Math. J. **56**, 655–667 (2008)
10. Gentili, G., Struppa, D.C.: On the multiplicity of zeroes of polynomials with quaternionic coefficients. Milan J. Math. **76**, 15–25 (2008)
11. Gordon, B., Motzkin, T.S.: On the zeros of polynomials over division rings. Trans. Am. Math. Soc. **116**, 218–226 (1965)
12. Huang, L., So, W.: Quadratic formulas for quaternions. Appl. Math. Lett. **15**, 533–540 (2002)
13. Janovská, D., Opfer, G.: A note on the computation of all zeros of simple quaternionic polynomials. SIAM J. Numer. Anal. **48**, 244–256 (2010)
14. Jia, Z., Cheng, X., Zhao, M.: A new method for roots of monic quaternionic quadratic polynomial. Comput. Math. Appl. **58**, 1852–1858 (2009)
15. Kakié, K.: The resultant of several homogeneous polynomials in two indeterminates. Proc. Am. Math. Soc. **54**, 1–7 (1976)
16. Kalantari, B.: Algorithms for quaternion polynomial root-finding. J. Complex. **29**, 302–322 (2013)
17. Lam, T.Y.: A First Course in Noncommutative Rings, 2nd edn. Springer, New York (2001)
18. Mignotte, M.: An inequality of the greatest roots of a polynomial. Elem. Math. **46**, 85–86 (1991)
19. Niven, I.: Equations in quaternions. Am. Math. Mon. **48**, 654–661 (1941)
20. Opfer, G.: Polynomials and Vandermonde matrices over the field of quaternions. Electron. Trans. Numer. Anal. **36**, 9–16 (2009)
21. Pogorui, A., Shapiro, M.V.: On the structure of the set of zeros of quaternionic polynomials. Complex Var. **49**(6), 379–389 (2004)
22. Serodio, R., Siu, L.: Zeros of quaternion polynomials. Appl. Math. Lett. **14**, 237–239 (2001)
23. Serodio, R., Pereira, E., Vitoria, J.: Computing the zeros of quaternion polynomials. Comput. Math. Appl. **42**, 1229–1237 (2001)
24. Topuridze, N.: On roots of quaternion polynomials. J. Math. Sci. **160**(6), 843–855 (2009)
25. Vardulakis, A.I.G., Stoyle, P.N.R.: Generalized resultant theorem. J. IMA **22**, 331–335 (1978)
26. Wedderburn, J.H.M.: On division algebras. Trans. AMS **22**, 129–135 (1921)
27. Zhang, F.: Quaternions and matrices of quaternions. Linear Algebra Appl. **251**, 21–57 (1997)

Mathematical Renormalization in Quantum Electrodynamics via Noncommutative Generating Series

G.H.E. Duchamp, V. Hoang Ngoc Minh, Ngo Quoc Hoan, K. Penson and P. Simonnet

Abstract In order to push the study of solutions of nonlinear differential equations involved in quantum electrodynamics (The present work is part of a series of papers devoted to the study of the renormalization of divergent polyzetas (at positive and at negative indices) via the factorization of the non commutative generating series of polylogarithms and of harmonic sums and via the effective construction of pairs of bases in duality in φ-deformed shuffle algebras. It is a sequel of [6] and its content was presented in several seminars and meetings, including the 66th and 74th Séminaire Lotharingien de Combinatoire.), we focus on combinatorial aspects of their renormalization at $\{0, 1, +\infty\}$.

Keywords Nonlinear differential equations · Divergent polyzetas · Bases in duality · Lyndon words · Monoidal factorization

1 Introduction

During the last century, the functional expansions were common in physics as well as in engineering and have been developed by Tomonaga, Schwinger and Feynman [19]

G.H.E. Duchamp (✉)
LIPN - UMR 7030, CNRS, 93430 Villetaneuse, France
e-mail: gheduchamp@gmail.com

V. Hoang Ngoc Minh
Université Lille II, 59024 Lille, France
e-mail: hoang@univ-lille2.fr

Ngo Quoc Hoan
Université Paris XIII, 93430 Villetaneuse, France
e-mail: quochoan_ngo@yahoo.com.vn

K. Penson
Université Paris VI - LPTMC, 75252 Paris Cedex 05, France
e-mail: penson@lptmc.jussieu.fr

P. Simonnet
Université de Corse, 20250 Corte, France
e-mail: simonnet@univ-corse.fr

I.S. Kotsireas and E. Martínez-Moro (eds.), *Applications of Computer Algebra*, Springer Proceedings in Mathematics & Statistics 198,
DOI 10.1007/978-3-319-56932-1_6

to represent the dynamical systems in quantum electrodynamics. The main difficulty of this approach is the divergence of these expansions at the singularity 0 or at $+\infty$ (see [2]) and leads to problems of *regularization* and *renormalization* which can be solved by combinatorial technics: Feynman diagrams [21] and their siblings [16, 48], noncommutative formal power series [23], trees [11].

Recently, in the same vein, and based on the one hand on the shuffle and quasi-shuffle bialgebras [6], the combinatorics of noncommutative formal power series was intensively amplified for the asymptotic analysis of dynamical systems with three regular singularities in[1] $\{0, 1, +\infty\}$; and, on the other hand with the monodromy and the Galois differential group of the Knizhnik–Zamolodchikov equation KZ_3 [41, 43] i.e., the following noncommutative evolution equation[2]

$$\frac{dG(z)}{dz} = \left(\frac{x_0}{z} + \frac{x_1}{1-z}\right)G(z),$$

the monoidal factorization facilitates mainly the renormalization and the computation of the associators[3] via the universal one, i.e. Φ_{KZ} of Drinfel'd [43].

In fact, these associators are noncommutative formal power series on two variables and regularize the Chen generating series of the differential forms admitting singularities at 0 or at 1 along the integration paths on the universal covering of $\mathbb{C}$ without points 0 and 1 (i.e. $\widetilde{\mathbb{C}\setminus\{0,1\}}$). Their coefficients are, up to a multiple of powers of $2i\pi$, polynomial on polyzetas, i.e. the following real numbers[4] [4, 45, 48, 56]

$$\zeta(s_1,\ldots,s_r) = \sum_{n_1>\cdots>n_r>0} \frac{1}{n_1^{s_1}\ldots n_r^{s_r}}, \text{ for } r \geq 1, s_1 \geq 2, s_2,\ldots,s_r \geq 1,$$

and these numbers admit a natural structure of algebra over the rational numbers deduced from the combinatorial aspects of the shuffle and quasi-shuffle Hopf algebras [33, 37, 55]. It is conjectured that this algebra is $\mathbb{N}$-graded.[5] More precisely, for $s_1 \geq 2, s_2,\ldots,s_r \geq 1$, the polyzeta $\zeta(s_1,\ldots,s_r)$ can be obtained as the limit of the polylogarithm [26, 33] $\mathrm{Li}_{s_1,\ldots,s_r}(z)$, for $z \to 1$, and of the harmonic sum $\mathrm{H}_{s_1,\ldots,s_r}(N)$, for $N \to +\infty$:

---

[1] Any differential equation with singularities in $\{a, b, c\}$ can be changed into a differential equation with singularities in $\{0, 1, +\infty\}$ via an homographic transformation.

[2] $x_0 := t_{1,2}/2i\pi$ and $x_1 := -t_{2,3}/2i\pi$ are noncommutative variables and $t_{1,2}, t_{2,3}$ belong to $\mathcal{T}_3 = \{t_{1,2}, t_{1,3}, t_{2,3}\}$ satisfying the infinitesimal 3-braid relations, i.e. $[t_{1,3}, t_{1,2} + t_{2,3}] = [t_{2,3}, t_{1,2} + t_{1,3}] = 0$.

[3] They were introduced in QFT by Drinfel'd and it plays an important role for the still open problem of the effective determination of the polynomial invariants of knots and links via Kontsevich's integral (see [7, 48]) and Φ_{KZ}, was obtained firstly, in [48], with explicit coefficients which are polyzetas and regularized polyzetas (see [43, 44] for the computation of the other involving *only* convergent polyzetas as local coordinates, and for algorithms regularizing divergent polyzetas).

[4] $s_1 + \cdots + s_r$ is the *weight* of $\zeta(s_1,\ldots,s_r)$, i.e. the weight of the composition $(s_1,\ldots,s_r)$.

[5] One of us wrote a tentative proof of this claim in [43, 44].

$$\mathrm{Li}_{s_1,\ldots,s_r}(z) = \sum_{n_1>\cdots>n_r>0} \frac{z^{n_1}}{n_1^{s_1}\ldots n_r^{s_r}} \text{ and } \mathrm{H}_{s_1,\ldots,s_r}(N) = \sum_{n_1>\cdots>n_r>0}^{N} \frac{1}{n_1^{s_1}\ldots n_r^{s_r}}.$$

Then, by a theorem of Abel, one has

$$\zeta(s_1,\ldots,s_r) = \lim_{z\to 1} \mathrm{Li}_{s_1,\ldots,s_r}(z) = \lim_{N\to+\infty} \mathrm{H}_{s_1,\ldots,s_r}(N).$$

Since the algebras of polylogarithms and of harmonic sums are isomorphic to the shuffle algebra $(\mathbb{Q}\langle X\rangle, ш, 1_{X^*})$ and quasi-shuffle algebra $(\mathbb{Q}\langle Y\rangle, ⧢, 1_{Y^*})$ respectively both admitting the Lyndon words $\mathscr{L}ynX$ over $X = \{x_0, x_1\}$ and $\mathscr{L}ynY$ over $Y = \{y_i\}_{i\geq 1}$, as (pure) transcendence bases (recalled in Sect. 2.1) one can use

- The (one-to-one) correspondence between the combinatorial compositions, the words[6] in Y^* and the words in $X^*x_1 + 1_{X^*}$, i.e.[7]

$$(\{1\}^k, s_{k+1},\ldots,s_r) \leftrightarrow y_1^k y_{s_{k+1}}\ldots y_{s_r} \underset{\pi_Y}{\overset{\pi_X}{\rightleftharpoons}} x_1^k x_0^{s_{k+1}-1}x_1\ldots x_0^{s_r-1}x_1. \quad (1)$$

- The ordering $x_1 \succ x_0$ and $y_1 \succ y_2 \succ \ldots$ over X and Y respectively.
- The transcendence basis $\{S_l\}_{l\in\mathscr{L}ynX}$ (resp. $\{\Sigma_l\}_{l\in\mathscr{L}ynY}$) of $(\mathbb{Q}\langle X\rangle, ш, 1_{X^*})$ (resp. $(\mathbb{Q}\langle Y\rangle, ⧢, 1_{Y^*})$) in duality[8] with $\{P_l\}_{l\in\mathscr{L}ynX}$ (resp. $\{\Pi_l\}_{l\in\mathscr{L}ynY}$), a basis of the Lie algebra of primitive elements of the bialgebra[9] $\mathscr{H}_ш = (\mathbb{Q}\langle X\rangle, \mathtt{conc}, 1_{X^*}, \Delta_ш, \varepsilon)$ (resp. $\mathscr{H}_⧢ = (\mathbb{Q}\langle Y\rangle, \mathtt{conc}, 1_{Y^*}, \Delta_⧢, \varepsilon)$) to factorize the following noncommutative generating series of polylogarithms, hormanic sums and polyzetas

$$\mathrm{L} = \prod_{l\in\mathscr{L}ynX} \exp(\mathrm{Li}_{S_l} P_l) \text{ and } \mathrm{H} = \prod_{l\in\mathscr{L}ynY} \exp(\mathrm{H}_{\Sigma_l}\Pi_l),$$

$$Z_ш = \prod_{\substack{l\in\mathscr{L}ynX\\ l\neq x_0,x_1}} \exp(\zeta(S_l)P_l) \text{ and } Z_⧢ = \prod_{\substack{l\in\mathscr{L}ynY\\ l\neq y_1}} \exp(\zeta(\Sigma_l)\Pi_l),$$

we then obtain two formal power series over Y, Z_1 and Z_2, such that

$$\lim_{z\to 1}\exp\left[y_1\log\frac{1}{1-z}\right]\pi_Y\mathrm{L}(z) = Z_1,\quad \lim_{N\to\infty}\exp\left[\sum_{k\geq 1}\mathrm{H}_{y_k}(N)\frac{(-y_1)^k}{k}\right]\mathrm{H}(N) = Z_2.$$

---

[6]Here, X^* (resp. Y^*) is the monoid generated by X (resp. Y) and its neutral element of is denoted by 1_{X^*} (resp. 1_{Y^*}).

[7]Here, π_Y is the adjoint of π_X for the canonical scalar products where π_X is the morphism of AAU $k\langle Y\rangle \to k\langle X\rangle$ defined by $\pi_X(y_k) = x_0^{k-1}x_1$.

[8]In a more precise way the S and Σ are the "Lyndon part" of the dual bases of the PBW expansions of the P and the Π respectively.

[9]ε is the "constant term" character.

Moreover, Z_1, Z_2 are equal and stand for the noncommutative generating series of $\{\zeta(w)\}_{w\in Y^*-y_1Y^*}$, or $\{\zeta(w)\}_{w\in x_0X^*x_1}$, as one has $Z_1 = Z_2 = \pi_Y Z_{\sqcup\!\sqcup}$ [42–44]. This allows, by extracting the coefficients of the noncommutative generating series, to explicit the counter-terms eliminating the divergence of $\{\mathrm{Li}_w\}_{w\in x_1X^*}$ and of $\{\mathrm{H}_w\}_{w\in y_1Y^*}$ and leads naturally to an equation connecting algebraic structures

$$\prod_{\substack{l\in\mathcal{L}ynY\\ l\neq y_1}}^{\searrow} \exp(\zeta(\Sigma_l)\Pi_l) = \exp\left[\sum_{k\geq 2} -\zeta(k)\frac{(-y_1)^k}{k}\right]\pi_Y \prod_{\substack{l\in\mathcal{L}ynX\\ l\neq x_0,x_1}}^{\searrow} \exp(\zeta(S_l)P_l). \quad (2)$$

Identity (2) allows to compute the Euler–MacLaurin constants and the Hadamard finite parts associated to divergent polyzetas $\{\zeta(w)\}_{w\in y_1Y^*}$ and, by identifying local coordinates, to describe the graded core of ker ζ by its *algebraic* generators.

In this paper, we will focus on the approach by noncommutative formal power series, adapted from [22, 23], and explain how some of the results of [42–44], allow to study the combinatorial aspects of the renormalization at the singularities in $\{0, 1, +\infty\}$ of the solutions of linear differential equations (see Example 1 below) as well as the solutions of nonlinear differential equations (see Examples 2 and 3 below) described in Sect. 3.2 and involved in quantum electrodynamics.

Example 1 (*Hypergeometric equation*) Let t_0, t_1, t_2 be parameters and

$$z(1-z)\ddot{y}(z) + [t_2 - (t_0 + t_1 + 1)z]\dot{y}(z) - t_0t_1y(z) = 0.$$

Let $q_1(z) = -y(z)$ and $q_2(z) = (1-z)\dot{y}(z)$. One has

$$\begin{pmatrix}\dot{q}_1\\ \dot{q}_2\end{pmatrix} = \left(\frac{M_0}{z} + \frac{M_1}{1-z}\right)\begin{pmatrix}q_1\\ q_2\end{pmatrix},$$

where M_0 and M_1 are the following matrices

$$M_0 = -\begin{pmatrix}0 & 0\\ t_0t_1 & t_2\end{pmatrix} \text{ and } M_1 = -\begin{pmatrix}0 & 1\\ 0 & t_2 - t_0 - t_1\end{pmatrix}.$$

Or equivalently, $\dot{q}(z) = A_0(q)\frac{1}{z} + A_1(q)\frac{1}{1-z}$ and $y(z) = -q_1(z)$ where A_0 and A_1 are the following parametrized linear vector fields

$$A_0 = -(t_0t_1q_1 + t_2q_2)\frac{\partial}{\partial q_2} \text{ and } A_1 = -q_1\frac{\partial}{\partial q_1} - (t_2 - t_0 - t_1)q_2\frac{\partial}{\partial q_2}.$$

acting by $\frac{\partial}{\partial q_1}(q) = \frac{\partial}{\partial q_1}\begin{pmatrix}q_1\\ q_2\end{pmatrix} = \begin{pmatrix}1\\ 0\end{pmatrix}$ and $\frac{\partial}{\partial q_2}(q) = \frac{\partial}{\partial q_2}\begin{pmatrix}q_1\\ q_2\end{pmatrix} = \begin{pmatrix}0\\ 1\end{pmatrix}$.

Example 2 (*Harmonic oscillator*) Let k_1, k_2 be parameters and

$$\dot{y}(z) + k_1 y(z) + k_2 y^2(z) = u_1(z).$$

which can be represented, with the same formalism as above, by the following state equations

$$\dot{q}(z) = A_0(q) + A_1(q)u_1(z) \text{ and } y(z) = q(z),$$

where A_0 and A_1 are the following vector fields $A_0 = -(k_1 q + k_2 q^2)\frac{\partial}{\partial q}$ and $A_1 = \frac{\partial}{\partial q}$.

Example 3 (*Duffing's equation*) Let a, b, c be parameters and

$$\ddot{y}(z) + a\dot{y}(z) + by(z) + cy^3(z) = u_1(z).$$

which can be represented by the following state equations

$$\dot{q}(z) = A_0(q) + A_1(q)u_1(z) \text{ and } y(z) = q_1(z),$$

where A_0 and A_1 are the following vector fields

$$A_0 = -(aq_2 + b^2 q_1 + cq_1^3)\frac{\partial}{\partial q_2} + q_2 \frac{\partial}{\partial q_1} \text{ and } A_1 = \frac{\partial}{\partial q_2}.$$

Example 4 (*Van der Pol oscillator*) Let γ, g be parameters and

$$\partial_z^2 x(z) - \gamma[1 + x(z)^2]\partial_z x(z) + x(z) = g\cos(\omega z)$$

which can be transformed into (where C is some constant of integration)

$$\partial_z x(z) = \gamma[1 + x(z)^2/3]x(z) - \int_{z_0}^{z} x(s)ds + \frac{g}{\omega}\sin(\omega z) + C.$$

Setting $y = \int_{z_0}^{z} x(s)ds$ and $u_1(z) = g\sin(\omega z)/\omega + C$, it leads then to

$$\partial_z^2 y(z) = \gamma[\partial_z y(z) + (\partial_z y(z))^3/3] - y(z) + u_1(z)$$

which can be represented by the following state equations (with $n = 2$)

$$\partial_z q(z) = [A_0 u_0(z) + A_1 u_1(z)](q) \text{ and } y(z) = q_1(z)$$

where A_0 and A_1 are the following vector fields

$$A_1 = \frac{\partial}{\partial q_2} \text{ and } A_0 = [\gamma(q_2 + q_2^3/3) - q_1]\frac{\partial}{\partial q_2} + q_2\frac{\partial}{\partial q_1}.$$

This approach by noncommutative formal power series is adequate for studying the algebraic combinatorial aspects of the asymptotic analysis at the singularities in $\{0, 1, +\infty\}$ for the nonlinear dynamical systems described in Sect. 3.2 because

- The polylogarithms form a basis of an infinite dimensional universal Picard–Vessiot extension by means of these differential equations [13, 41] and their algebra, isomorphic to the shuffle algebra, admits $\{\mathrm{Li}_l\}_{l\in\mathcal{L}ynX}$ as a transcendence basis,
- The harmonic sums generate the coefficients of the ordinary Taylor expansions of their solutions (when these expansions exist) [42] and their algebra is isomorphic to the quasi-shuffle algebra admitting $\{\mathrm{H}_l\}_{l\in\mathcal{L}ynY}$ as a transcendence basis,
- The polyzetas do appear as the fundamental arithmetical constants involved in the computations of the monodromies [33, 37], the Kummer type functional equations [34, 37], the asymptotic expansions of solutions [42, 43] and their algebra is freely generated by the polyzetas encoded by irreducible Lyndon words [43].

Hence, a lot of algorithms can be deduced from these facts and more general studies will be completed in [6, 13]. The organisation of this paper is the following

- In Sect. 2, we will give algebraic and analytic foundations, i.e. the combinatorial Hopf algebra of shuffles and the indiscernability respectively, for polyzetas.
- These will be exploited, in Sect. 3, to expand solutions, of nonlinear dynamical systems with singular inputs and their ordinary and *functional* differentiations.

2 Foundations of the Present Framework

2.1 Background about Combinatorics of Shuffle and Stuffle Bialgebras

2.1.1 Schützenberger's Monoidal Factorization

Let $\mathbb{Q}\langle X\rangle$ be equipped by the concatenation and the shuffle defined by

$$\forall w \in X^*,\quad w ⧢ 1_{X^*} = 1_{X^*} ⧢ w = w,$$
$$\forall x, y \in X, \forall u, v \in X^*,\quad xu ⧢ yv = x(u ⧢ yv) + y(xu ⧢ v),$$

or by their dual co-products, Δ_{conc} and $\Delta_{⧢}$, defined by, for $w \in X^*$ and $x \in X$,

$$\Delta_{\mathrm{conc}}(w) = \sum_{u,v\in X^*, uv=w} u \otimes v \text{ and } \Delta_{⧢}(x) = x \otimes 1 + 1 \otimes x,$$

$\Delta_{\sqcup\!\sqcup}$ is then extended to a conc-morphism $\mathbb{Q}\langle X\rangle \to \mathbb{Q}\langle X\rangle \otimes \mathbb{Q}\langle X\rangle$. These two comultiplications satisfy, for any $u, v, w \in X^*$,

$$\langle \Delta_{\mathrm{conc}}(w) \mid u \otimes v\rangle = \langle w \mid uv\rangle \text{ and } \langle \Delta_{\sqcup\!\sqcup}(w) \mid u \otimes v\rangle = \langle w \mid u \sqcup\!\sqcup v\rangle.$$

One gets two mutually dual bialgebras

$$\mathcal{H}_{\sqcup\!\sqcup} = (\mathbb{Q}\langle X\rangle, \mathrm{conc}, 1_{X^*}, \Delta_{\sqcup\!\sqcup}, \varepsilon), \quad \mathcal{H}_{\sqcup\!\sqcup}^{\vee} = (\mathbb{Q}\langle X\rangle, \sqcup\!\sqcup, 1_{X^*}, \Delta_{\mathrm{conc}}, \varepsilon).$$

After a theorem by Radford [49], the elements of $\mathcal{L}ynX$ form a transcendence basis of $(\mathbb{Q}\langle X\rangle, \sqcup\!\sqcup, 1_{X^*})$ and it can be completed then to the linear basis $\{w\}_{w\in X^*}$ which is auto-dual:

$$\forall v, v \in X^*, \quad \langle u \mid v\rangle = \delta_{u,v}. \tag{3}$$

But the elements $l \in \mathcal{L}ynX - X$ are not primitive, for $\Delta_{\sqcup\!\sqcup}$, and then $\mathcal{L}ynX$ does not constitute a basis for $\mathcal{L}ie_{\mathbb{Q}}\langle X\rangle$. Chen et al. [9] constructed $\{P_w\}_{w\in X^*}$, so-called the Poincaré–Birkhoff–Witt–Lyndon basis, for $\mathcal{U}(\mathcal{L}ie_{\mathbb{Q}}\langle X\rangle)$ as follows

$$\begin{aligned} P_x &= x \text{ for } x \in X,\\ P_l &= [P_s, P_r] \text{ for } l \in \mathcal{L}ynX, \text{ with standard factorization of } l = (s, r),\\ P_w &= P_{l_1}^{i_1} \ldots P_{l_k}^{i_k} \text{ for } w = l_1^{i_1} \ldots l_k^{i_k}, l_1 \succ \cdots \succ l_k, l_1 \ldots, l_k \in \mathcal{L}ynX. \end{aligned} \tag{4}$$

where here $\succ$ stands for the lexicographic (strict) ordering defined[10] by $x_0 \prec x_1$. Schützenberger constructed bases for $(\mathbb{Q}\langle X\rangle, \sqcup\!\sqcup)$ defined by duality as follows:

$$\forall u, v \in X^*, \quad \langle S_u \mid P_v\rangle = \delta_{u,v}$$

and obtained the transcendence and linear bases, $\{S_l\}_{l\in\mathcal{L}ynX}$, $\{S_w\}_{w\in X^*}$, as follows

$$\begin{aligned} S_l &= xS_u, && \text{for } l = xu \in \mathcal{L}ynX,\\ S_w &= \frac{S_{l_1}^{\sqcup\!\sqcup\, i_1} \sqcup\!\sqcup \ldots \sqcup\!\sqcup S_{l_k}^{\sqcup\!\sqcup\, i_k}}{i_1! \ldots i_k!} && \text{for } w = l_1^{i_1} \ldots l_k^{i_k}, l_1 \succ \ldots \succ l_k. \end{aligned}$$

After that, Mélançon and Reutenauer [52] proved that, for any $w \in X^*$,

$$P_w = w + \sum_{v\succ w, |v|_X=|w|_X} c_v v \text{ and } S_w = w + \sum_{v\prec w, |v|_X=|w|_X} d_v v. \tag{5}$$

where $|w|_X = (|w|_x)_{x\in X}$ is the family of all partial degrees (number of times a letter occurs in a word). In other words, the elements of the bases $\{S_w\}_{w\in X^*}$ and $\{P_w\}_{w\in X^*}$

---

[10] In here, the (lexicographic) order relation $\succ$ on X^* is defined by, for any $u, v \in X^*$, $u \succ v$ iff $u = vw$ with $w \in X^+$ else there are $w, w_1, w_2 \in X^*$ and $a \succ b \in X$ such that $u = waw_1$ and $v = wbw_2$.

are lower and upper triangular respectively and they are of multihomogeneous (all the monomials have the same partial degrees).

Example 5 (*of* $\{P_w\}_{w\in X^*}$ *and* $\{S_w\}_{w\in X^*}$, [30]) Let $X=\{x_0,x_1\}$ with $x_0\prec x_1$.

l	P_l	S_l
x_0	x_0	x_0
x_1	x_1	x_1
x_0x_1	$[x_0,x_1]$	x_0x_1
$x_0^2x_1$	$[x_0,[x_0,x_1]]$	$x_0^2x_1$
$x_0x_1^2$	$[[x_0,x_1],x_1]$	$x_0x_1^2$
$x_0^3x_1$	$[x_0,[x_0,[x_0,x_1]]]$	$x_0^3x_1$
$x_0^2x_1^2$	$[x_0,[[x_0,x_1],x_1]]$	$x_0^2x_1^2$
$x_0x_1^3$	$[[[x_0,x_1],x_1],x_1]$	$x_0x_1^3$
$x_0^4x_1$	$[x_0,[x_0,[x_0,[x_0,x_1]]]]$	$x_0^4x_1$
$x_0^3x_1^2$	$[x_0,[x_0,[[x_0,x_1],x_1]]]$	$x_0^3x_1^2$
$x_0^2x_1x_0x_1$	$[[x_0,[x_0,x_1]],[x_0,x_1]]$	$2x_0^3x_1^2+x_0^2x_1x_0x_1$
$x_0^2x_1^3$	$[x_0,[[[x_0,x_1],x_1],x_1]]$	$x_0^2x_1^3$
$x_0x_1x_0x_1^2$	$[[x_0,x_1],[[x_0,x_1],x_1]]$	$3x_0^2x_1^3+x_0x_1x_0x_1^2$
$x_0x_1^4$	$[[[[x_0,x_1],x_1],x_1],x_1]$	$x_0x_1^4$
$x_0^5x_1$	$[x_0,[x_0,[x_0,[x_0,[x_0,x_1]]]]]$	$x_0^5x_1$
$x_0^4x_1^2$	$[x_0,[x_0,[x_0,[[x_0,x_1],x_1]]]]$	$x_0^4x_1^2$
$x_0^3x_1x_0x_1$	$[x_0,[[x_0,[x_0,x_1]],[x_0,x_1]]]$	$2x_0^4x_1^2+x_0^3x_1x_0x_1$
$x_0^3x_1^3$	$[x_0,[x_0,[[[x_0,x_1],x_1],x_1]]]$	$x_0^3x_1^3$
$x_0^2x_1x_0x_1^2$	$[x_0,[[x_0,x_1],[[x_0,x_1],x_1]]]$	$3x_0^3x_1^3+x_0^2x_1x_0x_1^2$
$x_0^2x_1^2x_0x_1$	$[[x_0,[[x_0,x_1],x_1]],[x_0,x_1]]$	$6x_0^3x_1^3+3x_0^2x_1x_0x_1^2+x_0^2x_1^2x_0x_1$
$x_0^2x_1^4$	$[x_0,[[[[x_0,x_1],x_1],x_1],x_1]]$	$x_0^2x_1^4$
$x_0x_1x_0x_1^3$	$[[x_0,x_1],[[[x_0,x_1],x_1],x_1]]$	$4x_0^2x_1^4+x_0x_1x_0x_1^3$
$x_0x_1^5$	$[[[[[x_0,x_1],x_1],x_1],x_1],x_1]$	$x_0x_1^5$

Then, Schützenberger's factorization of the diagonal series $\mathcal{D}_X$ follows [52]

$$\mathcal{D}_X := \sum_{w\in X^*} w\otimes w = \sum_{w\in X^*} S_w\otimes P_w = \prod_{l\in\mathcal{L}ynX}^{\searrow} \exp(S_l\otimes P_l). \tag{6}$$

2.1.2 Extended Schützenberger's Monoidal Factorization

Let us define the commutative product over $\mathbb{Q}\langle Y\rangle$, denoted by μ, as follows

$$\forall y_n, y_m\in Y,\quad \mu(y_n,y_m)=y_{n+m},$$

or by its associated coproduct, Δ_μ, defined by

$$\forall y_n \in Y,\quad \Delta_\mu(y_n) = \sum_{i=1}^{n-1} y_i \otimes y_{n-i}$$

and satisfying, $\forall x, y, z \in Y,\ \langle \Delta_\mu \mid y \otimes z\rangle = \langle x \mid \mu(y,z)\rangle$. Let $\mathbb{Q}\langle Y\rangle$ be equipped by

1. The concatenation (or by its associated coproduct, Δ_{conc}).
2. The *shuffle* product, i.e. the commutative product defined by [23]

$$\forall w \in Y^*,\quad w \sqcup\!\sqcup\, 1_{Y^*} = 1_{Y^*} \sqcup\!\sqcup\, w = w,$$
$$\forall x, y \in Y, u, v \in Y^*,\quad xu \sqcup\!\sqcup\, yv = x(u \sqcup\!\sqcup\, yv) + y(xu \sqcup\!\sqcup\, v)$$

   or with its associated coproduct, $\Delta_{\sqcup\!\sqcup}$, defined, on the letters, by

$$\forall y_k \in Y,\quad \Delta_{\sqcup\!\sqcup}\, y_k = y_k \otimes 1 + 1 \otimes y_k$$

   and extended by morphism. It satisfies $\forall u, v, w \in Y^*,\ \langle \Delta_{\sqcup\!\sqcup} w \mid u \otimes v\rangle = \langle w \mid u \sqcup\!\sqcup\, v\rangle$.
3. The *quasi-shuffle* product, i.e. the commutative product defined by [46]

$$y_i u \uplus y_j v = y_i(u \uplus y_j v) + y_j(y_i u \uplus v) + \mu(y_i, y_j)(u \uplus v)$$

   or with its associated coproduct, $\Delta_{\uplus}$, defined, on the letters, by

$$\forall y_k \in Y,\quad \Delta_{\uplus}\, y_k = \Delta_{\sqcup\!\sqcup}\, y_k + \Delta_\mu y_k$$

   and extended by morphism. It satisfies $\forall u, v, w \in Y^*,\ \langle \Delta_{\uplus} w \mid u \otimes v\rangle = \langle w \mid u \uplus v\rangle$.

Hence, with the counit e defined by, for any $P \in \mathbb{Q}\langle Y\rangle$, $\mathsf{e}(P) = \langle P \mid 1_{Y^*}\rangle$, one gets two pairs of mutually dual bialgebras

$$\mathcal{H}_{\sqcup\!\sqcup} = (\mathbb{Q}\langle Y\rangle, \mathtt{conc}, 1_{Y^*}, \Delta_{\sqcup\!\sqcup}, \mathsf{e}) \text{ and } \mathcal{H}_{\sqcup\!\sqcup}^\vee = (\mathbb{Q}\langle Y\rangle, \sqcup\!\sqcup, 1_{Y^*}, \Delta_{\text{conc}}, \mathsf{e}),$$
$$\mathcal{H}_{\uplus} = (\mathbb{Q}\langle Y\rangle, \mathtt{conc}, 1_{Y^*}, \Delta_{\uplus}, \mathsf{e}) \text{ and } \mathcal{H}_{\uplus}^\vee = (\mathbb{Q}\langle Y\rangle, \uplus, 1_{Y^*}, \Delta_{\text{conc}}, \mathsf{e}).$$

By the CQMM theorem (see [6]), the connected $\mathbb{N}$-graded, co-commutative Hopf algebra $\mathcal{H}_{\sqcup\!\sqcup}$ is isomorphic to the enveloping algebra of the Lie algebra of its primitive elements which is equal to $\mathcal{L}ie_{\mathbb{Q}}\langle Y\rangle$:

$$\mathcal{H}_{\sqcup\!\sqcup} \cong \mathcal{U}(\mathcal{L}ie_{\mathbb{Q}}\langle Y\rangle) \text{ and } \mathcal{H}_{\sqcup\!\sqcup}^\vee \cong \mathcal{U}(\mathcal{L}ie_{\mathbb{Q}}\langle Y\rangle)^\vee.$$

Hence, let us consider [9]

1. The PBW–Lyndon basis $\{p_w\}_{w\in Y^*}$ for $\mathcal{U}(\mathcal{L}ie_{\mathbb{Q}}\langle Y\rangle)$ constructed recursively

$$\begin{cases} p_y = y & \text{for } y \in Y, \\ p_l = [p_s, p_r] & \text{for } l \in \mathcal{L}ynY, \text{ with the standard factorization } l = (s, r), \\ p_w = p_{l_1}^{i_1} \ldots p_{l_k}^{i_k} & \text{for } w = l_1^{i_1} \ldots l_k^{i_k}, l_1 \succ \ldots \succ l_k, l_1 \ldots, l_k \in \mathcal{L}ynY, \end{cases}$$

2. And, by duality,[11] the linear basis $\{s_w\}_{w \in Y^*}$ for $(\mathbb{Q}\langle Y \rangle, ш, 1_{Y^*})$, i.e.

$$\forall u, v \in Y^*, \ \langle p_u \mid s_v \rangle = \delta_{u,v}.$$

This basis can be computed recursively as follows [52]

$$\begin{cases} s_y = y, & \text{for } y \in Y, \\ s_l = y s_u, & \text{for } l = yu \in \mathcal{L}ynY, \\ s_w = \dfrac{s_{l_1}^{ш\, i_1} ш \ldots ш\, s_{l_k}^{ш\, i_k}}{i_1! \ldots i_k!} & \text{for } w = l_1^{i_1} \ldots l_k^{i_k}, l_1 \succ \ldots \succ l_k \in \mathcal{L}ynY. \end{cases}$$

As in (6), one also has Schützenberger's factorization for the diagonal series $\mathcal{D}_Y$

$$\mathcal{D}_Y := \sum_{w \in Y^*} w \otimes w = \sum_{w \in Y^*} s_w \otimes p_w = \prod_{l \in \mathcal{L}ynY}^{\searrow} \exp(s_l \otimes p_l).$$

Similarly, by the CQMM theorem, the connected $\mathbb{N}$-graded, co-commutative Hopf algebra $\mathcal{H}_{⧢}$ is isomorphic to the enveloping algebra of

$$\mathrm{Prim}(\mathcal{H}_{⧢}) = \mathrm{Im}(\pi_1) = \mathrm{span}_{\mathbb{Q}}\{\pi_1(w) | w \in Y^*\},$$

where, for any $w \in Y^*$, $\pi_1(w)$ is obtained as follows [6, 43]

$$\pi_1(w) = w + \sum_{k \geq 2} \frac{(-1)^{k-1}}{k} \sum_{u_1, \ldots, u_k \in Y^+} \langle w \mid u_1 ⧢ \ldots ⧢ u_k \rangle\, u_1 \ldots u_k. \tag{7}$$

Note that Equation (7) is equivalent to the following identity [6, 43, 44]

$$w = \sum_{k \geq 0} \frac{1}{k!} \sum_{u_1, \ldots, u_k \in Y^*} \langle w \mid u_1 ⧢ \ldots ⧢ u_k \rangle\, \pi_1(u_1) \ldots \pi_1(u_k). \tag{8}$$

In particular, for any $y_k \in Y$, we have successively [6, 43, 44]

[11] The dual family of a basis lies in the algebraic dual which is here the space of noncommutative series, but as the enveloping algebra under consideration is graded in finite dimensions (here by the multidegree), these series are in fact (multihomogeneous) polynomials.

$$\pi_1(y_k) = y_k + \sum_{l\geq 2} \frac{(-1)^{l-1}}{l} \sum_{\substack{j_1,\ldots,j_l\geq 1\\ j_1+\cdots+j_l=k}} y_{j_1}\ldots y_{j_l}, \tag{9}$$

$$y_n = \sum_{k\geq 1} \frac{1}{k!} \sum_{s'_1+\cdots+s'_k=n} \pi_1(y_{s'_1})\ldots\pi_1(y_{s'_k}) \tag{10}$$

Hence, by introducing the new alphabet $\bar{Y} = \{\bar{y}\}_{y\in Y} = \{\pi_1(y)\}_{y\in Y}$, one has

$$(\mathbb{Q}\langle\bar{Y}\rangle, \mathtt{conc}, 1_{\bar{Y}^*}, \Delta_{\sqcup\!\sqcup}) \cong (\mathbb{Q}\langle Y\rangle, \mathtt{conc}, 1_{Y^*}, \Delta_{\uplus})$$

as one can prove through (10) that the endomorphism $y \mapsto \bar{y}$ is, in fact, an isomorphism

$$\mathscr{H}_{\uplus} \cong \mathscr{U}(\mathscr{L}ie_{\mathbb{Q}}\langle\bar{Y}\rangle) \cong \mathscr{U}(\mathrm{Prim}(\mathscr{H}_{\uplus})),$$
$$\mathscr{H}_{\uplus}^{\vee} \cong \mathscr{U}(\mathscr{L}ie_{\mathbb{Q}}\langle\bar{Y}\rangle)^{\vee} \cong \mathscr{U}(\mathrm{Prim}(\mathscr{H}_{\uplus}))^{\vee}.$$

By considering

1. The PBW–Lyndon basis $\{\Pi_w\}_{w\in Y^*}$ for $\mathscr{U}(\mathrm{Prim}(\mathscr{H}_{\uplus}))$ constructed recursively as follows [43]

$$\begin{cases} \Pi_y = \pi_1(y) & \text{for } y \in Y,\\ \Pi_l = [\Pi_s, \Pi_r] & \text{for } l \in \mathscr{L}ynY, \text{ with the standard factorization } l = (s, r),\\ \Pi_w = \Pi_{l_1}^{i_1}\ldots\Pi_{l_k}^{i_k} & \text{for } w = l_1^{i_1}\ldots l_k^{i_k}, l_1 \succ \ldots \succ l_k, l_1\ldots, l_k \in \mathscr{L}ynY, \end{cases}$$

2. And, by duality, the linear basis $\{\Sigma_w\}_{w\in Y^*}$ for $(\mathbb{Q}\langle Y\rangle, \uplus, 1_{Y^*})$, i.e.

$$\forall u, v \in Y^*,\ \langle \Pi_u \mid \Sigma_v\rangle = \delta_{u,v}.$$

   This basis can be computed recursively as follows [5, 43]

$$\begin{cases} \Sigma_y = y, & \text{for } y \in Y,\\ \Sigma_l = \displaystyle\sum_{(!)} \frac{y_{s_{k_1}+\cdots+s_{k_i}}}{i!} \Sigma_{l_1\cdots l_n}, & \text{for } l = y_{s_1}\ldots y_{s_w} \in \mathscr{L}ynY,\\ \Sigma_w = \dfrac{\Sigma_{l_1}^{\uplus i_1} \uplus \ldots \uplus \Sigma_{l_k}^{\uplus i_k}}{i_1!\ldots i_k!}, & \text{for } w = l_1^{i_1}\ldots l_k^{i_k}, \text{ with } l_1 \succ \ldots \succ l_k \in \mathscr{L}ynY. \end{cases}$$

   In (!), the sum is taken over all subsequences $\{k_1, \ldots, k_i\} \subset \{1, \ldots, k\}$ and all Lyndon words $l_1 \succeq \cdots \succeq l_n$ such that $(y_{s_1}, \ldots, y_{s_k}) \stackrel{*}{\Leftarrow} (y_{s_{k_1}}, \ldots, y_{s_{k_i}}, l_1, \ldots, l_n)$, where $\stackrel{*}{\Leftarrow}$ denotes the transitive closure of the relation on standard sequences, denoted by $\Leftarrow$ (see [5]).

We also proved that, for any $w \in Y^*$, [6, 43, 44]

$$\Pi_w = w + \sum_{v \succ w, (v)=(w)} e_v v \text{ and } \Sigma_w = w + \sum_{v \prec w, (v)=(w)} f_v v. \tag{11}$$

In other words, the elements of the bases $\{\Sigma_w\}_{w\in Y^*}$ and $\{\Pi_w\}_{w\in Y^*}$ are lower and upper triangular respectively and they are of homogeneous in weight.

We also get the extended Schützenberger's factorization of $\mathscr{D}_Y$ [6, 43, 44]

$$\mathscr{D}_Y = \sum_{w\in Y^*} \Sigma_w \otimes \Pi_w = \prod_{l\in\mathscr{L}ynY}^{\searrow} \exp(\Sigma_l \otimes \Pi_l).$$

Example 6 (*of* $\{\Pi_w\}_{w\in Y^*}$ *and* $\{\Sigma_w\}_{w\in Y^*}$, [5])

l	Π_l	Σ_l
y_2	$y_2 - \frac{1}{2}y_1^2$	y_2
y_1^2	y_1^2	$\frac{1}{2}y_2 + y_1^2$
y_3	$y_3 - \frac{1}{2}y_1y_2 - \frac{1}{2}y_2y_1 + \frac{1}{3}y_1^3$	y_3
y_2y_1	$y_2y_1 - y_2y_1$	$\frac{1}{2}y_3 + y_2y_1$
y_1y_2	$y_2y_1 - \frac{1}{2}y_1^3$	y_1y_2
y_1^3	y_1^3	$\frac{1}{6}y_3 + \frac{1}{2}y_2y_1 + \frac{1}{2}y_1y_2 + y_1^3$
y_4	$y_4 - \frac{1}{2}y_1y_3 - \frac{1}{2}y_2^2 - \frac{1}{2}y_3y_1$ $+\frac{1}{3}y_1^2y_2 + \frac{1}{3}y_1y_2y_1 + \frac{1}{3}y_2y_1^2 - \frac{1}{4}y_1^4$	y_4
y_3y_1	$y_3y_1 - \frac{1}{2}y_2y_1^2 - y_1y_3 + \frac{1}{2}y_1^2y_2$	$\frac{1}{2}y_4 + y_3y_1$
y_2^2	$y_2^2 - \frac{1}{2}y_2y_1^2 - \frac{1}{2}y_1^2y_2 + \frac{1}{4}y_1^4$	$\frac{1}{2}y_4 + y_2^2$
$y_2y_1^2$	$y_2y_1^2 - 2\,y_1y_2y_1 + y_1^2y_2$	$\frac{1}{6}y_4 + \frac{1}{2}y_3y_1 + \frac{1}{2}y_2^2 + y_2y_1^2$
y_1y_3	$y_1y_3 - \frac{1}{2}y_1^2y_2 - \frac{1}{2}y_1y_2y_1 + \frac{1}{3}y_1^4$	$y_4 + y_3y_1 + y_1y_3$
$y_1y_2y_1$	$y_1y_2y_1 - y_1^2y_2$	$\frac{1}{2}y_4 + \frac{1}{2}y_3y_1 + y_2^2$ $+y_2y_1^2 + \frac{1}{2}y_1y_3 + y_1y_2y_1$
$y_1^2y_2$	$y_1^2y_2 - \frac{1}{2}y_1^4$	$\frac{1}{2}y_4 + y_3y_1 + y_2^2 + y_2y_1^2$ $+y_1y_3 + y_1y_2y_1 + y_1^2y_2$
y_1^4	y_1^4	$\frac{1}{24}y_4 + \frac{1}{6}y_3y_1 + \frac{1}{4}y_2^2 + \frac{1}{2}y_2y_1^2$ $+\frac{1}{6}y_1y_3 + \frac{1}{2}y_1y_2y_1 + \frac{1}{2}y_1^2y_2 + y_1^4$

2.2 *Indiscernability over a Class of Formal Power Series*

2.2.1 Residual Calculus and Representative Series

Definition 1 Let $S \in \mathbb{Q}\langle\langle X\rangle\rangle$ (resp. $\mathbb{Q}\langle X\rangle$) and let $P \in \mathbb{Q}\langle X\rangle$ (resp. $\mathbb{Q}\langle\langle X\rangle\rangle$). The left and right *residual* of S by P are respectively the formal power series $P \triangleright S$ and $S \triangleleft P$ in $\mathbb{Q}\langle\langle X\rangle\rangle$ defined by $\langle P \triangleright S \mid w\rangle = \langle S \mid wP\rangle$ (resp. $\langle S \triangleleft P \mid w\rangle = \langle S \mid Pw\rangle$).

For any $S \in \mathbb{Q}\langle\langle X\rangle\rangle$ (resp. $\mathbb{Q}\langle X\rangle$) and $P, Q \in \mathbb{Q}\langle X\rangle$ (resp. $\mathbb{Q}\langle\langle X\rangle\rangle$), we straightforwardly get $P \triangleright (Q \triangleright S) = PQ \triangleright S$, $(S \triangleleft P) \triangleleft Q = S \triangleleft PQ$ and $(P \triangleright S) \triangleleft Q = P \triangleright (S \triangleleft Q)$.

In case $x, y \in X$ and $w \in X^*$, we get[12] $x \triangleright (wy) = \delta_x^y w$ and $xw \triangleleft y = \delta_x^y w$.

Lemma 1 (Reconstruction lemma) *Let $S \in \mathbb{Q}\langle\langle X\rangle\rangle$. Then*

$$S = \langle S \mid 1_{X^*}\rangle + \sum_{x\in X} x(S \triangleleft x) = \langle S \mid 1_{X^*}\rangle + \sum_{x\in X} (x \triangleright S)x.$$

Theorem 1 *Let $\delta \in \mathfrak{Der}(\mathbb{Q}\langle X\rangle, ш, 1_{X^*})$ and $t \in \mathbb{Q}$. Moreover, we suppose that δ is locally nilpotent.*[13] *Then the family $(t\delta)^n/n!$ is summable and its sum, denoted $\exp(t\delta)$, is a one-parameter group of automorphisms of $(\mathbb{Q}\langle X\rangle, ш, 1_{X^*})$.*

Theorem 2 *Let L be a Lie series, i.e. $\Delta_ш(L) = L\hat{\otimes}1 + 1\hat{\otimes}L$. Let δ_L^r, δ_L^l be defined respectively by $\delta_L^r(P) := P \triangleleft L$, $\delta_L^l(P) := L \triangleright P$. Then δ_L^r, δ_L^l are locally nilpotent derivations of $(\mathbb{Q}\langle X\rangle, ш, 1_{X^*})$. Hence, $\exp(t\delta_L^r)$, $\exp(t\delta_L^l)$ are one-parameter groups of $Aut(\mathbb{Q}\langle X\rangle, ш, 1_{X^*})$ and $\exp(t\delta_L^r)P = P \triangleleft \exp(tL)$, $\exp(t\delta_L^l)P = \exp(tL) \triangleright P$.*

Example 7 Since $x_1\triangleright$ and $\triangleleft x_0$ are derivations and the polynomials $\{S_l\}_{l\in\mathcal{L}ynX-X}$ belong to $x_0\mathbb{Q}\langle X\rangle x_1$ then $x_1 \triangleright l = l \triangleleft x_0 = 0$ and $x_1 \triangleright \check{S}_l = \check{S}_l \triangleleft x_0 = 0$.

Theorem 3 *Let $S \in \mathbb{Q}\langle\langle X\rangle\rangle$. The following properties are equivalent:*

1. *The left $\mathbb{C}$-module $Res_g(S) = span\{w \triangleright S \mid w \in X^*\}$ is finite dimensional.*
2. *The right $\mathbb{C}$-module $Res_d(S) = span\{S \triangleleft w \mid w \in X^*\}$ is finite dimensional.*
3. *There are matrices $\lambda \in \mathcal{M}_{1,n}(\mathbb{Q})$, $\eta \in \mathcal{M}_{n,1}(\mathbb{Q})$ and $\mu : X^* \longrightarrow \mathcal{M}_{n,n}$, such that*

$$S = \sum_{w\in X^*} [\lambda\mu(w)\eta]\, w = \lambda \left(\prod_{l\in\mathcal{L}ynX}^{\searrow} e^{\mu(S_l)\, P_l} \right) \eta.$$

A series that satisfies the items of Theorem 3 will be called *representative (or rational) series*. This concept can be found in [1, 15, 18, 47]. The two first items are in [22, 27]. The third can be deduced from [8, 15] for example and it was used to factorize, for the first time, by Lyndon words, the output of bilinear and analytical dynamical systems respectively in [29, 30] and to study polylogarithms, hypergeometric functions and associated functions in [32, 36, 41]. The dimension of the orbit $Res_g(S)$ is equal to that of $Res_d(S)$, and to the minimal dimension of a representation satisfying the third point of Theorem 3. This rank is then equal to

[12] For any words $u, v \in X^*$, if $u = v$ then $\delta_u^v = 1$ else 0.

[13] $\phi \in End(V)$ is said to be locally nilpotent iff, for any $v \in V$, there exists $N \in \mathbb{N}$ s.t. $\phi^N(v) = 0$.

the rank of the Hankel matrix of S, i.e. the infinite matrix $(\langle S \mid uv\rangle)_{u,v\in X}$ indexed by $X^* \times X^*$ so called *Hankel rank*[14] of S [22, 27]. The triplet (λ, μ, η) is called a *linear representation* of S.[15] S is called *rational* if it belongs to the closure by scaling and by $+$, conc and star operation of proper elements.[16] Any noncommutative power series is representative if and only if it is rational [3, 54]. These rationality properties can be expressed in terms of differential operators in noncommutative geometry [15].

2.2.2 Background on continuity and indistinguishability

Definition 2 [28, 43] Let $\mathscr{H}$ be a class of series i.e. a subset of $\mathbb{C}\langle\langle X\rangle\rangle$ and $S \in \mathbb{C}\langle\langle X\rangle\rangle$.

a. The power series S is said to be *continuous* over $\mathscr{H}$ if for any $\Phi \in \mathscr{H}$, the sum $\sum_{w\in X^*}\langle S \mid w\rangle\langle \Phi \mid w\rangle$ is absolutely convergent, i.e. $\sum_{w\in X^*} |\langle S \mid w\rangle\langle \Phi \mid w\rangle| < +\infty$. This sum will be denoted by $\langle S \parallel \Phi\rangle$.
The set of continuous power series over $\mathscr{H}$ will be denoted by $\mathbb{C}^{\mathrm{cont}}\langle\langle X\rangle\rangle_{\mathscr{H}}$ (or simply $\mathbb{C}^{\mathrm{cont}}\langle\langle X\rangle\rangle$ if the context is clear).
b. S is said to be *indistinguishable* over $\mathscr{H}$ if and only if, for any $\Phi \in \mathscr{H}$, $\langle S \parallel \Phi\rangle = 0$.

Each series $S \in \mathbb{C}\langle\langle X\rangle\rangle$ then defines a (complex) measure μ_S over X^* determined by the charges $\langle S \mid w\rangle$. Here, X^* is considered as a discrete space and compactly supported continuous functions are exactly polynomials. The measure μ_S is a complex linear form over $\mathbb{C}\langle X\rangle$. It satisfies

$$\mu_S(P) = \langle S \mid P\rangle\ . \tag{12}$$

Proposition 1 [43] *Let $\mathscr{H} \subset \mathscr{H}_1 \subset \mathbb{C}\langle\langle X\rangle\rangle$ be two monoids (for the concatenation product), such that $\{e^{tx}\}_{x\in X}^{t\in\mathbb{C}} \subset \mathscr{H}$ ($x \in X$ is given), we suppose that $\mathscr{H}_1$ is closed*[17] *by $T \to |T|$. Let $S \in \mathbb{C}\langle\langle X\rangle\rangle$ be such that $\mathscr{H}_1 \subset \mathscr{L}^1(\mu_S)$ et $S \in \mathscr{H}^{\perp}$. Then for all $x \in X$, $S \triangleleft x, x \triangleright S \in \mathscr{H}^{\perp}$.*

Proof In this context, one has

$$\mathscr{L}^1(\mu_S) = \mathscr{L}^1(|\mu_S|) = \mathscr{L}^1(\mu_{|S|}) = \{T \in \mathbb{C}\langle\langle X\rangle\rangle | \sum_{w\in X^*} |\langle S \mid w\rangle\langle T \mid w\rangle| < +\infty\}\ . \tag{14}$$

[14] I.e. the dimension of span$\{S \triangleleft \Pi \mid \Pi \in \mathbb{C}\langle X\rangle\}$ (resp. span$\{\Pi \triangleright S \mid \Pi \in \mathbb{C}\langle X\rangle\}$).

[15] The minimal representation of S as being a representation of S of minimal dimension. It can be shown that all minimal representations are isomorphic (see [3]).

[16] For any *proper* series S, i.e. $\langle S \mid 1_{X^*}\rangle = 0$, the series $S^* = 1 + S + S^2 + \ldots$ is called "star of S".

[17] For all $S = \sum_{w\in X^*}\langle S \mid w\rangle w \in \mathbb{C}\langle\langle X\rangle\rangle$, we set

$$|S| := \sum_{w\in X*} |\langle S \mid w\rangle| w|. \tag{13}$$

We have to prove that, for all $x \in X$, $S \triangleleft x$ is indistinguishable over $\mathscr{H}$ (i.e. orthogonal to $\mathscr{H}$).

For convenience, we will note $\langle . \parallel\!| . \rangle$ the scalar product,[18] which indicates the absolute convergence within (14) i.e.

$$\langle S \parallel\!| T \rangle := \sum_{w \in X^*} |\langle S \mid w \rangle \langle T \mid w \rangle| \,. \tag{18}$$

For all $\Phi \in \mathscr{H}$, one has $e^{tx}|\Phi|,\ |\Phi| \in \mathscr{H}_1$, hence

$$\langle S \parallel\!| (e^x - 1)|\Phi| \rangle \leq \langle S \parallel\!| e^x|\Phi| \rangle + \langle S \parallel\!| |\Phi| \rangle < +\infty \,. \tag{19}$$

On remarks at once that, for the pointwise convergence topology,

$$\lim_{t \to 0_+} \frac{\exp(tx) - 1_{X^*}}{t} \Phi = x\,\Phi \tag{20}$$

and for all $t \in \mathbb{C}$, $|t| \leq 1$ et $w \in X^*$,

$$|\langle \frac{\exp(tx) - 1_{X^*}}{t} \Phi \mid w \rangle| = |\langle [\sum_{n \geq 1} \frac{t^{n-1}x^n}{n!}] \Phi \mid w \rangle| \leq \langle (e^x - 1)|\Phi| \mid w \rangle \,. \tag{21}$$

Then we can use the Dominated Convergence Theorem of Lebesgues [53] with $E = X^*$ ($\mathscr{B}$, being the σ-algebra generated by the finite subsets) and $\mu = \mu_S$ on one side, the family of functions $w \mapsto \langle \frac{\exp(t_n x) - 1_{X^*}}{t_n} \Phi \mid w \rangle$ (using a sequence $\{t_n \in]0, 1] \mid n \in \mathbb{N}_+\}$ which converges to 0 and the dominating function $w \mapsto \langle (e^x - 1)|\Phi| \mid w \rangle$ (21) on the other side. One then has

$$\begin{aligned}\langle S \triangleleft x \parallel\!| \Phi \rangle = \langle S \parallel\!| x\,\Phi \rangle = \mu_S(x\,\Phi) &= \int_{X^*} (\lim_{n \to +\infty} \frac{\exp(t_n x) - 1_{X^*}}{t_n} \Phi) d\mu_S \\ &= \lim_{n \to +\infty} \int_{X^*} (\frac{\exp(t_n x) - 1_{X^*}}{t_n} \Phi) d\mu_S = 0 \,.\end{aligned} \tag{22}$$

---

[18]Let $X = \{x\}$, consider $S := \sum_{n \geq 1} \frac{x^n}{n}$ and $T := \sum_{n \geq 0} (1-t)^n x^n$. then

$$\langle S \parallel\!| T \rangle := \sum_{w \in X^*} |\langle S \mid w \rangle \langle T \mid w \rangle| = \sum_{n \geq 1} |\langle S \mid x^n \rangle \langle T \mid x^n \rangle| \tag{15}$$

$$= \sum_{n \geq 1} |\frac{(1-t)^n}{n}| = \sum_{n \geq 1} \frac{|1-t|^n}{n} = -\log(1 - |1-t|),\ t \in]0, 2[\,. \tag{16}$$

$$= \begin{cases} -\log(t) = |\log(t)|, & t \in]\,0, 1] \\ -\log(2-t) = |\log(2-t)|, & t \in [1, 2[\end{cases} \tag{17}$$

2.3 Polylogarithms and Harmonic Sums

2.3.1 Structure of Polylogarithms and of Harmonic Sums

Let $\Omega := \mathbb{C} - (]-\infty, 0] \cup [1, +\infty[)$ and let $\mathscr{C} := \mathbb{C}[z, 1/z, 1/1-z]$. Note that the unit of $\mathscr{C}$ is denoted ,for the pointwise product, by $1_\Omega : \Omega \longrightarrow \mathbb{C}$ such that $z \longmapsto 1$.

One can check that $\mathrm{Li}_{s_1,\ldots,s_r}$ is obtained as the iterated integral over the differential forms $\omega_0(z) = dz/z$ and $\omega_1(z) = dz/(1-z)$ and along the path $0 \rightsquigarrow z$ [31]:

$$\mathrm{Li}_{s_1,\ldots,s_r}(z) = \alpha_0^z(x_0^{s_1-1}x_1\ldots x_0^{s_r-1}x_1) = \sum_{n_1>\cdots>n_r>0} \frac{z^{n_1}}{n_1^{s_1}\ldots n_r^{s_r}}. \tag{23}$$

By (1), $\mathrm{Li}_{s_1,\ldots,s_r}$ is then denoted also by $\mathrm{Li}_{x_0^{s_1-1}x_1\ldots x_0^{s_r-1}x_1}$ or $\mathrm{Li}_{y_{s_1}\ldots y_{s_r-1}}$ [32, 36, 37].

Example 8 (*of* $\mathrm{Li}_2 = \mathrm{Li}_{x_0x_1}$)

$$\alpha_0^z(x_0x_1) = \int_0^z \frac{ds}{s}\int_0^s \frac{dt}{1-t} = \int_0^z \frac{ds}{s}\int_0^s dt \sum_{k\geq 0} t^k = \sum_{k\geq 1}\int_0^z ds\frac{s^{k-1}}{k} = \sum_{k\geq 1}\frac{z^k}{k^2}.$$

The definition of polylogarithms is extended over the words $w \in X^*$ by putting $\mathrm{Li}_{x_0}(z) := \log(z)$. The $\{\mathrm{Li}_w\}_{w\in X^*}$ are $\mathscr{C}$-linearly independent [13, 33, 37] and then the following functions, for $v = y_{s_1}\ldots y_{s_r} \in Y^*$, are also $\mathbb{C}$-linearly independent [13, 40]

$$\mathrm{P}_v(z) := \frac{\mathrm{Li}_v(z)}{1-z} = \sum_{N\geq 0} \mathrm{H}_v(N)\, z^N, \text{ where } \mathrm{H}_v(N) := \sum_{N\geq n_1>\cdots>n_r>0} \frac{1}{n_1^{s_1}\ldots n_r^{s_r}}.$$

Proposition 2 ([40]) *By linearity, the following maps are isomorphisms of algebras*

$$\mathrm{P}_\bullet : (\mathbb{C}\langle Y\rangle, ⧢) \longrightarrow (\mathbb{C}\{\mathrm{P}_w\}_{w\in Y^*}, \odot), \quad u \longmapsto \mathrm{P}_u,$$
$$\mathrm{H}_\bullet : (\mathbb{C}\langle Y\rangle, ⧢) \longrightarrow (\mathbb{C}\{\mathrm{H}_w\}_{w\in Y^*}, .), \quad u \longmapsto \mathrm{H}_u = \{\mathrm{H}_u(N)\}_{N\geq 0}.$$

Theorem 4 ([40]) *The Hadamard $\mathscr{C}$-algebra of $\{\mathrm{P}_w\}_{w\in Y^*}$ can be identified with that of $\{\mathrm{P}_l\}_{l\in\mathscr{L}ynY}$. In the same way, the algebra of harmonic sums $\{\mathrm{H}_w\}_{w\in Y^*}$ with polynomial coefficients can be identified with that of $\{\mathrm{H}_l\}_{l\in\mathscr{L}ynY}$.*

Let L, P and H be the noncommutative generating series of respectively $\{\mathrm{Li}_w\}_{w\in X^*}$, $\{\mathrm{P}_w\}_{w\in X^*}$ and $\{\mathrm{H}_w(N)\}_{w\in Y^*}$, for $|z| < 1$ and $N > 1$ [33, 40]:

$$\mathrm{L}(z) = \sum_{w\in X^*} \mathrm{Li}_w(z)w; \quad \mathrm{P}(z) = \frac{\mathrm{L}(z)}{1-z}; \quad \mathrm{H}(N) = \sum_{w\in Y^*} \mathrm{H}_w(N)\, w. \tag{24}$$

Definition 3 (*Polylogarithms and harmonic sums at negative multi-indices*) For any $s_1, \ldots, s_r \in (\mathbb{N})^r$, let us define [17], for $|z| < 1$ and $N > 0$,

$$\mathrm{Li}_{-s_1,\ldots,-s_r}(z) := \sum_{n_1>\cdots>n_r>0} n_1^{s_1} \ldots n_r^{s_r} z^{n_1} \text{ and } \mathrm{H}_{-s_1,\ldots,-s_r}(N) := \sum_{N\geq n_1>\cdots>n_r>0} n_1^{s_1} \ldots n_r^{s_r}.$$

The ordinary generating series, $\mathrm{P}_{-s_1,\ldots,-s_r}(z)$, of $\{\mathrm{H}_{-s_1,\ldots,-s_r}(N)\}_{N\geq 0}$ is

$$\mathrm{P}_{-s_1,\ldots,-s_r}(z) := \sum_{N\geq 0} \mathrm{H}_{-s_1,\ldots,-s_r}(N)\, z^N = \frac{1}{1-z}\mathrm{Li}_{-s_1,\ldots,-s_r}(z).$$

Now, let[19] $Y_0 = Y \cup \{y_0\}$ and let Y_0^* denotes the free monoid generated by Y_0 admitting $1_{Y_0^*}$ as neutral element. As in (1), let us introduce another correspondence

$$(s_1, \ldots, s_r) \in \mathbb{N}^r \leftrightarrow y_{s_1} \ldots y_{s_r} \in Y_0^*.$$

In all the sequel, for some convenience, we will also adopt the following notations, for any $w = y_{s_1} \ldots y_{s_r} \in Y_0^*$,

$$\mathrm{Li}_w^- = \mathrm{Li}_{-s_1,\ldots,-s_r}; \quad \mathrm{P}_w^- = \mathrm{P}_{-s_1,\ldots,-s_r} \quad \text{and} \quad \mathrm{H}_w^- = \mathrm{H}_{-s_1,\ldots,-s_r}.$$

Example 9 ($\mathrm{Li}_{y_0^r}^-$ *and* $\mathrm{H}_{y_0^r}^-$) By Proposition (5), we have $\mathrm{Li}_{y_0^r}^- = \lambda^r$. Hence,

$$\frac{\mathrm{Li}_{y_0^r}^-(z)}{1-z} = \frac{z^r}{(1-z)^{r+1}} = \sum_{N\geq 0} \binom{N}{r} z^N \text{ and then } \mathrm{H}_{y_0^r}^-(N) = \binom{N}{r}.$$

Definition 4 With the convention $\mathrm{H}_{1_{Y_0^*}}^- = 1$, we put

$$\mathrm{L}^-(z) := \sum_{w\in Y_0^*} \mathrm{Li}_w^-(z) w; \quad \mathrm{P}^-(z) := \frac{\mathrm{L}^-(z)}{1-z}; \quad \mathrm{H}^-(N) := \sum_{w\in Y_0^*} \mathrm{H}_w^-(N) w.$$

Since, for $y_k \in Y, u \in Y^*$ (resp. $y_k \in Y_0, u \in Y_0^*$) and $N \geq 1$, one has $\mathrm{H}_{y_k u}(N) - \mathrm{H}_{y_k u}(N-1) = N^{-k}\mathrm{H}_u(N-1)$ (resp. $\mathrm{H}_{y_k u}^-(N) - \mathrm{H}_{y_k u}^-(N-1) = N^k \mathrm{H}_u^-(N-1)$). Then

Proposition 3 H *and* H^- *satisfy the following difference equations*

$$\mathrm{H}(N) = \Big(1_{Y^*} + \sum_{k\geq 1} \frac{y_k}{N^k}\Big)\mathrm{H}(N-1) = \prod_{n=1}^{N}\Big(1_{Y^*} + \sum_{k\geq 1} \frac{y_k}{n^k}\Big) = 1_{Y^*} + \sum_{w\in Y^*, |w|\geq N} \mathrm{H}_w(N)\, w,$$

$$\mathrm{H}^-(N) = \Big(1_{Y_0^*} + \sum_{k\geq 0} y_k N^k\Big)\mathrm{H}^-(N-1) = \prod_{n=1}^{N}\Big(1_{Y_0^*} + \sum_{k\geq 0} y_k n^k\Big) = 1_{Y_0^*} + \sum_{w\in Y_0^*, |w|\geq N} \mathrm{H}_w^-(N)\, w.$$

[19] with $y_0 \succ y_1$.

Hence, for any $w \in Y^*$ (resp. $w \in Y_0^*$), $\mathrm{H}_w(N)$ (resp. $\mathrm{H}_w^-(N)$) is of valuation N.

In all the sequel, the *length* and the *weight* of $u = y_{i_1} \dots y_{i_k} \in Y^*$ are defined respectively as the numbers $|u| = k$ and $(u) = i_1 + \cdots + i_k$.

Definition 5 Let $g, h \in \mathbb{Q}\langle\langle Y_0 \rangle\rangle [\![t]\!]$ be defined as follows (here, $|1_{Y_0^*}| = (1_{Y_0^*}) = 0$)

$$h(t) := \sum_{w \in Y_0^*} ((w) + |w|)! t^{(w)+|w|} w \text{ and } g(t) := \sum_{w \in Y_0^*} t^{(w)+|w|} w = \left(\sum_{y \in Y_0} t^{(y)+1} y \right)^* .$$

Remark 1 1. The generating series h is an extension of the Euler series $\sum_{n \geq 0} n! t^n$ and it can be obtained as Borel–Laplace transform of g.

2. The ordinary generating series $\mathscr{Y}(t) := 1 + \sum_{r \geq 0} y_r \, t^r$ and its inverse are group-like. The generating series $\Lambda(t) = \sum_{w \in Y_0^*} t^{(w)+|w|} w$ can be obtained from $1/\mathscr{Y}(t)$ by use the following change of alphabet $y_r \leftarrow t y_r$ it can be expressed as

$$g(t) = \left(1 - \sum_{r \geq 0} (-t y_r) \, t^r \right)^{-1} = \left(\sum_{r \geq 0} (-t y_r) \, t^r \right)^* .$$

Now, let us consider the following differential and integration operators acting on $\mathbb{C}\{\mathrm{Li}_w\}_{w \in X^*}$ which can be extended over $\mathscr{C}\{\mathrm{Li}_w\}_{w \in X^*}$ [41]:

$$\partial_z = d/dz, \; \theta_0 = z d/dz, \; \theta_1 = (1-z) d/dz, \iota_0 : \mathrm{Li}_w \longmapsto \mathrm{Li}_{x_0 w}, \; \iota_1 : \mathrm{Li}_w \longmapsto \mathrm{Li}_{x_1 w}$$

Let Θ and $\Im$ be monoid morphisms such that $\Theta(1_{X^*}) = \Im(1_{X^*}) = \mathrm{Id}$ and, for $x_i \in X, v \in X^*$, $\Theta(v x_i) = \Theta(v)\theta_i$ and $\Im(v x_i) = \Im(v)\iota_i$. Hence,

Proposition 4 *1. The operators $\{\theta_0, \theta_1, \iota_0, \iota_1\}$ satisfy in particular,*

$$\theta_1 + \theta_0 = [\theta_1, \theta_0] = \partial_z \text{ and } \forall k = 0, 1, \theta_k \iota_k = \mathrm{Id},$$
$$[\theta_0 \iota_1, \theta_1 \iota_0] = 0 \text{ and } (\theta_0 \iota_1)(\theta_1 \iota_0) = (\theta_1 \iota_0)(\theta_0 \iota_1) = \mathrm{Id}.$$

2. For any $w = y_{s_1} \dots y_{s_r} \in Y^*$ ($\pi_X(w) = x_0^{s_1 - 1} x_1 \dots x_0^{s_r - 1} x_1$) and $u = y_{t_1} \dots y_{t_r} \in Y_0^*$, we can rephrase Li_w, Li_u^- as follows

$$\mathrm{Li}_w = (\iota_0^{s_1 - 1} \iota_1 \dots \iota_0^{s_r - 1} \iota_1) 1_\Omega \text{ and } \mathrm{Li}_u^- = (\theta_0^{t_1 + 1} \iota_1 \dots \theta_0^{t_r + 1} \iota_1) 1_\Omega,$$
$$\theta_0 \, \mathrm{Li}_{x_0 \pi_X(w)} = \mathrm{Li}_{\pi_X(w)} \text{ and } \theta_1 \, \mathrm{Li}_{x_1 \pi_X(w)} = \mathrm{Li}_{\pi_X(w)},$$
$$\iota_0 \, \mathrm{Li}_{\pi_X(w)} = \mathrm{Li}_{x_0 \pi_X(w)} \text{ and } \iota_1 \, \mathrm{Li}_w = \mathrm{Li}_{x_1 \pi_X(w)} .$$

3. $\mathscr{C}\{\mathrm{Li}_w\}_{w \in X^*} \cong \mathscr{C} \otimes_{\mathbb{C}} \mathbb{C}\{\mathrm{Li}_w\}_{w \in X^*}$ is closed under of $\iota_0, \iota_1, \theta_0, \theta_1$.

4. Let $\lambda(z) := z/(1-z) \in \mathscr{C}$. Then λ and $1/\lambda$ are the eigenvalues of $\theta_0 \iota_1$ and $\theta_1 \iota_0$ within $\mathscr{C}\{\mathrm{Li}_w\}_{w \in X^*}$ respectively:

$$\forall f \in \mathscr{C}\{\mathrm{Li}_w\}_{w \in X^*}, (\theta_0 \iota_1) f = \lambda f \text{ and } (\theta_1 \iota_0) f = f/\lambda.$$

5. *For any $n \geq 0$ and $w \in X^*$, one has*[20]

$$\Theta(\widetilde{w})\,\mathrm{Li}_w = 1_\Omega \text{ and } \partial_z^n = \sum_{w \in X^n} (\Theta \otimes \Theta)\Delta_{\sqcup\!\sqcup}(w).$$

Proof The proofs are immediate.

Proposition 5 ([17])

1. *For any $w \in Y_0^*$, one has $\mathrm{Li}_w^-(z) = \lambda^{|w|}(z)A_w^-(z)(1-z)^{-(w)}$, where A_w^- is the extended Eulerian polynomial defined recursively as follows*

$$A_w^-(z) = \begin{cases} \displaystyle\sum_{k=0}^{n-1} A_{n,k} z^k & \text{if } w = y_k \in Y_0, \\ \displaystyle\sum_{i=0}^{s_1} \binom{s_1}{i} A_{y_i} A^-_{y_{(s_1+s_2-i)} y_{s_3} \ldots y_{s_r}} & \text{if } w = y_k u \in Y_0 Y_0^*, \end{cases}$$

and $A_{n,k}$ are Eulerian numbers satisfying $A_{n,k} = \sum_{j=0}^{k} (-1)^j \binom{n+1}{j} (k+1-j)^n$.

2. *For any $w \in Y^*$, let us define $\{G_w^-(n)\}_{n \in \mathbb{N}}$ by the following generating series*

$$\sum_{n \geq |w|} \frac{(n+1)!}{(n-|w|)!} G_w^-(n) z^n = \frac{\mathrm{Li}_w^-(z)}{1-z}.$$

Then $\mathrm{H}_w^-(N) = (N+1)N(N-1)\ldots(N-|w|+1)G_w^-(N)$.

3. *$\mathrm{Li}_w^-(z) \in \mathbb{Q}[(1-z)^{-1}] \subsetneq \mathscr{C}$ and $\mathrm{H}_w^-(N) \in \mathbb{Q}[N]$ of degree $|w| + (w)$.*

Example 10 [17](*Case of $r = 1$ by Maple*)

1. Since $A_n(z)/(1-z)^{n+1} = \sum_{j \geq 0} z^j (j+1)^n$ then $\mathrm{Li}_{y_n}^-(z) = zA_n(z)/(1-z)^{n+1}$ (see [20] for example). For example,

$$\begin{aligned}
\mathrm{Li}_{y_1}^-(z) &= z(1-z)^{-2} &&= -(1-z)^{-1} + (1-z)^{-2}. \\
\mathrm{Li}_{y_2}^-(z) &= z(z+1)(1-z)^{-3} &&= (1-z)^{-1} - 3(1-z)^{-2} + 2(1-z)^{-3}. \\
\mathrm{Li}_{y_3}^-(z) &= z(z^2+4z+1)(1-z)^{-4} &&= -(1-z)^{-1} + 7(1-z)^{-2} - 12(1-z)^{-3} + 6(1-z)^{-4}.
\end{aligned}$$

2. For any positive integer m, one has

$$\begin{aligned}
\mathrm{H}_{y_m}^-(N) &= \frac{1}{m+1} \sum_{k=0}^{m} \binom{m+1}{k} B_k (N+1)^{m+1-k} \\
&= \frac{1}{m+1} \sum_{k=1}^{m+1} \left[\sum_{l=0}^{m+1-k} \binom{m+1}{l} \binom{m+1-l}{k} B_l \right] N^l,
\end{aligned}$$

[20]For any $w = x_{i_1} \ldots x_{i_r} \in X^*$, we denote $\widetilde{w} = x_{i_r} \ldots x_{i_1}$.

where B_k is the kth Bernoulli's number given by its exponential generating series

$$\frac{t}{e^t - 1} = \sum_{k \geq 0} B_k \frac{t^k}{k!}.$$

For example, (recall that $B_0 = 1$, $B_1 = -1/2$, $B_2 = 1/6$, $B_3 = 0$, $B_4 = -1/30$),

$$\begin{aligned}
\mathrm{H}^-_{y_1}(N) &= \quad (N+1)^2/2 - (N+1)/2 &&= N(N+1)/2,\\
\mathrm{H}^-_{y_2}(N) &= (N+1)^3/3 - (N+1)^2/2 + (N+1)/6 &&= N(2N+1)(N+1)/6,\\
\mathrm{H}^-_{y_3}(N) &= (N+1)^4/4 - (N+1)^3/2 + (N+1)^2/4 &&= (N(N+1)/2)^2.
\end{aligned}$$

Example 11 (*Case of* $r = 2$ *by Maple*)

1. From what precedes, $\mathrm{Li}^-_{y_m y_n} = (\theta_0^{m+1}\iota_1)\,\mathrm{Li}^-_{y_n} = \theta_0^m(\theta_0\iota_1)\,\mathrm{Li}^-_{y_n}$. Since, by Example 9, we have $(\theta_0\iota_1)\,\mathrm{Li}^-_{y_n} = \mathrm{Li}^-_{y_0}\,\mathrm{Li}^-_{y_n}$ then $\mathrm{Li}^-_{y_m y_n} = \theta_0^m[\mathrm{Li}^-_{y_0}\,\mathrm{Li}^-_{y_n}] = \sum_{l=0}^m \binom{m}{l}\,\mathrm{Li}^-_{y_l}\,\mathrm{Li}^-_{y_{m+n-l}}$. For example,

$$\begin{aligned}
\mathrm{Li}^-_{y_1^2}(z) &= \mathrm{Li}^-_{y_0}(z)\,\mathrm{Li}^-_{y_2}(z) + (\mathrm{Li}^-_{y_1}(z))^2\\
&= -(1-z)^{-1} + 5(1-z)^{-2} - 7(1-z)^{-3} + 3(1-z)^{-4}\\
\mathrm{Li}^-_{y_2 y_1}(z) &= \mathrm{Li}^-_{y_0}(z)\,\mathrm{Li}^-_{y_3}(z) + 3\,\mathrm{Li}^-_{y_1}(z)\,\mathrm{Li}^-_{y_2}(z)\\
&= (1-z)^{-1} - 11(1-z)^{-2} + 31(1-z)^{-3} - 33(1-z)^{-4} + 12(1-z)^{-5},\\
\mathrm{Li}^-_{y_1 y_2}(z) &= \mathrm{Li}^-_{y_0}(z)\,\mathrm{Li}^-_{y_3}(z) + \mathrm{Li}^-_{y_1}(z)\,\mathrm{Li}^-_{y_2}(z)\\
&= (1-z)^{-1} - 9(1-z)^{-2} + 23(1-z)^{-3} - 23(1-z)^{-4} + 8(1-z)^{-5}.
\end{aligned}$$

2. For any positive integers m, n, one has

$$\mathrm{H}^-_{y_m y_n}(N) = \sum_{k_1=0}^{n} \sum_{k_2=0}^{m+n+1-k_1} \sum_{k_3=0}^{m+n+2-k_1-k_2} \frac{B_{k_1} B_{k_2}}{(n+1)(m+n+2-k_1)} \binom{n+1}{k_1}\binom{m+n+2-k_1}{k_2}\binom{m+n+2-k_1-k_2}{k_3} N^{k_3}.$$

For example,

$$\begin{aligned}
\mathrm{H}^-_{y_2 y_1}(N) &= N(N^2-1)(12N^2+15N+2)/120,\\
\mathrm{H}^-_{y_2^2}(N) &= N(N-1)(2N+1)(2N-1)(5N+6)(N+1)/360,\\
\mathrm{H}^-_{y_2 y_3}(N) &= N(N-1)(N+1)(30N^4+35N^3-33N^2-35N+2)/840,\\
\mathrm{H}^-_{y_2 y_4}(N) &= N(N-1)(N+1)(63N^5+72N^4-133N^3-138N^2+49N+30)/2520,\\
\mathrm{H}^-_{y_2 y_5}(N) &= N(N-1)(N+1)(280N^6+315N^5-920N^4-945N^3+802N^2+630N-108)/15120,\\
\mathrm{H}^-_{y_3^3}(N) &= N(N-1)(N+1)(21N^5+36N^4-21N^3-48N^2+8)/672,\\
\mathrm{H}^-_{5,6}(N) &= \frac{1}{2{,}162{,}160} N(N-1)(N+1)(23{,}760N^{10}+64{,}350N^9-109{,}620N^8\\
&\quad -386{,}100N^7+184{,}960N^6+920{,}205N^5-158{,}240N^4-1{,}036{,}035N^3\\
&\quad +97{,}444N^2+450{,}450N-16{,}956)).
\end{aligned}$$

Example 12 (*General case*)

1. One has, for any $y_{s_1}u = y_{s_1}\dots y_{s_r} \in Y_0^*$,

$$\mathrm{Li}^-_{y_{s_1}u} = \theta_0^{s_1}(\theta_0\iota_1)\,\mathrm{Li}^-_u = \theta_0^{s_1}(\lambda\,\mathrm{Li}^-_u) = \sum_{k_1=0}^{s_1}\binom{s_1}{k_1}(\theta_0^{k_1}\lambda)(\theta_0^{s_1-k_1}\,\mathrm{Li}^-_u),$$

$$\mathrm{Li}^-_{y_{s_1}\dots y_{s_r}} = \sum_{k_1=0}^{s_1}\sum_{k_2=0}^{s_1+s_2-k_1}\dots\sum_{k_r=0}^{\substack{(s_1+\dots+s_r)-\\(k_1+\dots+k_{r-1})}}\binom{s_1}{k_1}\binom{s_1+s_2-k_1}{k_2}\dots$$
$$\binom{s_1+\dots+s_r-k_1-\dots-k_{r-1}}{k_r}(\theta_0^{k_r}\lambda)(\theta_0^{k_2}\lambda)\dots(\theta_0^{k_r}\lambda).$$

Denoting $S_2(k_i, j)$ Stirling numbers of the second kind, one has

$$\forall i = 1,..,r,\ \ \theta_0^{k_i}\lambda(z) = \begin{cases} \lambda(z), \text{ if } k_i = 0, \\ \dfrac{1}{1-z}\displaystyle\sum_{j=1}^{k_i} S_2(k_i, j)j!\lambda^j(z), \text{ if } k_i > 0. \end{cases}$$

In particular, if $\omega \in Y^*$ then $(1-z)^{|w|}\,\mathrm{Li}^-_w(z)$ is polynomial of degree (w) in $\lambda(z)$.

2. We define, firstly, the *polynomials* $\{B_{y_{n_1}\dots y_{n_r}}(z)\}_{n_1,\dots,n_r\in\mathbb{N}}$ by their commutative exponential generating series as follows, for $z \in \mathbb{C}$,

$$\sum_{n_1,\dots,n_r\in\mathbb{N}} B_{y_{n_1}\dots y_{n_r}}(z)\frac{t_1^{n_1}\dots t_r^{n_r}}{n_1!\dots n_r!} = t_1\dots t_r e^{z(t_1+\dots+t_r)}\prod_{k=1}^{r}(e^{t_k+\dots+t_r}-1)^{-1},$$

or by the difference equation, for $n_1 \in \mathbb{N}_+$,

$$B_{y_{n_1}\dots y_{n_r}}(z+1) = B_{y_{n_1}\dots y_{n_r}}(z) + n_1 z^{n_1-1}B_{y_{n_2}\dots y_{n_r}}(z).$$

For any $w \in y_sY_0^*, s > 1$, we have $B_w(1) = B_w(0)$. Then let also, for any $1 \le k \le r$,

$$b_w := B_w(0) \text{ and } \beta_w(z) := B_w(z) - b_w.$$
$$b'_{y_k} := b_{y_k} \quad \text{and } b'_{y_{n_k}\dots y_{n_r}} := b_{y_{n_k}\dots y_{n_r}} - \sum_{j=0}^{r-1-k} b_{y_{n_{k+j+1}}\dots y_{n_r}} b'_{y_{n_k}\dots y_{n_{k+j}}}$$

Then we have the extended Faulhaber's identities

$$\beta_{y_{n_1}\cdots y_{n_r}}(N) = \sum_{k=1}^{r}\left(\prod_{i=1}^{k} n_i\right) b_{y_{n_{k+1}}\cdots y_{n_r}} \mathrm{H}^-_{y_{n_1-1}\cdots y_{n_k-1}}(N-1),$$

$$\mathrm{H}^-_{y_{n_1}\cdots y_{n_r}}(N) = \frac{\beta_{y_{n_1+1}\cdots y_{n_r+1}}(N+1) - \sum_{k=1}^{r-1} b'_{y_{n_{k+1}+1}\cdots y_{n_r+1}} \beta_{y_{n_1+1}\cdots y_{n_k+1}}(N+1)}{\prod_{i=1}^{r}(n_i+1)}.$$

Proposition 6 ([17]) *The following maps are morphisms of algebras*

$$\mathrm{H}^- : (\mathbb{C}\langle Y_0\rangle, \uplus) \longrightarrow (\mathbb{C}\{\mathrm{H}^-_w\}_{w\in Y_0^*}, .) \text{ and } \mathrm{P}^- : (\mathbb{C}\langle Y_0\rangle, \uplus) \longrightarrow (\mathbb{C}\{\mathrm{P}^- w\}_{w\in Y_0^*}, \odot).$$

Proof Recall that the quasi-symmetric functions on the variables $\mathbf{t} = \{t_i\}_{N\geq i\geq 1}$, i.e.

$$\mathrm{F}_{s_1,\ldots,s_r}(\mathbf{t}) = \mathrm{F}_{y_{s_1}\cdots y_{s_r}}(\mathbf{t}) = \sum_{n_1>\cdots>n_r>0} t_{n_1}^{s_1}\ldots t_{n_r}^{s_r}$$

satisfy the quasi-shuffle relation [52], i.e. for any $u, v \in Y_0^*$, $\mathrm{F}_{u\uplus v}(\mathbf{t}) = \mathrm{F}_u(\mathbf{t})\mathrm{F}_v(\mathbf{t})$.

Since $\mathrm{H}^-_{s_1,\ldots,s_r}(N)$ can be obtained by specializing, in $\mathrm{F}_{s_1,\ldots,s_r}(\mathbf{t})$, the variables $\mathbf{t}$ at

$$\forall 1 \leq i \leq N, t_i = i \text{ and } \forall i > N, t_i = 0$$

then H^- is a morphism of algebras. Therefore, P^- is also a morphism of algebras.

2.3.2 Global Renormalizations via Noncommutative Generating Series

By (2.3.2), L and H are images, by the tensor products Li ⊗Id and H ⊗ Id, of the diagonal series $\mathscr{D}_X$ and $\mathscr{D}_Y$ respectively. Then we get

Theorem 5 (Factorization of L and of H, [33, 37, 43]) *Let*

$$\mathrm{L}_{\mathrm{reg}} = \prod_{l\in\mathscr{L}ynX-X}^{\searrow} e^{\mathrm{Li}_{S_l}\, P_l} \text{ and } \mathrm{H}_{\mathrm{reg}}(N) = \prod_{l\in\mathscr{L}ynY-\{y_1\}}^{\searrow} e^{\mathrm{H}_{\check{\Sigma}_l}(N)\ \Sigma_l}.$$

Then $\mathrm{L}(z) = e^{-x_1\log(1-z)}\mathrm{L}_{\mathrm{reg}}(z)e^{x_0\log z}$ *and* $\mathrm{H}(N) = e^{\mathrm{H}_{y_1}(N)\ y_1}\mathrm{H}_{\mathrm{reg}}(N)$.

For any $l \in \mathscr{L}ynX - X$ (resp. $\mathscr{L}ynY - \{y_1\}$), the polynomial S_l (resp. Σ_l) is a finite combination of words in $x_0X^*x_1$ (resp. $Y^* - y_1Y^*$). Then we can state

Proposition 7 ([43]) *Let* $Z_{\sqcup\!\sqcup} := \mathrm{L}_{\mathrm{reg}}(1)$ *and* $Z_{\uplus} := \mathrm{H}_{\mathrm{reg}}(\infty)$. *Then* $Z_{\sqcup\!\sqcup}$ *and* $Z_{\uplus}$ *are group-like, for* $\Delta_{\sqcup\!\sqcup}$ *and* $\Delta_{\uplus}$ *respectively.*

Proposition 8 (Successive integrations and differentiations of L, [41, 43]) *We have, for any* $n \in \mathbb{N}$,

1. $\iota_0^n\mathrm{L} = x_0^n \triangleleft \mathrm{L}$ *and* $\iota_1^n\mathrm{L} = x_1^n \triangleleft \mathrm{L}$.

2. $\partial_z^n \mathrm{L} = D_n \mathrm{L}$ *and* $\theta_0^n \mathrm{L} = E_n \mathrm{L}$, *where*[21] *the polynomials* D_n *and* E_n *in* $\mathscr{C}\langle X\rangle$ *are*

$$D_n = \sum_{\mathrm{wgt}(\mathbf{r})=n} \sum_{w\in X^{\deg(\mathbf{r})}} \prod_{i=1}^{\deg(\mathbf{r})} \binom{\sum_{j=1}^{i} r_i + j - 1}{r_i} \tau_{\mathbf{r}}(w),$$

$$E_n = \sum_{\mathrm{wgt}(\mathbf{r})=n} \sum_{w\in X^{\deg(\mathbf{r})}} \prod_{i=1}^{\deg(\mathbf{r})} \binom{\sum_{j=1}^{i} r_i + j - 1}{r_i} \rho_{\mathbf{r}}(w),$$

and for any $w = x_{i_1}\cdots x_{i_k}$ *and* $\mathbf{r} = (r_1, \ldots, r_k)$ *of degree* $\deg(\mathbf{r}) = k$ *and of weight* $\mathrm{wgt}(\mathbf{r}) = k + r_1 + \cdots + r_k$, *the polynomials* $\tau_{\mathbf{r}}(w) = \tau_{r_1}(x_{i_1})\cdots\tau_{r_k}(x_{i_k})$ *and* $\rho_{\mathbf{r}}(w) = \rho_{r_1}(x_{i_1})\cdots\rho_{r_k}(x_{i_k})$ *are defined respectively by, for any* $r \in \mathbb{N}$,

$$\tau_r(x_0) = \partial_z^r \frac{x_0}{z} = \frac{-r!x_0}{(-z)^{r+1}} \text{ and } \tau_r(x_1) = \partial_z^r \frac{x_1}{1-z} = \frac{r!x_1}{(1-z)^{r+1}},$$

$$\rho_r(x_0) = \theta_0^r \frac{(-1)^{-1}x_0}{z} = 0 \text{ and } \rho_r(x_1) = \theta_0^r \frac{zx_1}{1-z} = \mathrm{Li}^-_{\pi_Y(x_0^{r-1}x_1)}(z)x_1.$$

Example 13 (*Coefficients of* $\theta_0^n\mathrm{L}$) Since, for any $u \in X^+$, $\theta_0 \mathrm{Li}_{x_0u} = \mathrm{Li}_u$ and $\theta_1 \mathrm{Li}_{x_0u} = \mathrm{Li}_0 \mathrm{Li}_u$, one obtains for example

- For any $n \geq 1$ and $w \in X^*$, one has $\theta_0^n \mathrm{Li}_{x_0^n w} = \mathrm{Li}_w$. Hence,

$$\theta_0 \mathrm{Li}_{x_1} = \mathrm{Li}_0,\ \theta_0^2 \mathrm{Li}_{x_1} = \mathrm{Li}^-_{\pi_Y(x_1)},\ \theta_0^3 \mathrm{Li}_{x_1} = \mathrm{Li}^-_{\pi_Y(x_0x_1)} \text{ and } \theta_0^4 \mathrm{Li}_{x_1} = \mathrm{Li}^-_{\pi_Y(x_0^2x_1)}.$$

- $\theta_0 \mathrm{Li}_{x_1^2} = \mathrm{Li}_0 \mathrm{Li}_{x_1}$, $\theta_0^2 \mathrm{Li}_{x_1^2} = \mathrm{Li}^-_{\pi_Y(x_1)} \mathrm{Li}_{x_1} + \mathrm{Li}_0^2$, $\theta_0^3 \mathrm{Li}_{x_1^2} = \mathrm{Li}^-_{\pi_Y(x_0x_1)} \mathrm{Li}_{x_1} + 3\ \mathrm{Li}^-_{\pi_Y(x_1)} \mathrm{Li}_0$ because

$$\forall k > 1,\ \theta_0^k \mathrm{Li}_{x_1^2} = \sum_{j=0}^{k-1} \binom{k-1}{j} \mathrm{Li}_{-j} \mathrm{Li}_{2+j-k}.$$

The noncommutative generating series L satisfies the differential equation

$$d\mathrm{L} = (x_0\omega_0 + x_1\omega_1)\mathrm{L} \tag{25}$$

with boundary condition

$$\mathrm{L}(z) \underset{z\to 0}{\widetilde{\ }} \exp(x_0 \log z) \quad\text{and}\quad \mathrm{L}(z) \underset{z\to 1}{\widetilde{\ }} \exp(-x_1 \log(1-z))\, Z_{ш}. \tag{26}$$

This implies that L is the exponential of a Lie series [33, 37]. Hence [41],

[21] Since $\theta_0 + \theta_1 = \partial_z$ then we also have $\theta_1^n\mathrm{L}(z) = [D_n(z) - E_n(z)]\mathrm{L}(z)$. The more general actions of $\{\Theta(w)\}_{w\in X^*}$ on L are more complicated to be expressed here.

$$\log \mathrm{L} = \sum_{k\geq 1} \frac{(-1)^{k-1}}{k} \sum_{u_1,\ldots,u_k \in X^+} \mathrm{Li}_{u_1 \sqcup\!\sqcup \ldots \sqcup\!\sqcup u_k}\; u_1 \ldots u_k = \sum_{w\in X^*} \mathrm{Li}_w\; \pi_1(w).$$

Theorem 6 ([41, 43])

1. *Let G, H be exponential solutions of (25). Then there exists a constant Lie series C such that $G = He^C$.*
2. *Let* $\mathrm{Gal}_{\mathbb{C}}(DE)$ *be the differential Galois group associated to the Drinfel'd equation. Then* $\mathrm{Gal}_{\mathbb{C}}(DE) = \{e^C \mid C \in \mathcal{L}ie_{\mathbb{C}}\langle\langle X\rangle\rangle\}$*, it contains the monodromy group defined by* $\mathcal{M}_0\mathrm{L} = \mathrm{L}\exp(2\mathrm{i}\pi\,\mathfrak{m}_0)$ *and* $\mathcal{M}_1\mathrm{L} = \mathrm{L}Z_{\sqcup\!\sqcup}^{-1}\exp(-2\mathrm{i}\pi x_1)Z_{\sqcup\!\sqcup} = \mathrm{L}\exp(2\mathrm{i}\pi\,\mathfrak{m}_1)$*, where* $\mathfrak{m}_0 = x_0$, $\mathfrak{m}_1 = \prod^{\searrow}_{l\in\mathcal{L}ynX-X}\exp(-\zeta(S_l)\,\mathrm{ad}_{P_l})(-x_1)$.

Then let us put[22] $\Lambda := \pi_Y \mathrm{L}$ and [42, 43]

$$\mathrm{Mono}(z) := e^{-(x_1+1)\log(1-z)} = \sum_{k\geq 0} \mathrm{P}_{y_1^k}(z)\; y_1^k \tag{27}$$

$$\mathrm{Const} := \sum_{k\geq 0} \mathrm{H}_{y_1^k}\; y_1^k = \exp\Big(-\sum_{k\geq 1} \mathrm{H}_{y_k} \frac{(-y_1)^k}{k}\Big), \tag{28}$$

$$B(y_1) := \exp\Big(\sum_{k\geq 1} \zeta(y_k) \frac{(-y_1)^k}{k}\Big), \tag{29}$$

and finally, $B'(y_1) := \exp(\gamma y_1)B(y_1)$. Hence, we get $\pi_Y \mathrm{P}(z) \underset{z\to 1}{\widetilde{\ }} \mathrm{Mono}(z)\pi_Y Z_{\sqcup\!\sqcup}$ and $\mathrm{H}(N) \underset{N\to+\infty}{\widetilde{\ }} \mathrm{Const}(N)\pi_Y Z_{\sqcup\!\sqcup}$ as a consequence of (27)–(28). Or equivalently,

Theorem 7 (First global renormalizations of divergent polyzetas, [42, 43])

$$\lim_{z\to 1} \exp\Big(-y_1 \log\frac{1}{1-z}\Big)\Lambda(z) = \lim_{N\to+\infty} \exp\Big(\sum_{k\geq 1} \mathrm{H}_{y_k}(N)\frac{(-y_1)^k}{k}\Big)\mathrm{H}(N) = \pi_Y Z_{\sqcup\!\sqcup}.$$

Theorem 8 ([12]) *For any $g \in \mathcal{C}\{\mathrm{P}_w\}_{w\in Y^*}$, there exist algorithmically computable coefficients $c_j, b_i \in \mathbb{C}$, $\alpha_j, \eta_i \in \mathbb{Z}$, $\beta_j, \kappa_i \in \mathbb{N}$ such that, at all orders*

$$g(z) \underset{z\to 1}{\widetilde{\ }} \sum_{j=0}^{+\infty} c_j (1-z)^{\alpha_j} \log^{\beta_j}(1-z), \quad \langle g(z) \mid z^n\rangle \underset{N\to+\infty}{\widetilde{\ }} \sum_{i=0}^{+\infty} b_i n^{\eta_i} \log^{\kappa_i}(n).$$

Theorem 8 means also that the $\{\mathrm{P}_w\}_{w\in Y^*}$ admit a full singular expansion, at 1, and then their ordinary Taylor coefficients, $\{\mathrm{H}_w\}_{w\in Y^*}$ admit a full asymptotic expansion, for $+\infty$. More precisely,

[22] Here, the coefficient $\langle B(y_1) \mid y_1^k\rangle$ corresponds to the Euler–Mac Laurin constant associated to $\langle \mathrm{Const}(N) \mid y_1^k\rangle$, i.e. the finite part of its asymptotic expansion in the scale of comparison $\{n^a \log^b(n)\}_{a\in\mathbb{Z}, b\in\mathbb{N}}$.

Corollary 1 *For any $w \in X^*$ and for any $k, i, j \in \mathbb{N}$, $k \geq 1$, there exists uniquely determined coefficients a_i, $b_{i,j}$ belonging to $\mathscr{Z}$; $\gamma_{\pi_Y(w)}$, α_i and $\beta_{i,j}$ belonging to the $\mathbb{Q}[\gamma]$-algebra generated by convergent polyzetas such that,*

$$\mathrm{Li}_w(z) = \sum_{i=1}^{|w|} a_i \log^i(1-z) + \langle Z_{ш} \mid w\rangle + \sum_{j=1}^{k}\sum_{i=0}^{|w|-1} b_{i,j} \frac{\log^i(1-z)}{(1-z)^{-j}} + \mathrm{o}_k^{(1)}((1-z)^k) \tag{30}$$

and, likely

$$\mathrm{H}_{\pi_Y(w)}(N) = \sum_{i=1}^{|w|} \alpha_i \log^i(N) + \gamma_{\pi_Y(w)} + \sum_{j=1}^{k}\sum_{i=0}^{|w|-1} \beta_{i,j} \frac{1}{N^j} \log^i(N) + \mathrm{o}_k^{(+\infty)}(N^{-k}). \tag{31}$$

Remark 2 1. The two expansions (30) and (31) are the asymptotic expansions of Li_w and H_w with respect to, respectively, the scales $\{(1-z)^n \log(1-z)^m\}_{n,m\geq 0}$ and $\{N^{-k}\log(N)^m\}_{k,m\geq 0}$.
2. In Eq. (30), the error term $\mathrm{o}_k^{(1)}((1-z)^k)$ can be put to the form $\mathrm{O}_k^{(1)}((1-z)^{k+\varepsilon})$ for any $\varepsilon \in]0,1[$.

More generally, by Theorem 6, we get

Proposition 9 ([43]) *For any commutative $\mathbb{Q}$-algebra A and for any Lie series $C \in \mathscr{L}ie_A\langle X\rangle$, we set $\overline{\mathrm{L}} = \mathrm{L}e^C$, $\overline{\Lambda} = \pi_Y \overline{\mathrm{L}}$ and $\overline{\mathrm{P}}(z) = (1-z)^{-1}\overline{\Lambda}(z)$, then*

1. $\overline{Z}_{ш} = Z_{ш} e^C$ *is group-like, for the co-product* $\Delta_{ш}$,
2. $\overline{\mathrm{L}}(z) \widetilde{_{z\to 1}} \exp(-x_1 \log(1-z))\, \overline{Z}_{ш}$,
3. $\overline{\mathrm{P}}(z) \widetilde{_{z\to 1}} \mathrm{Mono}(z)\pi_Y \overline{Z}_{ш}$,
4. $\overline{\mathrm{H}}(N) \widetilde{_{N\to\infty}} \mathrm{Const}(N)\pi_Y \overline{Z}_{ш}$,

where, for any $w \in Y^*$ and $N \geq 0$, one defines the coefficient $\langle \overline{\mathrm{H}}(N) \mid w\rangle$ of w in the power series $\overline{\mathrm{H}}(N)$ as the coefficient $\langle \overline{\mathrm{P}}_w(z) \mid z^N\rangle$ of z^N in the ordinary Taylor expansion of the polylogarithmic function $\overline{\mathrm{P}}_w(z)$.

By Proposition 9, we get successively

Proposition 10 ([39, 43]) *Let $\overline{\zeta}_{ш}$ and $\overline{\zeta}_{ш\!\!-}$ be the characters of respectively $(A\langle X\rangle, ш)$ and $(A\langle Y\rangle, ш\!\!-)$ satisfying $\overline{\zeta}_{ш}(x_0) = \overline{\zeta}_{ш}(x_1) = 0$ and $\overline{\zeta}_{ш\!\!-}(y_1) = 0$. Then*

$$\sum_{w\in Y^*} \overline{\zeta}_{ш}(w)\, w = \overline{Z}_{ш} = \prod_{l\in \mathscr{L}ynX - X}^{\searrow} \exp(\overline{\zeta}(S_l)\, P_l),$$

$$\sum_{w\in Y^*} \overline{\zeta}_{ш\!\!-}(w)\, w = \overline{Z}_{ш\!\!-} = \prod_{l\in \mathscr{L}ynY - \{y_1\}}^{\searrow} \exp(\overline{\zeta}(\Sigma_l)\, \Pi_l).$$

Proposition 11 ([43]) *Let $\{\overline{\gamma}_w\}_{w\in Y^*}$ be the Euler–Mac Laurin constants associated to $\{\overline{\mathrm{H}}_w(N)\}_{w\in Y^*}$. Let $\overline{Z}_\gamma$ be the noncommutative generating series of these constants. Then,*

1. *The following map realizes a character:*

$$\overline{\gamma}_\bullet : (A\langle Y\rangle, \; ⧢) \longrightarrow (\mathbb{R}, .), \quad w \longmapsto \langle \overline{\gamma}_\bullet \mid w\rangle = \overline{\gamma}_w.$$

2. *The noncommutative power series $\overline{Z}_\gamma$ is group-like, for $\Delta_{⧢}$.*
3. *There exists a group-like element $\overline{Z}_{⧢}$, for the co-product $\Delta_{⧢}$, such that*

$$\overline{Z}_\gamma = \sum_{w\in Y^*} \overline{\gamma}_w \, w = \exp(\gamma y_1)\overline{Z}_{⧢}.$$

By Theorem 7, Propositions 9 and 11, we also get

Proposition 12 ([43]) *For any $C \in \mathcal{L}ie_A\langle X\rangle$ such that $\overline{Z}_{ш} = Z_{ш} e^C$. Then*

$$\overline{Z}_\gamma = B(y_1)\pi_Y \overline{Z}_{ш}, \text{ or equivalently by cancellation, } \overline{Z}_{⧢} = B'(y_1)\pi_Y \overline{Z}_{ш},$$

where $B(y_1)$ and $B'(y_1)$ are given in (29).

By Proposition 9, the noncommutative generating series $\overline{Z}_{ш}$ and $\overline{Z}_{⧢}$ are group-like, for the co-product $\Delta_{ш}$ and $\Delta_{⧢}$ respectively. We also have

$$\overline{Z}_{ш} = \sum_{l\in\mathcal{L}ynX-X} \overline{\zeta}(S_l)\, P_l + \sum_{w\notin\mathcal{L}ynX-X} \overline{\zeta}_{ш}(S_w)\, P_w,$$

$$\overline{Z}_{⧢} = \sum_{l\in\mathcal{L}ynY-\{y_1\}} \overline{\zeta}(\Sigma_l)\, \Pi_l + \sum_{w\notin\mathcal{L}ynY-\{y_1\}} \overline{\zeta}_{⧢}(\Sigma_w)\, \Pi_w.$$

Hence, by Proposition 12, we deduce in particular,

$$\sum_{l\in\mathcal{L}ynY-\{y_1\}} \overline{\zeta}(\Sigma_l)\, \Pi_l + \ldots = B'(y_1)\Big(\sum_{l\in\mathcal{L}ynX-X} \overline{\zeta}(\pi_Y S_l)\, \pi_Y P_l + \ldots\Big).$$

The elements of $\{\pi_Y P_l\}_{l\in\mathcal{L}ynX}$ are decomposable in the linear basis $\{\Pi_w\}_{w\in Y^*}$ of $\mathcal{U}(\mathrm{Prim}(\mathcal{H}_{⧢}))$. Thus, by identification of local coordinates, i.e. the coefficients of $\{\Pi_l\}_{l\in\mathcal{L}ynY-\{y_1\}}$ in the basis $\{\Sigma_l\}_{l\in\mathcal{L}ynY-\{y_1\}}$, we get homogenous polynomial relations on polyzetas encoded by $\{\Sigma_l\}_{l\in\mathcal{L}ynY-\{y_1\}}$ [43].

Proposition 13 ([17]) *There exist A, B and $C \in \mathbb{Q}\langle Y_0\rangle$ such that*

$$\mathrm{L}^-(z)\widetilde{_{z\to 1}} A \odot g\Big(\frac{1}{1-z}\Big),\ \mathrm{P}^-(z)\widetilde{_{z\to 1}} B \odot \frac{1}{1-z} g\Big(\frac{1}{1-z}\Big),\ \mathrm{H}^-(N)\widetilde{_{N\to+\infty}} C \odot g(N).$$

where the series g, h were defined in the Definition 5.

Proof By Propositions 5, for $w = y_{s_1} \dots y_{s_r}$, there exists $a, b, c \in \mathbb{Q}$ such that

$$\mathrm{Li}^-_w(z) \widetilde{_{z\to 1}} \frac{a}{(1-z)^{|w|+(w)}}, \ \mathrm{P}^-_w(z) \widetilde{_{z\to 1}} \frac{b}{(1-z)^{|w|+(w)+1}}, \ \mathrm{H}^-_w(N) \widetilde{_{N\to+\infty}} cN^{|w|+(w)}.$$

Putting $\langle A \mid w\rangle = (-1)^{|w|}a$, $\langle B \mid w\rangle = (-1)^{|w|}b$, $\langle C \mid w\rangle = (-1)^{|w|}c$, it follows the expected results.

Proposition 14 ([17]) *For any $w \in Y_0^*$, there are non-zero constants, namely C^-_w and B^-_w, which only depend on w and r such that*

$$\lim_{N\to\infty} \frac{\mathrm{H}^-_w(N)}{N^{(w)+|w|}C^-_w} = 1, \text{ i.e. } \mathrm{H}^-_w(N) \widetilde{_{N\to+\infty}} N^{(w)+|w|}C^-_w,$$

$$\lim_{z\to 1^-} \frac{(1-z)^{(w)+|w|}\,\mathrm{Li}^-_w(z)}{B^-_w} = 1, \text{ i.e. } \mathrm{Li}^-_w(z) \widetilde{_{z\to 1}} \frac{N^{(w)+|w|}B^-_w}{(1-z)^{n+1}}.$$

Moreover, C^-_w and B^-_w are well determined by

$$C^-_w = \prod_{w=uv;\, v\neq 1_{Y_0^*}} \frac{1}{(v)+|v|} \in \mathbb{Q} \text{ and } B^-_w = ((w)+|w|)!C^-_w \in \mathbb{N}.$$

Example 14 (*of C^-_w and B^-_w*)

w	C^-_w	B^-_w	w	C^-_w	B^-_w
y_0	1	1	y_1y_2	1/15	8
y_1	1/2	1	y_2y_3	1/28	180
y_2	1/3	2	y_3y_4	1/49	8064
y_n	$1/(n+1)$	$n!$	y_my_n	$1/[(n+1)(m+n+2)]$	$n!m!\binom{m+n+1}{n+1}$
y_0^2	1/2	1	$y_2y_2y_3$	1/280	12960
y_0^n	$1/(n!)$	1	$y_2y_{10}y_1^2$	1/2160	9686476800
y_1^2	1/8	3	$y_2^2y_4y_3y_{11}$	1/2612736	416761182546508800000

Proposition 15 ([17]) *Let $u, v \in Y_0^*$. We get $\mathrm{H}^-_u\mathrm{H}^-_v = \mathrm{H}^-_{u \uplus v}$.*

Proof Let $w \in Y_0^*$ associated to $s = (s_1, \dots, s_k)$. The quasi-symmetric monomial functions on the commutative alphabet $t = \{t_i\}_{i\geq 1}$ are defined as follows

$$M_{1_{Y_0^*}}(t) = 1 \text{ and } M_w(t) = \sum_{n_1>\dots>n_k>0} t_{n_1}^{s_1} \dots t_{n_k}^{s_k},$$

For any $u, v \in Y_0^*$, we have $M_u(t)M_v(t) = M_{u \uplus v}(t)$. Then, the harmonic sum $\mathrm{H}^-_{s_1,\dots,s_k}(N)$ is obtained by specializing the indeterminates $t = \{t_i\}_{i\geq 1}$ from $M_w(t)$ as follows: $t_i = i$ for $1 \leq i \leq N$ and $t_i = 0$ for $N < i$.

Theorem 9 (Second global renormalizations of divergent polyzetas, [17])

1. The generating series H^- *is group-like and* $\log \mathrm{H}^-$ *is primitive. Moreover,*[23]

$$\lim_{N\to+\infty} g^{\odot-1}(N)\odot \mathrm{H}^-(N) = \lim_{z\to 1} h^{\odot-1}((1-z)^{-1})\odot \mathrm{L}^-(z) = C^-.$$

2. $\ker \mathrm{H}^-_\bullet$ *is a prime ideal of* $(\mathbb{Q}\langle Y_0\rangle, \,⧢\,)$, *i.e.* $\mathbb{Q}\langle Y_0\rangle \setminus \ker \mathrm{H}^-_\bullet$ *is closed by* ⧢.

Proof The first result is a consequence of the extended Friedrichs criterion [6, 43, 44] and the second is a consequence of Proposition 1.

Definition 6 For any $n\in\mathbb{N}_+$, let $\mathbb{P}_n := \mathrm{span}_{\mathbb{R}_+}\{w\in Y_0^*|(w)+|w|=n\}\setminus\{0\}$ be the blunt[24] convex cone generated by the set $\{w\in Y_0^*|(w)+|w|=n\}$.

By definition, $C^-_\bullet$ is linear on the set $\mathbb{P}_n$. For any $u,v\in Y_0^*$, one has $u ⧢ v = u \sqcup\!\sqcup v + \sum_{\substack{|w|<|u|+|v|\\(w)=(u)+(v)}} x_w w$ and the x_w's are positive. Moreover, for any w which belongs to the support of $\sum_{\substack{|w|<|u|+|v|\\(w)=(u)+(v)}} x_w w$, one has $(w)+|w|<(u)+(v)+|u|+|v|$, thus, by the definition of $C^-_\bullet$, one obtains

Corollary 2 *1. Let* $w,v\in Y_0^*$. *Then* $C^-_v C^-_w = C^-_{v\sqcup\!\sqcup w} = C^-_{v ⧢ w}$.

2. For any $P,Q\notin \ker \mathrm{H}^-_\bullet$, $C^-_P C^-_Q = C^-_{P ⧢ Q}$ *and* $\mathbb{Q}\langle Y_0\rangle\setminus \ker \mathrm{H}^-_\bullet$ *is a* ⧢ − *multiplicative monoid containing* Y_0^*.

Now, let us prove that $C^-_\bullet$ can be extended as a character, for ⧢, or equivalently, C^- is group-like (see the Freidrichs' criterion [52]) and then $\log C^-$ is primitive.

Lemma 2 *Let* $\mathscr{A}$ *is a* $\mathbb{R}$*-associative algebra with unit and let* $f:\bigsqcup_{n\geq 0}\mathbb{P}_n \longrightarrow \mathscr{A}$ *such that*

1. *For any* $u,v\in Y_0^*$, $f(u\sqcup\!\sqcup v)=f(u)f(v)$ *and* $f(1_{Y_0^*})=1_{\mathscr{A}}$.
2. *For any finite set* I, *one has* $f(\sum_{i\in I}\alpha_i w_i)=\sum_{i\in I}\alpha_i f(w_i)$ *where* $\sum_{i\in I}\alpha_i w_i\in \mathrm{P}_n$, $n\in\mathbb{N}$.

Then f *can be uniquely extended as a character i.e.* $S_f=\sum_{w\in Y_0^*} f(w)w$ *is group-like for* $\Delta_{\sqcup\!\sqcup}$.

Proof The linear span of $\mathbb{P}_n$ is the space of homogeneous polynomials of degree n, (*i.e.*, $\mathbb{P}_n-\mathbb{P}_n=\mathbb{R}_n\langle Y_0\rangle$), $\mathbb{P}_n$ being convex (and non-void), f extends uniquely, as a linear map, to $\mathbb{R}_n\langle Y_0\rangle$ and then, as a linear map, on $\oplus_{n\geq 0}\mathbb{R}_n\langle Y_0\rangle=\mathbb{R}\langle Y_0\rangle$. This linear extension is a morphism for the shuffle product as it is so on the (linear) generators Y_0^*.

By definition of f and S_f, it is immediate $\langle S_f \mid 1_{Y_0^*}\rangle = 1_{\mathscr{A}}$. One can check easily that $\Delta_{\sqcup\!\sqcup}(S_f)=S_f\otimes S_f$. Hence, S_f is group-like, for $\Delta_{\sqcup\!\sqcup}$.

---

[23]Here, the Hadamard product is denoted by $\odot$ and its dual law, the diagonal comultiplication is denoted by $\Delta_\odot$. The series g,h are defined in Definition 5.

[24]I.e. without zero or see Appendix A.

Corollary 3 *The noncommutative generating series C^- is group-like, for $\Delta_{\sqcup\!\sqcup}$.*

Proof It is a consequence of Lemma 2 and Corollary 2.

Example 15 (*of $C^-_{u \sqcup\!\sqcup v}$ and $C^-_{u \,\text{⊎}\, v}$*, [17]) Let $Y_0 = \{y_i\}_{i\geq 0}$ be an infinite alphabet.

u	C^-_u	v	C^-_v	$u \sqcup\!\sqcup v$	$C^-_{u \sqcup\!\sqcup v}$
y_0	1	y_0	1	$2y_0^2$	1
y_0^2	1/2				
y_1	1/2	y_2	1/3	$y_1y_2 + y_2y_1$	1/6
y_1y_2	1/15	y_2y_1	1/10		
y_m	$(m+1)^{-1}$	y_n	$(n+1)^{-1}$	$y_my_n + y_ny_m$	$[(m+1)(n+1)]^{-1}$
y_my_n	$\frac{(n+1)^{-1}}{(n+m+2)}$	y_ny_m	$\frac{(m+1)^{-1}}{(m+n+2)}$		
y_1	1/2	y_2y_5	1/54	$y_1y_2y_5 + y_2y_1y_5 + y_2y_5y_1$	1/108
$y_1y_2y_5$	1/594	$y_2y_1y_5$	1/528		
$y_2y_5y_1$	1/176				
y_0y_1	1/6	y_2y_3	1/28	$y_0y_1y_2y_3 + y_0y_2y_1y_3$ $+y_0y_2y_3y_1 + y_2y_3y_0y_1$ $+y_2y_0y_1y_3 + y_2y_0y_3y_1$	1/168
$y_0y_1y_2y_3$	1/2520	$y_0y_2y_1y_3$	1/2160		
$y_0y_2y_3y_1$	1/1080	$y_2y_3y_0y_1$	1/420		
$y_2y_0y_1y_3$	1/1680	$y_2y_0y_3y_1$	1/840		
y_ay_b	$\frac{(b+1)^{-1}}{(a+b+2)}$	y_cy_d	$\frac{(d+1)^{-1}}{(c+d+2)}$	$y_ay_by_cy_d + y_ay_cy_by_d$ $+y_ay_cy_dy_b + y_cy_dy_ay_b$ $+y_cy_ay_by_d + y_cy_ay_dy_b$	$\frac{(b+1)^{-1}(d+1)^{-1}}{(a+b+2)(c+d+2)}$

u	C^-_u	v	C^-_v	$u \,\text{⊎}\, v$	$C^-_{u \,\text{⊎}\, v}$
y_0	1	y_0	1	$2y_0^2 + y_0$	1
y_1	1/2	y_2	1/3	$y_1y_2 + y_2y_1 + y_3$	1/6
y_m	$(m+1)^{-1}$	y_n	$(n+1)^{-1}$	$y_my_n + y_ny_m + y_{n+m}$	$[(m+1)(n+1)]^{-1}$
y_1	1/2	y_2y_5	1/54	$y_1y_2y_5 + y_2y_1y_5 + y_2y_5y_1$ $+y_3y_5 + y_2y_6$	1/108
y_0y_1	1/6	y_2y_3	1/28	$y_0y_1y_2y_3 + y_0y_2y_1y_3$ $+y_0y_2y_3y_1 + y_2y_3y_0y_1$ $+y_2y_0y_1y_3 + y_2y_0y_3y_1 + y_0y_2y_4$ $+y_0y_3^2 + y_2y_3y_1 + y_2y_1y_3$ $+y_2y_0y_4 + y_2y_3y_1 + y_2y_4$	1/168
y_ay_b	$\frac{(b+1)^{-1}}{(a+b+2)}$	y_cy_d	$\frac{(d+1)^{-1}}{(c+d+2)}$	$y_ay_by_cy_d + y_ay_cy_by_d$ $+y_ay_cy_dy_b + y_cy_dy_ay_b + y_cy_ay_by_d$ $+y_cy_ay_dy_b + y_ay_cy_{b+d} + y_ay_{b+c}y_d$ $+y_cy_ay_{b+d} + y_cy_{a+d}y_b$ $+y_{a+c}y_by_d + y_{a+c}y_dy_b + y_{a+c}y_{b+d}$	$\frac{(b+1)^{-1}(d+1)^{-1}}{(a+b+2)(c+d+2)}$

In the above tables, it is clearly seen that $C^-_\bullet$ is linear on $\mathbb{P}_n$. For example, let $u = y_1$ and $v = y_2y_5$. Then $u \sqcup\!\sqcup v = y_1y_2y_5 + y_2y_1y_5 + y_2y_5y_1$. Hence, we get $C^-_{y_1y_2y_5} + C^-_{y_2y_1y_5} + C^-_{y_2y_5y_1} = \frac{1}{594} + \frac{1}{528} + \frac{1}{176} = \frac{1}{108} = C^-_{y_1}C^-_{y_2y_5} = C^-_{y_1 \sqcup\!\sqcup y_2y_5}$. Note that $y_1y_2y_5, y_2y_1y_5, y_2y_5y_1 \in \mathbb{P}_{11}$. But we have also $u \,\text{⊎}\, v = y_1y_2y_5 + y_2y_1y_5 + y_2y_5y_1 +$

$y_3y_5 + y_2y_6$. Moreover, $C^-_{y_1y_2y_5} + C^-_{y_2y_1y_5} + C^-_{y_2y_5y_1} + C^-_{y_3y_5} + C^-_{y_2y_6} = \frac{1}{108} + \frac{13}{420} \neq \frac{1}{108} = C^-_{y_1}C^-_{y_2y_5}$. However, from $y_3y_5, y_2y_6 \in \mathbb{P}_{10}$, we can conclude that

$$C^-_{y_1 \uplus y_2y_5} = C^-_{y_1y_2y_5+y_2y_1y_5+y_2y_5y_1+y_3y_5+y_2y_6} = C^-_{y_1y_2y_5+y_2y_1y_5+y_2y_5y_1} = 1/108 = C^-_{y_1}C^-_{y_2y_5}.$$

3 Polysystems and Differential Realization

3.1 Polysystems and Convergence Criterion

3.1.1 Estimates (from above) for Series

Here, $(\mathbb{K}, \|.\|)$ is a normed space.

Definition 7 ([28, 42, 43]) Let ξ, χ be real positive functions over X^*. Let $S \in \mathbb{K}\langle\langle X\rangle\rangle$.

1. S will be said *ξ-exponentially bounded from above* if it satisfies

$$\exists K \in \mathbb{R}_+, \exists n \in \mathbb{N}, \forall w \in X^{\geq n}, \quad \|\langle S \mid w\rangle\| \leq K\xi(w)/|w|!.$$

   We denote by $\mathbb{K}^{\xi-\mathrm{em}}\langle\langle X\rangle\rangle$ the set of formal power series in $\mathbb{K}\langle\langle X\rangle\rangle$ which are ξ-exponentially bounded from above.
2. S satisfies the *χ-growth condition* if it satisfies

$$\exists K \in \mathbb{R}_+, \exists n \in \mathbb{N}, \forall w \in X^{\geq n}, \quad \|\langle S \mid w\rangle\| \leq K\chi(w)|w|!.$$

   We denote by $\mathbb{K}^{\chi-\mathrm{gc}}\langle\langle X\rangle\rangle$ the set of formal power series in $\mathbb{K}\langle\langle X\rangle\rangle$ satisfying the χ-growth condition.

Lemma 3 ([28, 42, 43]) *If* $R = \sum_{w\in X^*} |w|!\, w$ *then* $\langle R^{\sqcup\!\sqcup 2} \mid w\rangle = \sum_{\substack{u,v\in X^*\\ \mathrm{supp}(u \sqcup\!\sqcup v)\ni w}} |u|!|v|! \leq 2^{|w|}|w|!$.

Proof One has

$$\sum_{\substack{u,v\in X^*\\ \mathrm{supp}(u \sqcup\!\sqcup v)\ni w}} |u|!|v|! = \sum_{k=0}^{|w|} \sum_{\substack{|u|=k,|v|=|w|-k\\ \mathrm{supp}(u \sqcup\!\sqcup v)\ni w}} k!(|w|-k)! = \sum_{k=0}^{|w|} \binom{|w|}{k} k!(|w|-k)! = \sum_{k=0}^{|w|} |w|!.$$

The last sum is equal to $(1 + |w|)|w|!$. By induction on $|w|$, one has $1 + |w| \leq 2^{|w|}$. Then the expected result follows.

Proposition 16 ([28, 42, 43]) *If S_1, S_2 satisfy the growth condition then $S_1 + S_2$, $S_1 \sqcup\!\sqcup S_2$ do also.*

Proof It is immediate for $S_1 + S_2$. Next, since $\|\langle S_i \mid w\rangle\| \leq K_i\chi_i(w)|w|!$ then

$$\langle S_1 \sqcup\!\sqcup S_2 \mid w\rangle = \sum_{\mathrm{supp}(u \sqcup\!\sqcup v)\ni w} \langle S_1 \mid u\rangle\langle S_2 \mid v\rangle,$$

$$\Rightarrow \quad \|\langle S_1 \sqcup\!\sqcup S_2 \mid w\rangle\| \leq K_1K_2 \sum_{\substack{u,v\in X^*\\ \mathrm{supp}(u \sqcup\!\sqcup v)\ni w}} (\chi_1(u)|u|!)(\chi_2(v)|v|!).$$

Let $K = K_1K_2$ and let χ be a real positive function over X^* such that, for any $w \in X^*$

$$\chi(w) = \max\{\chi_1(u)\chi_2(v) \mid u, v \in X^* \text{ and } supp(u \sqcup\!\sqcup v) \ni w\}.$$

With the notations in Lemma 3, we get $\|\langle S_1 \sqcup\!\sqcup S_2 \mid w\rangle\| \leq K\chi(w)\langle S_1R^{\sqcup\!\sqcup 2} \mid w\rangle$. Hence, $S_1 \sqcup\!\sqcup S_2$ satisfies the χ'-growth condition with $\chi'(w) = 2^{|w|}\chi(w)$.

Definition 8 ([28, 42, 43]) Let ξ be a real positive function defined over X^*, S will be said *ξ-exponentially continuous* if it is continuous over $\mathbb{K}^{\xi-\mathrm{em}}\langle\langle X\rangle\rangle$. The set of formal power series which are ξ-exponentially continuous is denoted by $\mathbb{K}^{\xi-ec}\langle\langle X\rangle\rangle$.

Lemma 4 ([28, 42, 43]) *For any real positive function ξ defined over X^*, we have $\mathbb{K}\langle X\rangle \subset \mathbb{K}^{\xi-ec}\langle\langle X\rangle\rangle$. Otherwise, for $\xi = 0$, we get $\mathbb{K}\langle X\rangle = \mathbb{K}^{0-ec}\langle\langle X\rangle\rangle$. Hence, any polynomial is 0-exponentially continuous.*

Proposition 17 ([28, 42, 43]) *Let ξ, χ be real positive functions over X^* and $P \in \mathbb{K}\langle X\rangle$.*

1. *Let $S \in \mathbb{K}^{\xi-\mathrm{em}}\langle\langle X\rangle\rangle$. The right residual of S by P belongs to $\mathbb{K}^{\xi-\mathrm{em}}\langle\langle X\rangle\rangle$.*
2. *Let $R \in \mathbb{K}^{\chi-\mathrm{gc}}\langle\langle X\rangle\rangle$. The concatenation SR belongs to $\mathbb{K}^{\chi-\mathrm{gc}}\langle\langle X\rangle\rangle$.*
3. *Moreover, if ξ and χ are morphisms over X^* satisfying $\sum_{x\in X}\chi(x)\xi(x) < 1$ then, for any $F \in \mathbb{K}^{\chi-\mathrm{gc}}\langle\langle X\rangle\rangle$, F is continuous over $\mathbb{K}^{\xi-\mathrm{em}}\langle\langle X\rangle\rangle$.*

Proof 1. Since $S \in \mathbb{K}^{\xi-\mathrm{em}}\langle\langle X\rangle\rangle$ then

$$\exists K \in \mathbb{R}_+, \exists n \in \mathbb{N}, \forall w \in X^{\geq n}, \quad \|\langle S \mid w\rangle\| \leq K\xi(w)/|w|!.$$

If $u \in \mathrm{supp}(P)$ then, for any $w \in X^*$, one has $\langle S \triangleleft u \mid w\rangle = \langle S \mid uw\rangle$ and $S \triangleleft u$ belongs to $\mathbb{K}^{\xi-\mathrm{em}}\langle\langle X\rangle\rangle$:

$$\exists K \in \mathbb{R}_+, \exists n \in \mathbb{N}, \forall w \in X^{\geq n}, \quad \|\langle S \triangleleft u \mid w\rangle\| \leq [K\xi(u)]\xi(w)/|w|!.$$

It follows that $S \triangleleft P$ is $\mathbb{K}^{\xi-\mathrm{em}}\langle\langle X\rangle\rangle$ by taking $K_1 = K\max_{u\in\mathrm{supp}(P)}\xi(u)$.
2. Since $R \in \mathbb{K}^{\chi-\mathrm{gc}}\langle\langle X\rangle\rangle$ then

$$\exists K \in \mathbb{R}_+, \exists n \in \mathbb{N}, \forall w \in X^{\geq n}, \quad \|\langle S \mid w\rangle\| \leq K\chi(w)|w|!.$$

Let $v \in \mathrm{supp}(P)$ such that $v \neq \varepsilon$. Since Rv belongs to $\mathbb{K}^{\chi-\mathrm{gc}}\langle\langle X\rangle\rangle$ and one has, for $w \in X^*$, $\langle Rv \mid w\rangle = \langle R \mid v \triangleright w\rangle$, i.e. there exists $K \in \mathbb{R}_+$, $n \in \mathbb{N}$ such that

$$\|\langle R \mid v \triangleright w\rangle\| \le K\chi(v \triangleright w)(|w| - |v|)! \le K|w|\chi(w)/\chi(v).$$

Note if $v \triangleright w = 0$ then $\langle Rv \mid w\rangle = 0$ and the previous conclusion holds. It follows that RP is $\mathbb{K}^{\chi-\mathrm{gc}}\langle\langle X\rangle\rangle$ by taking $K_2 = K \min_{v\in\mathrm{supp}(P)} \chi(v)^{-1}$.

3. Let ξ, χ be functions which satisfy the upper bound condition. The following quantity is well defined

$$\sum_{w\in X^*} \chi(w)\xi(w) = \Big(\sum_{x\in X} \chi(x)\xi(x)\Big)^*.$$

If $F \in \mathbb{K}^{\chi-\mathrm{gc}}\langle\langle X\rangle\rangle$, $C \in \mathbb{K}^{\xi-\mathrm{em}}\langle\langle X\rangle\rangle$ then there exist $K_i \in \mathbb{R}_+, n_i \in \mathbb{N}, i = 1, 2$ such that, for $w \in X^{\ge n_i}$, $\|\langle F \mid w\rangle\| \le K_1\chi(w)|w|!$ and $\|\langle C \mid w\rangle\| \le K_2\xi(w)/|w|!$. Thus,

$$\forall w \in X^*, |w| \ge \max\{n_1, n_2\}, \quad \|\langle F|w\rangle\langle C|w\rangle\| \le K_1K_2\chi(w)\xi(w),$$

$$\Rightarrow \sum_{w\in X^*} \|\langle F|w\rangle\langle C|w\rangle\| \le K_1K_2 \sum_{w\in X^*} \chi(w)\xi(w) = K_1K_2\Big(\sum_{x\in X} \chi(x)\xi(x)\Big)^*.$$

3.1.2 Upper Bounds *à la* Cauchy

The algebra of formal power series on commutative indeterminates $\{q_1, \ldots, q_n\}$ with coefficients in $\mathbb{C}$ is denoted by $\mathbb{C}[\![q_1, \ldots, q_n]\!]$.

Definition 9 ([28, 42, 43]) Let $f =\in \mathbb{C}[\![q_1, \ldots, q_n]\!]$. We set

$$\begin{array}{ll} E(f) & := \{\rho \in \mathbb{R}_+^n : \exists C_f \in \mathbb{R}_+ \text{ s.t. } \forall i_1, \ldots, i_n \ge 0, |f_{i_1,\ldots,i_n}|\rho_1^{i_1} \ldots \rho_n^{i_n} \le C_f\}. \\ \breve{E}(f) & : \text{ the interior of} E(f) \text{in } \mathbb{R}^n. \\ \mathrm{CV}(f) & := \{q \in \mathbb{C}^n : (|q_1|, \ldots, |q_n|) \in \breve{E}(f)\} : \quad \text{the convergence domain of } f. \end{array}$$

f is *convergent* if $\mathrm{CV}(f) \ne \emptyset$. Let $\mathcal{U} \subset \mathbb{C}^n$ be an open domain and $q \in \mathbb{C}^n$. f is convergent on q (resp. over $\mathcal{U}$) if $q \in \mathrm{CV}(f)$ (resp. $\mathcal{U} \subset \mathrm{CV}(f)$). We set $\mathbb{C}^{\mathrm{cv}}[\![q_1, \ldots, q_n]\!] := \{f \in \mathbb{C}[\![q_1, \ldots, q_n]\!] : \mathrm{CV}(f) \ne \emptyset\}$. Let $q \in \mathrm{CV}(f)$. There exist $C_f \in \mathbb{R}_+, \rho \in E(f), \bar\rho \in \breve{E}(f)$ such that $|q_1| < \bar\rho_1 < \rho_1, \ldots, |q_n| < \bar\rho_n < \rho_n$ and $|f_{i_1,\ldots,i_n}|\rho_1^{i_1} \ldots \rho_n^{i_n} \le C_f$ for any $i_1, \ldots, i_n \ge 0$.

The *convergence modulus* of f at q is $(C_f, \rho, \bar\rho)$.

Suppose $\mathrm{CV}(f) \ne \emptyset$ and let $q \in \mathrm{CV}(f)$. If $(C_f, \rho, \bar\rho)$ is a convergence modulus of f at q then $|f_{i_1,\ldots,i_n}q_1^{i_1} \ldots q_n^{i_n}| \le C_f(\bar\rho_1/\rho_1)^{i_1} \ldots (\bar\rho_1/\rho_1)^{i_n}$. Hence, at q, f is majored termwise by $C_f \prod_{k=0}^m (1 - \bar\rho_k/\rho_k)^{-1}$ and it is uniformly absolutely convergent in $\{q \in \mathbb{C}^n : |q_1| < \bar\rho, \ldots, |q_n| < \bar\rho\}$ which is open in $\mathbb{C}^n$. Thus, $\mathrm{CV}(f)$ is open in $\mathbb{C}^n$. Since the partial derivation $D_1^{j_1} \ldots D_n^{j_n} f$ is estimated by

$$\|D_1^{j_1} \dots D_n^{j_n} f\| \le C_f \frac{\partial^{j_1+\dots+j_n}}{\partial^{j_1}\bar{\rho}_1 \dots \partial^{j_n}\bar{\rho}_n} \prod_{k=0}^{m} \left(1 - \frac{\bar{\rho}_k}{\rho_k}\right)^{-1}.$$

Proposition 18 ([28, 42, 43]) *We have* $\mathrm{CV}(f) \subset \mathrm{CV}(D_1^{j_1} \dots D_n^{j_n} f)$.

Let $f \in \mathbb{C}^{\mathrm{cv}}[\![q_1, \dots, q_n]\!]$. Let $\{A_i\}_{i=0,1}$ be a polysystem defined as follows

$$A_i = \sum_{j=1}^{n} A_i^j(q) \frac{\partial}{\partial q_j}, \; \forall j = 1, \dots, n, \; A_i^j(q) \in \mathbb{C}^{\mathrm{cv}}[\![q_1, \dots, q_n]\!]. \tag{32}$$

Let $(\rho, \bar{\rho}, C_f), \{(\rho, \bar{\rho}, C_i)\}_{i=0,1}$ be convergence modulus at $q \in \mathrm{CV}(f) \bigcap_{i=0,1, j=1,\dots,n} \mathrm{CV}(A_i^j)$ of f and $\{A_i^j\}_{j=1,\dots,n}$. Let us consider the following monoid morphisms

$$\mathscr{A}(1_{X^*}) = \text{identity and } C(1_{X^*}) = 1, \tag{33}$$

$$\forall w = v x_i, x_i \in X, v \in X^*, \quad \mathscr{A}(w) = \mathscr{A}(v) A_i \text{ and } C(w) = C(v) C_i. \tag{34}$$

Lemma 5 ([24]) *For* $i = 0, 1$ *and* $j = 1, \dots, n$, *one has* $A_i \circ q_j = A_i^j$. *Hence,*

$$\forall i = 0, 1, \quad A_i = \sum_{j=1}^{n} (A_i \circ q_j) \frac{\partial}{\partial q_j}.$$

Lemma 6 ([23]) *For any word* w, $\mathscr{A}(w)$ *is continuous over* $\mathbb{C}^{\mathrm{cv}}[\![q_1, \dots, q_n]\!]$ *and, for any* $f, g \in \mathbb{C}^{\mathrm{cv}}[\![q_1, \dots, q_n]\!]$, *one has*

$$\mathscr{A}(w) \circ (fg) = \sum_{u,v \in X^*} \langle u \sqcup\!\sqcup v \mid w \rangle (\mathscr{A}(u) \circ f)(\mathscr{A}(v) \circ g).$$

These notations are extended, by linearity, to $\mathbb{K}\langle X \rangle$ and we will denote $\mathscr{A}(w) \circ f_{|q}$ the evaluation of $\mathscr{A}(w) \circ f$ at q.

Definition 10 ([23]) Let $f \in \mathbb{C}^{\mathrm{cv}}[\![q_1, \dots, q_n]\!]$. The generating series of the polysystem $\{A_i\}_{i=0,1}$ and of the observation f is given by

$$\sigma f := \sum_{w \in X^*} \mathscr{A}(w) \circ f \; w \quad \in \quad \mathbb{C}^{\mathrm{cv}}[\![q_1, \dots, q_n]\!]\langle\!\langle X \rangle\!\rangle.$$

Then the following generating series is called *Fliess generating series* of the polysystem $\{A_i\}_{i=0,1}$ and of the observation f at q:

$$\sigma f_{|_q} := \sum_{w \in X^*} \mathscr{A}(w) \circ f_{|q} \; w \quad \in \quad \mathbb{C}\langle\!\langle X \rangle\!\rangle.$$

Lemma 7 ([23]) *The map* $\sigma : (\mathbb{C}^{\rm cv}[\![q_1,\ldots,q_n]\!],.) \longrightarrow (\mathbb{C}^{\rm cv}[\![q_1,\ldots,q_n]\!]\langle\!\langle X\rangle\!\rangle, \sqcup\!\sqcup)$ *is an algebra morphism, i.e. for any* $f, g \in \mathbb{C}^{\rm cv}[\![q_1,\ldots,q_n]\!]$ *and* $\mu, \nu \in \mathbb{C}$*, one has* $\sigma(\nu f + \mu h) = \nu\sigma f + \mu\sigma g$ *and* $\sigma(fg) = \sigma f \sqcup\!\sqcup\, \sigma g$.

Lemma 8 ([24]) *For any* $w \in X^*$, $\sigma(\mathscr{A}(w) \circ f) = w \triangleright \sigma f \in \mathbb{C}^{\rm cv}[\![q_1,\ldots,q_n]\!]\langle\!\langle X\rangle\!\rangle$.

Theorem 10 ([28, 42, 43])

1. *Let* $\tau = \min_{1\le k\le n} \rho_k$ *and* $r = \max_{1\le k\le n} \bar{\rho}_k/\rho_k$. *We have*

$$\|\mathscr{A}(w) \circ f\| \le C_f \frac{(n+1)}{(1-r)^n} \frac{C(w)|w|!}{\binom{n+|w|-1}{|w|}} \left[\frac{n}{\tau(1-r)^{n+1}}\right]^{|w|}$$
$$\le C_f \frac{(n+1)}{(1-r)^n} C(w) \left[\frac{n}{\tau(1-r)^{n+1}}\right]^{|w|} |w|!.$$

2. *Let* $K = C_f(n+1)(1-r)^{-n}$ *and* χ *be the real positive function defined over* X^*:

$$\forall i = 0, 1, \ \ \chi(x_i) = C_i n(1-r)^{-(n+1)}/\tau.$$

Then[25] *the generating series* σf *of the polysystem* $\{A_i\}_{i=0,1}$ *and of the observation* f *satisfies the* χ*-growth condition.*

3.2 Polysystem and Nonlinear Differential Equation

3.2.1 Nonlinear Differential Equation (with Three Singularities)

Let us consider the singular inputs[26] $u_0(z) := z^{-1}$ and $u_1(z) := (1-z)^{-1}$, and

$$\begin{cases} y(z) &= f(q(z)), \\ \dot{q}(z) &= A_0(q)\, u_0(z) + A_1(q)\, u_1(z), \\ q(z_0) &= q_0, \end{cases} \tag{35}$$

where the state $q = (q_1,\ldots,q_n)$ belongs to a complex analytic manifold of dimension n, q_0 is the initial state, the observation f belongs to $\mathbb{C}^{\rm cv}[\![q_1,\ldots,q_n]\!]$ and $\{A_i\}_{i=0,1}$ is the polysystem defined on (32).

Definition 11 ([30]) The following power series is called *transport operator*[27] of the polysystem $\{A_i\}_{i=0,1}$ and of the observation f

---

[25] It is the same for the Fliess generating series $\sigma f_{|q}$ of $\{A_i\}_{i=0,1}$ and of f at q.

[26] These singular inputs are not included in the studies of Fliess motivated, in particular, by the renormalization of y at $+\infty$ [23, 24].

[27] It plays the rôle of the resolvent in Mathematics and the evolution operator in Physics.

$$\mathscr{T} := \sum_{w \in X^*} \alpha_{z_0}^{z}(w) \, \mathscr{A}(w).$$

By the factorization of the monoid by Lyndon words, we have [30]

$$\mathscr{T} = (\alpha_{z_0}^{z} \otimes \mathscr{A})\Big(\sum_{w \in X^*} w \otimes w\Big) = \prod_{l \in \mathscr{L}ynX} \exp[\alpha_{z_0}^{z}(S_l) \, \mathscr{A}(P_l)].$$

The Chen generating series along the path $z_0 \rightsquigarrow z$, associated to ω_0, ω_1 is

$$S_{z_0 \rightsquigarrow z} := \sum_{w \in X^*} \langle S \mid w \rangle \, w \text{ with } \langle S \mid w \rangle = \alpha_{z_0}^{z}(w) \tag{36}$$

which solves the differential equation (25) with the initial condition $S_{z_0 \rightsquigarrow z_0} = 1$. Thus, $S_{z_0 \rightsquigarrow z}$ and $\mathrm{L}(z)\mathrm{L}(z_0)^{-1}$ satisfy the same differential equation taking the same value at z_0 and $S_{z_0 \rightsquigarrow z} = \mathrm{L}(z)\mathrm{L}(z_0)^{-1}$. Any Chen generating series $S_{z_0 \rightsquigarrow z}$ is group like [50] and depends only on the homotopy class of $z_0 \rightsquigarrow z$ [10]. The product of $S_{z_1 \rightsquigarrow z_2}$ and $S_{z_0 \rightsquigarrow z_1}$ is $S_{z_0 \rightsquigarrow z_2} = S_{z_1 \rightsquigarrow z_2} S_{z_0 \rightsquigarrow z_1}$. Let $\varepsilon \in]0, 1[$ and $z_i = \varepsilon \exp(\mathrm{i}\beta_i)$, for $i = 0, 1$. We set $\beta = \beta_1 - \beta_0$. Let $\Gamma_0(\varepsilon, \beta_0)$ (resp. $\Gamma_1(\varepsilon, \beta_1)$) be the path turning around 0 (resp. 1) in the positive direction from z_0 to z_1. By induction on the length of w, one has $|\langle S_{\Gamma_i(\varepsilon,\beta)} \mid w \rangle| = (2\varepsilon)^{|w|_{x_i}} \beta^{|w|}/|w|!$, where $|w|$ denotes the length of w and $|w|_{x_i}$ denotes the number of occurrences of letter x_i in w, for $i = 0$ or 1. When ε tends to 0^+, these estimations yield $S_{\Gamma_i(\varepsilon,\beta)} = e^{\mathrm{i}\beta x_i} + o(\varepsilon)$. In particular, if $\Gamma_0(\varepsilon)$ (resp. $\Gamma_1(\varepsilon)$) is a circular path of radius ε turning around 0 (resp. 1) in the positive direction, starting at $z = \varepsilon$ (resp. $1 - \varepsilon$), then, by the noncommutative residue theorem [33, 37], we get

$$S_{\Gamma_0(\varepsilon)} = e^{2\mathrm{i}\pi x_0} + o(\varepsilon) \text{ and } S_{\Gamma_1(\varepsilon)} = e^{-2\mathrm{i}\pi x_1} + o(\varepsilon). \tag{37}$$

Finally, the asymptotic behaviors of L on (26) give [33, 37]

$$S_{\varepsilon \rightsquigarrow 1-\varepsilon} \underset{\varepsilon \to 0^+}{\widetilde{\ \ }} e^{-x_1 \log \varepsilon} Z_{\sqcup\!\sqcup} \, e^{-x_0 \log \varepsilon}. \tag{38}$$

In other terms, $Z_{\sqcup\!\sqcup}$ is the regularized Chen generating series $S_{\varepsilon \rightsquigarrow 1-\varepsilon}$ of differential forms ω_0 and ω_1: $Z_{\sqcup\!\sqcup}$ is the noncommutative generating series of the finite parts of the coefficients of the Chen generating series $e^{x_1 \log \varepsilon} \, S_{\varepsilon \rightsquigarrow 1-\varepsilon} \, e^{x_0 \log \varepsilon}$.

3.2.2 Asymptotic Behavior via Extended Fliess Fundamental Formula

Theorem 11 ([42, 43]) $y(z) = \mathscr{T} \circ f_{|q_0} = \langle \sigma f_{|q_0} \parallel S_{z_0 \rightsquigarrow z} \rangle$.

This extends then Fliess fundamental formula [23]. By Theorem 5, the expansions of the output y of nonlinear dynamical system with singular inputs follow

Corollary 4 (Combinatorics of Dyson series, [42, 43])

$$\begin{aligned} y(z) &= \sum_{w\in X^*} g_w(z)\ \mathscr{A}(w)\circ f_{|q_0} \\ &= \sum_{k\geq 0}\sum_{n_1,\ldots,n_k\geq 0} g_{x_0^{n_1}x_1\ldots x_0^{n_k}x_1}(z)\ \mathrm{ad}_{A_0}^{n_1}A_1\ldots\mathrm{ad}_{A_0}^{n_k}A_1 e^{\log z A_0}\circ f_{|q_0} \\ &= \prod_{l\in\mathscr{L}ynX} \exp\Big(g_{S_l}(z)\ \mathscr{A}(P_l)\circ f_{|q_0}\Big) \\ &= \exp\Big(\sum_{w\in X^*} g_w(z)\ \mathscr{A}(\pi_1(w))\circ f_{|q_0}\Big), \end{aligned}$$

where, for any word w in X^*, g_w belongs to the polylogarithm algebra.

Since $S_{z_0\rightsquigarrow z}=\mathrm{L}(z)\mathrm{L}(z_0)^{-1}$ and $\sigma f_{|q_0}, \mathrm{L}(z_0)^{-1}$ are invariant by $\partial_z=d/dz$ and $\theta_0=zd/dz$ then we get the nth order differentiation of y, with respect to ∂_z and θ_0:

$$\begin{aligned} \partial_z^n y(z) &= \langle\sigma f_{|q_0}\ \|\ \partial^n S_{z_0\rightsquigarrow z}\rangle = \langle\sigma f_{|q_0}\ \|\ \partial_z^n\mathrm{L}(z)\mathrm{L}(z_0)^{-1}\rangle, \\ \theta_0^n y(z) &= \langle\sigma f_{|q_0}\ \|\ \theta_0^n S_{z_0\rightsquigarrow z}\rangle = \langle\sigma f_{|q_0}\ \|\ \theta_0^n\mathrm{L}(z)\mathrm{L}(z_0)^{-1}\rangle. \end{aligned}$$

With the notations of Proposition 8, we get respectively

$$\begin{aligned} \partial_z^n y(z) &= \langle\sigma f_{|q_0}\ \|\ [D_n(z)\mathrm{L}(z)]\mathrm{L}(z_0)^{-1}\rangle = \langle\sigma f_{|q_0}\triangleleft D_n(z)\ \|\ \mathrm{L}(z)\mathrm{L}(z_0)^{-1}\rangle, \\ \theta_0^n y(z) &= \langle\sigma f_{|q_0}\ \|\ E_n(z)\mathrm{L}(z)]\mathrm{L}(z_0)^{-1}\rangle = \langle\sigma f_{|q_0}\triangleleft E_n(z)\ \|\ \mathrm{L}(z)\mathrm{L}(z_0)^{-1}\rangle. \end{aligned}$$

For $z_0=\varepsilon\to 0^+$, the asymptotic behavior and the renormalization at $z=1$ of $\partial_z^n y$ and $\theta_0^n y$ (or the asymptotic expansion and the renormalization of its Taylor coefficients at $+\infty$) are deduced from (38) and extend a little bit results of [42, 43]:

Corollary 5 (Asymptotic behavior of output, [42, 43])

1. *The n-order differentiation of the output y of the system (35) is a $\mathscr{C}$-combination of the elements g belonging to the polylogarithm algebra and,*[28] *for any $n\geq 0$,*

[28]Moreover, we get more out of this i.e. $\theta_1^n y(z)=\langle\sigma f_{|q_0}\ \|\ \theta_1^n S_{z_0\rightsquigarrow z}\rangle=\langle\sigma f_{|q_0}\ \|\ \theta_1^n\mathrm{L}(z)\mathrm{L}(z_0)^{-1}\rangle$. Therefore,

$$\theta_1^n y(z)=\langle\sigma f_{|q_0}\ \|\ [D_n(z)-E_n(z)]\mathrm{L}(z)\mathrm{L}(z_0)^{-1}\rangle=\langle\sigma f_{|q_0}\triangleleft[D_n(z)-E_n(z)]\ \|\ \mathrm{L}(z)\mathrm{L}(z_0)^{-1}\rangle.$$

Hence,

$$\theta_1^n y(1)\underset{\varepsilon\to 0^+}{\widetilde{\quad}}\sum_{w\in X^*}\langle\mathscr{A}(w)\circ f_{|q_0}\mid w\rangle\langle[D_n(1-\varepsilon)-E_n(1-\varepsilon)]e^{-x_1\log\varepsilon}Z_{\sqcup\!\sqcup}\,e^{-x_0\log\varepsilon}\mid w\rangle.$$

The actions of $\theta_0=u_0(z)^{-1}d/dz$ and $\theta_1=u_1(z)^{-1}d/dz$ over y are equivalent to those of the residuals of $\sigma f_{|q_0}$ by respectively x_0 and x_1. They correspond to *functional* differentiations [25] while $\partial_z=d/dz$ is the ordinary differentiation and is equivalent to the residual by x_0+x_1.

$$\partial_z^n y(1) \underset{\varepsilon \to 0^+}{\widetilde{\quad}} \sum_{w \in X^*} \langle \mathscr{A}(w) \circ f_{|_{q_0}} \mid w \rangle \langle D_n(1-\varepsilon)\, e^{-x_1 \log \varepsilon}\, Z_{\sqcup\!\sqcup}\, e^{-x_0 \log \varepsilon} \mid w \rangle,$$

$$\theta_0^n y(1) \underset{\varepsilon \to 0^+}{\widetilde{\quad}} \sum_{w \in X^*} \langle \mathscr{A}(w) \circ f_{|_{q_0}} \mid w \rangle \langle E_n(1-\varepsilon)\, e^{-x_1 \log \varepsilon}\, Z_{\sqcup\!\sqcup}\, e^{-x_0 \log \varepsilon} \mid w \rangle.$$

2. *If the ordinary Taylor expansions of $\partial_z^n y$ and $\theta_0^n y$ exist then the coefficients of these expansions belong to the algebra of harmonic sums and there exist algorithmically computable coefficients $a_i, a_i' \in \mathbb{Z}, b_i, b_i' \in \mathbb{N}, c_i, c_i' \in \mathscr{Z}[\gamma]$ such that*

$$\partial_z^n y(z) = \sum_{k \geq 0} d_k z^n \text{ and } d_k \underset{k \to \infty}{\widetilde{\quad}} \sum_{i \geq 0} c_i k^{a_i} \log^{b_i} k,$$

$$\theta_0^n y(z) = \sum_{k \geq 0} t_k z^k \text{ and } t_k \underset{k \to \infty}{\widetilde{\quad}} \sum_{i \geq 0} c_i' k^{a_i'} \log^{b_i'} k.$$

3.3 Differential Realization

3.3.1 Differential Realization

Definition 12 ([24]) The *Lie rank* of a formal power series $S \in \mathbb{K}\langle\langle X \rangle\rangle$ is the dimension of the vector space generated by

$$\{S \triangleleft \Pi \mid \Pi \in \mathscr{L}ie_{\mathbb{K}}\langle X \rangle\}, \text{ or respectively by } \{\Pi \triangleright S \mid \Pi \in \mathscr{L}ie_{\mathbb{K}}\langle X \rangle\}.$$

Definition 13 ([51]) Let $S \in \mathbb{K}\langle\langle X \rangle\rangle$ and let us put $\mathrm{Ann}(S) := \{\Pi \in \mathscr{L}ie_{\mathbb{K}}\langle X \rangle \mid S \triangleleft \Pi = 0\}$, and $\mathrm{Ann}^{\perp}(S) := \{Q \in (\mathbb{K}\langle\langle X \rangle\rangle, \sqcup\!\sqcup) \mid Q \triangleleft \mathrm{Ann}(S) = 0\}$.

It is immediate that $\mathrm{Ann}^{\perp}(S) \ni S$. It follows then (see [24, 51] and Lemma 7),

Lemma 9 ([24]) *Let $S \in \mathbb{K}\langle\langle X \rangle\rangle$. Then*

1. *If S is of finite Lie rank, d, then the dimension of* $\mathrm{Ann}^{\perp}(S)$ *is d.*
2. *For any Q_1 and $Q_2 \in \mathrm{Ann}^{\perp}(S)$, one has $Q_1 \sqcup\!\sqcup Q_2 \in \mathrm{Ann}^{\perp}(S)$.*
3. *For any $P \in \mathbb{K}\langle X \rangle$ and $Q_1 \in \mathrm{Ann}^{\perp}(S)$, one has $P \triangleright Q_1 \in \mathrm{Ann}^{\perp}(S)$.*

Definition 14 The formal power series $S \in \mathbb{K}\langle\langle X \rangle\rangle$ is *differentially produced* if there exist an integer d, a power series $f \in \mathbb{K}[\![\bar{q}_1, \ldots, \bar{q}_d]\!]$, a homomorphism $\mathscr{A}$ from X^* to the algebra of differential operators generated by

$$\mathscr{A}(x_i) = \sum_{j=1}^{d} A_i^j(\bar{q}_1, \ldots, \bar{q}_d) \frac{\partial}{\partial \bar{q}_j}, \text{ where } \forall j = 1, \ldots, d,\ A_i^j(\bar{q}_1, \ldots, \bar{q}_d) \in \mathbb{K}[\![\bar{q}_1, \ldots, \bar{q}_d]\!]$$

such that, for any $w \in X^*$, one has $\langle S \mid w \rangle = \mathscr{A}(w) \circ f_{|_0}$.

The pair $(\mathscr{A}, f)$ is called the *differential representation* of S of dimension d.

Proposition 19 ([51]) *Let $S \in \mathbb{K}\langle\langle X\rangle\rangle$. If S is differentially produced then it satisfies the growth condition and its Lie rank is finite.*

Proof Let $(\mathscr{A}, f)$ be a differential representation of S of dimension d. Then, by the notations of Definition 10, we get $\sigma f_{|_0} = S = \sum_{w\in X^*} (\mathscr{A}(w) \circ f)_{|_0}\, w$. We put

$$\forall j = 1, \ldots, d,\ \ T_j = \sum_{w\in X^*} \frac{\partial(\mathscr{A}(w) \circ f)}{\partial \bar{q}_j}\, w.$$

Firstly, by Theorem 10, the generating series σf satisfies the growth condition. Secondly, for any $\Pi \in \mathscr{L}ie_{\mathbb{K}}\langle X\rangle$ and for any $w \in X^*$, one has

$$\langle \sigma f \triangleleft \Pi \mid w\rangle = \langle \sigma f \mid \Pi w\rangle = \mathscr{A}(\Pi w) \circ f = \mathscr{A}(\Pi) \circ (\mathscr{A}(w) \circ f).$$

Since $\mathscr{A}(\Pi)$ is a derivation over $\mathbb{K}[\![\bar{q}_1, \ldots, \bar{q}_d]\!]$:

$$\begin{aligned} \mathscr{A}(\Pi) &= \sum_{j=1}^{d} (\mathscr{A}(\Pi) \circ \bar{q}_j) \frac{\partial}{\partial \bar{q}_j}, \\ \Rightarrow \quad \mathscr{A}(\Pi) \circ (\mathscr{A}(w) \circ f) &= \sum_{j=1}^{d} (\mathscr{A}(\Pi) \circ \bar{q}_j) \frac{\partial(\mathscr{A}(w) \circ f)}{\partial \bar{q}_j} \end{aligned}$$

then we deduce that

$$\begin{aligned} \forall w \in X^*, \quad \langle \sigma f \triangleleft \Pi \mid w\rangle &= \sum_{j=1}^{d} (\mathscr{A}(\Pi) \circ \bar{q}_j) \langle T_j \mid w\rangle, \\ \Longleftrightarrow \quad \sigma f \triangleleft \Pi &= \sum_{j=1}^{d} (\mathscr{A}(\Pi) \circ \bar{q}_j)\ T_j. \end{aligned}$$

This means that $\sigma f \triangleleft \Pi$ is a $\mathbb{K}$-linear combination of $\{T_j\}_{j=1,\ldots,d}$ and the dimension of the vector space $\mathrm{span}\{\sigma f \triangleleft \Pi \mid \Pi \in \mathscr{L}ie_{\mathbb{K}}\langle X\rangle\}$ is less than or equal to d.

3.3.2 Fliess' Local Realization Theorem

Proposition 20 ([51]) *Let $S \in \mathbb{K}\langle\langle X\rangle\rangle$ with Lie rank d. Then there exists a basis $S_1, \ldots, S_d \in \mathbb{K}\langle\langle X\rangle\rangle$ of $(\mathrm{Ann}^{\perp}(S), ш) \cong (\mathbb{K}[\![S_1, \ldots, S_d]\!], ш)$ such that the S_i's are proper and for any $R \in \mathrm{Ann}^{\perp}(S)$, one has*

$$R = \sum_{i_1,\ldots,i_d \geq 0} \frac{r_{i_1,\ldots,i_n}}{i_1! \ldots i_d!} S_1^{ш i_1} ш \ldots ш S_d^{ш i_d}, \text{ where } r_{0,\ldots,0} = \langle R \mid 1_{X^*}\rangle, r_{i_1,\ldots,i_d} \in \mathbb{K}.$$

Proof By Lemma 9, such a basis exists. More precisely, since the Lie rank of S is d then there exist $P_1, \ldots, P_d \in \mathcal{L}ie_{\mathbb{K}}\langle X\rangle$ such that $S \triangleleft P_1, \ldots, S \triangleleft P_d \in (\mathbb{K}\langle\langle X\rangle\rangle, ⧢)$ are $\mathbb{K}$-linearly independent. By duality, there exists $S_1, \ldots, S_d \in (\mathbb{K}\langle\langle X\rangle\rangle, ⧢)$ such that

$$\forall i, j = 1, \ldots, d, \quad \langle S_i \mid P_j\rangle = \delta_{i,j}, \text{ and } R = \prod_{i=1}^{d} \exp(S_i\ P_i).$$

Expanding this product, one obtains, via PBW theorem, the expected expression for the coefficients $\{r_{i_1,\ldots,i_d} = \langle R \mid P_1^{i_1} \ldots P_d^{i_d}\rangle\}_{i_1,\ldots,i_d \geq 0}$. Hence, $(\mathrm{Ann}^{\perp}(S), ⧢)$ is generated by $S_1, \ldots, S_d$.

With the notations of Proposition 20, one has

Corollary 6 *1. If $S \in \mathbb{K}[S_1, \ldots, S_d]$ then, for any $i = 0, 1$ and for any $j = 1, \ldots, d$, one has $x_i \triangleright S \in \mathrm{Ann}^{\perp}(S) = \mathbb{K}[S_1, \ldots, S_d]$.*

2. The power series S satisfies the growth condition if and only if, for any $i = 1, \ldots, d$, S_i also satisfies the growth condition.

Proof Assume there exists $j \in [1, \ldots, d]$ such that S_j does not satisfy the growth condition. Since $S \in \mathrm{Ann}^{\perp}(S)$ then using the decomposition of S on $S_1, \ldots, S_d$, one obtains a contradiction with the fact that S satisfies the growth condition.

Conversely, using Proposition 16, we get the expected results.

Theorem 12 ([24]) *The formal power series $S \in \mathbb{K}\langle\langle X\rangle\rangle$ is differentially produced if and only if its Lie rank is finite and if it satisfies the χ-growth condition.*

Proof By Proposition 19, one gets a direct proof. Conversely, since the Lie rank of S equals d then by Proposition 20, setting $\sigma f_{|0} = S$ and, for $j = 1, \ldots, d, \sigma\bar{q}_i = S_i$,

1. We choose the observation f as follows

$$f(\bar{q}_1, \ldots, \bar{q}_d) = \sum_{i_1,\ldots,i_d \geq 0} \frac{r_{i_1,\ldots,i_n}}{i_1! \ldots i_d!} \bar{q}_1^{i_1} \ldots \bar{q}_d^{i_d} \in \mathbb{K}[\![\bar{q}_1, \ldots, \bar{q}_d]\!]$$

such that

$$\sigma f_{|0}(\bar{q}_1, \ldots, \bar{q}_d) = \sum_{i_1,\ldots,i_d \geq 0} \frac{r_{i_1,\ldots,i_n}}{i_1! \ldots i_d!} (\sigma\bar{q}_1)^{⧢ i_1} ⧢ \ldots ⧢ (\sigma\bar{q}_d)^{⧢ i_d},$$

2. It follows that, for $i = 0, 1$ and for $j = 1, \ldots, d$, the residual $x_i \triangleright \sigma\bar{q}_j$ belongs to $\mathrm{Ann}^{\perp}(\sigma f_{|0})$ (see also Lemma 9),
3. Since σf satisfies the χ-growth condition then, the generating series $\sigma\bar{q}_j$ and $x_i \triangleright \sigma\bar{q}_j$ (for $i = 0, 1$ and for $j = 1, \ldots, d$) verify also the growth condition. We then take (see Lemma 8)

$$\forall i = 0, 1, \forall j = 1, \ldots, d, \quad \sigma A_j^i(\bar{q}_1, \ldots, \bar{q}_d) = x_i \triangleright \sigma\bar{q}_j,$$

by expressing σA^i_j on the basis $\{\sigma\bar{q}_i\}_{i=1,\ldots,d}$ of $\mathrm{Ann}^{\perp}(\sigma f_{|0})$,

4. The homomorphism $\mathscr{A}$ is then determined as follows

$$\forall i = 0, 1,\quad \mathscr{A}(x_i) = \sum_{j=0}^{d} A^i_j(\bar{q}_1, \ldots, \bar{q}_d)\frac{\partial}{\partial \bar{q}_j},$$

where, by Lemma 5, one has $A^i_j(\bar{q}_1, \ldots, \bar{q}_d) = \mathscr{A}(x_i) \circ \bar{q}_j$.

Thus, $(\mathscr{A}, f)$ provides a differential representation[29] of dimension d of S.

Moreover, one also has the following

Theorem 13 ([24]) *Let $S \in \mathbb{K}\langle\langle X\rangle\rangle$ be a differentially produced formal power series. Let $(\mathscr{A}, f)$ and $(\mathscr{A}', f')$ be two differential representations of dimension n of S. There exist a continuous and convergent automorphism h of $\mathbb{K}$ such that*

$$\forall w \in X^*, \forall g \in \mathbb{K},\quad h(\mathscr{A}(w) \circ g) = \mathscr{A}'(w) \circ (h(g)) \text{ and } f' = h(f).$$

Since any rational power series satisfies the growth condition and its Lie rank is less than or equal to its Hankel rank which is finite [24] then

Corollary 7 *Any rational power series and any polynomial over X with coefficients in $\mathbb{K}$ are differentially produced.*

References

1. Abe, E.: Hopf Algebra. Cambridge University Press, Cambridge (1980)
2. Bender, C.M., Brody, D.C., Meister, B.K.: Quantum field theory of partitions. J. Math. Phys. **40**, 3239–3245 (1999)
3. Berstel, J., Reutenauer, C.: Rational Series and Their Languages. Springer, Berlin (1988)
4. Borwein, J.M., Bradley, D.M., Broadhurst, D.J., Lisonek, P.: Special values of multiple polylogarithms. Trans. Am. Math. Soc. **353**, 907–941 (2000)
5. Bui, C., Duchamp, G.H.E., Hoang Ngoc Minh, V.: Schützenberger's factorization on the (completed) Hopf algebra of q-stuffle product. ACM Commun. Comput. Algebra **47**(3/4), 90–91 (2013)
6. Bui, V.C., Duchamp, G.H.E., Hoang Ngoc Minh, V., Tollu, C., Ngo, Q.H.: (Pure) transcendence bases in ϕ-deformed shuffle bialgebras. Seminaire Lotharingien de Combinatoire, Université Louis Pasteur. arXiv:1507.01089 [cs.SC] (2015)
7. Cartier, P.: Développements récents sur les groupes de tresses. Applications à la topologie et à l'algèbre, Sém BOURBAKI, 42^{ème}(716). Springer, New York (1989–1990)
8. Chari, R., Pressley, A.: A Guide to Quantum Group. Cambridge University Press, Cambridge (1994)
9. Chen, K.T., Fox, R.H., Lyndon, R.C.: Free differential calculus IV, the quotient groups of the lower central series. Ann. Math. **68**, 81–95 (1958)
10. Chen, K.T.: Iterated path integrals. Bull. Am. Math. Soc. **83**, 831–879 (1977)

[29] In [24, 51], the reader can find the discussion on the *minimal* differential representation.

11. Connes, A., Kreimer, D.: Hopf algebras, renormalization and noncommutative geometry. Commun. Math. Phys. **199**, 203–242 (1998)
12. Costermans, C., Enjalbert, J.Y., Hoang Ngoc Minh, V.: Algorithmic and combinatorial aspects of multiple harmonic sums. In: Discrete Mathematics & Theoretical Computer Science Proceedings (2005)
13. Deneufchâtel, M., Duchamp, G.H.E., Hoang Ngoc Minh, V., Solomon, A.I.: Independence of Hyperlogarithms over Function Fields via Algebraic Combinatorics. Lecture Notes in Computer Science, vol. 6742, pp. 127–139. Springer, Berlin (2011)
14. Deneufchâtel, M., Duchamp, G.H.E., Hoang Ngoc Minh, V.: Radford bases and Schützenberger's Factorizations. arXiv:1111.6759 (2011)
15. Duchamp, G.H.E., Reutenauer, C.: Un critère de rationalité provenant de la géométrie noncommutative. Invent. Math. **128**(3), 613–622 (1997)
16. Duchamp, G.H.E., Hoang Ngoc Minh, V., Solomon, A.I., Goodenough, S.: An interface between physics and number theory. J. Phys. **284**(1), 012–023 (2011)
17. Duchamp, G.H.E., Hoang Ngoc Minh, V., Hoan, N.Q.: Harmonic sums and polylogarithms at negative multi-indices. J. Symb. Comput. (2016)
18. Duchamp, G.H.E., Tollu, C.: Sweedler's duals and Schützenberger's calculus. IN: Conference on Combinatorics and Physics. arXiv: 0712.0125v3
19. Dyson, F.J.: The radiation theories of Tomonaga, Schwinger and Feynman. Phys. Rev. **75**, 486–502 (1949)
20. Foata, D., Schützenberger, M.P.: Théorie Géométrique des Polynômes Eulériens. Lecture Notes in Mathematics, vol. 138. Springer, Berlin (1979)
21. Feynman, R.P., Hibbs, A.R.: Quantum Mechanics and Path Integrals. Wiley, New York (1965)
22. Fliess, M.: Matrices de Hankel. J. Math. Pures Appl. **53**, 197–222 (1974)
23. Fliess, M.: Fonctionnelles causales non linéaires et indéterminées non commutatives. Bull. SMF **109**, 3–40 (1981)
24. Fliess, M.: Réalisation locale des systèmes non linéaires, algèbres de Lie filtrées transitives et séries génératrices non commutatives. Invent. Math. **71**(3), 521–537 (1983)
25. Fliess, M.: Vers une notion de dérivation fonctionnelle causale. Annales de l'institut Henri Poincar (C) Analyse non linaire **3**(1), 67–76 (1986)
26. Goncharov, A.B.: Multiple polylogarithms, cyclotomy and modular complexes. Math. Res. Lett. **5**, 497–516 (1998)
27. Hespel, C.: Une étude des séries formelles noncommutatives pour l'Approximation et l'Identification des systèmes dynamiques. Thèse docteur d'état, Université Lille 1 (1998)
28. Hoang Ngoc Minh, V.: Contribution au développement d'outils informatiques pour résoudre des problèmes d'automatique non linéaire. Thèse, Lille (1990)
29. Hoang Ngoc Minh, V.: Input/Output behaviour of nonlinear control systems: about exact and approximated computations. In: IMACS-IFAC Symposium, Lille, Mai (1991)
30. Hoang Ngoc Minh, V., Jacob, G., Oussous, N.: Input/output behaviour of nonlinear control systems: rational approximations, nilpotent structural approximations. Analysis of controlled dynamical systems. In: Bonnard, B., Bride, B., Gauthier, J.P., Kupka, I. (eds.) Progress in Systems and Control Theory, pp. 253–262. Birkhäuser, Boston (1991)
31. Hoang Ngoc Minh, V.: Summations of polylogarithms via evaluation transform. Math. Comput. Simul. **1336**, 707–728 (1996)
32. Hoang Ngoc Minh, V.: Fonctions de Dirichlet d'ordre n et de paramètre t. Discret. Math. **180**, 221–242 (1998)
33. Hoang Ngoc Minh, V., Petitot, M., Van der Hoeven, J.: Polylogarithms and Shuffle algebra. In: Proceedings of FPSAC'98 (1998)
34. Hoang Ngoc Minh, V., Petitot, M., Van der Hoeven, J.: L'algèbre des polylogarithmes par les séries génératrices. In: Proceedings of FPSAC'99 (1999)
35. Hoang Ngoc Minh, V.: Calcul symbolique non commutatif: aspects combinatoires des fonctions spéciales et des nombres spéciaux. HDR, Lille (2000)
36. Hoang Ngoc Minh, V., Jacob, G.: Symbolic integration of meromorphic differential systems via Dirichlet functions. Discret. Math. **210**, 87–116 (2000)

37. Hoang Ngoc Minh, V., Jacob, G., Oussous, N.E., Petitot, M.: Aspects combinatoires des polylogarithmes et des sommes d'Euler-Zagier. J. Électron. Sémin. Lothar. Combin., Article ID B43e (2000)
38. Hoang Ngoc Minh, V., Petitot, M.: Lyndon words, polylogarithmic functions and the Riemann ζ function. Discret. Math. **217**, 273–292 (2000)
39. Hoang Ngoc Minh, V., Jacob, G., Oussous, N.E., Petitot, M.: De l'algèbre des ζ de Riemann multivariées l'algèbre des ζ de Hurwitz multivariées. J. Électron. Sémin. Lothar. Combin. **44**, Art. B44i (2001)
40. Hoang Ngoc Minh, V.: Finite polyzêtas, Poly-Bernoulli numbers, identities of polyzêtas and noncommutative rational power series. In: Proceedings of 4th International Conference on Words, pp. 232–250 (2003)
41. Hoang Ngoc Minh, V.: Differential Galois groups and noncommutative generating series of polylogarithms. In: Automata, Combinatorics and Geometry, 7th World Multi-conference on Systemics, Cybernetics and Informatics, Florida (2003)
42. Hoang Ngoc Minh, V.: Algebraic combinatoric aspects of asymptotic analysis of nonlinear dynamical system with singular inputs. Acta Acad. Aboensis **B67**(2), 117–126 (2007)
43. Hoang Ngoc Minh, V.: On a conjecture by Pierre Cartier about a group of associators. Acta Math. Vietnam. **3**, 39 (2013)
44. Hoang Ngoc Minh, V.: Structure of polyzetas and Lyndon words. Vietnam. Math. J. (2013). doi:10.1007/10013-013-0047-x
45. Hoffman, M.: The multiple harmonic series. Pac. J. Math. **152**(2), 275–290 (1992)
46. Hoffman, M.: The algebra of multiple harmonic series. J. Algebra **194**, 477–495 (1997)
47. Hochschild, G.: The Structure of Lie Groups. Holden-Day, San Francisco (1965)
48. Lê, T.Q.T., Murakami, J.: Kontsevich's integral for Kauffman polynomial. Nagoya Math **142**, 39–65 (1996)
49. Radford, D.E.: A natural ring basis for shuffle algebra and an application to group schemes. J. Algebra **58**, 432–454 (1979)
50. Ree, R.: Lie elements and an algebra associated with shuffles. Ann. Math. **68**, 210–220 (1985)
51. Reutenauer, C: The local realisation of generating series of finite Lie rank. In: Algebraic and Geometric Methods in Nonlinear Control Theory, pp. 33–43. Reidel, Dordrecht (1986)
52. Reutenauer, C.: Free Lie Algebras. London Mathematical Society Monographs. Clarendon Press, Oxford (1993)
53. Royden, H.L.: Real Analysis. 3rd ed., (1988)
54. Schützenberger, M.P.: On the definition of a family of automata. Inf. Control **4**, 245–270 (1961)
55. Viennot, G.: Algèbres de Lie Libres et Monoïdes Libres, vol. 691. Lecture Notes in Mathematics, pp. 94–112. Springer, Berlin (1978)
56. Zagier, D.: Values of zeta functions and their applications. In: First European Congress of Mathematics, vol. 2, pp. 497–512. Birkhäuser, Boston (1994)

The Root Lattice A_2 in the Construction of Substitution Tilings and Singular Hypersurfaces

Juan García Escudero

Abstract The analysis of the critical points of a one-parameter family of polynomials allows us to define sets of pseudolines in the fundamental region of the affine Weyl group associated with the root lattice A_2. The pseudolines are transformed into configurations of lines containing the prototiles of substitution tilings with n-fold symmetry. The configurations of lines have been used recently to obtain hypersurfaces with many singularities. Calabi–Yau threefolds can be constructed from resolutions of some of the singular hypersurfaces.

Keywords Calabi–Yau threefolds · Singular hypersurfaces · Substitution tilings

1 Introduction

In [14, 15] we have shown that special types of simple arrangements of d lines are related to a class of bivariate polynomials $\hat{J}_{d,\tau}(x, y)$ having many critical points with few critical values. The polynomials have been used in the construction of algebraic surfaces with many A and D singularities [13, 14].

Tilings exhibiting non-crystallographic symmetries have been significant in the past decades in the field of quasicrystals. The root lattice A_4 was considered in [1] to generate planar tilings with tenfold symmetry by projection methods. An arrangement of pseudolines is a collection of curves topologically equivalent to lines (pseudolines are also called topological lines) such that any two of them intersect exactly once. If no three of them meet in a common point then the arrangement is said to be simple. The analysis of the critical points of $\hat{J}_{d,\tau}(x, y)$ allows us to define pseudoline arrangements inside the fundamental region of the affine Weyl group associated with the root lattice A_2, which can be transformed into simple and simplicial (all the bounded cells are triangles) arrangements of lines containing the triangular prototiles of substitution tilings with n-fold symmetry [10]. Topological invariants

J.G. Escudero (✉)
Facultad de Ciencias, Universidad de Oviedo, 33007 Oviedo, Spain
e-mail: jjge@uniovi.es

I.S. Kotsireas and E. Martínez-Moro (eds.), *Applications of Computer Algebra*, Springer Proceedings in Mathematics & Statistics 198,
DOI 10.1007/978-3-319-56932-1_7

of tiling spaces connected with the simplicial arrangements have been studied in [11, 12], where we have shown that there are fivefold and ninefold symmetry tiling spaces having minimal first cohomology groups, a property that distinguish them from others with the same symmetries. Random tilings can be generated from both the line and the pseudoline configurations [10].

On the other hand, by following Hirzebruch's methods [20, 21] applied to special line configurations associated with $\hat{J}_{d,\tau}(x, y)$, threefolds with trivial canonical bundle and absolute value of the Euler number not large but different from zero can be obtained. Calabi–Yau threefolds associated with higher dimensional root lattices like A_4 were studied in [23]. In this paper we use Mathematica [33] and Singular [19] computer algebra systems.

2 On a One-Parameter Family of Bivariate Polynomials

The folding polynomials of degree d, obtained in the study of the generalisation of Chebyshev polynomials in two variables [22, 24, 25], associated with the affine Weyl group of the root lattice A_2, are defined for $(z, w) \in \mathbf{C}^2$ as $F_d^{A_2}(z, w) := j_{1,d}(z, w) + j_{2,d}(z, w)$, where $j_{2,d}(z, w) = j_{1,d}(w, z)$ and $j_{1,d}(z, w)$ satisfies the recursion relation

$$j_{1,d}(z, w) = z j_{1,d-1}(z, w) - w j_{1,d-2}(z, w) + j_{1,d-3}(z, w) \tag{1}$$

with $j_{1,1}(z, w) = z$, $j_{1,2}(z, w) = z^2 - 2w$, $j_{1,3}(z, w) = z^3 - 3zw + 3$.

We consider, for $(x, y) \in \mathbf{R}^2$ and $\tau \in \mathbf{R}$, the one-parameter family of degree d polynomials with real coefficients

$$\hat{J}_{d,\tau}(x, y) := e^{i(\tau+\frac{2\pi}{3})}\widetilde{j}_{1,d}(x, y) + e^{-i(\tau+\frac{2\pi}{3})}\widetilde{j}_{2,d}(x, y) + 2\cos 3\tau \tag{2}$$

where $\widetilde{j}_{1,d}(x, y) = j_{1,d}(x + iy, x - iy)$, $\widetilde{j}_{2,d}(x, y) = \widetilde{j}^{*}_{1,d}(x, y)$ and the superscript asterisk stands for complex conjugation.

The map $\widetilde{h}(u, v) : \mathbf{R}^2 \to \mathbf{R}^2$, is defined by $\widetilde{h}(u, v) := (\cos(2\pi(u + v)) + \cos(2\pi u) + \cos(2\pi v), \sin(2\pi(u + v)) - \sin(2\pi u) - \sin(2\pi v)) = (x, y)$. A basis of simple roots $\{\alpha_1, \alpha_2\}$ for the root lattice A_2 is $\alpha_1 = (2, 0)$, $\alpha_2 = (-1, \sqrt{3})$. In [14] we showed that the critical points with critical value $\zeta = 0$ in the (u, v) plane are situated in the downscaled root lattice with basis $\{\frac{\alpha_1}{6d}, \frac{\alpha_1+\alpha_2}{6d}\}$. They determine a set of d pseudolines, whose images under $\widetilde{h}$ are the lines in the (x, y) plane whose union is used to define $\hat{J}_{d,0}(x, y)$. The pseudolines represent also periodic billiard trajectories in the region $u - v \geq 0$, $u + 2v \geq 0$, $2u + v \leq 1$, which is the fundamental region, denoted by Δ, of the affine Weyl group $\widetilde{W}(A_2)$. The images of $\partial\Delta$ and the pseudolines under $\widetilde{h}$ are the deltoid and its tangents respectively. We can extend the results in [14] by taking into account the positions of the corresponding critical points of $\hat{J}_{d,\tau}(x, y)$. We get the lines $L_{d,\nu,\tau}(x, y) = 0$, $\nu = -\lfloor\frac{d-2}{2}\rfloor, -\lfloor\frac{d-2}{2}\rfloor + 1, \ldots, \lfloor\frac{d+1}{2}\rfloor$

with $L_{d,\nu,\tau}(x, y) := y - (x - \cos(\frac{2\pi}{d}(\frac{6\nu-1}{6} - \frac{\tau}{\pi})))\tan(\frac{\pi}{d}(\frac{6\nu-1}{6} - \frac{\tau}{\pi})) + \sin(\frac{2\pi}{d}(\frac{6\nu-1}{6} - \frac{\tau}{\pi}))$. The polynomials are the union of the lines, up to the normalising factor $\lambda_{d,\tau} = e^{i(\tau+\frac{2\pi}{3}+\frac{d\pi}{2})} + e^{-i(\tau+\frac{2\pi}{3}+\frac{d\pi}{2})}$ corresponding to the term y^d for $\tau + \frac{2\pi}{3} + \frac{d\pi}{2} \neq (2m+1)\frac{\pi}{2}$, or $\lambda_{d,\tau} = (-1)^m 2d$ corresponding to xy^{d-1} when $\tau = (6m - 3d - 1)\frac{\pi}{6}$ (in this case the line $L_{d,\nu,\tau}(x, y) = 0$ parallel to the y-axis is interpreted as the line $x = -1$):

$$\hat{J}_{d,\tau}(x, y) = \lambda_{d,\tau} \prod_{\nu} L_{d,\nu,\tau}(x, y) \tag{3}$$

The following theorem describes some properties of the critical points of $\hat{J}_{d,\tau}(x, y)$ [15]:

Theorem 1 *The critical points and critical values of $\hat{J}_{d,\tau}(x, y)$ have the following properties:*
(p1) The critical values are

$$\zeta = 0, \zeta_M(\tau) = 6\cos\tau + 2\cos 3\tau, \zeta_{m1}(\tau) = \zeta_M(\tau - \frac{2\pi}{3}), \zeta_{m2}(\tau) = \zeta_M(\tau + \frac{2\pi}{3}).$$

(p2) The critical points with critical values $\zeta_M(\tau)$, $\zeta_{m1}(\tau)$ and $\zeta_{m2}(\tau)$ have the same coordinates $\forall \tau \in \mathbf{R}$.
(p3) At the points $\tau \in \Lambda_\Sigma := \{k\frac{\pi}{3}, k \in \mathbf{Z}, 0 \leq k \leq 5\}$, either $\hat{J}_{d,\tau}(x, y)$ or $-\hat{J}_{d,\tau}(x, y)$ have all the maxima with critical value 8 and all the minima with value -1.
(p4) All the maxima have values $3\sqrt{3}$ and all the minima $-3\sqrt{3}$ at $\tau \in \Lambda_S := \{(2k+1)\frac{\pi}{6}, k \in \mathbf{Z}, 0 \leq k \leq 5\}$.
(p5) For each $\tau \in \Lambda_3 := [0, 2\pi) \setminus \{k\frac{\pi}{6}, k \in \mathbf{Z}, 0 \leq k \leq 11\}$, $\hat{J}_{d,\tau}(x, y)$ has critical values $\zeta = 0$ and ζ_m, $m = 1, 2, 3$, with $0 < |\zeta_3| < 1 < |\zeta_2| < 3\sqrt{3} < |\zeta_1| < 8$ and $sgn(\zeta_1) \neq sgn(\zeta_2) = sgn(\zeta_3)$.

3 Substitution Tilings

The critical points of $\hat{J}_{d,\tau}(x, y)$ are the images under $\widetilde{h}$ of the critical points of $H_{d,\tau}(u, v) = 2\cos(2\pi du - \frac{2\pi}{3} - \tau) + 2\cos(2\pi dv - \frac{2\pi}{3} - \tau) + 2\cos(2\pi d(u+v) + \frac{2\pi}{3} + \tau)$, with $(u, v) \in \Delta \setminus \partial\Delta$ (when $\tau \in \Lambda_S$, some critical points with critical value 0 are situated in $\partial\Delta$). A direct computation of the critical values of $H_{d,\tau}(u, v) + 2\cos 3\tau$ leads to the following cases ($k, l \in \mathbf{Z}$) [15]:

(a) $\zeta_M(\tau) = 6\cos\tau + 2\cos 3\tau$; $u = \frac{3k+1}{3d}$, $v = \frac{3l+1}{3d}$.
(b) $\zeta_{m1}(\tau) = 6\cos(\tau - \frac{2\pi}{3}) + 2\cos 3\tau$; $u = \frac{3k+2}{3d}$, $v = \frac{3l+2}{3d}$.
(c) $\zeta_{m2}(\tau) = 6\cos(\tau + \frac{2\pi}{3}) + 2\cos 3\tau$; $u = \frac{k}{d}$, $v = \frac{l}{d}$.
(d) $\zeta = 0$;
  (d1) $u = \frac{6k-1}{6d} - \frac{\tau}{\pi d}$, $v = \frac{6l-1}{6d} - \frac{\tau}{\pi d}$;

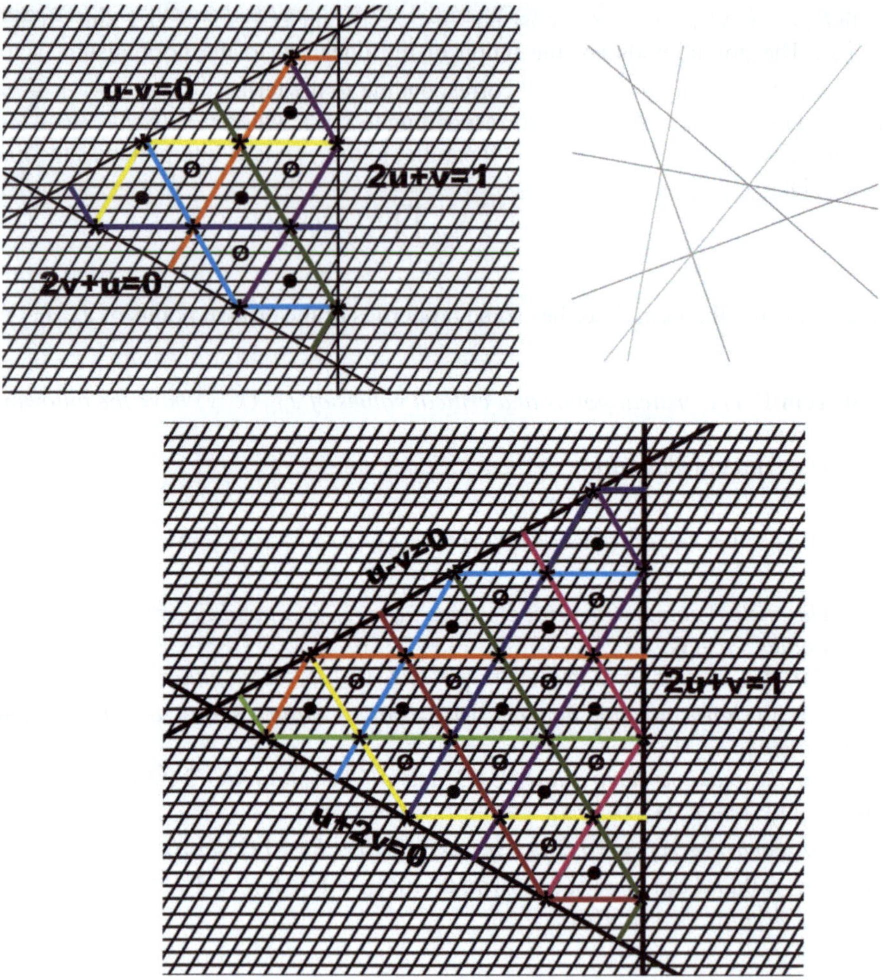

Fig. 1 Pseudolines in the (u, v) plane, represented in *color*, inside Δ for d = 6 (*top, left*) and d = 9 (*bottom*). Lines in the (x, y) plane for $d = 6$ (*top, right*) as images of the pseudolines under $\widetilde{h}$

(d2) $u = \frac{6k-1}{6d} - \frac{\tau}{\pi d}$, $v = \frac{3l+1}{3d} + \frac{2\tau}{\pi d}$;
(d3) $u = \frac{3k+1}{3d} + \frac{2\tau}{\pi d}$, $v = \frac{6l-1}{6d} - \frac{\tau}{\pi d}$.

Critical points of $H_{d,\pi/6}(u, v) + 2$ with critical values $3\sqrt{3}, 0, -3\sqrt{3}$ are represented by $\circ, *, \bullet$ respectively in Fig. 1. The critical points with critical value 0 define a set of pseudolines represented in color. The lines $u = c$, c being a constant, denoted by $l_{u=c}$, are transformed under $\widetilde{h}$ into $y = (x - \cos(2\pi c))\tan(\pi c) - \sin(2\pi c)$ [14]. If two lines l_1, l_2 are transformed under $\widetilde{h}$ into the same line, then we write $l_1 \sim l_2$ and the pseudoline is $l_1 \cup l_2$. We have the following properties:

$$l_{u=c} \sim l_{v=c}, l_{u+v=c} \sim l_{u=-c} \ (\mathrm{mod}\ 6d), l_{u+v=c} \sim l_{v=-c} \ (\mathrm{mod}\ 6d).$$

For $d = 6, \tau = \pi/6$ the critical points are (up to the factor $1/36$)

$$\zeta_M(\pi/6) = 3\sqrt{3} : \{(14, -4), (8, 2), (14, 2)\} \subset \Delta \setminus \partial\Delta,$$

$$\zeta_{m2}(\pi/6) = -3\sqrt{3} : \{(18, -6), (12, 0), (6, 0), (12, 6)\} \subset \Delta \setminus \partial\Delta,$$

$$\zeta_{m1}(\pi/6) = 0 : \{(10, -2), (16, -2), (10, 2)\} \subset \Delta \setminus \partial\Delta,$$

$$\zeta = 0 : \{(22, -8), (16, -8), (4, -2), (4, 4), (16, 4), (10, 10)\} \subset \partial\Delta$$

Having in mind the coordinates of the critical points with critical value zero, we get the following 6 pseudolines connecting those points (Fig. 1, top, left):

$$l_{u=4} \cup l_{v=4}, l_{u=10} \cup l_{v=10}, l_{u=16} \cup l_{u+v=20},$$

$$l_{u=22} \cup l_{u+v=14}, l_{v=-8} \cup l_{u+v=8}, l_{v=-2} \cup l_{u+v=2}$$

The images of the pseudolines under $\widetilde{h}$ in the (x, y) plane are simple arrangements of lines, denoted by Σ^d_D, Σ^d_C, when $\tau \in \Lambda_\Sigma$, and simplicial arrangements of lines S^d_D, S^d_C if $\tau \in \Lambda_S$ (the subindex C denotes cyclic symmetry). The arrangement shown in Fig. 1 (top, right) is S^6_C. In Fig. 2 (left) we can see Σ^9_C (discontinuous lines) and S^9_C (continuous lines) superimposed, the last one being the image of the configuration

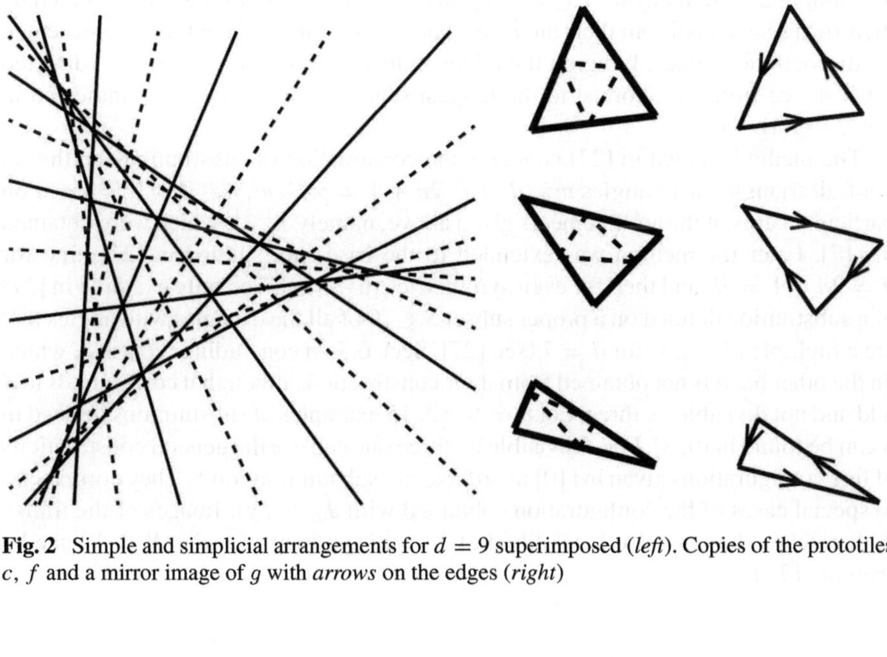

Fig. 2 Simple and simplicial arrangements for $d = 9$ superimposed (*left*). Copies of the prototiles c, f and a mirror image of g with *arrows* on the edges (*right*)

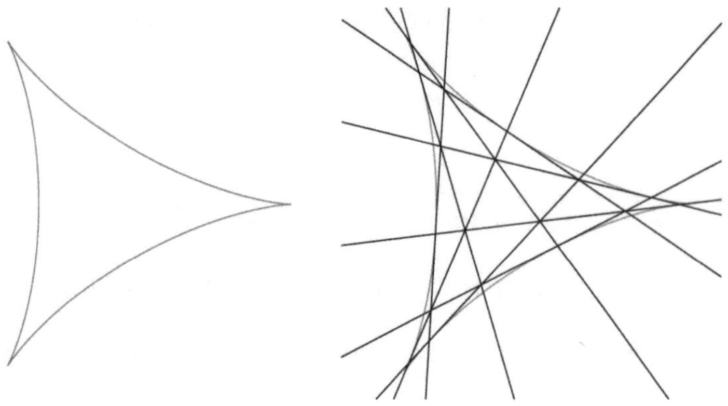

Fig. 3 The deltoid (*left*). The *lines* of the simplicial arrangement S_C^9 are tangents of the deltoid (*right*)

of pseudolines in Fig. 1 (bottom). The deltoid and S_C^9 are represented in Fig. 3. The prototiles (minimal set of tiles such that each tile in the tiling is congruent to one of those in the prototile set) for a wide class of tilings can be obtained from the simple or the simplicial arrangements. A substitution or inflation rule determines how to replace each prototile with a patch of tiles. Iteration of the substitution rules gives, in the limit, a substitution tiling. The possible patches of tiles necessary for the derivation of the substitution rules are included in the simplicial arrangements. However, in order to get the inflation rules a decoration by arrows on the prototile edges is needed. In Fig. 2 (left) we see that when we represent the union of simple and simplicial arrangements (S_C^d and Σ_C^9 in this case), the prototiles are decorated by their own scaled copies in their interiors. The vertices of the scaled tiles, represented by discontinuous lines, lying on the edges induce a decoration by arrows (directed for instance from the shortest to the longest segment) on the edges as indicated in Fig. 2 (right).

The method studied in [27] concerns the construction of substitutions on the set T of all triangles with angles $m\pi/d$, $d = 2n + 1 \neq 3l, l, m, n \in \mathbf{Z}^+$. It is based on particular cases of the arrangements given above, namely S_D^{2n+1}, which were obtained in [17]. Later the method was extended to the cases not studied in [27], first for $d = 2n + 1 = 3l$, and then for even symmetries [6–8, 10]. The only example in [27] of a substitution defined on a proper subset $S \subset T$ of all the triangles with angles that are a multiple of π/d is for $d = 7$ (see [27], Sect. 6.3 on concluding remarks), which on the other hand is not obtained from their construction, although it corresponds to d odd and not divisible by three. For $d = 9$, 12, 15 examples of substitutions defined in S can be found in [6, 8]. For d divisible by three one can use the general constructions of line configurations given in [10] in order to get substitutions on S. They correspond to special cases of the configurations obtained with $\hat{J}_{d,\tau}(x, y)$. Images of the tilings with patches showing also local dihedral $2d$-fold symmetry for $d = 8, 9, 12$ can be seen in [12, 16].

There are several methods to get non-deterministic structures. The method of composition of inflation rules, also called multisubstitutions, consists in applying the same inflation rule to each tile in a given inflation step. Examples with only one prototile set were treated in [27] and with different prototile sets in [7, 8].

A different type of structures are the random substitution tilings, which are characterised by the fact that one can apply at each inflation step several substitution rules to each tile. Examples of edge-to-edge random substitutions in the plane were obtained in [9]. Now, we show how tile rearrangements in the inflation rules with PV inflation factors give random tilings for $d = 5$ and $d = 9$.

Inflation factors that are unit PV numbers for $d = 9$ are $\mu_2\mu_4, \mu_4, \mu_4^2$, where $\mu_k := \frac{\sin(k\pi/d)}{\sin(\pi/d)}$, with minimal polynomials $1 + 3x - 6x^2 + x^3$, $1 - 3x^2 + x^3$ and $-1 + 6x - 9x^2 + x^3$ respectively. The inflation rules for the sets with 3 and 7 prototiles are denoted by Φ_m, Φ'_m, respectively, if the inflation factor is μ_m. In Fig. 4a

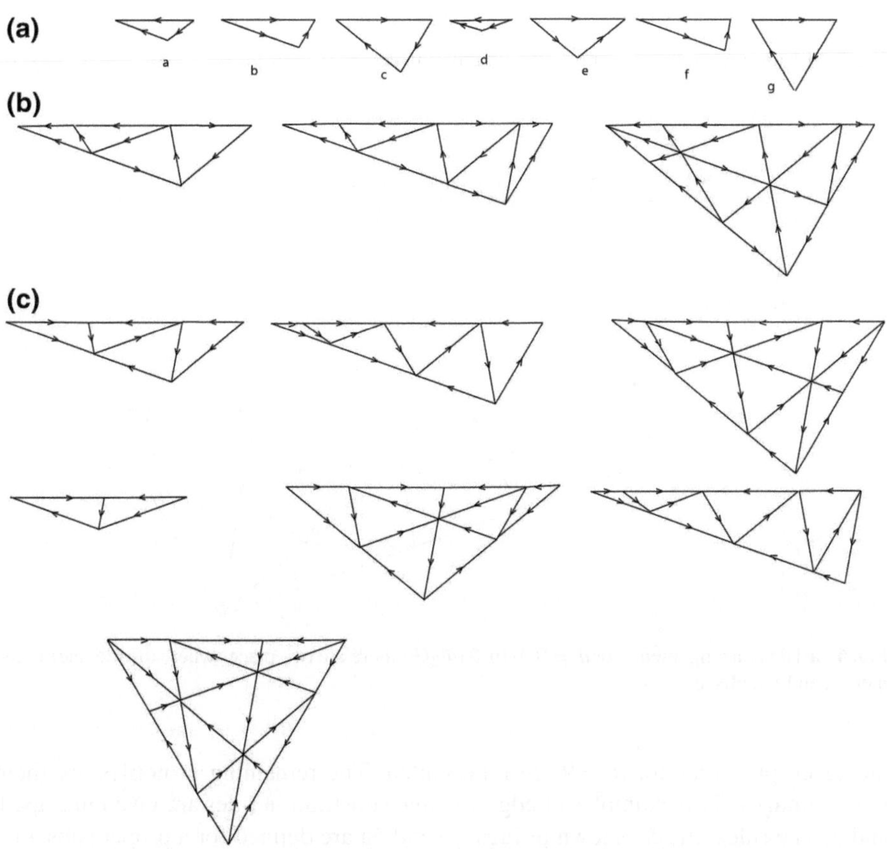

Fig. 4 **a** The two sets of prototiles for $d = 9$ are $\{a, b, c\}$ and $\{a, b, c, d, e, f, g\}$. **b** Inflation rules Φ_4 with inflation factor μ_4 for $\{a, b, c\}$. **c** Inflation rules Φ'_4 with inflation factor μ_4 for $\{a, b, c, d, e, f, g\}$

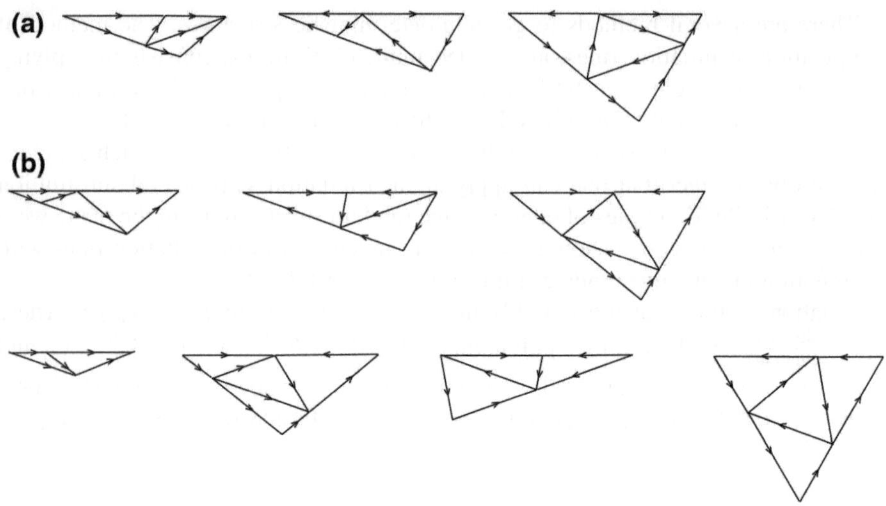

Fig. 5 Inflation rules Φ_2, Φ_2' for $d = 9$ with inflation factor μ_2 for **a** $\{a, b, c\}$, **b** $\{a, b, c, d, e, f, g\}$

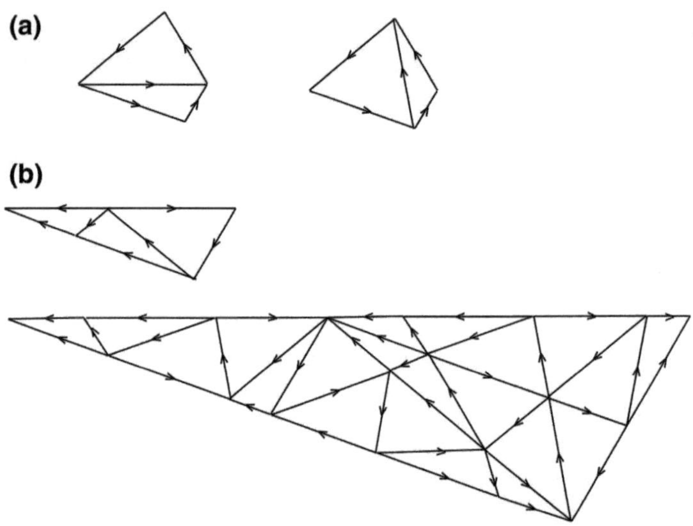

Fig. 6 **a** Tile rearrangements for $d = 9$. **b** In $\Phi_4\Phi_2(b)$ there are two places where the tile rearrangements can be realised

the seven prototiles for $d = 9$ are represented. The remaining prototiles are their mirror images. Two examples of edge-to-edge substitution rules are given in Figs. 4 and 5. The rules Φ_4, Φ_2 shown in Figs. 4b and 5a are defined for a proper subset of 3 prototiles, whereas the rules Φ_4', Φ_2' are defined for 7 prototiles (Figs. 4c and 5b). In order to get random tilings we first generate the pattern $(\Phi_4\Phi_2)^{k-1}$, we apply tile rearrangements as indicated in Fig. 6a to get a series of patterns and then we end by

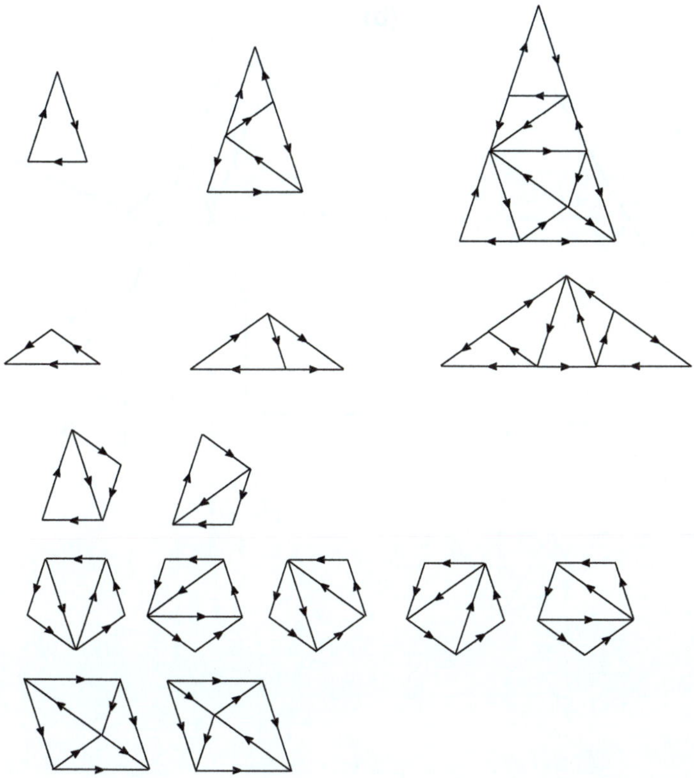

Fig. 7 The substitution rules Φ_-^2 (*top*) and tile rearrangements in the inflation rules (*bottom*)

applying to them $\Phi_4'\Phi_2'$. The tile rearrangements in this case produce the tile g not belonging to $\{a, b, c\}$ therefore we can not get random tilings with only 3 prototiles (only multisubstitutions).

For $d = 5$ two basic substitution rules Φ_+, Φ_-, their compositions and randomisations have been studied in [11]. The prototiles are two golden triangles A, B appearing in the Robinson decomposition of the Penrose tiling. The two successive inflation steps to get Φ_-^2 are shown in Fig. 7 (top). The tile rearrangements corresponding to the inflation rules of Φ_-^2 are given in Fig. 7 (bottom). They can be used in this case to obtain random tilings: we get 12 different rules for the tile A and 2 for B. The analysis of Φ_+^2 shows that no rearrangements are possible in this case. Random tilings having an inflation factor, that is the golden number squared have been obtained also recently in [18], where an algorithm for generating inflation rules by computer is given. Several examples for $d = 7$ defined on a proper subset of T are given in [18].

Proper subsets of triangular prototiles appear in the simple arrangements forming $\hat{J}_{d,\tau}(x, y)$ with $\tau \in \Lambda_3$. In this cases, the polynomials have three non-zero critical values: one for the critical points inside the non-triangular cells of the arrangement

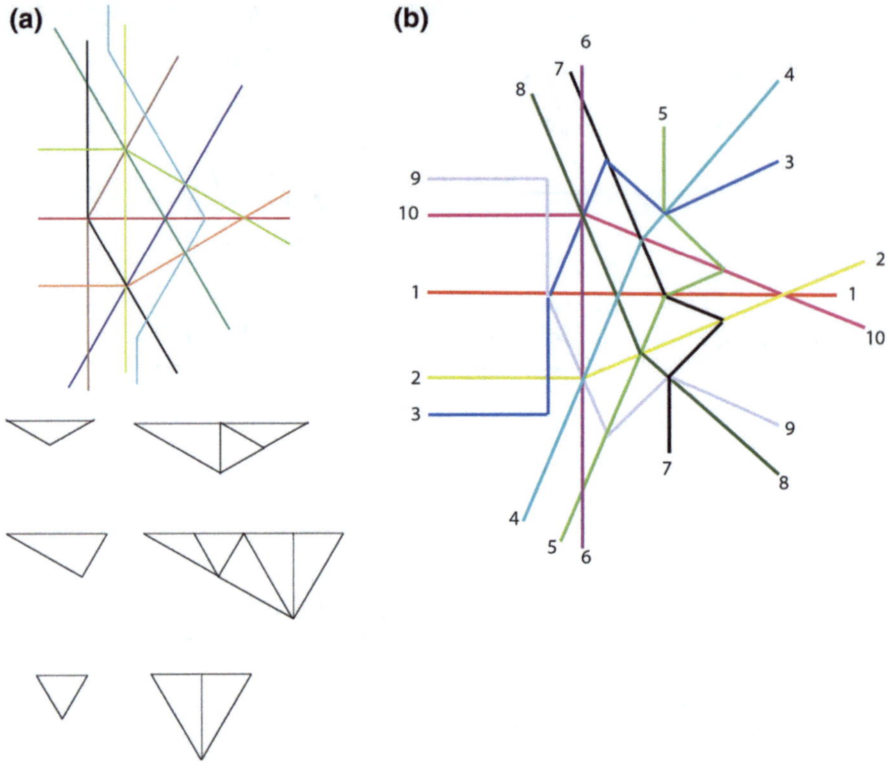

Fig. 8 Other pseudoline arrangements for random tilings: **a** d = 6, **b** d = 8

and the other two for the critical points inside two subsets of triangles with different sizes [15].

In [9] hexagonal and octagonal random substitution tilings were generated. The prototiles and inflation rules can be obtained from other types of simplicial pseudoline arrangements. Nine pseudolines are necessary for the arrangement corresponding to the hexagonal tilings (Fig. 8a), which has three triangular prototiles (see also a substitution rule in Fig. 8a). For octagonal tilings, the arrangement has ten pseudolines containing four prototiles (Fig. 8b). There are two different substitution rules which may be combined in order to obtain random tilings in the sense explained above.

4 Singular Algebraic Surfaces

The construction of algebraic surfaces with many singularities given in [13, 14], is based on $\hat{J}_{d,0}(x, y)$. By varying τ in $\hat{J}_{d,\tau}(x, y)$, which are solutions of a second-order

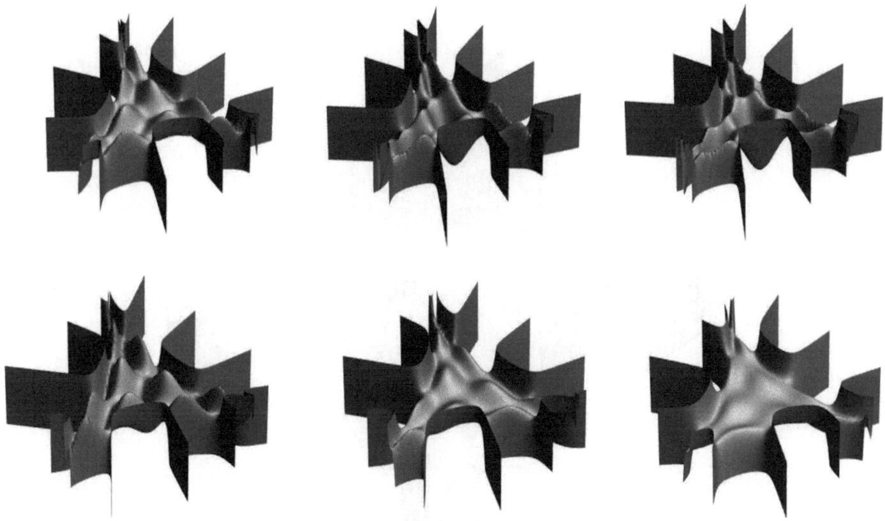

Fig. 9 Images of $G_{9,\tau}(x, y)$ for $\tau = k\pi/54$, $k = 0, 13, 17, 30, 40, 47$

linear partial differential equation [15], we can get transformations between surfaces with different number of singularities.

We consider the family of polynomials :

$$G_{d,\tau}(x, y) := (\hat{J}_{d,\tau}(\eta_{d,n+1}, 0) - \hat{J}_{d,\tau}(\eta_{d,n}, 0))\hat{J}_{d,\tau}(x, y) =$$

$$2(\cos(\tau + 2\pi/3))(\widetilde{j}_{1,d}(\eta_{d,n+1}, 0) - \widetilde{j}_{1,d}(\eta_{d,n}, 0))\hat{J}_{d,\tau}(x, y)$$

where $\eta_{d,m} := 2\cos\frac{d-m}{d}\pi + 1$. In Fig. 9 we can see the representation of $G_{9,\tau}(x, y)$ for several values of τ.

A picture of one period of $\hat{J}_{9,\tau}(x, 0) = 2\widetilde{j}_{1,9}(x, 0)\cos(\tau + 2\pi/3) + 2\cos 3\tau$ in terms of τ and x can be seen in Fig. 10. The polynomial

$$\hat{T}_d(x) := \frac{\hat{J}_{d,\tau}(x, 0) - \hat{J}_{d,\tau}(\eta_{d,n+1}, 0)}{\hat{J}_{d,\tau}(\eta_{d,n+1}, 0) - \hat{J}_{d,\tau}(\eta_{d,n}, 0)} = \frac{\widetilde{j}_{1,d}(x, 0) - \widetilde{j}_{1,d}(\eta_{d,n+1}, 0)}{\widetilde{j}_{1,d}(\eta_{d,n+1}, 0) - \widetilde{j}_{1,d}(\eta_{d,n}, 0)}$$

is a normalised Chebyshev polynomial $\hat{T}_d(x) = -\frac{T_d(\frac{x-1}{2})+1}{2}$, where $T_d(x)$ is the Chebyshev polynomial with critical values -1 and 1.

By varying τ, the family of surfaces with affine equations

$$G_{d,\tau}(x, y) - \hat{J}_{d,\tau}(z, 0) + \hat{J}_{d,\tau}(\eta_{d,n+1}, 0) = 0$$

describe transformations between the real variants of the Chmutov surfaces [2, 3] and the surfaces obtained in [14] which have more singularities. In Fig. 11 we have

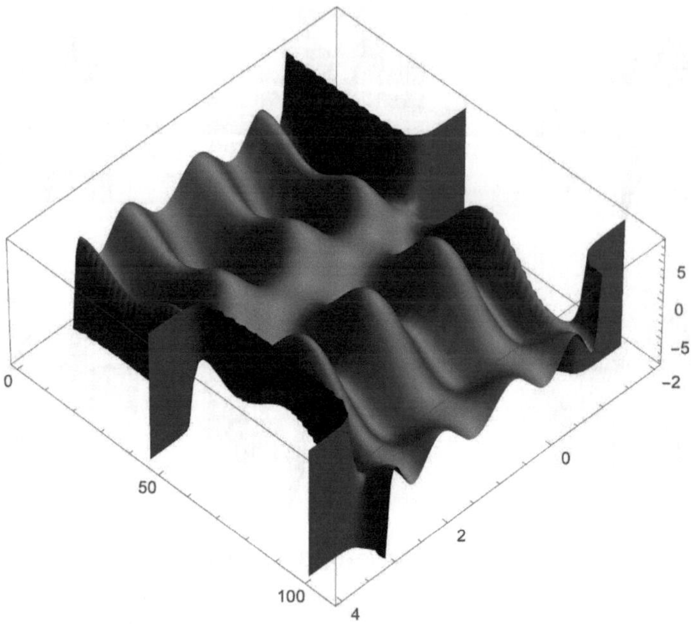

Fig. 10 Representation of one period of $\hat{J}_{9,\tau}(x, 0)$ as a function of τ and x

represented the surfaces corresponding to $d = 6$ for eight increasing values of τ. The first surface (top, left) has 59 real nodes and the fifth (third row, left) is equivalent to the real variant of the Chmutov surface with 57 real nodes. Transformations between the nodal threefolds constructed with $\hat{J}_{d,\tau}(x, y)$ can be obtained along the same lines. The computation of the number of nodes for $\tau = 0$ can be done by using Lemma 1 in [14]:

Lemma 1 *The real polynomial $\hat{J}_{d,0}(x, y)$ has $\binom{d}{2}$ real critical points with critical value 0. The number of real points with critical value 8 is $\frac{d(d-3)}{6}$ if $d = 0$ mod 3, and $\frac{(d-1)(d-2)}{6}$ otherwise. The number of real critical points with critical value -1 is $\frac{d^2}{3} - d + 1$ for $d = 0$ mod 3, and $\frac{(d-1)(d-2)}{3}$ otherwise.*

A direct consequence of this Lemma is the following result for nodal threefolds (also called conifolds):

Theorem 2 *The threefold in $\mathbf{P}^4(\mathbf{C})$ defined by the homogenization of the equation*

$$\hat{J}_{d,0}(u, v) - \hat{J}_{d,0}(z, w) = 0$$

has the following number of nodal singularities:

$$\frac{1}{18}(18 - 36d + 39d^2 - 24d^3 + 7d^4)$$

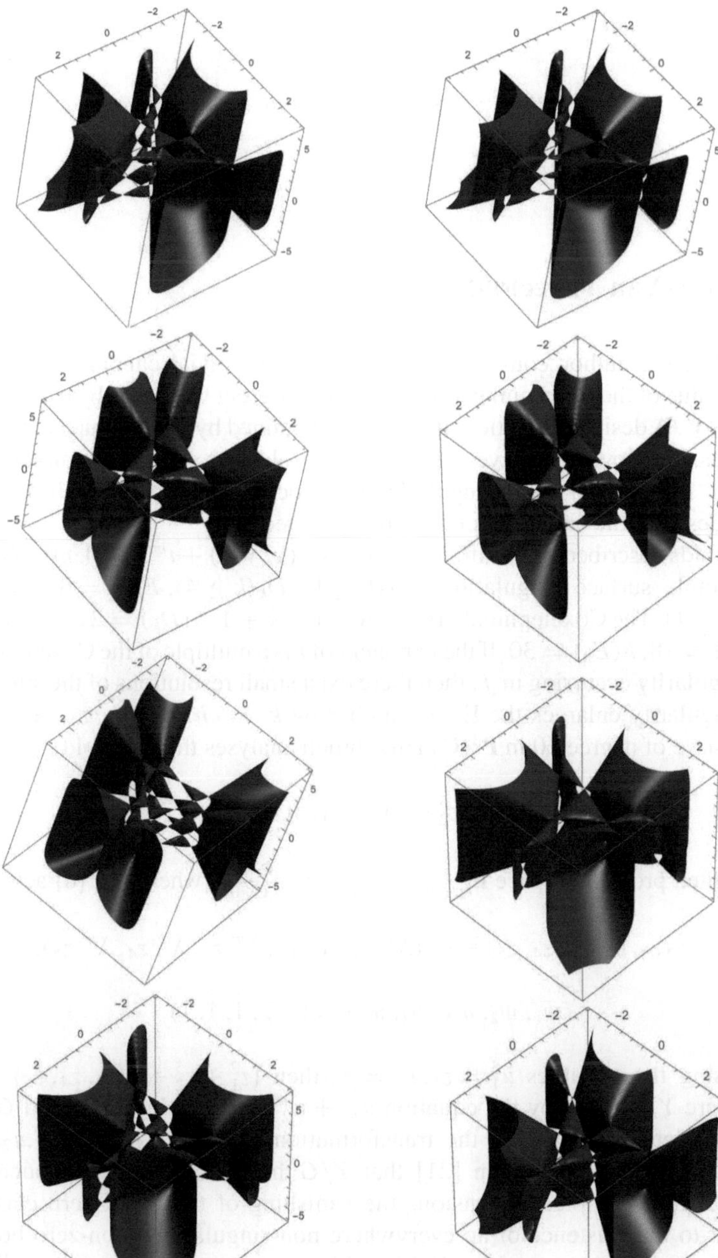

Fig. 11 Transformation between degree six singular surfaces when τ varies

if $d \equiv 0 \bmod 3$, *and*

$$\frac{1}{18}(10 - 30d + 37d^2 - 24d^3 + 7d^4)$$

otherwise.

5 Calabi–Yau Threefolds

In [20, 21], the author constructed threefolds with trivial canonical bundle and absolute value of the Euler number not large but different from zero by using singular threefolds V. A desingularization of V may be obtained by blowing up along the singular locus, but blowing up gives a smooth threefold with a different canonical class in general. However, if the desingularization process is based on small resolutions of the nodes then the canonical class is not changed.

Threefolds described by the affine equations $f(x, y, z) + t^{m \cdot h} = 0, m \in \mathbf{N}$, where f has simple surface singularities $A_k(k \geq 1), D_k(k \geq 4), E_k(k = 6, 7, 8)$, were studied in [21]. The Coxeter numbers are $h(A_k) = k + 1, h(D_k) = 2k - 2, h(E_6) = 12, h(E_7) = 18, h(E_8) = 30$. If the exponent of t is a multiple of the Coxeter number h of a singularity occurring in f, then there exist small resolutions of the singularity. Every singularity enlarges the Euler number by $k \cdot m \cdot h$. If $g(z_3, z_4, z_5) = 0$ is a smooth curve of degree 10 in $\mathbf{P}^2(\mathbf{C})$ Hirzebruch analyses the threefold

$$z_1^2 + z_2^5 + g(z_3, z_4, z_5) = 0$$

in a weighted projective space $\mathbf{P}^4_{(w_1,w_2,w_3,w_4,w_5)} := \frac{\mathbf{C}^5\setminus\{0\}}{\mathbf{C}\setminus\{0\}}$, where $\mathbf{C} \setminus \{0\}$ acts by

$$\lambda : (z_1, z_2, z_3, z_4, z_5) \longmapsto (\lambda^{w_1} z_1, \lambda^{w_2} z_2, \lambda^{w_3} z_3, \lambda^{w_4} z_4, \lambda^{w_5} z_5),$$

$$(w_1, w_2, w_3, w_4, w_5) = (5, 2, 1, 1, 1)$$

By choosing the variables $u_1^5 = z_1, u_2^2 = z_2$ then $\{z_1^2 + z_2^5 + g(z_3, z_4, z_5) = 0\} = Y/G$, where Y is given by the equation $u_1^{10} + u_2^{10} + g = 0$ in $\mathbf{P}^4(\mathbf{C})$ and G is the group of order 10 formed by the transformations $(u_1, u_2) \to (\alpha.u_1, \beta.u_2)$, with $\alpha^5 = 1, \beta^2 = 1$. It is shown in [21] that Y/G has trivial canonical bundle and $\chi = -288$. In n complex dimensions the vanishing of the first Chern class c_1 is equivalent to the existence of an everywhere non-singular and non-zero holomorphic $(n, 0)$-form. This type of manifolds, with $g(z_3, z_4, z_5) = z_3^{10} + z_4^{10} + z_5^{10}$, were studied from this point of view in [29].

If $g = 0$ has n_k singularities of types $A_k, k = 1, 4$, then the Euler number in the resolution increases by $5n_1 + 20n_4$. In the example analysed in [21], $g(z_3, z_4, z_5) = (z_3^5 + z_4^5 + z_5^5)^2 - 4(z_3^5 z_4^5 + z_3^5 z_5^5 + z_4^5 z_5^5)$. This is a particular case of the curves studied in the Lemma in [28], p.311, where it is shown that $(z_3^n + z_4^n + z_5^n)^2 - 4(z_3^n z_4^n +$

$z_3^n z_5^n + z_4^n z_5^n) = 0$ has $3n$ A_{n-1} singularities. The degree 10 projective curve $g = 0$ has then 15 A_4 singularities and the Euler number of the small resolutions of the threefold is $-288 + 15 \cdot 20 = 12$.

We can get Calabi–Yau threefolds with this method using $\hat{J}_{10,\tau}(x, y)$. For $\tau \in \Lambda_\Sigma \cup \Lambda_3$, the small resolutions of the threefold with g obtained homogenising $\hat{J}_{10,\tau}(x, y)$, which has 45 nodes, have Euler number $-288 + 5 \cdot 45 = -63$ and $c_1 = 0$.

The coefficients of $\hat{J}_{d,\tau}(x, y)$ are not rational for generic τ. However, we can get polynomials with integer coefficients for certain values of τ. For instance, if $\tau = \frac{(6m+1)\pi}{12}$, $m \in \mathbf{Z}$ we have $\sqrt{2}\cos\frac{(2m+3)\pi}{4} = (-1)^{\lfloor\frac{m+2}{2}\rfloor}$, $\sqrt{2}\sin\frac{(2m+3)\pi}{4} = (-1)^{\lfloor\frac{m+1}{2}\rfloor}$ and $\sqrt{2}\cos\frac{(6m+1)\pi}{4} = (-1)^{\lfloor\frac{m}{2}\rfloor}$. In this case $\frac{1}{\sqrt{2}}\hat{J}_{d,\tau}(x, y)$ is a polynomial with integer coefficients.

Another method for constructing Calabi–Yau threefolds in [21] is based on double coverings of $\mathbf{P}^3(\mathbf{C})$ branched along an octic surface [4], which is allowed to have singularities. The octic curve we use is $-\frac{1}{\sqrt{2}}\hat{J}_{8,\frac{\pi}{12}}(x, y)$. We add a normalised Chebyshev polynomial and we get a surface with 133 nodes. We can check that the Tjurina number is $T = 133$ with the following Singular code (short input is used, e.g. $3x^2 - x^3$ is denoted by $3x2 - x3$):

```
LIB "sing.lib";
ring R= 0, (x,y,z), dp ; //affine ring, 0 is the characteristic of the ground field
poly f= -1-8x+12x2+24x3-30x4-8x5+20x6-8x7+x8+8y+24xy-24x2y
-72x3y+56x4y+40x5y-40x6y+8x7y-12y2+24xy2-12x2y2-112x3y2+20x4y2
+72x5y2-28x6y2-24y3-56xy3-48x2y3+80x3y3+40x4y3-56x5y3+34y4
+24xy4-20x2y4+40x3y4+70x4y4+24y5+40xy5+72x2y5+56x3y5-20y6
-40xy6-28x2y6-8y7-8xy7+y8-(2-32z2+160z4-256z6+128z8);
ideal sl= jacob(f),f; //the singular locus
vdim(std(sl)); //Total Tjurina number
```

A double covering of $\mathbf{P}^3(\mathbf{C})$ branched along this octic surface (also called octic double solid) has Euler number $-296 + 2 \cdot 133 = -30$.

Quintic threefolds in $\mathbf{P}^4(\mathbf{C})$ with $c_1 = 0$ were also studied in [21]. If $p(x, y) = 0$ is the equation of the curve of degree 5 in the real (x, y)-plane given by a simple configuration of five lines having a regular pentagon in the centre, the threefold in $\mathbf{P}^4(\mathbf{C})$ obtained by homogenising the affine equation $p(u, v) - p(z, w) = 0$ has 126 nodes. The small resolutions of all the nodes of the Hirzebruch quintic threefold have Euler number $-200 + 2 \cdot 126 = 52$.

Now, we consider

$$f(x, y) := \frac{1}{\sqrt{2}}\hat{J}_{5,\frac{\pi}{12}}(x, y) = 1 + 5x - 5x^2 - 5x^3 + 5x^4 - x^5 - 5y - 10xy + 5x^2y$$

$$+10x^3y - 5x^4y + 5y^2 - 5xy^2 + 10x^3y^2 + 5y^3 + 10xy^3 + 10x^2y^3 - 5y^4 - 5xy^4 - y^5$$

Let M be the threefold in $\mathbf{P}^4(\mathbf{C})$ defined by the homogenization of the equation

$$f(u, v) - f(z, w) = 0$$

The conifold M has 112 nodal singularities [15]. The small resolution $\widetilde{M}$ of all the nodes of M has Euler characteristic $\chi = 24$.

The method for computing the Hodge numbers $h^{i,j}$ in [26, 31, 32] is based on a theorem of Weil–Deligne on the eigenvalues of Frobenius and the Lefschetz fixed-point formula. One way to find $h^{1,1}$ is to count the points of a small resolution of the threefold over an appropriate finite field $\mathbf{F}_p$.

In order to count points in finite fields we used Processing, an open-source programming language based on Java. The program for this purpose uses only loops and elementary integer arithmetic. The analysis of $\#M(\mathbf{F}_p)$, the number of points of M over $\mathbf{F}_p$, do not give a precise answer to the problem of determining $h^{1,1}$ for low values of the primes of good reduction p. We found a unique answer for $p = 601$ where $\#M(\mathbf{F}_{601}) = 223, 916, 358$. The result was also checked with the C++ programming language. Having in mind the number of nodes and the rulings of their tangent cones which are rational over $\mathbf{F}_{601}$, and also the number of independent quintics which do not vanish in the whole set of nodes, we obtained the Betti number $h^2(\widetilde{M}) = 19$. The Hodge numbers of $\widetilde{M}$ are, therefore, $h^{1,1} = 19$, $h^{2,1} = 7$.

The small resolutions are no longer projective in general. According to [32] there are projective resolutions if all local divisors of all singularities can be extended to global smooth divisors. An open question concerns the existence of projective resolutions for the examples presented in this section. Even if there are no projective resolutions, the threefolds might be of interest from the point of view of their applications in physics. Non-Kähler Calabi–Yau threefolds [30] have been considered in the past years in the context of string theory.

Another point of interest, which is related to the Langlands program, is the study of the modularity of the Calabi–Yau threefolds defined over the rationals. The modularity of non-rigid Calabi–Yau threefolds has been proved only for special cases [5, 23, 34].

Acknowledgements This work was partially supported by Consejería de Educación del Principado de Asturias, Spain (UO-15-INVES-38).

References

1. Baake, M., Kramer, P., Schlottmann, M., Zeidler, D.: Planar patterns with fivefold symmetry as sections of periodic structures in 4-space. Int. J. Mod. Phys. B **4**, 2217–2268 (1990)
2. Breske, S., Labs, O., van Straten, D.: Real Line Arrangements and Surfaces with Many Real Nodes. Geometric Modeling and Algebraic Geometry. Springer, New York (2008)
3. Chmutov, S.V.: Examples of projective surfaces with many singularities. J. Algebr. Geom. **1**, 191–196 (1992)
4. Clemens, C.H.: Double solids. Adv. Math. **47**, 107–230 (1983)
5. Dieulefait, L., Pacetti, A., Schütt, M.: Modularity of the Consani-Scholten quintic. With an appendix by José Burgos Gil and Ariel Pacetti. Doc. Math. J. DMV **17**, 953–987 (2012)
6. Escudero, J.G.: Quasicrystal tilings with nine-fold and fifteen-fold symmetry and their Bragg Spectra. J. Phys. Soc. Jpn. **67**, 71–77 (1998)
7. Escudero, J.G.: Composition of inflation rules for aperiodic structures with nine-fold symmetry. Int. J. Mod. Phys. B **13**, 363–373 (1999)

8. Escudero, J.G.: ET0L-systems for composite dodecagonal quasicrystal patterns. Int. J. Mod. Phys. B **15**, 1165–1175 (2001)
9. Escudero, J.G.: Configurational entropy for stone-inflation hexagonal and octagonal patterns. Int. J. Mod. Phys. B. **18**, 1595–1602 (2004)
10. Escudero, J.G.: Random tilings of spherical 3-manifolds. J. Geom. Phys. **58**, 1451–1464 (2008)
11. Escudero, J.G.: Randomness and topological invariants in pentagonal tiling spaces. Discrete Dyn. Nat. Soc. Article ID 946913 (2011)
12. Escudero, J.G.: Substitutions with vanishing rotationally invariant first cohomology. Discrete Dyn. Nat. Soc. Article ID 818549 (2012)
13. Escudero, J.G.: Arrangements of real lines and surfaces with A and D singularities. Exp. Math. **23**, 482–491 (2014)
14. Escudero, J.G.: A construction of algebraic surfaces with many real nodes. Ann. Mat. Pura Appl. **195**, 571–583 (2016)
15. Escudero, J.G.: Threefolds from solutions of a partial differential equation. Exp. Math. **26**(2), 189–196 (2017)
16. Escudero, J.G., García, J.G.: Non-periodic tessellations with unit and non-unit Pisot inflation factors. J. Phys. A **38**, 6525–6543 (2005)
17. Füredi, Z., Palásti, I.: Arrangements of lines with a large number of triangles. Proc. Am. Math. Soc. **92**, 561–566 (1984)
18. Gähler, F., Kwan, E.E., Maloney, G.R.: A computer search for planar substitution tilings with n-fold rotational symmetry. Discrete Comput. Geom. **53**(2), 445–465 (2015)
19. Greuel, G.M., Pfister, G.: A Singular Introduction to Commutative Algebra. Springer, New York (2008)
20. Hirzebruch, F.: Some examples of algebraic surfaces. In: Proceedings 21st Summer Research Institute Australian Mathematical Society. Contemporary Mathematics, vol. 9, pp. 55–71. American Mathematical Society, Providence (1982)
21. Hirzebruch, F.: Some Examples of Threefolds with Trivial Canonical Bundle. Gesammelte Abhandlungen, Bd. II. Springer, New York (1987)
22. Hoffman, M.E., Withers, D.: Generalized Chebyshev polynomials associated with affine Weyl groups. Trans. Am. Math. Soc. **282**, 555–575 (1988)
23. Hulek, K., Verril, H.: On modularity of rigid and non-rigid Calabi-Yau varieties associated to the root lattice A_4. Nagoya Math. J. **179**(2), 103–146 (2005)
24. Koornwinder, T.H.: Orthogonal polynomials in two variables which are eigenfunctions of two algebraically independent partial differential operators. III. Neder. Akad. Wet. Proc. Ser. A **77**, 357–369 (1974)
25. Lidl, R.: Tschebyscheffpolynome in mehreren Variablen. J. Reine Angew. Math. **273**, 178–198 (1975)
26. Meyer, C.: Modular Calabi-Yau Threefolds, Fields Institute Monographs. American Mathematical Society, Providence (2005)
27. Nischke, K.P., Danzer, L.: A construction of inflation rules based on n-fold symmetry. Discret. Comput. Geom. **15**, 221–236 (1996)
28. Persson, U.: Horikawa surfaces with maximal Picard numbers. Math. Ann. **259**, 287–312 (1982)
29. Strominger, A., Witten, E.: New manifolds for superstring compactification. Commun. Math. Phys. **101**, 341–361 (1985)
30. Tseng, L.S., Yau, S.T.: Non Kähler Calabi-Yau manifolds. String-Math 2011. In: Proceedings of Symposia in Pure Mathematics, vol. 85, pp. 241–253. American Mathematical Society, Providence (2012)
31. Van Geemen, B., Werner, J.: Nodal quintics in P^4. Arithmetics of Complex Manifolds. Lect. Notes Math. **1399**, 48–59 (1989)
32. Werner, J., Van Geemen, B.: New examples of three-folds with $c_1 = 0$. Math. Z. **203**, 211–225 (1990)
33. Wolfram, S.: Mathematica. Addison-Wesley Publishing Co., Boston (1991)
34. Yui, N.: Modularity of Calabi-Yau varieties: 2011 and beyond. Arithmetic and geometry of $K3$ surfaces and Calabi-Yau threefolds. Fields Institute Communications, vol. 67, pp. 101–139. Springer, New York (2013)

Finding Eigenvalues of Self-maps with the Kronecker Canonical Form

Marc Ethier, Grzegorz Jabłoński and Marian Mrozek

Abstract Recent research has examined how to study the topological features of a continuous self-map by means of the persistence of the eigenspaces, for given eigenvalues, of the endomorphism induced in homology over a field. This raised the question of how to select dynamically significant eigenvalues. The present paper aims to answer this question, giving an algorithm that computes the persistence of eigenspaces for every eigenvalue simultaneously, also expressing said eigenspaces as direct sums of "finite" and "singular" subspaces.

Keywords Computational topology · Persistent homology · Self-maps · Matrix pencils · Kronecker canonical form

1 Introduction

The theory of persistent homology [2, 6] has proved in the past two decades to be a very useful tool in several branches of applied mathematics and computer science. In [1], a novel application of persistence to the computational analysis of dynami-

---

This research is supported by the Toposys project FP7-ICT-318493-STREP, by the Polish MNiSzW grant Nr 2621/7.PR/12/2013/2, and by the National Science Centre (Poland) grant DEC-2013/09/N/ST6/02995

M. Ethier (✉) · G. Jabłoński · M. Mrozek
Division of Computational Mathematics, Faculty of Mathematics and Computer Science, Jagiellonian University, Kraków, Poland
e-mail: methier@ustboniface.ca

M. Mrozek
e-mail: marian.mrozek@uj.edu.pl

M. Ethier
University of Saint-Boniface, Winnipeg, MB, Canada

G. Jabłoński
Institute of Science and Technology Austria, Klosterneuburg, Austria
e-mail: grzegorz.jablonski@ist.ac.at

I.S. Kotsireas and E. Martínez-Moro (eds.), *Applications of Computer Algebra*, Springer Proceedings in Mathematics & Statistics 198,
DOI 10.1007/978-3-319-56932-1_8

cal systems is introduced. Building upon the concept of *towers* in a given category (a tower in the category of modules or vector spaces is equivalent to a *persistence module* as defined in [6]), the authors define the tower of eigenspaces for an endomorphism of a tower of (finite dimensional) vector spaces. When these vector spaces are obtained as the homology over a field $\mathbb{F}$ of a filtration representing the underlying topological space, and the endomorphism is the map induced in homology by a self-map of said topological space, the eigenvectors are homology classes invariant under the self-map and provide a first step towards understanding the persistence of this map.

When the self-map is expanding, there is no guarantee that the image of a homology class by the endomorphism is in the filtration at the same step or even at any step. To overcome this difficulty, the authors of [1] adapted persistent homology to the study of a self-map by using two towers of vector spaces, which are equivalent to persistence modules indexed over integer numbers: (Y_i, η_i), a tower of homology spaces obtained from a filtration of the underlying topological space, and (X_i, ξ_i), a tower of homology spaces obtained by restricting domains such that maps induced by the self-map are simplicial. The morphisms $\varphi_i: X_i \to Y_i$, $\psi_i: X_i \to Y_i$ are obtained, respectively, from the self-map and from the inclusion map. In [1], the *eigenspace for pairs* $E_t(\varphi, \psi)$ was constructed by defining, for every $t \in \mathbb{F}$,

$$\overline{E}_t(\varphi, \psi) = \ker(\varphi - t\psi)$$

and then quotienting out the common kernel of φ and ψ, that is,

$$E_t(\varphi, \psi) = \overline{E}_t(\varphi, \psi)/(\ker \varphi \cap \ker \psi). \tag{1}$$

Nevertheless, despite quotienting out the common kernel of φ and ψ, it may happen that $E_t(\varphi, \psi)$ is non-trivial for every $t \in \mathbb{F}$, a phenomenon that was termed "abundance of eigenvalues" in [1]. This difficulty in finding the eigenvalues for the pair (φ, ψ), and in identifying them as dynamically significant, leads to the question whether there exists a way to compute the eigenspace towers for a pair of morphisms, for all eigenvalues simultaneously. The present article aims to answer this question in the affirmative, providing an algorithm to extract eigenvectors for every eigenvalue all at once. In addition, using the theory of the Kronecker canonical form for matrix pencils (a generalization of the Jordan form to polynomial matrices of the form $tB - A$), the eigenspace for every eigenvalue can be expressed as the direct sum of a "finite" and a "singular" part, the latter of which being associated with the abundance of eigenvalues phenomenon. We believe that the dynamically significant eigenvectors are contained in the former, finite part.

In Sect. 2, we reintroduce the concept of the Kronecker canonical form along with invariant polynomials of polynomial matrices, which while belonging to classical theories in linear algebra, appear not to be part of the common mathematical knowledge. Section 3 is dedicated to the algorithm to extract eigenvectors, as well as generalized eigenvectors, for all eigenvalues simultaneously. Section 4 shows numerical examples.

2 Kronecker Canonical Form

By the term *linear matrix pencil*, or simply *matrix pencil*, we refer to the polynomial matrix $tB - A$, where $A, B \in M^{m\times n}(\mathbb{F})$ and $\mathbb{F}$ is a fixed field. Fix a value $\widehat{t} \in \mathbb{F}$; if the equation

$$(\widehat{t}B - A)x = 0$$

possesses a nonzero solution $x \in \mathbb{F}^n$, then x is said to be an *eigenvector* for the *eigenvalue* $\widehat{t}$. In addition, if there is a finite sequence $x_1, x_2, \ldots, x_k \in \mathbb{F}^n$ of nonzero vectors such that the system

$$\begin{aligned}(\widehat{t}B - A)x_1 &= 0,\\ (\widehat{t}B - A)x_2 &= Bx_1,\\ &\vdots\\ (\widehat{t}B - A)x_k &= Bx_{k-1}\end{aligned}$$

has a solution, then this sequence is called a *sequence of generalized eigenvectors* for the eigenvalue $\widehat{t}$. Let $tB_1 - A_1$ and $tB_2 - A_2$ be two $m \times n$ pencils; if there exist invertible matrices $Q \in M^{m\times m}(\mathbb{F})$, $R \in M^{n\times n}(\mathbb{F})$ such that $Q^{-1}(tB_1 - A_1)R = tB_2 - A_2$, then the pencils are said to be *similar*.

In order to study the eigenstructure of the pencil $tB - A$, that is find its eigenvalues and the dimension of its eigenspaces and generalized eigenspaces, and hence to extract (generalized) eigenvectors, we recall the classical concepts of invariant polynomials and of Kronecker indices of matrix pencils.

We first start by considering a particular type of pencil. Call the *rank* of a pencil, rank $(tB - A)$, the largest integer k such that there exist non-vanishing $k \times k$ minors of $tB - A$. If a pencil $tB - A$ is square ($B, A \in M^{n\times n}(\mathbb{F})$) and has rank n, it is said to be *regular*. If it is non-square, or if it is $n \times n$ square but its rank is strictly lower than n, it is said to be *singular*. We will additionally say that a pencil has *full row rank* (respectively *full column rank*) if its rank equals its number of rows (respectively its number of columns).

Let us recall the well-known rational canonical form and primary rational canonical form of a square matrix.

Definition 1 For $p(t) = c_0 + c_1t + c_2t^2 + \ldots + c_{k-1}t^{k-1} + t^k$ a monic polynomial, the $k \times k$ matrix

$$C(p) = \begin{bmatrix} 0 & 0 & \cdots & 0 & -c_0 \\ 1 & 0 & \cdots & 0 & -c_1 \\ 0 & 1 & \cdots & 0 & -c_2 \\ \vdots & \vdots & \ddots & \vdots & \vdots \\ 0 & 0 & \cdots & 1 & -c_{k-1} \end{bmatrix}$$

is called the *companion matrix* of d.

Theorem 2 [4, Theorem 11.17] *Let T be a square matrix, then T is similar to a unique square matrix*

$$\operatorname{diag}\{C(d_1), C(d_2), \ldots, C(d_s)\} \tag{2}$$

where $C(d_i)$ is the companion matrix of a non-constant monic polynomial d_i and $d_1|d_2|\ldots|d_s$.

Theorem 3 [4, Theorem 11.20] *Let T be a square matrix, then T is similar to a square matrix*

$$\operatorname{diag}\{C(p_1), C(p_2), \ldots, C(p_r)\} \tag{3}$$

where each $p_i = q_i^{s_i}$ is a power of a monic prime polynomial q_i, and $C(p_i)$ its companion matrix. This matrix is uniquely determined up to the order of the blocks $C(p_i)$ on the diagonal.

We refer to the form (2) as the *rational canonical form* of T, and to the form (3) as the *primary rational canonical form* of T.

Proposition 4 *Every regular pencil $tB - A$ over $\mathbb{F}$ is similar to a pencil in the form*

$$\operatorname{diag}\{tN - I_{r_1}, tI_{r_2} - C\} \tag{4}$$

where N is the direct sum of nilpotent companion matrices, C is a square matrix in rational canonical form, and I_{r_1} and I_{r_2} are identity matrices of the given size.

Proof This proof proceeds similarly to the proof of [3, Chapter 12, Theorem 3]. If $tB - A$ is regular, then there exists $\widehat{t} \in \mathbb{F}$ such that $\widehat{t}B - A$ has full rank. Call $\widetilde{A}$ the matrix $-(\widehat{t}B - A)$, then

$$\begin{aligned} tB - A &= (t - \widehat{t})\, B - \widetilde{A} \\ \Rightarrow \widetilde{A}^{-1}\,(tB - A) &= (t - \widehat{t})\, \widetilde{A}^{-1} B - I. \end{aligned}$$

We can write the primary rational canonical form of $\widetilde{A}^{-1}B$ by ordering the blocks such that the block corresponding to t^{r_1} for $r_1 > 0$, if it exists, is in the top left. The pencil $tB - A$ is thus similar (in the sense for pencils given above) to

$$(t - \widehat{t})\, \operatorname{diag}\{C_0, C_1\} - I = \operatorname{diag}\left\{t\, C_0 - (I_{r_1} + \widehat{t}\, C_0), (t\, C_1 - (I_{r_2} + \widehat{t}\, C_1)\right\}$$

where C_0 is the companion matrix of $p(t) = t^{r_1}$. Since $I_{r_1} + \widehat{t}\, C_0$ is invertible, and so is C_1, we can left-multiply the above pencil by $\operatorname{diag}\left\{(I_{r_1} + \widehat{t}\, C_0)^{-1}, C_1^{-1}\right\}$, yielding

$$\operatorname{diag}\left\{t\,(I_{r_1} + \widehat{t}\, C_0)^{-1}\, C_0 - I_{r_1}, t\, I_{r_2} - C_1^{-1}\,(I_{r_2} + \widehat{t}\, C_1)\right\}.$$

The result is obtained by putting the matrices $(I_{r_1} + \widehat{t}\, C_0)^{-1}\, C_0$ and $C_1^{-1}\,(I_{r_2} + \widehat{t}\, C_1)$ into their rational canonical forms, respectively N and C. ■

Matrix N in (4) is a block diagonal matrix, whose blocks are nilpotent companion matrices $N_i, i = 1, \dots, l$. Each such matrix is the companion of the polynomial t^{k_i}, $k_i \geq 1$, and so N has only 0 as eigenvalue. We say that $tB - A$ possesses l *infinite elementary divisors*, whose orders are $k_1, k_2, \dots, k_l$.

We also encountered in (4) a matrix C in rational canonical form, that is

$$C = \operatorname{diag}\{C(d_1), C(d_2), \dots, C(d_s)\}$$

with $d_1|d_2|\dots|d_s$. These polynomials are referred to as the *invariant polynomials* of the pencil $tB - A$. We will refer to the eigenstructure of C as the *finite* eigenstructure of the pencil. From Proposition 4 and the fact that $tN - I_{r_1}$ has no eigenvalue, we see that t is an eigenvalue of a regular pencil if and only if it is a root of one of its invariant polynomials, with the dimension of its eigenspace being the number of such invariant polynomials. In [3, Chapter 6], the classical algorithm to put a polynomial matrix into Smith normal form is shown to yield a diagonal matrix in canonical form, whose first diagonal elements are ones followed by the invariant polynomials of the matrix, with zero rows at the bottom and zero columns at the right. A regular pencil is of full rank and can, therefore, not have zero rows or columns, so the classical Smith normal form algorithm provides invertible matrices $Q(t)$, $R(t)$ such that

$$Q(t)^{-1}\,(tB - A)\,R(t) = \operatorname{diag}\{1, \dots, 1, d_1, \dots, d_s\}.$$

We easily see that if $R(t) = [y_1(t)\,y_2(t)\dots y_{n-s}(t)\,x_1(t)\,x_2(t)\dots x_s(t)]$, and if $\widehat{t}$ is a root of polynomial d_i, then

$$(\widehat{t}B - A)\,x_i(\widehat{t}) = 0.$$

Since $R(t)$ is invertible, its columns are linearly independent for every value t. Therefore, if $\widehat{t}$ is a root of more than one invariant polynomial, we can find the same number of linearly independent eigenvectors.

Now, consider a general $m \times n$ pencil $tB - A$. We can study solutions of

$$\forall t \in \mathbb{F} \quad (tB - A)\,x(t) = 0, \tag{5}$$

where $x : \mathbb{F} \to \mathbb{F}^n$ is the variable. If there exists a linear dependence over $\mathbb{F}[t]$ between the columns of $tB - A$, then there exists a polynomial solution of Eq. (5) which we call a *polynomial eigenvector* for the pencil. Write such a solution as

$$x(t) = x_0 + t\,x_1 + t^2\,x_2 + \dots + t^{\varepsilon}\,x_{\varepsilon},\ \varepsilon \geq 0 \tag{6}$$

with x_i, $i = 0, \dots, \varepsilon$ vectors in $\mathbb{F}^n$, and $x_\varepsilon \neq 0$, where ε is the degree of the polynomial eigenvector. Without loss of generality, we can assume that $x_0 \neq 0$. Indeed, suppose that $x_0 = x_1 = \dots = x_{k-1} = 0$ and $x_k \neq 0$ for $k \leq \varepsilon$ in Eq. (6). Then we can factor out t^k, leaving

$$t^k\,(tB - A)\,(x_k + t\,x_{k+1} + t^2\,x_{k+1} + \dots + t^{\varepsilon-k}\,x_\varepsilon) = 0,$$

that is, $x_k + t\, x_{k+1} + t^2\, x_{k+1} + \cdots + t^{\varepsilon - k}\, x_\varepsilon$ is a new polynomial eigenvector with nonzero constant term. Therefore, if $x(t)$ is a polynomial eigenvector of $tB - A$ with nonzero constant term, then for every $\widehat{t} \in \mathbb{F}$, $x(\widehat{t})$ is an eigenvector of $tB - A$ for eigenvalue $\widehat{t}$.

Theorem 5 [3, Chapter 12, Theorem 4] *Suppose that ε is the smallest positive integer such that the pencil $tB - A$ possesses a polynomial solution (6) of degree $\varepsilon > 0$. Then the pencil is similar to*

$$\begin{bmatrix} L_\varepsilon & 0 \\ 0 & t\widehat{B} - \widehat{A} \end{bmatrix}$$

where

$$L_\varepsilon = \begin{bmatrix} t & -1 & & \\ & \ddots & \ddots & \\ & & t & -1 \end{bmatrix} \tag{7}$$

is a bidiagonal pencil of dimension $\varepsilon \times (\varepsilon + 1)$, known as a column Kronecker block *of* index ε, *and $t\widehat{B} - \widehat{A}$ has no polynomial eigenvector analogous to (6) of degree less than ε.*

Theorem 5 is also valid in the case where $\varepsilon = 0$, in which case a "0×1" block L_0 means a column of zeros to the left of $t\widehat{B} - \widehat{A}$.

Proposition 6 *A vector $x_0 \in \ker A \cap \ker B$ if and only if $x(t) = x_0$ is a polynomial eigenvector of $tB - A$ of degree* 0.

Proof If $x_0 \in \ker A \cap \ker B$, then obviously $(tB - A)\, x_0 = 0$. Now suppose that $(tB - A)\, x_0 = 0$, then for every $t \in \mathbb{F}$, $A\, x_0 = t\, B\, x_0$. Since $A\, x_0$ and $B\, x_0$ are elements of $\mathbb{F}$, then this can only be true if $A\, x_0 = B\, x_0 = 0$. ■

The last theorem in this section concerns a decomposition of the pencil $tB - A$:

Theorem 7 *Any $m \times n$ pencil $tB - A$ over $\mathbb{F}$ is similar to the pencil*

$$\operatorname{diag}\{L_{\varepsilon_1}, \ldots, L_{\varepsilon_p}, L^T_{\eta_1}, \ldots, L^T_{\eta_q}, t\overline{B} - \overline{A}\}$$

where $t\overline{B} - \overline{A}$ is a regular and therefore square pencil.

Proof Repeatedly Applying Theorem 5, we may successively extract from $tB - A$ Kronecker blocks of nonincreasing index until we end up with the following decomposition: $tB - A$ is similar to

$$\operatorname{diag}\{L_{\varepsilon_1}, L_{\varepsilon_2}, \ldots, L_{\varepsilon_p}, t\widehat{B} - \widehat{A}\}$$

where the columns of $t\widehat{B} - \widehat{A}$ are linearly independent and the blocks L_{ε_i} may be ordered in a way that $0 \le \varepsilon_1 \le \varepsilon_2 \le \cdots \le \varepsilon_p$. At this point, $t\widehat{B} - \widehat{A}$ may still have a linearly dependent set of rows, in which case it would possess *left polynomial eigenvectors* $y(t)$ such that $y(t)\,(t\widehat{B} - \widehat{A}) = 0$. This is obviously equivalent to $(t\widehat{B}^T - \widehat{A}^T)\,y^T(t) = 0$, and therefore Theorem 5 can now be applied to this transposed subpencil, yielding *row Kronecker blocks* $L^T_{\eta_j}$, $j = 1, \ldots, q$.

Since we already know the decomposition $t\overline{B} - \overline{A}$ of (4), this completes the presentation of the *Kronecker canonical form* of a pencil. We will more precisely call this form the *rational Kronecker canonical form* since it includes a matrix in rational canonical form; the classical Kronecker canonical form is a generalization of the Jordan form and therefore is only guaranteed to exist when working with an algebraically closed field.

Let us now show an example.

Example 8 Consider the following pencil over $\mathbb{Q}$:

$$tB - A = \begin{bmatrix} -t & 0 & 0 & 1 & 0 & 0 & 0 & 0 \\ t-1 & 0 & t-1 & t-1 & 0 & -t & 1 & 0 \\ -1 & 0 & 0 & t & 0 & 0 & 0 & 0 \\ 0 & -t-1 & 1 & 0 & 0 & 0 & 0 & 0 \\ 0 & 0 & 1 & t+1 & t+1 & 0 & 0 & 0 \\ 0 & 0 & -1 & -t-1 & -t-1 & 0 & 0 & 0 \\ 0 & 0 & 0 & 0 & 0 & 0 & 0 & -1 \end{bmatrix}.$$

We can show that this pencil has column Kronecker indices $\varepsilon_1 = \varepsilon_2 = 1$, row Kronecker index $\eta_1 = 0$, one infinite elementary divisor of order 1 and invariant polynomials $t + 1$ and $t^2 - 1$. Therefore, the rational Kronecker canonical form of $tB - A$ is

$$\begin{bmatrix} t & -1 & & & & & & \\ & & t & -1 & & & & \\ & & & & & & & \\ & & & & -1 & & & \\ & & & & & t+1 & & \\ & & & & & & t & -1 \\ & & & & & & -1 & t \end{bmatrix}.$$

We leave to the reader to verify that the following transition matrices put $tB - A$ into this canonical form:

$$Q^{-1} = \begin{bmatrix} 0 & 0 & 0 & -1 & 0 & 0 & 0 \\ 0 & -1 & 0 & 0 & 0 & 0 & 0 \\ 0 & 0 & 0 & 0 & 1 & 1 & 0 \\ 0 & 0 & 0 & 0 & 0 & 0 & 1 \\ 0 & 0 & 0 & -1 & 1 & 0 & 0 \\ 0 & 0 & 1 & 0 & 0 & 0 & 0 \\ -1 & 0 & 0 & 0 & 0 & 0 & 0 \end{bmatrix},$$

$$R = \begin{bmatrix} 0 & 0 & 0 & 0 & 0 & 0 & 0 & 1 \\ 1 & 0 & 0 & 0 & 0 & 0 & 0 & 0 \\ 1 & 1 & 0 & 0 & 0 & 0 & 0 & 0 \\ 0 & 0 & 0 & 0 & 0 & 0 & 1 & 0 \\ -1 & 0 & 0 & 0 & 0 & 1 & -1 & 0 \\ 1 & 1 & 1 & 0 & 0 & 0 & 1 & 1 \\ 1 & 1 & 0 & 1 & 0 & 0 & 1 & 1 \\ 0 & 0 & 0 & 0 & 1 & 0 & 0 & 0 \end{bmatrix}.$$

3 Algorithm

Van Dooren [5] introduced an algorithm to transform a pencil $tB - A$ into a form from which its column Kronecker indices can be computed, and the associated polynomial eigenvectors are easily extracted. This is done by successively column- and row-reducing subpencils of $tB - A$. In the following algorithm, indices denote step number except for zero and identity matrices, where they denote dimension.

Algorithm 9

Input: $tB - A$
$j := 1;\ m_1 := m;\ n_1 := n;$
$A_{1,1} := A;\ B_{1,1} := B;\ Q^{-1} := I_m;\ R := I_n;$
while (true)
  if $B_{j,j}$ has n_j linearly independent columns
    $l := j - 1;$
    return $Q^{-1}, R;$
  $\left[B_{j+1} \middle| 0_{m_j \times s_j} \right] := B_{j,j}\ R_j;$
  Let R_j be obtained through column reduction algorithm on $B_{j,j}$
  $\left[A_{j+1} \middle| A_j \right] := A_{j,j}\ R_j;$
  for $i = 1$ **to** $j - 1$ **do**
    (* Update other blocks in column j *)
    $\left[B_{j+1,i} \middle| B_{j,i} \right] := B_{j,i}\ R_j;$
    $\left[A_{j+1,i} \middle| A_{j,i} \right] := A_{j,i}\ R_j;$
  (* Update transition matrix R *)
  $$R := R \left[\begin{array}{c|c} R_j & 0_{n_j \times (n-n_j)} \\ \hline 0_{(n-n_j) \times n_j} & I_{n-n_j} \end{array} \right];$$
  $$\left[\begin{array}{c} 0_{(m_j - r_j) \times s_j} \\ \hline A_{j,j} \end{array} \right] := Q_j^{-1}\ A_j;$$
  Let Q_j^{-1} be obtained through row reduction algorithm on A_j
    and permutation so zero rows are on top
  $$\left[\begin{array}{c} A_{j+1,j+1} \\ \hline A_{j+1,j} \end{array} \right] := Q_j^{-1}\ A_{j+1};\ \left[\begin{array}{c} B_{j+1,j+1} \\ \hline B_{j+1,j} \end{array} \right] := Q_j^{-1}\ B_{j+1};$$
  (* Update transition matrix Q^{-1} *)

$$Q^{-1} := \left[\begin{array}{c|c} Q_j^{-1} & 0_{m_j\times(m-m_j)} \\ \hline 0_{(m-m_j)\times m_j} & I_{m-m_j} \end{array}\right] Q^{-1};$$
$m_{j+1} := m_j - r_j;\ n_{j+1} := n_j - s_j;$
$j := j+1;$

Theorem 10 *Algorithm 9 stops when $B_{l+1,l+1}$ has full column rank. At this point, the output are the matrices Q^{-1} and R such that $Q^{-1}(tB - A)R$ is the following block lower triangular matrix:*

$$\begin{array}{c} \left[\begin{array}{c|c|c|c|c} tB_{l+1,l+1} - A_{l+1,l+1} & 0 & \cdots & 0 & 0 \\ \hline tB_{l+1,l} - A_{l+1,l} & -A_{l,l} & \cdots & 0 & 0 \\ \hline \vdots & \ddots & \ddots & \vdots & \vdots \\ \hline tB_{l+1,2} - A_{l+1,2} & tB_{l,2} - A_{l,2} & \cdots & -A_{2,2} & 0 \\ \hline tB_{l+1,1} - A_{l+1,1} & tB_{l,1} - A_{l,1} & \cdots & tB_{2,1} - A_{2,1} & -A_{1,1} \end{array}\right] \begin{array}{l} m_{l+1} \\ r_l \\ \vdots \\ r_2 \\ r_1 \end{array} \\ \begin{array}{ccccc} n_{l+1} & \quad s_l & \quad \cdots & \quad s_2 & \quad s_1 \end{array} \end{array} \tag{8}$$

where the $A_{j,j}$'s have full row rank r_j for $j = 1, \ldots, l$, and the $B_{j,j-1}$'s have full column rank s_j for $j = 2, \ldots, l$. Some of the r_j's can equal 0.

Proof Form (8) is a direct consequence of the algorithm. Indeed, the initial form of the pencil is

$$tB_{1,1} - A_{1,1},$$

and at step j, the left block of columns,

$$\left[\begin{array}{c} tB_{j,j} - A_{j,j} \\ \hline tB_{j,j-1} - A_{j,j-1} \\ \hline \vdots \\ \hline tB_{j,1} - A_{j,1} \end{array}\right]$$

is the only part of the pencil to change, being transformed by multiplying on the right by R_j and on the left by

$$\left[\begin{array}{cc} Q_j^{-1} & \\ & I_{m-m_j} \end{array}\right]$$

into

$$\left[\begin{array}{c|c} tB_{j+1,j+1} - A_{j+1,j+1} & 0_{(m_j-r_j)\times s_j} \\ \hline tB_{j+1,j} - A_{j+1,j} & -A_{j,j} \\ \hline tB_{j+1,j-1} - A_{j+1,j-1} & tB_{j,j-1} - A_{j,j-1} \\ \hline \vdots & \vdots \\ \hline tB_{j+1,1} - A_{j+1,1} & tB_{j,1} - A_{j,1} \end{array}\right].$$

The $A_{j,j}$ blocks, $j = 1, \ldots, l$, have full row rank r_j, being obtained from the nonzero rows of a row-reduced matrix. In addition, at every step j, the block B_{j+1} created has full column rank, being obtained from the nonzero columns of a column-reduced matrix. Multiplying it on the left by Q_j^{-1} yields

$$\left[\frac{B_{j+1,j+1}}{B_{j+1,j}}\right]. \tag{9}$$

If $B_{j+1,j+1}$ has full column rank, then the algorithm stops and this block becomes the upper left block $B_{l+1,l+1}$. Otherwise, the block (9) is multiplied on the right by R_{j+1}, yielding

$$\left[\begin{array}{c|c} B_{j+2} & 0 \\ \hline B_{j+2,j} & B_{j+1,j} \end{array}\right] = Q_j^{-1}\, B_{j+1}\, R_{j+1}.$$

Since B_{j+1} has full column rank, then $B_{j+1,j}$ also does. This is true for $j = 1, \ldots, l-1$, proving the theorem. ■

From the row and column ranks r_j and s_j, we then compute (putting $s_{l+1} := 0$)

$$\begin{aligned} e_j &:= s_j - r_j \geq 0 \text{ for } j = 1, \ldots, l; \\ d_j &:= r_j - s_{j+1} \geq 0 \text{ for } j = 1, \ldots, l. \end{aligned}$$

As shown in [5, Proposition 4.3], the indices d_j and e_j fully determine the infinite elementary divisors and the column Kronecker indices, respectively. More precisely, they tell us that $tB - A$ has d_j infinite elementary divisors of degree j, $j = 1, \ldots, l$, and e_j column Kronecker blocks L_{j-1} of size $(j-1) \times j$, $j = 1, \ldots, l$. The pencil $tB_{l+1,l+1} - A_{l+1,l+1}$ additionally contains the finite structure of the original pencil.

In [5] a dual algorithm is also described. It extracts the infinite elementary divisors and row Kronecker indices of $tB - A$. Here, let us recall that if B is an identity matrix, that is for the classical eigenproblem for a square matrix A, there exists a natural isomorphism between the left and right eigenspaces, and generalized eigenspaces, of A. Indeed, the left generalized eigenspace of A (equivalently the generalized eigenspace of A^T) for every given eigenvalue is the dual space of its (right) generalized eigenspace. This natural isomorphism breaks down in the case of matrix pencils since column and row Kronecker indices are completely independent of each other, but it is possible to retain it by quotienting out vectors from the column (respectively row) Kronecker structure from the eigenspace (respectively left eigenspace).

When the pencil has been put into form (8), we can further use the fact that $B_{l+1,l+1}$ has full column rank, as do the blocks $B_{i,i-1}$ for $i = 2, \ldots, l$, and that $A_{i,i}$ has full row rank for $i = 1, \ldots, l$, to zero out the majority of subdiagonal blocks in the following way.

Algorithm 11

Input: Q^{-1}, R, $Q^{-1}\,(tB - A)\,R$ from Algorithm 9
for $i = 1$ **to** l

(* Zero out block $B_{l+1,l+1-i}$ *)
Find X such that $B_{l+1,l+1-i} = X\, B_{l+1,l+1}$;
$B_{l+1,l+1-i} := 0$;
$A_{l+1,l+1-i} := A_{l+1,l+1-i} - X\, A_{l+1,l+1}$;

$$Q^{-1} := \begin{bmatrix} I_{m_{l+1}} & & & \\ & \ddots & & \\ -X & & I_{r_{l+1-i}} & \\ & & & \ddots \\ & & & & I_{r_1} \end{bmatrix} Q^{-1};$$

for $j = 1$ **to** $i - 2$
(* Zero out block $B_{l+1-j,l+1-i}$ *)
Find Z such that $B_{l+1-j,l+1-i} = Z\, B_{l+1-j,l-j}$;
$B_{l+1-j,l+1-i} := 0$;
$A_{l-j,l+1-i} := A_{l-j,l+1-i} - Z\, A_{l-j,l-j}$;

$$Q^{-1} := \begin{bmatrix} I_{m_{l+1}} & & & & & \\ & \ddots & & & & \\ & & I_{r_{l-j}} & & & \\ & & & \ddots & & \\ & & -Z & & I_{r_{l+1-i}} & \\ & & & & & \ddots \\ & & & & & & I_{r_1} \end{bmatrix} Q^{-1};$$

for $j = 1$ **to** i
(* Zero out block $A_{l+1-i+j,l+1-i}$ *)
Find Y such that $A_{l+1-i+j,l+1-i} = A_{l+1-i,l+1-i}\, Y$;
$A_{l+1-i+j,l+1-i} := 0$;
for $k = 1$ **to** $l - i$
$A_{l+1-i+j,k} := A_{l+1-i+j,k} - A_{l+1-i,l+1-i-k}\, Y$;
$B_{l+1-i+j,k} := B_{l+1-i+j,k} - B_{l+1-i,l+1-i-k}\, Y$;

$$R := R \begin{bmatrix} I_{n_{l+1}} & & & & & \\ & \ddots & & & & \\ & & I_{s_{l+1-i+j}} & & & \\ & & & \ddots & & \\ & & -Y & & I_{s_{l+1-i}} & \\ & & & & & \ddots \\ & & & & & & I_{s_1} \end{bmatrix};$$

At this point, having reused the names of the blocks, $Q^{-1}\,(tB - A)\,R$ equals

$$\left[\begin{array}{c|c|c|c|c|c} tB_{l+1,l+1} - A_{l+1,l+1} & 0 & 0 & \cdots & 0 & 0 \\ \hline 0 & -A_{l,l} & 0 & \cdots & 0 & 0 \\ \hline 0 & tB_{l,l-1} & -A_{l-1,l-1} & \cdots & 0 & 0 \\ \hline 0 & 0 & tB_{l-1,l-2} & \ddots & \vdots & \vdots \\ \hline \vdots & \vdots & \vdots & \ddots & -A_{2,2} & 0 \\ \hline 0 & 0 & 0 & \cdots & tB_{2,1} & -A_{1,1} \end{array}\right] \begin{array}{l} m_{l+1} \\ r_l \\ r_{l-1} \\ \vdots \\ r_2 \\ r_1 \end{array} \qquad (10)$$
$$\begin{array}{cccccc} n_{l+1} & s_l & s_{l-1} & \cdots & s_2 & s_1 \end{array}$$

Note that the blocks $A_{i,i}$ have $s_i - r_i = e_i$ zero columns, which is also the number of Kronecker blocks of index $i-1$, each of which corresponds to a polynomial eigenvector of degree $i-1$. Therefore, using the blocks $A_{i,i}$ to zero out the blocks $tB_{i+1,i}$, $i = 1$ going up to $l-1$ in this order, will expose zero columns in the pencil.

Algorithm 12

Input: R, $Q^{-1}(tB - A)R$ from Algorithm 11
for $i = 1$ **to** $l-1$
  Find Y such that $B_{i+1,i} = A_{i,i}Y$;
  $B_{i+1,i} := 0$;

$$R := R \begin{bmatrix} I_{n_{l+1}} & & & & \\ & \ddots & & & \\ & & I_{s_{i+1}} & & \\ & & tY & I_{s_i} & \\ & & & \ddots & \\ & & & & I_{s_1} \end{bmatrix};$$

Note that R is now a matrix over $\mathbb{F}[t]$, and for every zero column of $Q^{-1}(tB - A)R$, we find a column $x(t) \in \mathbb{F}[t]^n$ of R which is a polynomial eigenvector of $tB - A$. In addition, Algorithm 12 ensures that the degrees of such columns of R are equal to the column Kronecker indices of $tB - A$.

Furthermore, since the block $tB_{l+1,l+1} - A_{l+1,l+1}$ contains the whole finite structure of the pencil, we can at this point (also updating R) put it into Smith normal form, whose non-constant diagonal elements will be the invariant polynomials of $tB - A$. Here as well, we expose a column in the pencil that is zero except for one entry, an invariant polynomial of $tB - A$. When evaluated at a root t_0 of this polynomial, the corresponding column of R is an eigenvector for eigenvalue t_0.

When working on $\mathbb{Q}$, the rational roots of a polynomial with integer (or rational) coefficients can be obtained with the following well-known theorem:

Theorem 13 (Rational Root Theorem) *Let*

$$a_n x^n + a_{n-1}x^{n-1} + \ldots + a_1 x + a_0 = 0 \qquad (11)$$

be a polynomial equation with integer coefficients, and suppose that $a_n \neq 0$, $a_0 \neq 0$. *Then every rational root* p/q *of (11), where* p, q *are relatively prime, has the property that* $p|a_0$ *and* $q|a_n$.

Applying the previous theorem to the invariant polynomials of $tB - A$ over $\mathbb{Q}$ (multiplying by an integer if necessary) allows one to find every rational eigenvalue.

Note that in case left eigenvectors are required, the dual algorithm of [5] can be used instead of Algorithm 9, followed by a dual "row" version of Algorithms 11 and 12, keeping track of the left transition matrix Q^T.

Example 14 Consider again the pencil of Example 8. Applying Algorithm 9 yields

$$\left[\begin{array}{ccc|cc|ccc} t+1 & 0 & 0 & 0 & 0 & 0 & 0 & 0 \\ -t-1 & 0 & 0 & 0 & 0 & 0 & 0 & 0 \\ t & -t & -1 & 0 & 0 & 0 & 0 & 0 \\ 1 & -1 & -t & 0 & 0 & 0 & 0 & 0 \\ \hline 0 & 0 & t-1 & t-1 & 0 & -1 & 1 & 0 \\ -1 & -t & t-1 & t & t+1 & 0 & 1 & 0 \\ 0 & 0 & 0 & 0 & 0 & 0 & 0 & -1 \end{array}\right] \begin{array}{l} \\ \\ m_3 \\ r_2 \\ \\ r_1 \\ \\ \end{array}, \qquad \begin{array}{ccc} n_3 & s_2 & s_1 \end{array}$$

so we can verify the presence of $s_2 - r_2 = 2$ column Kronecker blocks of index 1, and $r_1 - s_2 = 1$ infinite elementary divisor of order 1. Applying Algorithms 11, 12 and the Smith normal form algorithm, we obtain

$$tB - A \sim \left[\begin{array}{ccc|cc|ccc} 1 & 0 & 0 & & & & & \\ 0 & t+1 & 0 & & & & & \\ 0 & 0 & t^2-1 & & & & & \\ 0 & 0 & 0 & & & & & \\ \hline & & & 0 & 0 & -1 & 0 & 0 \\ & & & 0 & 0 & 0 & 1 & 0 \\ & & & 0 & 0 & 0 & 0 & -1 \end{array}\right]$$

$$R = \begin{bmatrix} 0 & 0 & 1 & 0 & 0 & 0 & 0 & 0 \\ 0 & 1 & -t & 0 & -1 & 0 & 0 & 0 \\ -1 & 0 & -t-1 & 0 & -t-1 & 1 & 1 & 0 \\ 1 & 0 & t & 0 & 0 & 0 & 0 & 0 \\ 0 & 0 & 0 & 0 & 1 & 0 & 0 & 0 \\ 0 & 0 & -1 & -1 & -t-1 & 1 & 1 & 0 \\ -1 & 0 & -t & -t & -t-1 & 0 & 1 & 0 \\ 0 & 0 & 0 & 0 & 0 & 0 & 0 & 1 \end{bmatrix}$$

verifying that $t + 1$ and $t^2 - 1$ are invariant polynomials of this pencil. We conclude that

$$x_1(t) = [0, 0, 0, 0, 0, -1, -t, 0]^T; \quad x_2(t) = [0, -1, -t-1, 0, 1, -t-1, -t-1, 0]^T$$

are polynomial eigenvectors,

$$x_3 = [0, 1, 0, 0, 0, 0, 0, 0]^T$$

is an eigenvector of $tB - A$ for eigenvalue -1, and

$$x_4(t) = [1, -t, -t - 1, t, 0, -1, -t, 0]^T$$

is a vector that can be evaluated at ± 1 to yield an eigenvector for each of these two eigenvalues.

We note that, as can been seen in Example 14, our algorithm allows us to identify, for every eigenvector for a given eigenvalue, whether it originates from the singular structure of the pencil or not. Since we believe that the singular structure is not associated with topologically significant eigenvectors, this identification is useful in applications.

Let us now discuss the computation of generalized eigenvectors of a pencil. Algorithm 12 provides polynomial eigenvectors of degree equal to the column Kronecker indices. A pencil whose Kronecker structure has one index ε possesses a sequence of $\varepsilon + 1$ generalized eigenvectors. To see this, consider the Kronecker block L_ε in (7). It can easily be checked that for every field value t, the sequence

$$\begin{bmatrix} 1 \\ t \\ t^2 \\ \vdots \\ t^{\varepsilon-1} \\ t^{\varepsilon} \end{bmatrix}, \begin{bmatrix} 0 \\ -1 \\ -2\,t \\ \vdots \\ -(\varepsilon-1)\,t^{\varepsilon-2} \\ -\varepsilon\,t^{\varepsilon-1} \end{bmatrix}, \ldots, \begin{bmatrix} 0 \\ 0 \\ 0 \\ \vdots \\ (-1)^{\varepsilon-1} \\ (-1)^{\varepsilon-1}\,\varepsilon\,t \end{bmatrix}, \begin{bmatrix} 0 \\ 0 \\ 0 \\ \vdots \\ 0 \\ (-1)^{\varepsilon} \end{bmatrix}$$

is a sequence of $\varepsilon + 1$ linearly independent generalized eigenvectors of L_ε. We also notice that the above sequence of polynomial vectors has been obtained by formal differentiation over the ring $\mathbb{F}[t]$ of the first vector. This is a ring homomorphism denoted $\frac{\mathrm{d}}{\mathrm{d}t}$ (or with prime notation) with the property that

$$\frac{\mathrm{d}}{\mathrm{d}t} t^k = k\,t^{k-1} \text{ for } k \in \mathbb{N}, \text{ and } \frac{\mathrm{d}}{\mathrm{d}t} c = 0, c \in \mathbb{F}.$$

If we denote $x(t) = [1, t, t^2, \ldots, t^{\varepsilon-1}, t^{\varepsilon}]^T$, then the above sequence is

$$x(t), -x'(t), \frac{1}{2}\,x''(t), \ldots, \frac{(-1)^{\varepsilon-1}}{(\varepsilon-1)!}\,x^{(\varepsilon-1)}(t), \frac{(-1)^{\varepsilon}}{\varepsilon!}\,x^{\varepsilon}(t).$$

This property generalizes to other pencils. Suppose that $tB - A$ possesses a polynomial eigenvector $x(t)$ of degree ε, as obtained for example by Algorithm 12. Then $x(t)$ satisfies Eq. (5). Formal differentiation verifies the chain rule, and so we can apply it repeatedly to both sides of this equation:

$$\begin{aligned}(tB - A)x'(t) &= -Bx(t),\\ (tB - A)x''(t) &= -2Bx'(t),\\ &\vdots\\ (tB - A)x^{(\varepsilon)}(t) &= -\varepsilon Bx^{(\varepsilon-1)}(t).\end{aligned}$$

From this it can easily be seen that $((-1)^k/k!\,x^{(k)}(t))$, $k = 0, \ldots, \varepsilon$ is a sequence of $\varepsilon + 1$ linearly independent generalized eigenvectors for $tB - A$. The $(\varepsilon + 1)$st derivative of $x(t)$ is the zero vector and therefore not linearly independent. The same procedure can also be applied to the columns of R in the output of Algorithm 12, say $x(t)$, that correspond to invariant polynomials of $tB - A$ in the sense that

$$Q^{-1}\,(tB - A)\,x(t) = d(t)\,e$$

for d an invariant polynomial and e a vector of the canonical basis of $\mathbb{F}^m$. Indeed, if t_0 is a root of d, we can write $d(t) = (t - t_0)^{k+1}\,r_0(t)$ for a certain $k \geq 0$, where r does not have t_0 as a root. Then

$$(tB - A)\,x(t) = (t - t_0)^{k+1}\,r_0(t)\,Q\,e.$$

Proposition 15 *For $i \leq k$, applying formal differentiation i times to both sides of the previous equation yields*

$$(tB - A)x^{(i)}(t) = -iBx^{(i-1)}(t) + (t - t_0)^{k+1-i}r_i(t)Q\,e \qquad (12)$$

where $r_i(t)$ is another polynomial such that $r_i(t_0) \neq 0$.

Proof Suppose, for $0 \leq i \leq k - 1$, that (12) holds. Then, applying formal differentiation on both sides, we obtain

$$(tB - A)\,x^{(i+1)}(t) = -(i+1)B^{(i)}(t) + (t - t_0)^{k-i}\left((k - i + 1)\,r_i(t) + (t - t_0)\,r_i'(t)\right)Q\,e.$$

We can fix $r_{i+1} = (k - i + 1)\,r_i(t) + (t - t_0)\,r_i'(t)$; it is obvious that t_0 is not a root of this polynomial. ■

Evaluating the previous sequence at t_0, we find that $((-1)^i/i!\,x^{(i)}(t_0))$, $i = 0, \ldots, k$ provides us with a sequence of generalized eigenvectors for eigenvalue t_0.

Example 16 In Example 14, the vectors

$$x_1(t) = [0, 0, 0, 0, 0, -1, -t, 0]^T; \quad x_2(t) = [0, -1, -t - 1, 0, 1, -t - 1, -t - 1, 0]^T$$

are eigenvectors of $tB - A$ for every field value, and so they are also generalized eigenvectors for every field value. The vectors

$$-x_1'(t) = [0, 0, 0, 0, 0, 0, 1, 0]^T; \quad -x_2'(t) = [0, 0, 1, 0, 0, 1, 1, 0]^T$$

are also generalized eigenvectors for every field value.

4 Numerical Results

We studied a map on a cloud of 100 points, taken in $S^1 \subset \mathbb{C}$ and then subjected to Gaussian noise with standard deviation varying from $\sigma = 0$ to 0.30. The image of each point z is taken to be the closest point to z^2, so the map is angle-doubling with noise. It is expected that we should find in homology H_1, computed over the field $\mathbb{Z}_{19}$, an eigenvector of long persistence for eigenvalue $t = 2$, but that stronger noise may make it harder to distinguish. Figure 1 shows the persistence barcodes for the eigenvector associated with $t = 2$ along a filtration of complexes indexed with parameter value ε. Since we can identify, at every step along the filtration, whether the eigenvector originates from the singular structure of the pencil or not, we can code the bar with the following colours: red when it does originate from the singular structure, and blue when it does not. It can be seen that as the noise level is increased,

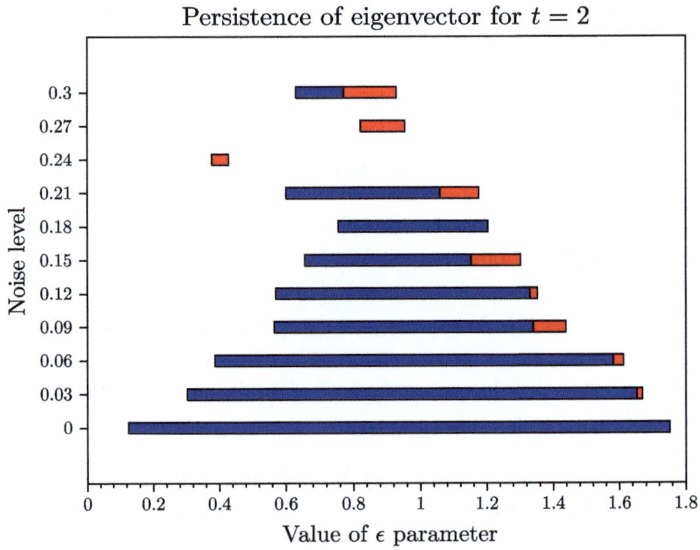

Fig. 1 Persistence of the longest lasting eigenvector associated with $t = 2$ in H_1 persistence over $\mathbb{Z}_{19}$ for several noise levels of a cloud of sample points on S^1, subject to the map $z \mapsto z^2$. Bar is *red* for vectors from singular structure, *blue* otherwise

the persistence of this eigenvector tends to become shorter, being born later and dying earlier, and additionally the eigenvector becomes "degenerate" (associated with the singular structure of the pencil) for a longer term.

Our 3D example uses a map on the torus constructed in the following way. Consider the square $[0, 1]^2$, identifying its left and right edges, as well as its top and bottom edges. Take a randomly selected sample of 200 points on this square, and build the map sending each point (x, y) to the closest point to $A[x, y]^T$, for the 2×2 matrix

$$A = \begin{bmatrix} 0 & 1 \\ 1 & 0 \end{bmatrix},$$

which has eigenvalues 1 and -1. In Fig. 2 we show persistence barcodes in H_1 homology over the field $\mathbb{Q}$ for eigenvalues 0, 1 and -1 for this test case. Here too

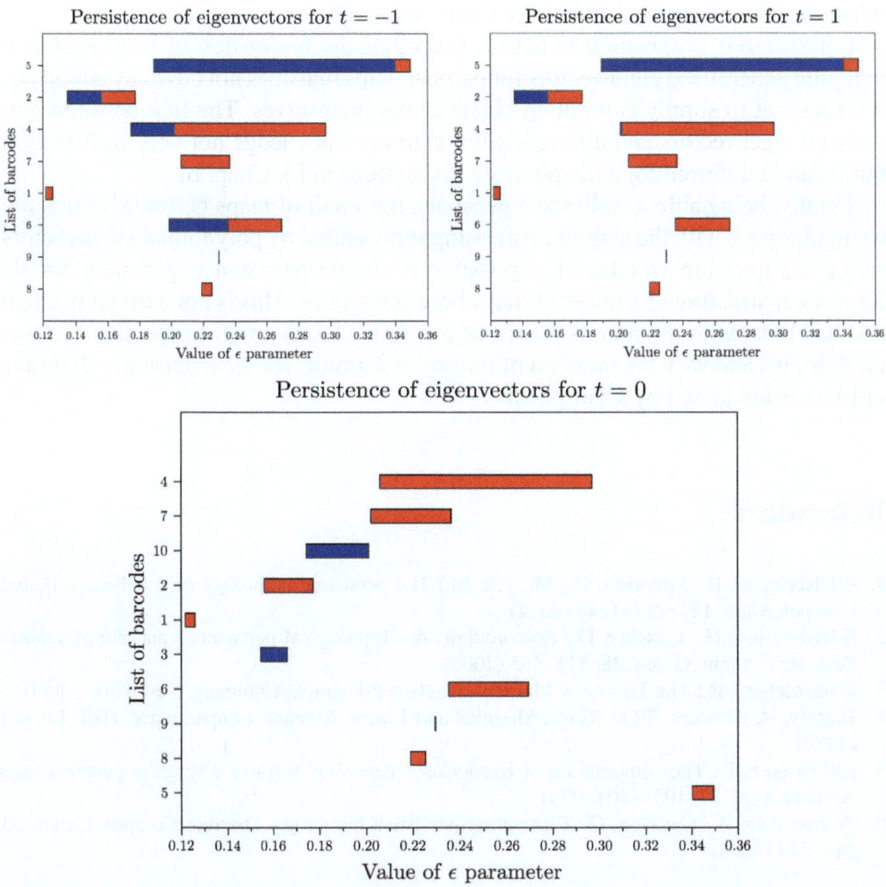

Fig. 2 Persistence barcodes for eigenvalues $t = -1$, $t = 1$ and $t = 0$ in H_1 persistence over $\mathbb{Q}$ for matrix A on the torus. Numbering is arbitrary. Bar is *red* for vectors from singular structure, *blue* otherwise

the bars are colour-coded red if the vector comes from the singular structure of the pencil, and blue if it comes from its finite structure. We notice several long-lasting vectors, but only two of those, one for eigenvalue 1 and one for eigenvalue -1, have a long life as non-singular vectors.

5 Conclusion

Algorithms 9, 11 and 12 positively answer the question asked in [1], on whether it is possible to compute the eigenspace towers of a pair of morphisms between two towers of vector spaces for all eigenvalues simultaneously. This is a necessary condition in applications, where candidate eigenvalues for long-lasting eigenvectors are not and cannot be known. It also makes it possible to study towers of eigenspaces when the spaces are over an infinite field such as $\mathbb{Q}$.

Furthermore, Proposition 15 and the preceding discussion describe a procedure to compute generalized eigenvectors for pairs of maps that does not have any added cost with respect to simply computing eigenvectors themselves. The link between generalized eigenvectors and differentiation is to our knowledge not very well-known, but it can be inferred for example from discussions in [3, Chap. 6].

Finally, being able to split the eigenspace for a pair of maps between a finite and a singular part, with the singular part being represented by polynomial eigenvectors, raises the question whether it is possible to define persistence generally for the Kronecker structure of a tower of maps between spaces. This is not a trivial problem and has links with the non-existence of a simple classification for persistence over modules [6] and with the problem of finding constraints for the persistence diagrams of two towers joined by a morphism.

References

1. Edelsbrunner, H., Jabłoński, G., Mrozek, M.: The persistent homology of a self-map. Found. Comput. Math. **15**, 1213–1244 (2014)
2. Edelsbrunner, H., Letscher, D., Zomorodian, A.: Topological persistence and simplification. Discrete Comput. Geom. **28**, 511–533 (2002)
3. Gantmacher F.R.: The Theory of Matrices. Chelsea Publishing Company, New York (1959)
4. Hartley, B., Hawkes, T.O.: Rings, Modules and Linear Algebra. Chapman and Hall, London (1970)
5. van Dooren, P.: The computation of Kronecker's canonical form of a singular pencil. Linear Algebra Appl. **27**, 103–140 (1979)
6. Zomorodian, A., Carlsson, G.: Computing persistent homology. Discrete Comput. Geom. **33**, 249–274 (2005)

Algorithm for Predicting Mathematical Formulae from Linear Strings for Mathematical Inputs

Tetsuo Fukui

Abstract Recently, computer-aided assessment (CAA) systems have been used for mathematics education, with some CAA systems capable of assessing learners' answers using mathematical expressions. However, the standard input method for mathematics education systems is cumbersome for novice learners. In 2011, we proposed a new mathematical input method that allowed users to input mathematical expressions through an interactive conversion of mathematical expressions from colloquial-style linear strings in WYSIWYG. In this study, we propose a predictive algorithm to improve the input efficiency of this conversion process by using machine learning to determine the score parameters with a structured perceptron similar to natural language processing. In our experimental evaluation, with a training dataset comprising 700 formulae, the prediction accuracy was 96.2% for the top ten ranking by stable score parameter learning; this accuracy is sufficient for a mathematical input interface system.

Keywords Mathematical input interface · Predictive algorithm · Machine learning · Mathematical formula editor

1 Introduction

In recent years, computer-aided assessment (CAA) systems have been used for the purpose of mathematics education. Some CAA systems enable users to directly enter mathematical expressions such that their answers can be evaluated automatically by using a computer algebra system (CAS). These systems have also been used to provide instructions to students at universities. However, the procedure through which answers are entered into the system, using a standard input method for mathematics education, is still cumbersome [10, 11].

T. Fukui (✉)
Mukogawa Women's University, Nishinomiya, Japan
e-mail: fukui@mukogawa-u.ac.jp

I.S. Kotsireas and E. Martínez-Moro (eds.), *Applications of Computer Algebra*, Springer Proceedings in Mathematics & Statistics 198,
DOI 10.1007/978-3-319-56932-1_9

In 2011, we proposed a new mathematical input method through the conversion of colloquial-style mathematical text (string) [1, 3]. This method is similar to those used for inputting Japanese characters in many operating systems. In this system, the list of candidate characters and symbols corresponding to the desired mathematical expression, as obtained through the user input, is displayed in WYSIWYG format; once all elements required to be included by the user are selected, the process of formatting of the expressions is complete. This method enables the user to input almost any mathematical expression without having to learn a new language or syntax [12]. However, the disadvantage of the above-mentioned method is that the user has to convert each element in the colloquial-style mathematical string proceeding from left to right in order [13].

This study aims to address this shortcoming by improving the input efficiency of such systems through intelligent predictive conversion of a linear mathematical string to an entire expression instead of converting each element individually.

2 Related Works

In this section, we describe related works on natural language processing, along with other predictive inputs for mathematical formulae using an N-gram model.

Input-word prediction has been studied since the 1980s in the field of natural language processing. Input characters are usually predicted for a word unit [5]. An N-gram model is typically used to predict text entries in popular probabilistic language models. For example, one typical system for word prediction, Reactive Keyboard, uses an N-gram model for augmentative and alternative communication (AAC) [7]. In such systems, a tree is built for prediction, where each alphabetical character corresponds to a node. Priority is assigned to each node based on the number of occurrences in the N-gram. When a user inputs characters, the system matches them with tree nodes, and the words in the child nodes of the matched node are provided as proposed predictions. A structured perceptron in machine learning for natural language processing has been used to input Japanese characters since the 1990s. As explained in Sect. 4.1, Algorithm 1 is similar to machine learning. It uses a structured perceptron for natural language processing [9]. However, mathematical formulae have tree structures, rather than the sentential chain structures of natural language. Indeed, none of the above-mentioned methods consider the structure of a sentence; however, our method considers the structure of mathematical formulae.

Structure-based user interfaces for inputting mathematical formulae are popular. They enable users to format a desired mathematical formula on a PC in WYSIWYG by selecting an icon corresponding to the structure of the expression. User do so using a GUI template, e.g., a fraction bar and an exponent form, into which the mathematical elements can be entered. Hijikata et al. from Osaka University improved the input efficiency of mathematical formulae by proposing an algorithm for predicting mathematical elements using an N-gram model [6]. However, their proposal is nevertheless a structure-based interface in the sense that users must understand the entire structure of a desired mathematical formula before selecting the corresponding icons.

By contrast, our predictive conversion method predicts such mathematical structures from a linear string of the mathematical formulae, rendering it significantly different from structure-based input methods.

3 Predictive Conversion

In this section, we define the linear string of a mathematical expression to be input by the user and describe the design of an intelligent predictive conversion system of such linear strings in Sect. 3.2. In Sect. 3.3, we formulate a predictive algorithm by using machine learning.

3.1 Linear String Rules

The rules for a linear mathematical string for a mathematical expression are described as follows:

> Set the key letters (or words) corresponding to the elements of a mathematical expression linearly in the order of the colloquial (or reading) style, without considering two-dimensional placement and delimiters.

In other words, a key letter (or word) consists of the ASCII code(s) corresponding to the initial or the clipped form (such as the LaTeX -form) of the objective mathematical symbol. Therefore, a single key often supports many mathematical symbols. For example, when a user wants to input α^2, the linear string is denoted by "a2", where "a" represents the "alpha" symbol and it is unnecessary to include the power sign (i.e., the caret letter (^)). In the case of $\frac{1}{\alpha^2+3}$, the linear string is denoted by "1/a2 + 3", where it is not necessary to put the denominator (which is generally the operand of an operator) in parentheses, because those are never printed.

Other representative category cases are shown in Table 1. For example, the linear string for $e^{\pi x}$ is only denoted by "epx". However, the linear string of the expressions $e_p x$, e^{px}, $e^{\pi} x$ are also denoted by "epx". Hence, there are some ambiguities for representing linear strings using these rules.

3.2 Design of an Intelligent Predictive Conversion System

In this paper, we propose a predictive algorithm to convert a linear string s into the most suitable mathematical expression y_p. For prediction purposes, we devise a method through which each candidate to be selected would be ranked in terms of

Table 1 Examples of mathematical expressions using linear string rules

Category	Linear strings	Math formulae
Variable	a	a or α
Polynomial	3x2+4x+1	$3x^2 + 4x + 1$
Fraction	2/3	$\frac{2}{3}$
Equation	(x–1/2)2=x2–x+1/4	$(x - \frac{1}{2})^2 = x^2 - x + \frac{1}{4}$
Square root	root3	$\sqrt{3}$
Trigonometric	sin2x	$\sin^2 x$
Logarithm	log10x	$\log_{10} x$
Exponent	epx	$e^{\pi x}$
Summation	sumk=1nk2	$\sum_{k=1}^{n} k^2$
Integral	intabfdx	$\int_a^b f dx$

its suitability. Our method uses a function Score(y) to assign a score proportional to the probability of occurrence of the mathematical expression y, which would enable us to predict the candidate y_p by using Eq. (1) as being the most suitable expression with the maximum score. Here, $Y(s)$ in Eq. (1) represents the totality of all possible mathematical expressions converted from s.

$$y_p \text{ s.t. } \mathrm{Score}(y_p) = \max\{\mathrm{Score}(y) | y \in Y(s)\} \tag{1}$$

A mathematical expression consists of mathematical symbols, such as numbers, variables, and *operators*,[1] together with the operating relations between an operator and an element. Therefore, we decided to represent a mathematical expression by a tree structure consisting of nodes and edges corresponding to the symbols and operating relations, respectively.

First, all node elements of the mathematical expressions are classified into nine categories, as listed in Table 2 in this mathematical conversion system. Therefore, a node element is characterized by (k, e, t), where k is the key letter (or word) of the mathematical symbol e that belongs to type $t (= N, V, P, A, B_L, B_R, C, Q, R,$ or $T)$ in Table 2. For example, the number 2 is characterized as ("2",2,N) and similarly a variable x as ("x", x, V) and as for the Greek letter α, it can either be characterized as ("$alpha$", α, V) or ("a", α, V). In the case of an operator, the character ("/", $\frac{\Delta_1}{\Delta_2}$, C) represents a fractional symbol with input key "/", where Δ_1, Δ_2 represents arbitrary operands.

In this study, a total of 510 mathematical symbols and 597 operators in node element table $\mathscr{D}$ are implemented by our prototype system.

[1]In this article, "operator" is used in the sense of operating on, i.e. performing actions on elements in terms of their arrangements for two-dimensional mathematical notation.

Table 2 Nine types of mathematical expressive structures

Math element	Codes of type	Examples ($\Delta_1, \Delta_2, \Delta_3$ represent operands)
Number	N	$2,\ 128$
Variable, symbol	V	$x,\ \alpha$
Prefix unary operator	P	$\sqrt{\Delta_1},\ \sin \Delta_1$
Postfix unary operator	A	Δ_1'
Bracket	B_L, B_R	(Δ_1)
Infix binary operator	C	$\Delta_1 + \Delta_2,\ \frac{\Delta_1}{\Delta_2}$
Prefix binary operator	Q	$\log_{\Delta_1} \Delta_2$
Prefix ternary operator	R	$\int_{\Delta_1}^{\Delta_2} \Delta_3$
Infix ternary operator	T	$\Delta_1 \xrightarrow{\Delta_2} \Delta_3$

The totality $Y(s)$ of the mathematical expressions converted from s is calculated by using the following procedure **Proc. 1–Proc. 3** (cf. [2, 4]) referring to node element table $\mathscr{D}$.

Proc. 1 A linear string s is separated in the group of keywords defined in Eq. (2) by using the parser in this system. All possible key separation vectors $(k_1, k_2, \cdots, k_K)$ are obtained by matching every part of s with a key in $\mathscr{D}$.

$$s = k_1 \uplus k_2 \uplus \cdots k_K \text{ where } (k_i, v_i, t_i) \in \mathscr{D}, i = 1, ..., K \tag{2}$$

Proc. 2 Predictive expressive structures are fixed by analyzing all key separation vectors of s and comparing the nine types of structures provided in Table 2.

Proc. 3 From the fixed structures corresponding to the operating relations between the nodes, we obtain $Y(s)$ by applying all possible combinations of mathematical elements belonging to each keyword in $\mathscr{D}$.

Complexity of $Y(s)$

Generally, the number of elements in $Y(s)$, denoted by $n(Y(s))$, becomes enormous corresponding to the increase in the length of s. For example, because the key letter "a" corresponds to seven symbols, namely $Y(\text{"}a\text{"})=\{a, \alpha, \mathrm{a}, \mathbf{a}, \boldsymbol{a}, \mathsf{a}, \aleph\}$, and the invisible times between a and b corresponds to $Y(\text{"}ab\text{"}) = \{ab, a^b, a_b, {}^a b, {}_a b\}$, then $n(Y(\text{"}abc\text{"}))=7^3 \times 5^2 = 8575$. However, for the purpose of a mathematical input interface, it is enough to calculate the N-best high score candidates in $Y(s)$ as shown in Eq. (1). Therefore, for improving the efficiency of calculations we obtain the N-best candidates in $Y(s)$ as follows:

1. In **Proc. 1**, all the key separation vectors $(k_1, k_2, \cdots, k_K)$ of s are sorted in ascending order of the number K in Eq. (2), i.e. in an order starting from higher probability.
2. In **Proc. 2,** we set upper limit L of the number of loops for breaking down all the possible calculations of the predictive expressive structures.

3. In **Proc. 3**, to obtain the N-best candidates in $Y(s)$, we apply only the N-best mathematical elements for operand expressions related to an operator instead of all possible combinations.

3.3 Predictive Algorithm

Let us assume that the probability of occurrence of a certain mathematical element is proportional to its frequency of use. Then, the probability of occurrence of mathematical expression y, which is possibly converted from a given string s, is estimated from the total score of all the mathematical elements included in y. Given the numbering of each element from 1 to F_{total}, which is the total number of elements, let θ_f be the score of the $f(=1, \cdots, F_{total})$-th element, and let $x_f(y)$ be the number of times the f-th element is included in y. Then, Score(y) in Eq. (1) is estimated by Eq. (3), where $\boldsymbol{\theta}^T = (\theta_1, \cdots, \theta_{F_{total}})$ denotes the score vector and $\mathbf{X} = (x_f(y)),\ f = 1, \cdots, F_{total}$ is the F_{total}-dimensional vector.

$$h_\theta\left(\mathbf{X}(y)\right) = \boldsymbol{\theta}^T \cdot \mathbf{X}(y) = \sum_{f=1}^{F_{total}} \theta_f x_f(y) \tag{3}$$

Equation (3) is in agreement with the hypothesis function of linear regression, and $\mathbf{X}(y)$ is referred to as the characteristic vector of y. To solve our linear regression problem and predict the probability of occurrence of a mathematical expression, we conduct supervised machine learning on the m elements of a training dataset $\{(s_1, y_1), (s_2, y_2), \cdots, (s_m, y_m)\}$. Our learning algorithm to obtain the optimized score vector is performed through the following four-step procedure:

Step 1 Initialization: $\boldsymbol{\theta} = \mathbf{0}$, $i = 1$
Step 2 Decision regarding a candidate: y_p s.t. $h_\theta\left(\mathbf{X}(y_p)\right) = \max\{h_\theta\left(\mathbf{X}(y)\right) | y \in Y(s_i)\}$
Step 3 Training parameter: if($y_p \neq y_i$) {

$$\begin{array}{ll} \theta_f := \theta_f + 1 & \text{for } \{f \leq F_{total} | x_f(y_i) > 0\} \\ \theta_{\bar{f}} := \theta_{\bar{f}} - 1 & \text{for } \{\bar{f} \leq F_{total} | x_{\bar{f}}(y_p) > 0\} \end{array} \tag{4}$$

}
Step 4 if($i < m$){ i=i+1; go to **Step 2** for repetition.}
else { Output $\boldsymbol{\theta}$ and end.}

This learning algorithm is very simple, and similar to machine learning using a structured perceptron for natural language processing [9].

4 Main Algorithm

In this section, we experimentally investigate the prediction accuracy by using the algorithm described in the previous section. Then, we discuss the results of the evaluation in Sect. 4.1 and propose the main algorithm of this study in Sect. 4.2.

4.1 Experimental Evaluation

We examine the prediction accuracy using two score learning parameter sets on an evaluation dataset $\mathscr{E} = \{(s_i, y_i) | i = 0, \ldots, 799\}$ containing 800 mathematical formulae from a mathematics textbook [8]. As the scope of the evaluation dataset $\mathscr{E}$, we adopted the mathematical subjects: "Quadratic-polynomials, -equations, -inequalities and -functions" that are studied in the tenth grade in Japan. The dataset $\mathscr{E}$ has generated manually with our previous system [3] in the order of appearance from the textbook by choosing individual expressions y_i with length of s_i, which is less than 16. Some samples of the dataset $\mathscr{E}$ are shown in Table 3.

Two parameter sets of $\boldsymbol{\theta}$ for scoring were trained by using the following two algorithms programmed in Java on a desktop computer (MacOS 10.9, 3.2 GHz Intel core i3, 8 GB memory):

Algorithm 1 **Step 1–Step 4**, using Eq. (4).
Algorithm 2 **Step 1–Step 4**, with **Step 3** using

$$\begin{aligned} \theta_f &:= \theta_f + 2 \quad \text{for } \{f \leq F_{total} | x_f(y_i) > 0\} \\ \theta_{\bar{f}} &:= \theta_{\bar{f}} - 1 \quad \text{for } \{\bar{f} \leq F_{total} | x_{\bar{f}}(y_p) > 0\} \end{aligned} \tag{5}$$

in place of Eq. (4).

Table 3 Samples of the evaluation dataset $\mathscr{E}$

Input strings (s_i)	Length of s_i	Formulae (y_i)
a/=0	4	$a \neq 0$
A (3,–2)	7	A $(3, -2)$
3 <= y <= 7	7	$3 \leqq y \leqq 7$
7/9=0.7.	8	$\frac{7}{9} = 0.\dot{7}$
root32=[3]	10	$\sqrt{3^2} = \|3\|$
y=1/2x2–2x–1	12	$y = \frac{1}{2}x^2 - 2x - 1$
(a4)3=a4*3=a12	14	$\left(a^4\right)^3 = a^{4\times 3} = a^{12}$
3x2y4*(–2x4y)3	14	$3x^2y^4 \times \left(-2x^4y\right)^3$

Table 4 Prediction accuracy using Algorithms 1 and 2

Training number	Best 1 (%)		Best 3 (%)		Best 10 (%)		Correct score	
	Algo. 1	Algo. 2	Algo. 1	Algo. 2	Algo. 1	Algo. 2	Algo. 1	Algo. 2
0	25.9 (3.8)	25.9 (3.8)	41.3 (4.4)	41.3 (4.4)	52.3 (4.3)	52.3 (4.3)	2.8 (0.1)	2.8 (0.1)
100	62.7 (14.7)	53.3 (14.6)	75.5 (9.2)	82.5 (6.4)	81.5 (6.8)	88.5 (4.3)	15.4 (1.2)	307.0 (58.1)
200	75.6 (6.6)	60.3 (5.0)	82.7 (5.0)	86.1 (4.2)	86.6 (4.3)	91.7 (3.2)	18.0 (1.5)	568.9 (99.4)
300	79.3 (4.1)	64.1 (5.1)	85.2 (4.3)	89.1 (3.2)	88.1 (4.3)	93.8 (2.9)	20.4 (1.9)	964.3 (186.8)
400	79.2 (3.8)	67.7 (5.7)	85.1 (3.8)	90.1 (3.1)	88.2 (3.5)	94.1 (3.1)	21.0 (2.2)	1103.4 (75.2)
500	80.0 (4.4)	67.6 (5.7)	86.7 (4.0)	90.6 (2.9)	89.5 (3.3)	94.5 (2.8)	23.1 (2.2)	1290.6 (99.8)
600	79.5 (3.7)	69.1 (4.6)	85.9 (3.4)	90.8 (2.7)	89.2 (3.8)	94.3 (2.5)	22.4 (2.4)	1492.2 (106.5)
700	79.1 (5.7)	68.5 (6.0)	85.7 (5.3)	91.1 (2.5)	89.2 (4.2)	95.0 (2.5)	22.9 (1.7)	1692.9 (114.7)

Numbers within parentheses denote the *SD*

Fig. 1 The result by Algorithm 1 (training number–prediction accuracy)

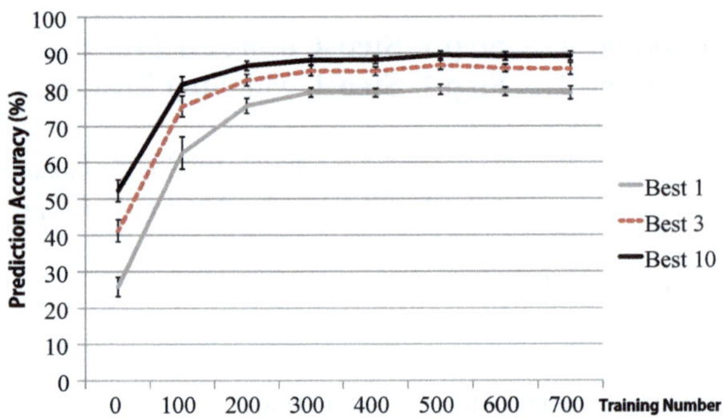

In the experimental evaluation, we measured the proportion of correct predictions from among 100 test datasets after learning the parameters through Algorithms 1 and 2 using a training dataset consisting of 700 formulae by eightfold cross-validation.

The machine learning results using Algorithms 1 and 2 are given in Table 4 for each training number. By using Algorithm 1, the prediction accuracy of "Best 1" is about 79.1% after being trained 700 times. In the top ten ranking ("Best 10"), it achieves about 89.2%. Figure 1 shows the change in the prediction accuracy as a result of Algorithm 1 for each training number.

On the other hand, the result obtained by using Algorithm 2 with another learning weight shows that the prediction accuracy of "Best 1" is approximately 68.5% after being trained 700 times. It achieves about 95.0% in the top ten ranking. The change in the prediction accuracy as a result of Algorithm 2 is shown in Fig. 2.

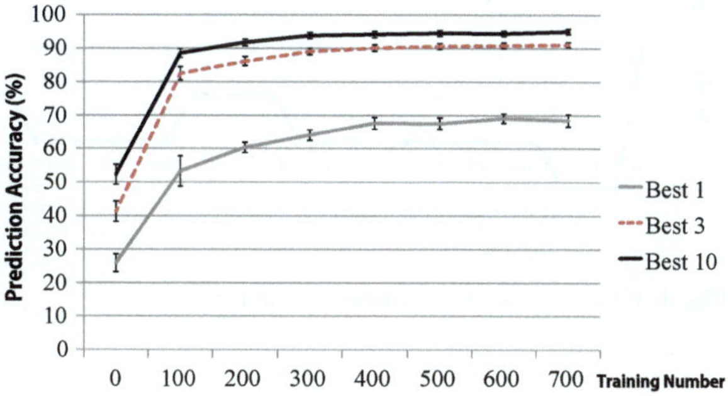

Fig. 2 The result by Algorithm 2 (training number–prediction accuracy)

4.2 Discussion

The mean scores for the correct expressions ("correct score" in short) in the test dataset for each training number are shown in the fifth column of Table 4 and illustrated in Fig. 3. The prediction accuracy of "Best 1" by using Algorithm 1 becomes sufficiently high, i.e., approximately 80%, with the mean correct score approximately equal to 23 after being trained 500 times. However, this is disadvantageous for a mathematical input interface, because the correct expressions out of the top ten ranking are more than 10%. One of the causes of this 10% leak is because the priorities of some correct expressions are not reflected in their occurrence frequency. In the case when two different candidates belonging to the same key appear from the training data, e.g., the pair a and α and the pair p and π, their scores change into a positive value from a negative value or vice versa. This means that even if a candidate with negative score occurred many times in $\mathscr{E}$, it has lower priority than the one with zero score because the increase and decrease in the weights of the score in Eq. (4) are mutually the same. For example, changes in score parameters (a and α) are shown in Fig. 4.

To avoid such problems, we have modified Algorithm 2 such that the increase in weight for the correct candidate is greater than the decrease in weight for the incorrect one as shown in Eq. (5). From the results of experimental evaluation of the

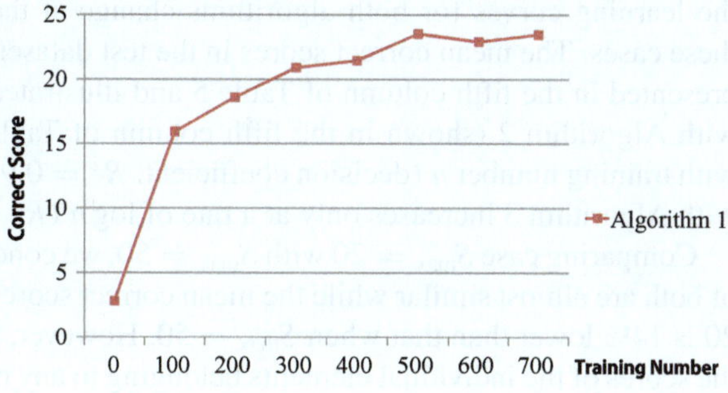

Fig. 3 Change in correct score given by Algorithm 1

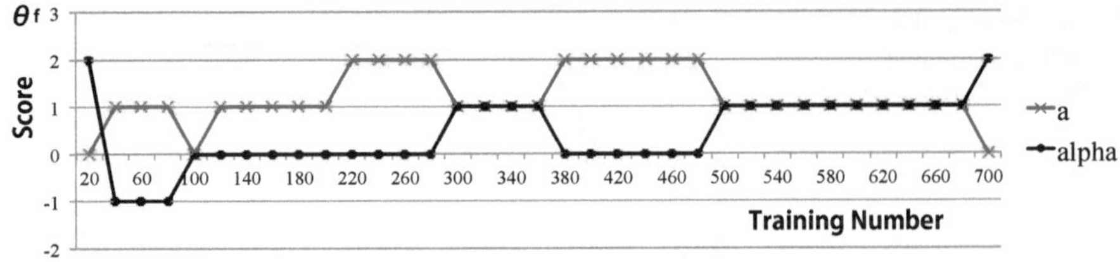

Fig. 4 Change in score parameters (a and α)

prediction accuracy by using Algorithm 2, the ratio of correct expressions from the top ten ranking is less than 5%, which is sufficient for a mathematical input interface system. However, we remark that the score parameter continues to increase while Algorithm 2 is learning.

In this study, we propose the following algorithm, Algorithm 3, to overcome the problems encountered in Algorithm 2.

Algorithm 3 **Step 1–Step 4**, where **Step 3** using

$$\begin{aligned} \text{if}(\theta_f < S_{\max})\{\theta_f &:= \theta_f + 2 \quad \text{for } \{f \le F_{total} | x_f(y_i) > 0\}\} \\ \theta_{\bar{f}} &:= \theta_{\bar{f}} - 1 \quad \text{for } \{\bar{f} \le F_{total} | x_{\bar{f}}(y_p) > 0\} \end{aligned} \tag{6}$$

in place of Eq. (5).

Here, $S_{\max}$ in Eq. (6) is a suitable upper bound for any mathematical element score. Because the result of Algorithm 1 provides good precision with a mean score of approximately 23, we set the upper bound $S_{\max}$ to 20 for any mathematical element score θ_f.

The machine learning results for Algorithm 3 for the case $S_{\max} = 20$ are given in Table 5 for various sizes of the training dataset. It can be seen that the accuracy of "Best 1" with Algorithm 3 was approximately 68.3% after being trained 700 times. This algorithm achieved an accuracy of 90.5% for the top three ranking, and 96.2% for the top ten ranking. With a training set of size 700, there is no statistically significant difference (at the 5% level) between the results for Algorithm 2 and those for Algorithm 3 for the "Best 1," "Best 3," or "Best 10" cases. Additionally, the learning curves for both algorithms change at the same skill rate for each of these cases. The mean correct scores in the test dataset for each training number are presented in the fifth column of Table 5 and illustrated in Fig. 5. The correct score with Algorithm 2 (shown in the fifth column of Table 4) increases proportionally with training number n (decision coefficient: $R^2 = 0.98$); however, the correct score with Algorithm 3 increases only at a rate of $\log n$ ($R^2 = 0.96$).

Comparing case $S_{\max} = 20$ with $S_{\max} = 50$, we conclude that precision properties of both are almost similar while the mean correct score for the test data when $S_{\max} = 20$ is 14% lower than that when $S_{\max} = 50$. However, if we set $S_{\max}$ to less than 20, the scores of the individual elements belonging to any one key are not much different

Table 5 Prediction accuracy using Algorithm 3

Training no.	Best 1 (%)	Best 3 (%)	Best 10 (%)	Correct score
0	25.9 (3.8)	41.3 (4.4)	52.3 (4.3)	2.8 (0.1)
100	54.2 (13.8)	82.6 (6.4)	88.7 (4.3)	283.0 (74.0)
200	65.7 (6.8)	87.7 (3.4)	93.0 (2.7)	428.6 (110.6)
300	69.5 (6.1)	88.3 (3.1)	94.0 (3.0)	494.1 (134.7)
400	67.9 (6.3)	88.8 (2.4)	94.3 (2.8)	536.7 (148.8)
500	69.2 (5.6)	89.8 (3.0)	95.2 (2.7)	566.2 (162.7)
600	70.6 (5.2)	90.9 (2.7)	95.9 (2.5)	590.0 (169.4)
700	68.3 (6.1)	90.5 (2.8)	96.2 (2.3)	608.0 (180.0)

Numbers within parentheses denote the *SD*

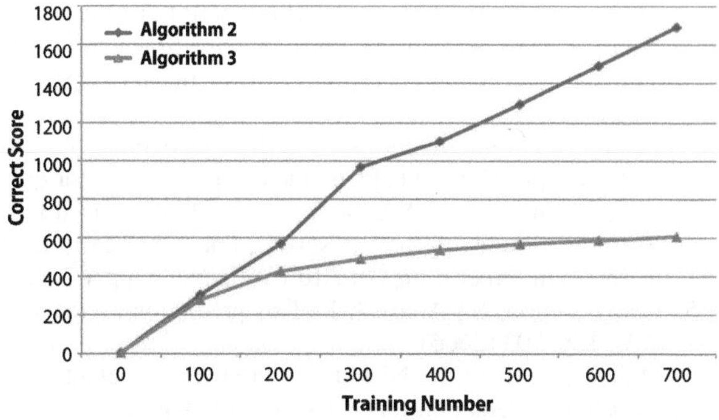

Fig. 5 Change in correct score given by Algorithms 2 and 3

in the machine training because the maximum number of elements belonging to any one key is equal to 20 in our key dictionary $\mathscr{D}$. Therefore, we propose $S_{\max} = 20$ to be the most suitable value in this study.

5 Conclusion and Future Work

In this paper, we proposed a predictive algorithm with an accuracy of 96.2% for the top ten ranking by improving upon a previously proposed algorithm in terms of a structured perceptron for stable score parameter learning. The mean CPU time for predicting each mathematical expression with corresponding linear string of length less than 16 obtained from a mathematics textbook was 0.44 s (SD=0.61).

Because the linear strings for mathematical expressions are easily recognized from both handwritten image data and voice data for such mathematical expressions, it is possible that the desired mathematical expression is predicted with high accuracy from such linear strings by using this predictive algorithm. We believe that there is a possibility to apply this predictive algorithm to not only a mathematical input method on a PC with the keyboard but also for recognizing handwritten mathematical expressions and voice for mathematical expressions.

Finally, the most important avenues for future research are to reduce the time for prediction and develop an intelligent mathematical input interface by implementing our proposed predictive algorithm.

Acknowledgements This work was supported by JSPS KAKENHI Grant Number 26330413.

References

1. Fukui, T.: An intelligent method of interactive user interface for digitalized mathematical expressions. RIMS Kokyuroku **1780**, 160–171 (2012) (in Japanese)
2. Fukui, T.: The performance of interactive user interface for digitalized mathematical expressions using an intelligent formatting from linear strings. RIMS Kokyuroku **1785**, 32–44 (2012). (in Japanese)
3. Fukui, T.: An intelligent user interface technology for easy formatting of digitalized mathematical expressions–a mathematical expression editor on web-browser. Interaction 2013 IPSJ symposium, No.1, 2EX13-50, pp. 537–540 (2013) (in Japanese)
4. Fukui, T.: Prediction for converting linear strings to mathematical formulae using machine learning. In: Proceedings of ARG WI2, No. 6, pp. 67–72 (2015) (in Japanese)
5. Garay-Vitoria, N., Abascal, J.: Text prediction systems: a survey. Univers. Access. Inf. Soc. **4**(3), 188–203 (2006)
6. Hijikata, Y., Horie, K., Nishida, S.: Predictive input interface of mathematical formulas Human-Computer Interaction-INTERACT2013, Vol. 8117 of the series Lecture Notes in Computer Science. Springer, New York (2013)
7. Hunnicutt, S.: Input and output alternative in word prediction. STL/QPRS **28**(2–3), 15–29 (1987)
8. Iidaka, S., Matsumoto, Y., et al.: Mathematics I, **001**, TOKYO SHOSEKI (2012) (in Japanese)
9. Manning, C.D., Scheutze, H.: Foundations of Statistical Natural Language Processing. The MIT Press, London (2012)
10. Pollanen, M., Wisniewski, T., Yu, X.: XPRESS: a novice interface for the real-time communication of mathematical expressions, In: Proceedings of MathUI (2007)
11. Sangwin, CJ.: Computer aided assessment of mathematics using STACK, In: Proceedings of ICME, vol. 12 (2012)
12. Shirai, S., Fukui, T.: Development and evaluation of a web-based drill system to master basic math formulae using a new interactive math input method, Mathematical Software—ICMS2014, vol. 8592 of the Series Lecture Notes in Computer Science, pp. 621–628. Springer, New York (2014)
13. Shirai, S., Fukui, T.: Improvement in the input of mathematical formulae into STACK using interactive methodology. Comput. Educ. **37**, 85–90 (2014) (in Japanese)

Algebraic Modelling of Covering Arrays

Bernhard Garn and Dimitris E. Simos

Abstract We introduce a novel technique to model and compute binary covering arrays, discrete combinatorial structures, based on a tuple-level modelling and using methods arising from linear algebra, commutative algebra and symbolic computation. Concrete instances of covering arrays for given parameters will arise as points in a generated variety of a system of multivariate polynomial equations with Gröbner Bases playing an important role.

Keywords Covering arrays · Algebraic modelling · Symbolic computation · Algorithms

1 Introduction

Over the last few years, covering arrays have seen a new application in a novel branch of software testing called combinatorial testing [30]. Traditional applications of especially orthogonal and covering arrays lie in a field known as Design of Experiments, however, a change of perspective led to the supremacy usage of covering arrays in combinatorial testing. A change of focus when moving from classical Design of Experiments to software testing makes the coverage notion of covering arrays particularly appealing and has been shown beneficial in software testing practice [29].

The computation of optimal covering arrays is a well-known NP-hard problem [35], which has been attacked with multiple techniques from various fields (see Sect. 2.3). For a comprehensive treatment of the different kinds of methodologies for the construction of covering arrays, we refer the interested reader to the recent survey of [37].

Contribution. In this paper, we propose a Computational Algebra formalism to provide an algebraic modelling that leads to the construction of covering arrays. Our

B. Garn · D.E. Simos (✉)
SBA Research, 1040 Vienna, Austria
e-mail: dsimos@sba-research.org

B. Garn
e-mail: bgarn@sba-research.org

I.S. Kotsireas and E. Martínez-Moro (eds.), *Applications of Computer Algebra*,
Springer Proceedings in Mathematics & Statistics 198,
DOI 10.1007/978-3-319-56932-1_10

formalism is based on a translation of the coverage property of covering arrays to zeros of multivariate polynomials and the subsequent computation of varieties arising from systems of such polynomials, incorporating the theory of Gröbner bases. For given binary covering array configuration, our approach is able to generate all possible solutions, if any exist or show that there are no solutions at all.

This paper is structured as follows. Section 2 describes the necessary theoretical background and related problems for covering arrays. In Sect. 3 we give our algebraic modelling and present detailed examples illustrating our algorithmic approaches for the construction of covering arrays in Sect. 4. Finally, in Sect. 5 we compare our approach to some greedy (algorithmic) approaches and conclude the paper in Sect. 6 with directions for future research.

2 Related Problems and Algorithms for Covering Arrays

In this section, we first review the necessary definitions for covering arrays and some of their properties. Afterwards, we reformulate various problems related to the computation and construction of covering arrays, referencing past works where needed. Finally, we describe related algorithmic approaches, for solving the aforementioned problems, and we make special mention to the most efficient of them.

2.1 *Preliminaries for Covering Arrays*

We give below the necessary definitions for the notions of covering arrays used in this paper, taken from [12].

Definition 1 (*CA*) A *covering array* $\mathsf{CA}_\lambda\,(N; t, k, v)$ is an $N \times k$ array. In every $N \times t$ subarray, each t-tuple occurs at least λ times. Then t is the strength of the coverage of interactions, k is the number of components (degree) and v is the number of symbols for each component (order). Only the case when $\lambda = 1$ is treated; the subscript is then omitted in the notation. The size N is omitted when inessential in the context.

Definition 2 (*MCA*) A *mixed-level covering array* $\mathsf{MCA}_\lambda\,(N; t, k, (v_1, v_2, \ldots, v_k))$ is an $N \times k$ array. Let $\{i_1, \ldots, i_t\} \subseteq \{1, \ldots, k\}$, and consider the subarray of size $N \times t$ obtained by selecting columns $i_1, \ldots, i_t$ of the MCA. There are $\prod_{i=1}^{t} v_i$ distinct t-tuples that could appear as rows, and an MCA requires that each appear at least once. $\mathsf{CAN}\,(t, k, (v_1, v_2, \ldots, v_k))$ denotes the smallest N for which such a mixed covering array exists.

Definition 3 (*MCA and CA configuration*) A *configuration* C for a mixed-level covering array is a tuple $(t, k, (v_1, v_2, \ldots, v_k))$. When $v_1 = v_2 = \ldots = v_k = v$, then the specified MCA is in fact a CA and we denote its configuration simply by (t, k, v).

We will, however, denote and use (mixed-level) covering arrays M as their transpose $M^\top$ in their matrix notation, following the terminology used in [16]. The later used statement "a matrix M is compatible with a MCA configuration $C = (t, k, (v_1, \dots, v_k))$" is to be understood as that the matrix M has k columns and that its elements in the i-th column arise either from the set $\{0, \dots, v_i - 1\}$ or constitute variables which take values exactly in $\{0, \dots, v_i - 1\}$.

2.2 Problems for Covering Arrays

In this section we present different problems that arise in the generation and computation of covering arrays. We reformulate some of the problems found in [10, 16, 31, 32] to a proper computational or decisional version (in terms of computational complexity) and introduce some more that will be needed in the course of our work. Since for any given MCA configuration it is possible to give immediately at least one mixed-level covering array—the Cartesian product—the most important challenge lies in the construction of optimal or near optimal mixed-level covering arrays. In particular, considerable effort has been put into developing theoretical upper and lower bounds for the covering array numbers CAN $(t, k, (v_1, v_2, \dots, v_k))$, as well as explicit constructions which we list in Sect. 2.3.

Problem 1 (*Decisional Existence*) For given MCA configuration C and given $N \in \mathbb{N}$, decide whether a MCA for the configuration C with N rows exists.

Problem 2 (*Computational Existence (one solution), version 1*) For given MCA configuration C and given $N \in \mathbb{N}$, construct one MCA for the configuration C with exactly N rows or terminate with an error.

Problem 3 (*Computational Existence (one solution), version 2*) For given MCA configuration C and given $N \in \mathbb{N}$ with $\text{CAN}(C) \leq N$, construct one MCA for the configuration C with exactly N rows.

Problem 4 (*Computational Existence (all solutions), version 1*) For given MCA configuration C and given $N \in \mathbb{N}$, construct all MCAs for the configuration C with exactly N rows or terminate with an error.

Problem 5 (*Computational Existence (all solutions), version 2*) For given MCA configuration C and given $N \in \mathbb{N}$ with $\text{CAN}(C) \leq N$, construct all MCAs for the configuration C with exactly N rows.

Problem 6 (*Decisional Parameter Extension*) Given a MCA M of strength t and given an alphabet size v, decide whether it is possible to extend the given matrix M with one additional column corresponding to a new parameter taking v values such that the extended matrix constitutes a MCA of strength t without adding additional rows.

Problem 7 (*Computational Parameter Extension* (*one solution*), *version 1*) Given a MCA of strength t, given an alphabet size v, construct one new additional column such that the extended matrix constitutes a MCA of strength t with the additional parameter taking v values without adding new rows or terminate with an error.

Problem 8 (*Computational Parameter Extension* (*one solution*), *version 2*) Given a MCA of strength t, given an alphabet size v and assume an affirmative parameter extension decision, construct one new additional column such that the extended matrix constitutes a MCA of strength t with the additional parameter taking v values without adding new rows.

Problem 9 (*Computational Parameter Extension* (*all solutions*), *version 1*) Given a MCA of strength t, given an alphabet size v, construct all possible new additional columns such that the extended matrices constitute MCAs of strength t with the additional parameter taking v values without adding new rows or terminate with an error.

Problem 10 (*Computational Parameter Extension* (*all solutions*), *version 2*) Given a MCA of strength t, given an alphabet size v and assume an affirmative parameter extension decision, construct all possible new additional columns such that the extended matrices constitute MCAs of strength t with the additional parameter taking v values without adding new rows.

Problem 11 (*Decisional Vertical Extension*) Given a MCA configuration C, a compatible matrix M and an integer r, decide whether it is possible to extend the given matrix M with exactly r rows, such that after the extension the new matrix constitutes a MCA for the given configuration C.

Problem 12 (*Computational Vertical Extension* (*one solution*), *version 1*) Given a MCA configuration C, a compatible matrix M and an integer r, construct one vertical extension for M of exactly r rows such that the extended matrix constitutes a MCA for the given configuration C or terminate with an error.

Problem 13 (*Computational Vertical Extension* (*one solution*), *version 2*) Given a MCA configuration C, a compatible matrix M, an integer r and assume an affirmative vertical extension decision for r, construct one vertical extension of exactly r rows such that the extended matrix constitutes a MCA for the given configuration C.

Problem 14 (*Computational Vertical Extension* (*all solutions*), *version 1*) Given a MCA configuration C, a compatible matrix M and an integer r, construct all possible vertical extensions for M of exactly r rows such that the extended matrices constitute MCAs for the given configuration C or terminate with an error.

Problem 15 (*Computational Vertical Extension* (*all solutions*), *version 2*) Given a MCA configuration C, a compatible matrix M, an integer r and assume an affirmative vertical extension decision for r, construct all possible vertical extensions of exactly r rows such that the extended matrices constitute MCAs for the given configuration C.

Problem 16 (*Decisional Minimal Vertical Extension*) Given a MCA configuration C, compatible matrix M and integer r, decide whether r is the least positive integer such that an r vertical extension of M for C is possible.

Problem 17 (*Computational Minimal Vertical Extension*) Given a MCA configuration C and compatible matrix M, construct the least positive integer r such that there is an affirmative minimal vertical extension decision for r.

Problem 18 (*Decisional Coverage Verification*) Given a MCA configuration C and a compatible matrix M, decide whether it constitutes a MCA for the given configuration C.

2.3 Related Algorithms for Covering Arrays

We give some references to works describing algorithms and techniques for constructing covering arrays. We do not aim to provide a comprehensive, or by all means complete, treatment of the subject, as this is not the purpose of the present paper. We are merely interested in giving a flavour of the many different approaches used, in order to exhibit that while covering arrays are specialized types of combinatorial structures there has been a great interest on applying algorithmic techniques for their construction.

For binary covering arrays of strength $t = 2$ the exact value of CAN $(2, k, 2)$ and an explicit construction for an optimal covering array are known [23]. In general, exact methods will return a covering array with CAN rows, i.e., they construct the optimal number of rows [5, 17, 40]. Practically successful greedy constructions include [6, 8]. A branch and bound approach was presented in [39]. Although metaheuristic methods do not provide any guarantees regarding the quality of the generated solution, their application to the generation of covering arrays has been successful in practice, including simulated annealing [1, 38] and Tabu search [15, 33]. Last but not least, algorithms arising in the field of discrete mathematics include constructions based on cyclotomy [11] and linear feedback shift registers [34].

Finally, we make special mention to the In-Parameter-Order-General (IPOG) [31, 32] family of algorithms for constructing covering arrays, since it is among the most popular algorithmic solutions for constructing covering arrays today and is heavily used by practitioners of combinatorial testing. It is developed jointly at the University of Texas at Arlington and NIST. These algorithms are implemented in a software called ACTS, which constructs competitive quality mixed-level covering arrays. Given a configuration C of a CA, the IPOG strategy for constructing a strength $t = 2$ covering array works as follows: it starts by constructing a strength $t = 2$ covering array for the first two parameters, which can be done easily by constructing the Cartesian product (i.e., ordered pairs of values from domain of the first parameter and the domain of the second parameter). Then, this matrix is extended to a matrix which will be a strength $t = 2$ covering array for the first three parameters. This strategy is continued until all parameters have been considered and as such a covering array with strength $t = 2$ for all parameters has been constructed. The extension step to also cover an additional parameter is performed in two independent steps:

Step 1 In the *horizontal extension step*, each row of the matrix is extended from an i-tuple to an $i + 1$-tuple, and an appropriate value in the domain of the $i + 1$-th parameter is chosen.

Step 2 In the *vertical extension step*, new rows are added to the matrix until the desired t-wise coverage is achieved.

3 Algebraic Tuple Modelling

In this section, we establish the connection between the notion of t-wise coverage of covering arrays and algebraic techniques. In particular, we show how to model, enforce and construct binary strength two covering arrays as defined in Sect. 2.

A brief, informal description of our approach can be given as follows: for given covering array configuration $C = (2, k, 2)$, and depending on the covering array problem considered (see Sect. 2.2), we construct a matrix where some entries are variables x_ι from a suitable multivariate polynomial ring over the field of rational numbers $\mathbb{Q}$. The later introduced concept of row-selectors (see Sect. 3.2, Definition 5) with specific entries and transformations of coverage conditions into an algebraic formulation serve to arrive at a multivariate system of equations over a specific multivariate polynomial ring. Subsequently, we rely on the theory of Gröbner Bases to compute the variety. In the case that there are solutions, each point in the computed variety corresponds to a matrix which constitutes a covering array for the given initial configuration C. Below we show how to construct an algebraic system that incorporates all the necessary conditions.

Different discrete structures have been studied with the help of (generalized) linear systems and Gröbner Bases, such as correlated sequences [27, 36], design matrices [24–26, 28] and linear codes [3]. However, to the best of our knowledge no such algebraic approach has been devoted thus far to the construction and computation of covering arrays.

3.1 Binary Conditions

We are interested in binary covering arrays, and therefore we have to ensure that all entries in the considered matrices are either zero or one. For all variables x_ι in a matrix, we enforce the binary condition via an equation of the following form:

$$x_\iota \,(x_\iota - 1) \tag{1}$$

3.2 Coverage Equations

Definition 4 Let P be a ring and $a, b \in P$. We say that the triple (P, a, b) has the *pairwise binary tuple distinguishing property*, if and only if,

1. P is a unary ring.
2. P is an integral domain.
3. The elements of the set $\{0, a, b, a + b\}$ are pairwise different.

We start with inferring a "distinguisher" for coverage properties for pairwise coverage, meaning we are interested in a function defined on pairs (i.e., two tuples) such that we can infer coverage properties based on the function value. We begin with a fundamental (trivial) observation.

Remark 1 Assume that P is a ring, $a, b \in P$ and that (P, a, b) has the pairwise binary tuple distinguishing property. Then the following equations involving standard inner products of vectors x and y of length η defined over P,

$$\langle x, y \rangle := x \cdot y := \sum_{r=1}^{\eta} x_r y_r,$$

are true statements (we denote by $^\top$ the transpose of a matrix):

$$\left(a, b\right) \cdot \left(0, 0\right)^\top = 0 \tag{2a}$$
$$\left(a, b\right) \cdot \left(1, 0\right)^\top = a \tag{2b}$$
$$\left(a, b\right) \cdot \left(0, 1\right)^\top = b \tag{2c}$$
$$\left(a, b\right) \cdot \left(1, 1\right)^\top = a + b \tag{2d}$$

From the computation results in Eqs. (2a)–(2d) and the pairwise binary tuple distinguishing property assumption, we conclude that the results are pairwise different, and therefore we can use the evaluation of the standard inner product of a tuple having only zero and one as elements with the vector $\left(a, b\right)$ to determine a tuple from the set $\{(0, 0), (1, 0), (0, 1), (1, 1)\}$ uniquely. We formulate this observation in the following Lemma:

Lemma 1 *Assume that P is a ring, $a, b \in P$, and that (P, a, b) has the pairwise binary tuple distinguishing property. Further assume that $\left(t_1, t_2\right)$ is a tuple where each entry is either zero or one. Then, the following statements hold:*

$$\left(t_1, t_2\right) = \left(0, 0\right) \Longleftrightarrow \left(a, b\right) \cdot \left(t_1, t_2\right)^\top = 0 \tag{3a}$$
$$\left(t_1, t_2\right) = \left(1, 0\right) \Longleftrightarrow \left(a, b\right) \cdot \left(t_1, t_2\right)^\top - a = 0 \tag{3b}$$
$$\left(t_1, t_2\right) = \left(0, 1\right) \Longleftrightarrow \left(a, b\right) \cdot \left(t_1, t_2\right)^\top - b = 0 \tag{3c}$$
$$\left(t_1, t_2\right) = \left(1, 1\right) \Longleftrightarrow \left(a, b\right) \cdot \left(t_1, t_2\right)^\top - a - b = 0 \tag{3d}$$

Proof The computations in Remark 1 yield a bijection φ from the set

$$\{(0,0), (1,0), (0,1), (1,1)\} \tag{4}$$

onto the set $\{0, a, b, a+b\}$. For all four cases, the direction "$\Rightarrow$" follows from the computation in Remark 1, and the direction "$\Leftarrow$" from the fact that each element in the set $\{0, a, b, a+b\}$ has exactly one pre-image under the function φ. □

It is the defining property that given a non-empty finite product of elements of an integral domain, the product is zero if and only if at least one factor is zero. This paves the way for making a connection between the notion of coverage in the theory of covering arrays and zeros of finite non-empty products of elements in certain multivariate polynomial rings.

To broaden the technique presented so far to be able to reason about matrices, in Definition 5 below introduced concept of row-selectors established the transformation of the statement "for any selection of t distinct rows" into our algebraic setting using linear operations.

Remark 2 The definition of mixed-level covering arrays in Definition 2 speaks about the selection of any t distinct parameters, which corresponds to the selection of columns in a matrix since they correspond to parameters of the system. In this section, we will almost always work with the transpose of such a matrix and are therefore interested in the selection of t distinct rows.

Definition 5 For $i, j, k \in \mathbb{N}$, $k \geq 2$, $1 \leq i < j \leq k$, let the function

$$e^{k,i,j} : P \times P \longrightarrow P^{1\times k} \tag{5}$$

map a pair of elements from a ring P to a row vector of length k over P, where the first component is mapped to the i-th position, the second component to the j-th position in the vector, and all other entries in the vector are zero.
We call these functions *row-selectors*.

Example 1 Following the terminology in Definition 5 and assuming $a, b \in P$, we will be particularly interested in $e^{k,i,j}(a,b)$, for example

$$e^{9,2,5}(a,b) = (0, a, 0, 0, b, 0, 0, 0, 0). \tag{6}$$

Remark 3 Let P be a ring and $\varepsilon_i, \varepsilon_j \in P$. For any matrix M defined over P with $2 \leq k$ rows and given $i, j \in \mathbb{N}$ such that $1 \leq i < j \leq k$, let the matrix $\widetilde{M}$ consist of the vertical concatenation of the i-th and j-th row in this order of the matrix M. Then

$$e^{k,i,j}(\varepsilon_i, \varepsilon_j)\, M = \left(\varepsilon_i,\ \varepsilon_j\right)\, \widetilde{M}. \tag{7}$$

Example 2 Following the terminology in Definition 5 and Example 1, we consider the matrix

$$M = \begin{pmatrix} 0\,0\,1\,1\,1\,0 \\ 0\,1\,0\,1\,1\,0 \\ 1\,0\,0\,1\,0\,1 \\ 0\,1\,1\,0\,0\,1 \\ 0\,1\,0\,1\,0\,1 \\ 0\,0\,1\,1\,0\,1 \\ 1\,0\,0\,0\,1\,1 \\ 1\,0\,1\,1\,0\,0 \\ 1\,1\,0\,1\,0\,0 \end{pmatrix}, \tag{8}$$

and the row-selector $e^{9,2,5}(a, b) = (0, a, 0, 0, b, 0, 0, 0, 0)$. Simple computation yields that

$$\left(0,\ a,\ 0,\ 0,\ b,\ 0,\ 0,\ 0,\ 0\right) \begin{pmatrix} 0\,0\,1\,1\,1\,0 \\ 0\,1\,0\,1\,1\,0 \\ 1\,0\,0\,1\,0\,1 \\ 0\,1\,1\,0\,0\,1 \\ 0\,1\,0\,1\,0\,1 \\ 0\,0\,1\,1\,0\,1 \\ 1\,0\,0\,0\,1\,1 \\ 1\,0\,1\,1\,0\,0 \\ 1\,1\,0\,1\,0\,0 \end{pmatrix} = \left(0,\ a+b,\ 0,\ a+b,\ a,\ b\right), \tag{9a}$$

$$\left(a,\ b\right) \begin{pmatrix} 0\,1\,0\,1\,1\,0 \\ 0\,1\,0\,1\,0\,1 \end{pmatrix} = \left(0,\ a+b,\ 0,\ a+b,\ a,\ b\right), \tag{9b}$$

illustrating the observations given in Remark 3.

Theorem 1 *Assume that P is a ring, $a, b \in P$, and that (P, a, b) has the pairwise binary tuple distinguishing property. For any given $k \times N$ matrix M defined over P containing only zero and one as entries, and any $1 \leq i < j \leq k$, let $\widetilde{M}$ denote the vertical concatenation of the i-th and j-th row in this order of the matrix M. Let s_a denote the $1 \times N$ row vector with all components equal to a, and let*

$$h = e^{k,i,j}(a, b)M - s_a. \tag{10}$$

Then, the following statements are equivalent:

1. *The tuple $(1, 0)^\top$ appears at least once as a column in the matrix $\widetilde{M}$.*
2. *The vector h contains at least one component equal to zero.*
3. $\prod_{\ell=1}^{N} h_\ell = 0$.

A similar statement holds for the tuples $(0,\ 0)^\top$, $(0,\ 1)^\top$, and $(1,\ 1)^\top$.

Proof The equivalence of 1 and 2 follows from Lemma 1. The equivalence of 2 and 3 follows from the defining property of an integral domain.

Example 3 The conditions of Theorem 1 and the appearances of tuples can be observed in Eqs. (9a) and (9b).

Definition 6 We call the equations appearing in Theorem 1, point 3, *coverage equations* and, using the notation from Theorem 1, define the following notation for $1 \leq i < j \leq k, \tau \in \{(0,0)^\top, (1,0)^\top, (0,1)^\top, (1,1)^\top\}$:

$$\mathrm{coveq}_{\tau}^{(i,j)}(M) = \prod_{\ell=1}^{N} h_\ell. \tag{11}$$

Remark 4 The coverage equations are modelled after the coverage conditions appearing in the definition of covering arrays. The coverage equations are formulated in such a way that they are semantically equivalent to the pairwise coverage conditions for binary covering arrays. This statement is the main result of this section and will be formulated in Corollary 1.

Corollary 1 *Assume that P is a ring, $a, b \in P$, and that (P, a, b) has the pairwise binary tuple distinguishing property. Let M be a $k \times N$ matrix with $2 \leq k$ defined over P, containing only zero and one as entries.*

Then, the statements 1, 2 and 3 are equivalent.

1. *For every selection of two different rows of M, each possible binary tuple appears at least once as a column of the selected $2 \times N$ submatrix of M.*
2. *$M^\top$ is a covering array for the configuration $(2, k, 2)$ in the sense of Definition 1, i.e., the strength two coverage conditions of covering arrays hold for $M^\top$.*
3. *For every i and j with the property $1 \leq i < j \leq k$ and for all $\tau \in \{(0,0)^\top, (1,0)^\top, (0,1)^\top, (1,1)^\top\}$:*

$$\mathrm{coveq}_{\tau}^{(i,j)}(M) = 0. \tag{12}$$

Proof The equivalence of 1 and 2 follows from Definition 1. The equivalence of 1 and 3 follows from Theorem 1.

Based on the previous algebraic modelling, we can now describe the different computational and/or decisional problems of covering arrays from Sect. 2.2 as related problems found in multilinear algebra. In particular, through our *algebraic modelling* the problem(s) of constructing and computing covering arrays can be formulated as instance(s) of algebraic systems, where each solution of them corresponds to a covering array.

3.3 Candidate Matrices

Depending on the problem under consideration (see Sect. 2.2), we obtain a matrix which in some entries has variables x_ι appearing. Assume that there are exactly γ variables appearing in the matrix. In order to apply our distinguisher-based approach, we will regard this matrix as being defined over the multivariate polynomial ring $\mathbb{Q}\left[x_1, x_2, \ldots, x_\gamma, a, b\right]$ in $\gamma + 2$ variables over the rational field $\mathbb{Q}$. A detailed description of the used polynomial ring is given in Sect. 3.4. In the next lemma, we prove that the results of Sect. 3.2 do hold for the case of candidate matrices:

Lemma 2 *Let P be a multivariate polynomial ring in at least two variables defined over the rational field and assume that a and b are two different indeterminates, then (P, a, b) has the pairwise binary tuple distinguishing property.*

Proof The requirements for the pairwise binary tuple distinguishing property hold because of the properties of P.

Using Lemma 2 and candidate matrices, possibly with variables in some entries, we can now compute all binary conditions and coverage equations according to the previous two subsections and use them to reason about coverage statements concerning these matrices. We would like to explicitly point out that there are no binary conditions computed for a and b.

3.4 Treating the Variables

We speak of those variables appearing in a matrix as X-variables $(x_1, \ldots, x_\gamma)$, whereas we think of a and b as A-variables. So far, all matrices are defined over the multivariate polynomial ring

$$\mathbb{Q}\left[x_1, x_2, \ldots, x_\gamma, a, b\right]$$

in $\gamma + 2$ variables. Concerning the A-variables, they do not appear in the solutions of the modelled matrices. Note that all X-variables take values in $\{0, 1\}$. Since we are only interested in the solutions w.r.t. X-components, we want to project the variety in the subspace spanned by X.

Gathering all the polynomials mentioned in this section, we obtain an algebraic description. This algebraic description is an ideal in $\mathbb{Q}\left[x_1, x_2, \ldots, x_\gamma, a, b\right]$, which we call the *coverage ideal* of the candidate matrix. From the theory of Gröbner bases we know that the Gröbner basis is a full description of an ideal, but has a better form than a random set of generators (as the ones we obtained by our analysis of the problem).

3.5 Solving of the Systems

We rely on the theory of Gröbner bases, we give the reference to their initial proposition in [7], whereby a comprehensive account can be found in [2]. There exist efficient algorithms for computing Gröbner bases, such as the F4 [13] and F5 algorithms [14], among others.

Given a set of equations forming a coverage ideal I in $\mathbb{Q}\left[x_1, x_2, \ldots, x_\gamma, a, b\right]$, to solve the system, we first choose random values for a and b. We evaluate the polynomials in this set with the chosen values for a and b and interpret them as elements of $R = \mathbb{Q}\left[x_1, x_2, \ldots, x_\gamma\right]$. These equations define an ideal I_R restricted to R, whereas we compute a Gröbner Basis (GB) of this ideal in the MAGMA computer algebra system [4]. When $\mathrm{GB}(I_R) \neq \{1\}$ then the resulting variety will entail all points (solutions of the algebraic system) that correspond to actual covering arrays upon replacing the values of X-variables into the entries of the candidate matrices. Otherwise, when $\mathrm{GB}(I_R) = \{1\}$ there is no solution to the *specific* algebraic system, even though we cannot exclude the possibility that another random replacement of A-variables will result in a non-empty variety. However, we have not observed such a case in our experiments. We would like to note that whenever a nontrivial variety is obtained, the corresponding matrices are covering arrays by Corollary 1.

4 Algorithmic Approaches to the Covering Array Problems

We would like to point out that most constructions in this section operate on the transpose of a covering array, i.e. meaning that rows are corresponding to parameters.

4.1 An Algorithmic Approach to the Vertical Extension Problems

In the first example we present how to extend a given matrix, which is compatible with but not a covering array for a covering array configuration C, with one additional column such that all missing tuples will appear in the extended matrix (relates to Problems 11, 12, 13, 16).

Specifically, we consider the configuration $C = (2, 2, 2)$, i.e. two binary parameters for strength two, and the following matrix:

$$M = \begin{pmatrix} 0 & 1 & 0 \\ 0 & 0 & 1 \end{pmatrix}. \tag{13}$$

We extend with one additional column, therefore we will be working in $P = \mathbb{Q}\left[x_1, x_2, a, b\right]$, i.e. in the multivariate polynomial ring in four variables over the rational field. The extension column consists of the first two indeterminates, $\left(x_1\ x_2\right)^\top$, and is added at the end of the given matrix M:

$$Mext = \begin{pmatrix} 0 & 1 & 0 & x_1 \\ 0 & 0 & 1 & x_2 \end{pmatrix}. \tag{14}$$

We select the first and second row of the matrix $Mext$ and create the coverage equations. We start with deriving the coverage equation for the $(0,\ 0)^\top$ tuple.

$$v_{00} = (a,\ b) \begin{pmatrix} 0 & 1 & 0 & x_1 \\ 0 & 0 & 1 & x_2 \end{pmatrix} = \tag{15a}$$

$$(0,\ a,\ b,\ x_1 a + x_2 b). \tag{15b}$$

Taking the product of the elements of the vector v_{00} leads to the first coverage equation $coveq^{(1,2)}_{(0,0)^\top}(Mext) = 0$. As the tuple $(0,\ 0)^\top$ appears in the matrix M, we expect the respective coverage equation to hold which it does as can derived from Eq. (15b). Similarly, the coverage equations for the tuples $(1,\ 0)^\top$ and $(0,\ 1)^\top$ hold as well. The only nontrivial coverage equation arises for the $(1,\ 1)^\top$ tuple:

$$-x_1 a^3 b - x_1 a^2 b^2 - x_2 a^2 b^2 - x_2 a b^3 + a^3 b + 2a^2 b^2 + ab^3. \tag{16}$$

Next, we add the binary conditions for the variables x_1 and x_2:

$$x_1^2 - x_1,\ x_2^2 - x_2. \tag{17}$$

We now substitute random values for the A-variables

$$arand = -13400/112, \tag{18a}$$

$$brand = 290349/125, \tag{18b}$$

and denote by `alleqnoab` the set consisting of the polynomials occurring in the only nontrivial coverage equation (cf. Eq. (16)) and in the binary conditions (cf. Eq. (17)). All further computations take place in $R = \mathbb{Q}[x_1, x_2]$. We compute the Gröbner Basis of the following ideal in MAGMA:

```
I_R = ideal < R | alleqnoab >,
```

where `I_R` denotes the restricted ideal I_R as defined in Sect. 3.5. The following computation returns the Gröbner Basis polynomials (where *rank* refers to the number of variables in a multivariate polynomial ring) in MAGMA:

Ideal of Polynomial ring of rank 2 over Rational Field, Lexicographical Order, Variables x_1, x_2, Dimension 0, Groebner basis:

$$(x_1 - 1, x_2 - 1) \tag{19}$$

The variety consists of only one point, corresponding to the tuple $(1,\ 1)^\top$. Substituting this solution into the extended matrix $Mext$ leads to the transpose of a covering array in the sense of Definition 1 for the configuration (2, 2, 2):

$$\begin{pmatrix} 0 & 1 & 0 & 1 \\ 0 & 0 & 1 & 1 \end{pmatrix}. \tag{20}$$

The pseudocode for the Vertical Extension procedure, that has been used in this example, is given in Algorithm 1.

Algorithm 1 Vertical Extension

```
procedure VERT-EXT(M)
Require: matrix M                                   ▷ rows corresponding to parameters
    k ← NumberOfRows(M)
    P ← ℚ[x_1, ..., x_k, a, b]
    E ← (x_1, ..., x_k)^T
    Mext ← HorizontalConcatenation(M, E)
    eqall ← ∅
    for i = 1, 2, ..., k do
        for j = i + 1, ..., k do
            for τ ∈ {(0, 0)^T, (1, 0)^T, (0, 1)^T, (1, 1)^T} do
                eqall ← eqall ∪ {coveq_τ^(i,j) (Mext)}
            end for
        end for
    end for

    SetOfBinaryConditions ← Compute binary equations
    eqall ← eqall ∪ SetOfBinaryConditions
    Randomly replace a and b in eqall
    Regard polynomials in eqall as elements of a set s over R = ℚ[x_1, ..., x_k]
    I_R ← ideal < R|s >
    GB ← GröbnerBasis(I_R)
    if GB ≠ {1} then
        V ← Variety(GB)
        Print "Non-empty set of solutions (CAs) found."
        return V
    else
        Print "No solutions found."
        return ∅
    end if
end procedure
```

4.2 An Algorithmic Approach to the Parameter Extension Problems

In this example we are given a covering array with k parameters and we want to extend it to a covering array with one additional parameter by extending the given matrix with a new row and in particular without adding more columns (relates to Problems 6, 7, 8). Consider the following matrix M, which is a covering array for the configuration $C = (2, 2, 2)$:

$$M^\top = \begin{pmatrix} 0 & 1 & 0 & 1 \\ 0 & 0 & 1 & 1 \end{pmatrix}. \tag{21}$$

We will add a row vector of length four to the matrix, whose entries are the first four indeterminates of the multivariate polynomial ring $P = \mathbb{Q}[x_1, x_2, x_3, x_4, a, b]$, leading to the matrix

$$Mext = \begin{pmatrix} 0 & 1 & 0 & 1 \\ 0 & 0 & 1 & 1 \\ x_1 & x_2 & x_3 & x_4 \end{pmatrix}. \tag{22}$$

The next step is to derive the coverage equations. We begin by selecting the first and second row of the matrix $Mext$, the respective row-selector vector is $e^{3,1,2}(a,b) = \big(a, b, 0\big)$. All resulting coverage equations for this pair of selected rows hold, since by choice we started with a matrix that is already a covering array of strength two for two parameters. Therefore, we only have to consider row-selection pairs which include the newly added row. The four coverage equations for the selection of the first and third row are as follows:

$$x_1x_2x_3x_4b^4 + x_1x_2x_3ab^3 + x_1x_3x_4ab^3 + x_1x_3a^2b^2 \tag{23a}$$

$$\begin{aligned} &x_1x_2x_3x_4b^4 + x_1x_2x_3ab^3 - x_1x_2x_3b^4 - x_1x_2x_4b^4 - x_1x_2ab^3 + x_1x_2b^4 + \\ &\quad x_1x_3x_4ab^3 - x_1x_3x_4b^4 + x_1x_3a^2b^2 - 2x_1x_3ab^3 + x_1x_3b^4 - x_1x_4ab^3 + \\ &\quad x_1x_4b^4 - x_1a^2b^2 + 2x_1ab^3 - x_1b^4 - x_2x_3x_4b^4 - x_2x_3ab^3 + x_2x_3b^4 + \\ &\quad x_2x_4b^4 + x_2ab^3 - x_2b^4 - x_3x_4ab^3 + x_3x_4b^4 - x_3a^2b^2 + \\ &\quad 2x_3ab^3 - x_3b^4 + x_4ab^3 - x_4b^4 + a^2b^2 - 2ab^3 + b^4 \end{aligned} \tag{23b}$$

$$x_1x_2x_3x_4b^4 - x_1x_2x_4ab^3 - x_2x_3x_4ab^3 + x_2x_4a^2b^2 \tag{23c}$$

$$\begin{aligned} &x_1x_2x_3x_4b^4 - x_1x_2x_3b^4 - x_1x_2x_4ab^3 - x_1x_2x_4b^4 + x_1x_2ab^3 + x_1x_2b^4 - \\ &\quad x_1x_3x_4b^4 + x_1x_3b^4 + x_1x_4ab^3 + x_1x_4b^4 - x_1ab^3 - x_1b^4 - \\ &\quad x_2x_3x_4ab^3 - x_2x_3x_4b^4 + x_2x_3ab^3 + x_2x_3b^4 + x_2x_4a^2b^2 + \\ &\quad 2x_2x_4ab^3 + x_2x_4b^4 - x_2a^2b^2 - 2x_2ab^3 - x_2b^4 + x_3x_4ab^3 + x_3x_4b^4 - \\ &\quad x_3ab^3 - x_3b^4 - x_4a^2b^2 - 2x_4ab^3 - x_4b^4 + a^2b^2 + 2ab^3 + b^4. \end{aligned} \tag{23d}$$

We follow again the random replacement approach for the indeterminates a and b and substitute the random values into the equations and interpret them as members of $R = \mathbb{Q}[x_1, x_2, x_3, x_4]$. In MAGMA, we compute a Gröbner Basis of the respective ideal and the following computation is returned (where *rank* refers to the number of variables in a multivariate polynomial ring):

Ideal of Polynomial ring of rank 4 over Rational Field, Lexicographical Order, Variables: x_1, x_2, x_3, x_4, *Dimension 0, Groebner basis:*

$$(x_1 - x_4, x_2 + x_4 - 1, x_3 + x_4 - 1, x_4^2 - x_4) \tag{24}$$

The variety consists of the following two points,

```
(<0, 1, 1, 0> , <1, 0, 0, 1>)
```

meaning that there are two possible ways to extend to a covering array with three binary parameters of strength two while using the given matrix M as a "seed".

The pseudocode for the Parameter Extension procedure, that has been used in this example, is given in Algorithm 2.

Algorithm 2 Parameter Extension

procedure PARA- EXT(M)
Require: covering array M ▷ rows corresponding to parameters
  $N \leftarrow$ `NumberOfColumns`(M)
  $k \leftarrow$ `NumberOfRows`(M)
  $P \leftarrow \mathbb{Q}[x_1, \ldots, x_N, a, b]$
  $E \leftarrow (x_1, \ldots, x_N)$
  $Mext \leftarrow$ `VerticalConcatenation`(M, E)
  $j \leftarrow k + 1$
  $eqall \leftarrow \emptyset$
  for $i = 1, 2, \ldots, k$ **do**
    for $\tau \in \{(0, 0)^\top, (1, 0)^\top, (0, 1)^\top, (1, 1)^\top\}$ **do**
      $eqall \leftarrow eqall \cup \{coveq_\tau^{(i,j)}(Mext)\}$
    end for
  end for

  SetOfBinaryConditions $\leftarrow$ Compute binary equations
  $eqall \leftarrow eqall \cup$ SetOfBinaryConditions
  Randomly replace a and b in $eqall$
  Regard polynomials in $eqall$ as elements of a set s over $R = \mathbb{Q}[x_1, \ldots, x_N]$
  $I_R \leftarrow$ `ideal` $< R|s >$
  $GB \leftarrow$ `GröbnerBasis`(I_R)
  if $GB \neq \{1\}$ **then**
    $V \leftarrow$ `Variety`(GB)
    Print "Parameter extension successful."
    return V
  else
    Print "Parameter extension not possible."
    return $\emptyset$
  end if
end procedure

4.3 An Algorithmic Approach to the Computational Existence of Covering Arrays

Given a configuration $C = (t, k, v)$ of a covering array with a chosen value of k (i.e., number of parameters), one may "guess" the number of rows N required for a covering array in the sense of Definition 1 for C (relates to Problems 1, 2, 3, 18). Clearly, in the case $\mathsf{CAN}(C) \leq N$ there is at least one solution, whereas in the case $\mathsf{CAN}(C) > N$ there are no solutions. In the first case, our algebraic modelling provides the means to actually construct such a matrix. The idea is detailed in an approach called *Guess*. There exists a 4×2 covering array for the configuration $(2, 2, 2)$, and assume that we "guess" that there exists a 4×3 matrix which forms a covering array for the configuration $(2, 3, 2)$. In contrast to Sect. 4.2, the Guess approach constructs the complete matrix. While in this example an exhaustive search-based approach is still feasible, this might no longer be the case for a greater number of parameters or rows. Continuing the example, we will work in

$$P = \mathbb{Q}[x_1, x_2, x_3, x_4, x_5, x_6, x_7, x_8, x_9, x_{10}, x_{11}, x_{12}, a, b],$$

while the initialization of the candidate matrix is described in the following MAGMA code:

```
S:=RationalField();
P:=PolynomialRing(S,k*N+2);
R:=PolynomialRing(S,k*N);
M := ZeroMatrix(P,k,N);
for i in [1..k] do
    for j in [1..N] do
        M[i][j] := P.((i-1)*N+j); // P_i variable is P.i in MAGMA
    end for;
end for;
```

MAGMA code for Guess candidate matrix generation.

In the next step, we create all coverage equations and all binary conditions. It follows that in the Guess approach, there arise

$$\binom{k}{2} \cdot 2^2 + kN \tag{25}$$

equations in total. In our example, we have 12 coverage equations and 12 binary conditions, yielding a total of 24 equations. So far, these equations are defined over the polynomial ring P in 14 variables.

Again, we choose random values for a and b, evaluate the polynomials and interpret the resulting polynomials as elements of

$$R = \mathbb{Q}[x_1, x_2, x_3, x_4, x_5, x_6, x_7, x_8, x_9, x_{10}, x_{11}, x_{12}].$$

We compute a Gröbner Basis of the respective ideal comprised the new 24 equations in MAGMA and the computation returns that the basis consists in 17 polynomials.

The corresponding variety has 48 points, which means we have computed 48 different 3×4 matrices, those transposes constitute covering arrays in the sense of Definition 1 of strength two for three binary parameters. With an independent exhaustive search and simple tuple counting approach, we have verified that the transposes of these 48 matrices are in fact *all* 4×3 matrices, which constitute covering arrays in the sense of Definition 1 of strength two for three binary parameters.

The pseudocode for the Guess procedure, that has been used in this example, is given in Algorithm 3.

Algorithm 3 Guess

```
procedure GUESS(k, N)
Require: k ∈ ℕ                                        ▷ k is the number of parameters
Require: N ∈ ℕ                                        ▷ N is the number of columns
    A ← k × N matrix over ℚ[x_1, ..., x_{kN}, a, b] with entries x_ι
    eqall ← ∅
    for i = 1, 2, ..., k do
        for j = i + 1, ..., k do
            for τ ∈ {(0, 0)^T, (1, 0)^T, (0, 1)^T, (1, 1)^T} do
                eqall ← eqall ∪ {coveq_τ^(i,j) (A)}
            end for
        end for
    end for

    SetOfBinaryConditions ← Compute binary equations
    eqall ← eqall ∪ SetOfBinaryConditions
    Randomly replace a and b in eqall
    Regard polynomials in eqall as elements of a set s over R = ℚ[x_1, ..., x_{kN}]
    I_R ← ideal < R|s >
    GB ← GröbnerBasis(I_R)
    if GB ≠ {1} then
        V ← Variety(GB)
        Print "Non-empty set of solutions (CAs) found."
        return V
    else
        Print "No solutions found."
        return ∅
    end if
end procedure
```

---

5 Comparison with Greedy Algorithms

In this section, we give some cases where our method compares favourably to the IPOG algorithm, one of the most known greedy algorithms. These cases are merely used for illustration of the method's potential rather than a benchmark, as this is not the scope of the current paper.

The NIST tables of covering arrays [22] are a publicly accessible source of covering arrays for various covering array configurations that have been constructed using the IPOG-F algorithm. This archive serves also to support practitioners of combinatorial testing by providing additional resources.

At [21], a covering array with 6 rows for 9 binary parameters is available. We took this covering array and applied our Parameter Extension procedure. By doing so, we were able to successfully extend the chosen initial matrix in two ways to a covering array for 10 binary parameters of strength two, while keeping 6 rows. The best covering array for the configuration (2, 10, 2) provided at the NIST tables is a matrix with 8 rows at [18]. The two new matrices (in the sense of Definition 1) are given below:

$$\begin{pmatrix} 0&0&1&0&0&0&1&1&1&0 \\ 0&1&0&1&1&0&0&0&1&0 \\ 1&0&0&1&0&1&0&1&0&0 \\ 1&1&1&0&1&1&0&1&1&1 \\ 1&1&0&0&0&0&1&0&0&1 \\ 0&0&1&1&1&1&1&0&0&1 \end{pmatrix}$$

$$\begin{pmatrix} 0&0&1&0&0&0&1&1&1&1 \\ 0&1&0&1&1&0&0&0&1&1 \\ 1&0&0&1&0&1&0&1&0&1 \\ 1&1&1&0&1&1&0&1&1&0 \\ 1&1&0&0&0&0&1&0&0&0 \\ 0&0&1&1&1&1&1&0&0&0 \end{pmatrix}$$

Continuing in this direction, we start with a covering array for 16 binary parameters of strength two and 8 rows available at [19]. This covering array is given in a specialized format, where the described matrix contains entries which are undefined so as to indicate that the algorithm determined during its construction that these entries in the matrix are irrelevant from the standpoint of ensuring the pairwise coverage property. In a preprocessing step, we replaced these entries with zeros and denote the resulting matrix as $\hat{M}$. Again, we applied our Parameter Extension procedure to the matrix $\hat{M}$, yielding twelve possible extensions while keeping 8 rows. The best covering array for the configuration (2, 17, 2) provided at the NIST tables is a matrix with 10 rows available at [20].

Finally, a table listing the best known sizes of binary covering arrays of strength two is available at [9].

6 Conclusion and Future Work

In this paper, we provided some ideas on how to model covering arrays from an algebraic perspective with a close connection to computer algebra. Since we considered only binary strength two covering arrays, an immediate future work is the extension of this approach to higher strengths, parameters with order at least three, and extensions to related structures such as mixed-level and variable-strength covering arrays. Finally, the structure of the generated system of equations can be studied further.

Acknowledgements The research presented in the paper has been funded in part by the Austrian Research Promotion Agency (FFG) under Grant 851205 (SPLIT—Security ProtocoL Interaction Testing in practice) and the Austrian COMET K1 Program (FFG).

References

1. Avila-George, H., Torres-Jimenez, J., Hernández, V.: New bounds for ternary covering arrays using a parallel simulated annealing. Math. Probl. Eng. (2012)
2. Becker, T., Weispfenning, V.: Gröbner bases. In: A Computational Approach to Commutative Algebra. Graduate Studies in Mathematics, vol. 141. Springer, New York (1993)
3. Borges-Quintana, M., Borges-Trenard, M.A., Fitzpatrick, P., Martínez-Moro, E.: Gröbner bases and combinatorics for binary codes. Appl. Algebra Eng. Commun. Comput. **19**(5), 393–411 (2008)
4. Bosma, W., Cannon, J., Playoust, C.: The Magma algebra system. I. The user language. J. Symb. Comput. **24**(3-4), 235–265 (1997). Computational algebra and number theory (London, 1993)
5. Bracho-Rios, J., Torres-Jimenez, J., Rodriguez-Tello, E.: A new backtracking algorithm for constructing binary covering arrays of variable strength. In: MICAI 2009: Advances in Artificial Intelligence, pp. 397–407. Springer, New York (2009)
6. Bryce, R.C., Colbourn, C.J.: The density algorithm for pairwise interaction testing. Softw. Test. Verif. Reliab. **17**(3), 159–182 (2007)
7. Buchberger, B.: Bruno Buchberger's phd thesis 1965: an algorithm for finding the basis elements of the residue class ring of a zero dimensional polynomial ideal. J. Symb. Comput. **41**, 475–511 (2006). doi:10.1016/j.jsc.2005.09.007
8. Cohen, D.M., Dalal, S.R., Fredman, M.L., Patton, G.C.: The AETG system: an approach to testing based on combinatorial design. IEEE Trans. Softw. Eng. **23**(7), 437–444 (1997)
9. Colbourn, C.: Table for CAN(2,k,2) for k up to 20000. http://www.public.asu.edu/~ccolbou/src/tabby/2-2-ca.html. Accessed 31 Dec 2015
10. Colbourn, C.J.: Combinatorial aspects of covering arrays. Le Mathematiche **LIX**(I–II), pp. 125–172 (2004)
11. Colbourn, C.J.: Covering arrays from cyclotomy. Des. Codes Cryptogr. **55**(2–3), 201–219 (2010)
12. Colbourn, C.J., Dinitz, J.H.: Handbook of Combinatorial Designs. CRC Press, Boca Raton (2006)
13. Faugere, J.C.: A new efficient algorithm for computing Gröbner bases (f 4). J. Pure Appl. Algebr. **139**(1), 61–88 (1999)
14. Faugère, J.C.: A new efficient algorithm for computing grÖbner bases without reduction to zero (f5). In: Proceedings of the 2002 International Symposium on Symbolic and Algebraic Computation (ISSAC '02), pp. 75–83. ACM, New York (2002). doi:10.1145/780506.780516

15. Gonzalez-Hernandez, L., Rangel-Valdez, N., Torres-Jimenez, J.: Construction of mixed covering arrays of strengths 2 through 6 using a tabu search approach. Discret. Math. Algorithm Appl. **4**(03), 1250,033 (2012)
16. Hartman, A., Raskin, L.: Problems and algorithms for covering arrays. Discret. Math. **284**(1), 149–156 (2004)
17. Hnich, B., Prestwich, S.D., Selensky, E., Smith, B.M.: Constraint models for the covering test problem. Constraints **11**(2–3), 199–219 (2006)
18. IPOG-F: CA(2,10,2). http://math.nist.gov/coveringarrays/ipof/cas/t=2/v=2/ca.2.2^10.txt.zip. Accessed 31 Dec 2015
19. IPOG-F: CA(2,16,2). http://math.nist.gov/coveringarrays/ipof/cas/t=2/v=2/ca.2.2^16.txt.zip. Accessed 31 Dec 2015
20. IPOG-F: CA(2,17,2). http://math.nist.gov/coveringarrays/ipof/cas/t=2/v=2/ca.2.2^17.txt.zip. Accessed 31 Dec 2015
21. IPOG-F: CA(2,9,2). http://math.nist.gov/coveringarrays/ipof/cas/t=2/v=2/ca.2.2^9.txt.zip. Accessed 31 Dec 2015
22. ITL, N.: Covering array tables. http://math.nist.gov/coveringarrays/. Accessed 31 Dec 2015
23. Kleitman, D.J., Spencer, J.: Families of k-independent sets. Discret. Math. **6**(3), 255–262 (1973)
24. Kotsireas, I., Koukouvinos, C., Seberry, J.: Hadamard ideals and hadamard matrices with circulant core. J. Combin. Math. Combin. Comput. **57**, 47–63 (2006)
25. Kotsireas, I., Koukouvinos, C., Seberry, J.: Hadamard ideals and Hadamard matrices with two circulant cores. Eur. J. Comb. **27**(5), 658–668 (2006)
26. Kotsireas, I.S., Kutsia, T., Simos, D.E.: Constructing orthogonal designs in powers of two: Gröbner bases meet equational unification. In: 26th International Conference on Rewriting Techniques and Applications (RTA), June 29 to July 1, 2015, Warsaw, pp. 241–256 (2015)
27. Koukouvinos, C., Simos, D.E., Zafeirakopoulos, Z.: An algebraic framework for extending orthogonal designs. In: ISSAC '11: Abstracts of Poster Presentations of the 36th International Symposium on Symbolic and Algebraic Computation, ACM Communications in Computer Algebra, vol. 45, pp. 123–124 (2011)
28. Koukouvinos, C., Simos, D.E., Zafeirakopoulos, Z.: A Gröbner bases method for complementary sequences. In: Proceedings of Applications of Computer Algebra (ACA), p. 255. Málaga (2013)
29. Kuhn, R., Kacker, R., Lei, Y., Hunter, J.: Combinatorial software testing. Computer **8**, 94–96 (2009)
30. Kuhn, D.R., Kacker, R.N., Lei, Y.: Introduction to Combinatorial Testing. CRC Press, Boca Raton (2013)
31. Lei, Y., Tai, K.C.: In-parameter-order: a test generation strategy for pairwise testing. In: Third IEEE International Proceedings High-Assurance Systems Engineering Symposium, pp. 254–261 (1998)
32. Lei, Y., Kacker, R., Kuhn, D.R., Okun, V., Lawrence, J.: IPOG: a general strategy for t-way software testing. In: 14th Annual IEEE International Conference and Workshops on the Engineering of Computer-Based Systems (ECBS'07), pp. 549–556 (2007)
33. Nurmela, K.J.: Upper bounds for covering arrays by tabu search. Discret. Appl. Math. **138**(1), 143–152 (2004)
34. Raaphorst, S., Moura, L., Stevens, B.: A construction for strength-3 covering arrays from linear feedback shift register sequences. Des. Codes Cryptogr. **73**(3), 949–968 (2014)
35. Seroussi, G., Bshouty, N.H.: Vector sets for exhaustive testing of logic circuits. IEEE Trans. Inf. Theory **34**(3), 513–522 (1988)
36. Shorin, V.V., Loidreau, P.: Application of Groebner bases techniques for searching new sequences with good periodic correlation properties. In: Proceedings International Symposium on Information Theory (ISIT), pp. 1196–1200 (2005)
37. Torres-Jimenez, J., Izquierdo-Marquez, I.: Survey of covering arrays. In: 15th International Symposium on Symbolic and Numeric Algorithms for Scientific Computing (SYNASC), pp. 20–27 (2013)

38. Torres-Jimenez, J., Rodriguez-Tello, E.: New bounds for binary covering arrays using simulated annealing. Inf. Sci. **185**(1), 137–152 (2012)
39. Torres-Jimenez, J., Izquierdo-Marquez, I., Gonzalez-Gomez, A., Avila-George, H.: A branch & bound algorithm to derive a direct construction for binary covering arrays. In: Advances in Artificial Intelligence and Soft Computing, pp. 158–177. Springer, New York (2015)
40. Yan, J., Zhang, J.: Backtracking algorithms and search heuristics to generate test suites for combinatorial testing. In: 30th Annual International Computer Software and Applications Conference (COMPSAC'06), vol. 1, pp. 385–394 (2006)

Applications of Signatures Curves to Characterize Melanomas and Moles

Anna Grim and Chehrzad Shakiban

Abstract In this paper, we focus on the application of an Euclidean invariant curve, called the signature curve, formed by taking curvature and derivative of curvature with respect to arc length of a closed curve, $\Sigma = \{(\kappa(t), \kappa_s(t))\}$ to analyze the contour of melanomas and moles. We calculate the signature curves of the contours of the skin lesions to detect asymmetry, boundary irregularity and diameter size of the skin lesions. By analyzing the signature curves of 60 benign moles and 60 melanomas, we show that the benign and malignant lesions have different global and local symmetry patterns in their signature curves. We will also demonstrate that the regular moles show a high degree of global symmetry, whereas melanomas exhibit multiple types of local symmetry that are embedded within their signature curves. We then turn our attention to the *C* aspect of the ABCD method by analyzing the color of melanomas and moles. Finally, we use ROC Analysis, a key statistical tool, to analyze the performance of our method.

Keywords Curvature · Derivative · Signature curves

1 Introduction

Noninvasive diagnosis of melanoma persists as a challenge for dermatologists because of the structural differences between moles and melanomas are often indistinguishable to the human eye. Melanoma, the most serious type of skin cancer, develops in the cells that produce melanin—the pigment that gives skin its color [1]. The cancerous skin lesion is capable of spreading throughout the body, making it difficult to treat in advanced cases. In addition, visual similarities between melanoma and mole make diagnosis difficult, and often require the use of an invasive skin biopsy. Dermatologists often use the ABCD method (Fig. 1) to determine the necessity of a skin biopsy. This research focuses on *B* and *C* aspects of the ABCD method.

A. Grim · C. Shakiban (✉)
University of St. Thomas, St. Paul, MN 55105, USA
e-mail: cshakiban@stthomas.edu

I.S. Kotsireas and E. Martínez-Moro (eds.), *Applications of Computer Algebra*, Springer Proceedings in Mathematics & Statistics 198,
DOI 10.1007/978-3-319-56932-1_11

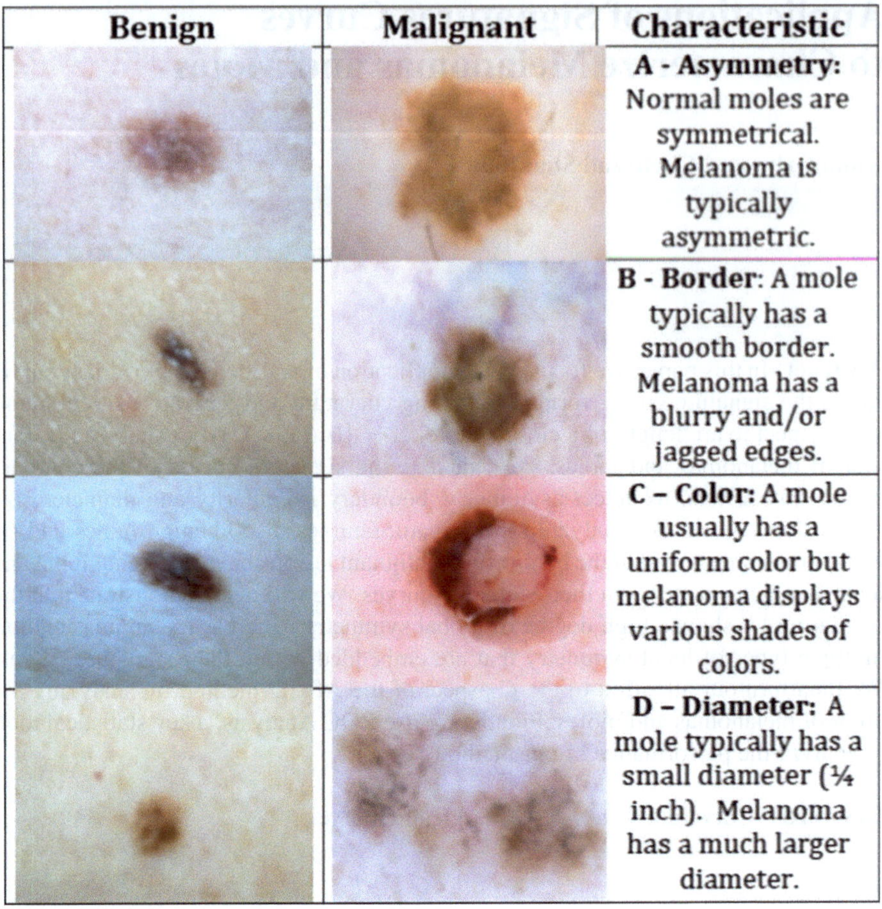

Fig. 1 ABCD of melanomas and moles

In this paper, we focus on the application of an Euclidean invariant curve, called the *signature curve*, formed by taking curvature and derivative of curvature with respect to arc length of a closed curve, $\Sigma = \{(\kappa(t), \kappa_s(t))\}$ to analyze the contour of melanomas and moles. We calculate the signature curves of the contours of the skin lesions to detect asymmetry, boundary irregularity and diameter size of the skin lesions. By analyzing the signature curves of 60 benign moles and 60 melanomas, we show that the benign and malignant lesions have different global and local symmetry patterns in their signature curves. We will also demonstrate that the regular moles show a high degree of global symmetry, whereas melanomas exhibit multiple types of local symmetry that are embedded within their signature curves. We then turn our attention to the *C* aspect of the ABCD method by analyzing the color of melanomas and moles. Finally, we use receiver operating characteristic (ROC) analysis, a key statistical tool, to analyze the performance of our method.

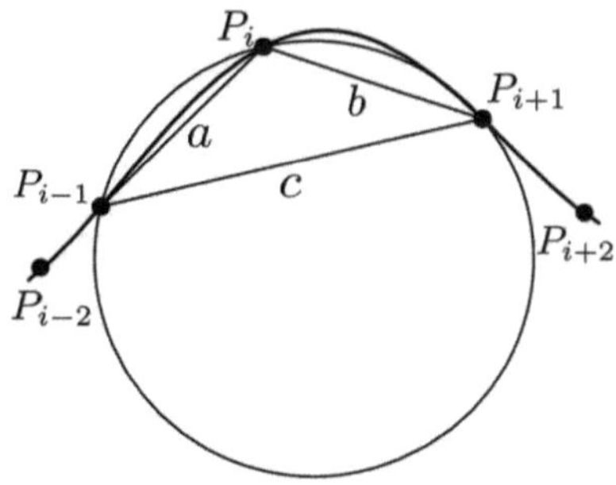

Fig. 2 Approximate curvature at an arbitrary point

1.1 Signature Curves

Signature curves are the fundamental tool in the methodology, we use in distinguishing melanomas and moles, due to their invariance properties in the Euclidean plane [8]. In order to calculate the signature curve of a closed curve in the Euclidean space, first we must parameterize the curve $\mathcal{C} = (x(t), y(t))$. The signature curve Σ corresponding to $\mathcal{C}$ is given by $\Sigma = \{(\kappa(t), \kappa_s(t))\}$, where κ is curvature and κ_s is the derivative of curvature with respect to arc length. A reformation of a theorem by Élie Cartan states that

Theorem 1 *Two smooth and nondegenerate curves $\mathcal{C}$ and $\bar{\mathcal{C}}$ can be mapped to each other by a proper Euclidean transformation, g, i.e., $\bar{\mathcal{C}} = g\mathcal{C}$, if and only if their signature curves are identical: $\Sigma = \bar{\Sigma}$,* [3, 5, 14].

This indicates that the signature curve can provide an efficient mechanism for recognizing both exact and approximate Euclidean symmetries of objects.

In order to calculate the signature curve of the contours of the images of the lesions, we can use a numerical method proposed in [2, 3]. Our goal is to approximate the curvature of $\mathcal{C}$ in a Euclidean invariant manner. This requires the approximation to depend only on the distances $d(P_i, P_j)$ between mesh points. Because the curvature is a second order differential function, the simplest approximation will require three mesh points. With this in mind, we now derive the basic approximation formula for the curvature (Fig. 2).

The approximate curvature $\tilde{\kappa}$ at an arbitrary point $P_i \in \mathcal{C}$ with respect to arc length we choose the points $P_{i-1}, P_{i+1} \in \mathcal{C}$, forming the triangle P_{i-1}, P_i, P_{i+1} illustrated in Fig. 1. Let Δ represent the signed area of this triangle and let $s = \frac{1}{2}(a+b+c)$ denote its semi-perimeter, so that $\Delta = \pm\sqrt{s(s-a)(s-b)(s-c)}$. We apply Heron's formula to compute the radius of the circle passing through the points P_{i-1}, P_i, P_{i+1}, leading to the formula of

$$\tilde{\kappa}(P_{i-1}, P_i, P_{i+1}) = 4\frac{\Delta}{abc} = \pm 4\frac{\sqrt{s(s-a)(s-b)(s-c)}}{abc}. \tag{1}$$

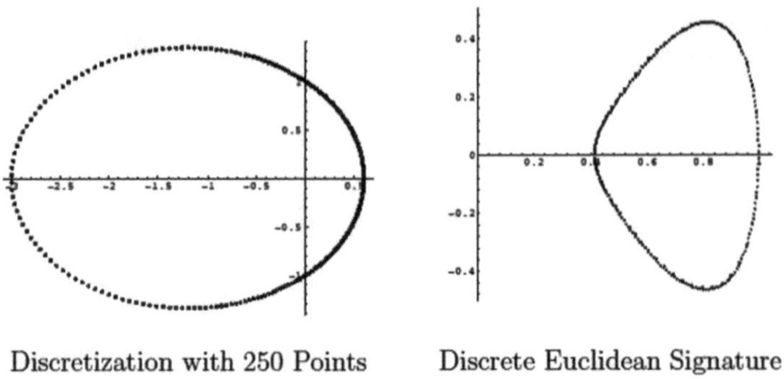

Fig. 3 Discrete Euclidean signature curve of an ellipse

The derivative of curvature is calculated in a similar manner by calculating the approximate curvature at the points $P_{i-1}, P_{i+1} \in C$. Therefore, the approximate derivative of curvature $\tilde{\kappa}_s$ at point P_i can be calculated by

$$\tilde{\kappa}_s\left(P_{i-2}, P_{i-1}, P_i, P_{i+1}, P_{i+2}\right) = \frac{\tilde{\kappa}(P_i, P_{i+1}, P_{i+2}) - \tilde{\kappa}(P_{i-2}, P_{i-1}, P_i)}{\mathbf{d}(P_{i+1}, P_{i-1})}, \qquad (2)$$

where $\mathbf{d}\left(P_{i+1}, P_{i-1}\right)$ is the Euclidean distance between P_{i+1} and P_{i-1}. Equations (1) and (2) are used to obtain an approximation for the signature curve. Therefore, Σ can be graphed by using

$$\{\tilde{\kappa}\left(P_{i-1}, P_i, P_{i+1}\right), \tilde{\kappa}_s\left(P_{i-2}, P_{i-1}, P_i, P_{i+1}, P_{i+2}\right)\}.$$

Figure 3 illustrates the graph of the discrete Euclidean signature for an ellipse with 250 mesh points.

1.2 Calculation of Signature Curves of Melanomas and Moles

The process for finding signature curves of melanomas and moles begins with a number of high resolution images. An active contour segmentation program in MATLAB processes the image and determines the outermost boundary of the skin lesion.

The output of this program consists of the cartesian coordinates (x_i, y_i) of n set of points $\{P_1, P_2, P_3, \ldots, P_n\}$. This data set is then exported as a text file to Mathematica where the discrete closed curve obtained by these points is smoothed several times using a smoothing spline algorithm in order to robustly calculate the signature curve of the lesion. This process is illustrated in Figs. 4 and 5. As the images we consider are not uniformly sized, a method of scaling is required to ensure that data we obtain from the boundary of the images we use are comparable.

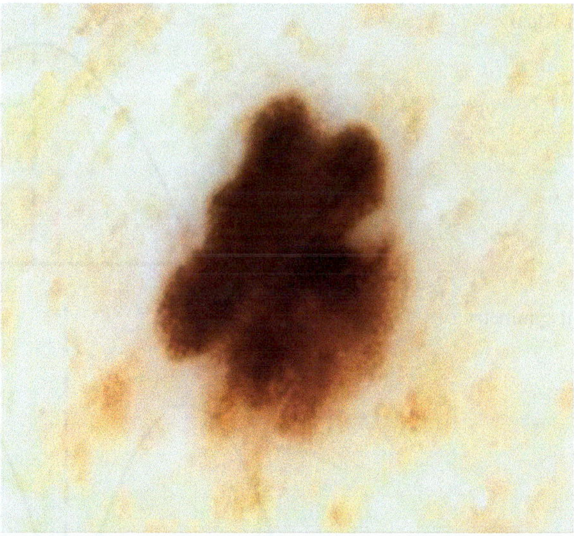

Fig. 4 Image of a melanoma

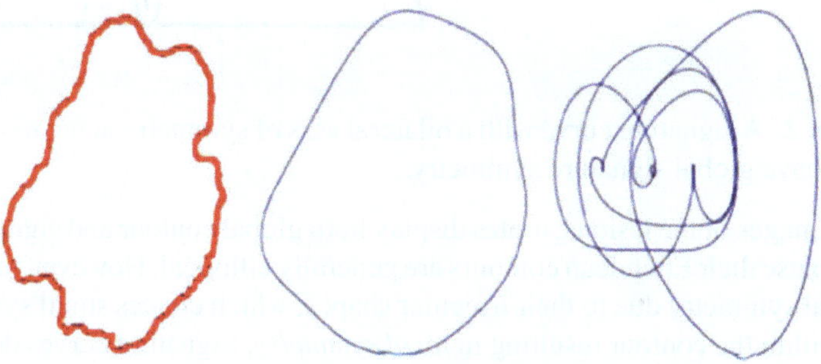

Fig. 5 Boundary of the melanoma is smoothed and signature curve calculated

1.3 Symmetry of Signature Curves

Signature curves encode the curvature complexity of a contour and accentuate global and local symmetry patterns. For example, a symmetrical contour such as an ellipse has a double overlapping signature curve due to its reflective *global contour symmetry*.

Definition 1 A contour possessing a bilateral axis of symmetry is said to have global contour symmetry.

Although contours could also have rotational, reflectional or translational symmetry, in this paper, we only concentrate on reflectional symmetry. Thus, we also define:

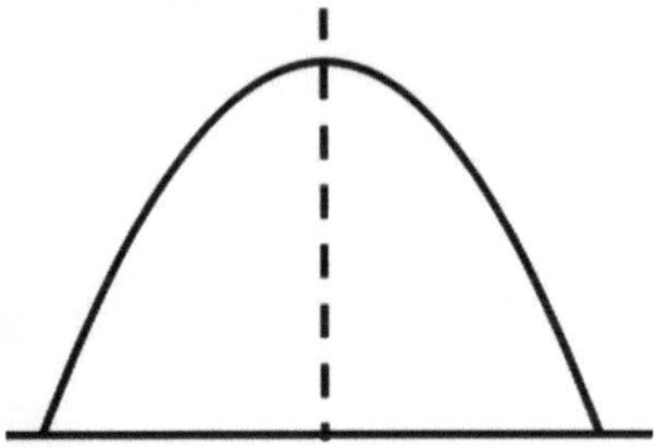

Fig. 6 Local individual symmetry

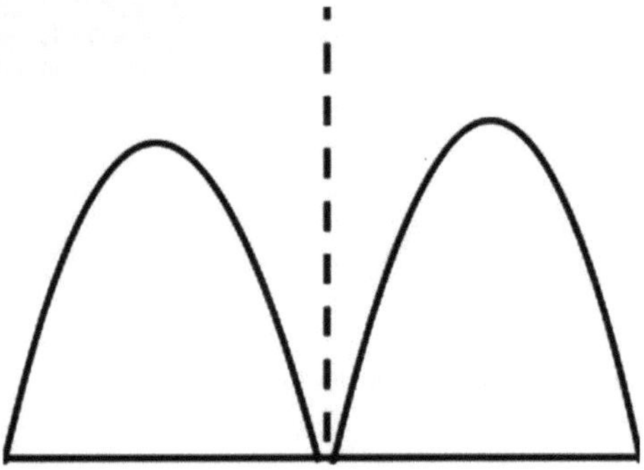

Fig. 7 Local joint symmetry

Definition 2 A signature curve with a bilateral axis of symmetry at the κ- or κ_s-axis, is said to have global signature symmetry.

In the images of the lesions, moles display both global contour and signature symmetry because their Euclidean contours are generally elliptical. However, melanomas lack global symmetry due to their irregular shapes, which creates small symmetrical regions within the contour resulting in *local symmetry*. Signature curves detect local symmetry as signature arcs that are symmetrical across either the κ- or κ_s-axis. We introduce the following definitions.

Definition 3 A signature arc with a bilateral axis of symmetry is said to have local individual symmetry.

The axis of symmetry passes perpendicularly through the midpoint of the horizontal axis connecting the initial and final points of the signature arc, as seen in Fig. 6.

Definition 4 A reflective symmetry between two distinct signature arcs is said to have local joint symmetry.

The axis of symmetry is equidistant from the arcs and perpendicular to the horizontal axis connecting the initial and final points of both signature arcs, as illustrated in Fig. 7.

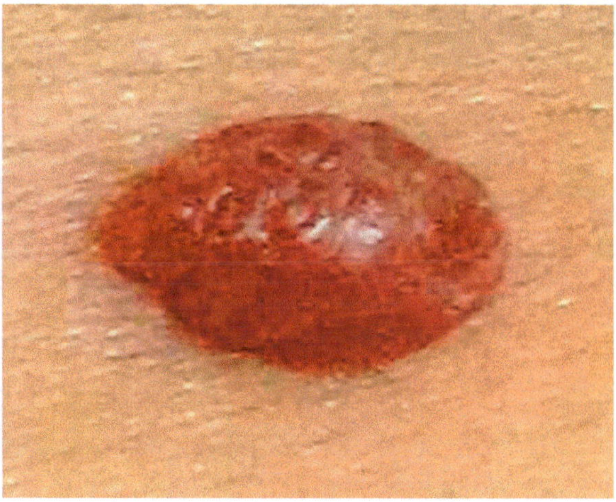

Fig. 8 Benign tumor contour

2 Signature Methodology

2.1 *Zero Curvature Points*

In this study, we say that a point along the contour, where either $\kappa(t) = 0$ or $\kappa_s(t) = 0$ is a *zero curvature point*. Zero curvature points are identified by detecting a change in sign of $\kappa(t)$ or $\kappa_s(t)$ caused by S crossing the κ- or κ_s-axis. The range R of zero curvature points on each respective axis is $R_\kappa = \max\{\kappa_s(t)\} - \min\{\kappa_s(t)\}$, where $\kappa(t) = 0$, and $R_{\kappa_s} = \max\{\kappa(t)\} - \min\{\kappa(t)\}$, where $\kappa_s(t) = 0$. The density of zero curvature points on each axis is calculated as

$$\rho_\kappa = \frac{R_\kappa}{\eta_{\kappa_s}} \quad \text{and} \quad \rho_{\kappa_s} = \frac{R_{\kappa_s}}{\eta_\kappa},$$

where η is the number of zero curvature points on the respective axis.

2.2 *Global Symmetry*

Benign contours tend to be approximately globally symmetrical with several axes of symmetry as seen in Fig. 8. The contour's corresponding signature in Fig. 9 has a nearly double overlapping signature curve due to the global symmetry. Therefore, we developed two methods referred to as *global contour* and *signature symmetry*.

Fig. 9 Signature curve

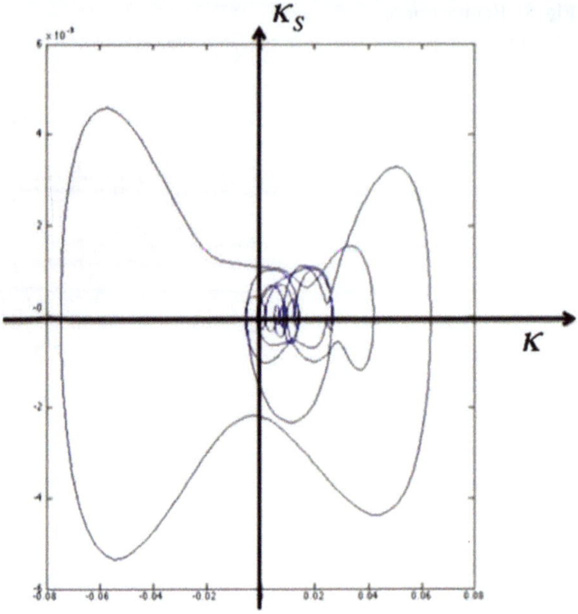

2.2.1 Global Contour Symmetry

Each contour, is translated so that its centroid is coincident with the origin. At the beginning of each symmetry calculation iteration, the contour is rotated $\Delta\theta = \frac{5\pi}{180}$ radians. The rotation increment $\Delta\theta$ was selected because it is relatively "small" and computationally efficient. The contour C is then partitioned between the sets C^+ and C^- defined to be

$$C^+ = \{(x(t), y(t): y(t) \geq 0\} \quad \text{and} \quad C^- = \{(x(t), y(t)) : y(t) < 0\} \tag{3}$$

where the points are denoted by $(x^+(t), y^+(t))$ when $(x(t), y(t)) \in C^+$ and similarly by $(x^-(t), y^-(t))$ when $(x(t), y(t)) \in C^-$. For the m points in C^+ and n points in C^-, the cumulative magnitudes $\|v_i^+\|$ and $\|v_i^-\|$ are calculated. Although we could sum the distribution by using the first point as our initial point and continuing successively, this can be problematic if, for example, the first point is an outlier. To circumvent this possibility, we will reorder the distributions so that the centroid is the initial point and each successive point alternates between the left and right side of the centroid. So now, we have distributions with nontrivial ordering and will proceed to sum the distributions by calculating the cumulative distance magnitude of each point. The *cumulative magnitude* is recursively defined, where the magnitude of a point is added to the summation of all preceding point's magnitudes, with $\|v_0^+\|$ being the magnitude of the first point

$$\|v_i^+\| = \sqrt{(x_i^+(t))^2 + (y_i^+(t))^2} + \sum_{m=0}^{i-1} \|v_m^+\|.$$

The magnitudes are compiled into a vector $\hat{v}$, where the magnitudes from $\|v^+\|$ are negated and so

$$\hat{v} = \{-\|v_0^+\|, \ldots, -\|v_m^+\|, \|v_0^-\|, \ldots, \|v_n^-\|\}.$$

The symmetry of the distribution $\hat{v}$ is quantified by calculating skewness δ [7], using the formula,

$$\delta = \frac{\frac{1}{m+n}\sum_{p=1}^{m+n} v_p^3}{\left(\frac{1}{m+n}\sum_{p=1}^{m+n} v_p^2\right)^{3/2}}. \tag{4}$$

The symmetry algorithm is repeated for 37 iterations for each of the $\Delta\theta = \frac{5\pi}{180}$ rotations of the contour through $\theta \in [0, \pi]$.

2.3 *Local Symmetry*

The signature curve of a melanoma tends to be symmetrical across both the κ-and κ_s-axis as illustrated in Fig. 11. The local symmetry is due an irregular contour such as in Fig. 10. Consequently, local joint and individual symmetry were quantified with respect to each axis using the following symmetry algorithm.

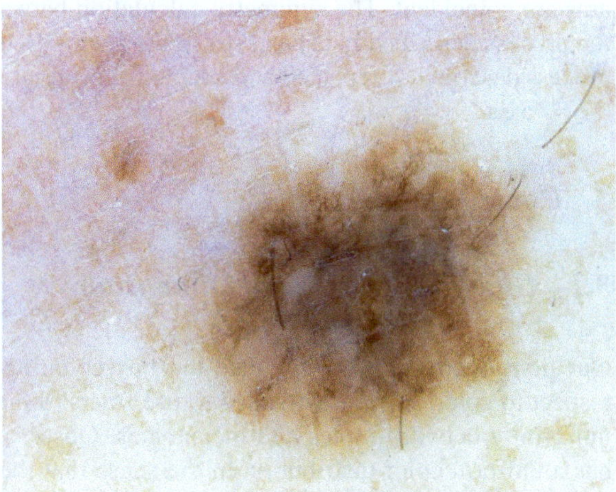

Fig. 10 Melanoma contour

2.3.1 Local Individual Symmetry

For individual symmetry, S is divided by the κ-axis so that $\kappa_s(t) = 0$ only at the initial and final points of an arc L. First, we let the point (x_c, y_c) be the midpoint of our signature arc, so that it lies on the bilateral axis dividing the signature arc. Now, we partition the signature into the sets L^+ and L^- defined to be

$$L^+ = \{(\kappa(t), \kappa_s(t)) : \kappa(t) < x_c\} \quad \text{and} \quad L^- = \{(\kappa(t), \kappa_s(t)) : \kappa(t) \geq x_c\}. \tag{5}$$

Further denote the points in L^+ by $(\kappa^+(t), \kappa_s^+(t))$ and similarly the points in L^- by $(\kappa^-(t), \kappa_s^-(t))$. For all points in L^+ and L^-, the cumulative distance between each point and the midpoint are calculated so that for an arbitrary point the cumulative distance is

$$\|v_i^+\| = \sqrt{(\kappa_i^+(t) - x_m)^2 + (\kappa_{s_i}^+(t) - y_m)^2} + \sum_{m=0}^{i-1} \|v_m^+\|. \tag{6}$$

The magnitudes are compiled into a vector $\hat{v}$ as previously described and the symmetry of the distribution is calculated with Eq. (4). This process is repeated and adapted appropriately for when S is segmented by the κ_s-axis.

2.3.2 Local Joint Symmetry

For local joint symmetry, S is divided by the κ-axis so that $\kappa_s(t) = 0$ at only the initial and final point of an arc L. Two distinct arcs L^+ and L^- are selected from S, then translated so that they are aligned as in Fig. 7 with (x_c, y_c) as the point where the arcs are coincident. The symmetry calculation between the two arcs is identical to the process described in (5) and (6) so the skewness can be calculated as in (4). The process described is repeated and adapted appropriately for when S is segmented by the κ_s-axis.

3 Color Methodology

3.1 Global Color Fractal Dimension

Our first algorithm is to calculate the box counting dimension of an image with respect to color. We are motivated to use fractal dimension by the contrast in color uniformity between moles and melanomas. Observe that in Fig. 8, a typical mole has a uniform color distribution in the sense that the pigment is the same through the entire lesion. In contrast, a typical melanoma, as seen in Figs. 4 and 8, has a nonuniform color distribution.

In short, we calculate the global color mean, then iteratively partition the lesion and calculate the fractal dimension from the count of subsets of the partition with

Fig. 11 Signature curve

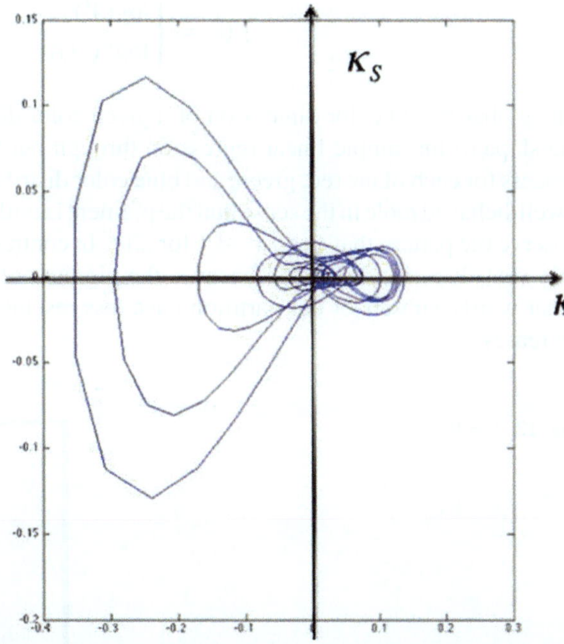

a mean color value sufficiently close the global color mean. We will refer to this measure as the *global color fractal dimension*. In fact, we calculate three fractal dimensions corresponding to each color in the image's RGB color distribution. Each color distribution is stored in an $n \times m$ matrix, which we denote Ψ_p such that $p = r, g$, or b, corresponding to red, green, or blue. An entry in Ψ_p is denoted by $\Psi(i, j)$ to indicate its position at the ith row and jth column in Ψ_p. First, we calculate the global color mean M of Ψ_p by

$$M = \frac{1}{mn} \sum_{i=1}^{m} \sum_{j=1}^{n} \Psi_p(i, j). \tag{7}$$

We then partition Ψ_p into 4^t submatrices, where $t$corresponds to the iteration count. In Figs. 12, 13, and 14, we show the partition of the image, where each cell represents a submatrix of Ψ_p with a size of $\frac{n}{2^t} \times \frac{m}{2^t}$.

On the tth iteration, we have a set of 4^t submatrices and calculate the local mean μ of each submatrix as in (7) to obtain the set $M_t = \{\mu_1, \ldots, \mu_{4^t}\}$. Let $\chi(t) = \#\mu_p$ with $p \in 1, \ldots, 4^t$ such that each μ_p satisfies $|M - \mu_p| < \epsilon$, where $\epsilon = 0.01, 0.03$, and 0.05. We use these three values for ϵ in order to enhance the accuracy of our diagnostic algorithm. Now, we introduce the discrete function $f(t)$ such that

$$f(t) = \begin{cases} \log(4^t) \\ \log(\chi(t)) \end{cases} .$$

The global fractal color dimension of a given color distribution Ψ_p is calculated as the slope of the simple linear regression through our function $f(t)$. We repeat this process for each of the red, green, and blue color distributions over five iterations. For a well-behaved mole in the sense that the pigment is uniform throughout the lesion, we observe the pattern that $\chi(t)/4^t \approx 1$ for all t. In contrast, for melanomas we observe that $\chi(t)/4^t > \chi(t+1)/4^{t+1}$ because the pigment of the lesion is nonuniform. In other words, subsets of our partition have less resemblance to the entire lesion as t increases.

Fig. 12 $t = 0$

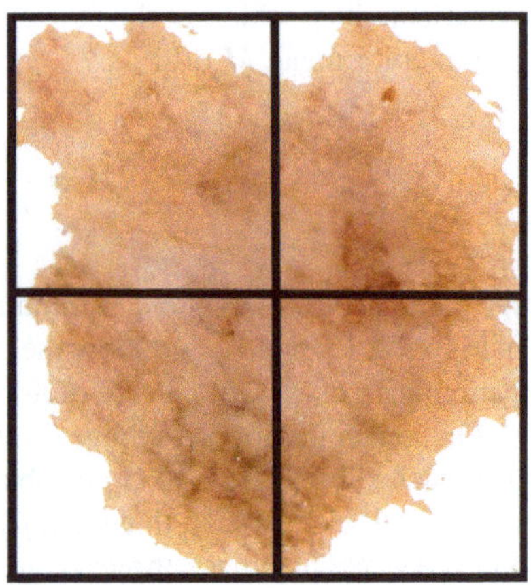

Fig. 13 $t = 1$

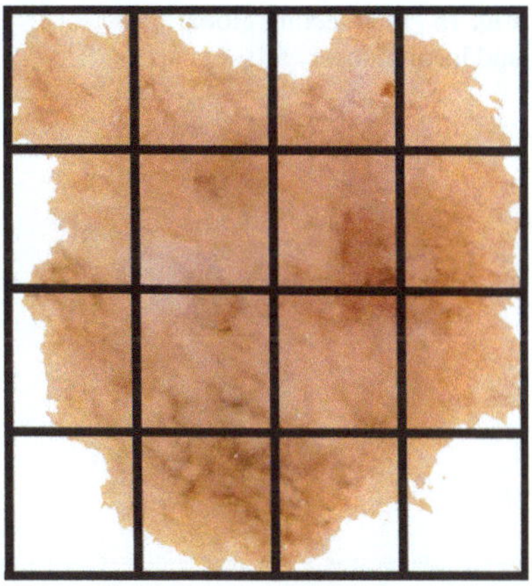

Fig. 14 $t = 2$

3.2 Global Plane Method

Now, we present an alternative geometric method for distinguishing between moles and melanomas. For any image of a lesion, consider a given red, green, or blue color distribution of that image with size $n \times m$. The distribution can be represented as a surface $\Phi(x, y) \in \mathbb{R}^3$. The surface has a characteristic shape corresponding to whether the lesion is a mole or melanoma. For a well behaved mole, the surface is relatively flat because the color is uniform throughout the lesion. In addition, the surface is smooth because the color changes continuously throughout the lesion. In contrast, the surface generated from a melanoma has many discontinuities in color and has a larger degree of variation. To distinguish these shapes, we calculate the global least squares plane through $\Phi(x, y)$, then partition $\Phi(x, y)$ into subsurfaces and calculate the corresponding least squares plane. Our metric for distinguishing between moles and melanomas is obtained by measuring how parallel the subsurfaces' least squares plane is to the global plane.

We begin by calculating the least squares plane through the surface defined by $\Phi(x, y)$ and determining its normal vector $\mathbf{n}_\Phi$. This plane will be referred to as the global plane through the surface of $\Phi(x, y)$. We proceed by uniformly partitioning $\Phi(x, y)$ into subsurfaces as in the global color fractal dimension algorithm. On the tth iteration, the surface is partitioned into 4^t subsurfaces and let $\phi_p(x, y)$ denote a subsurface with $p \in 1, \ldots, 4^t$. For each subsurface, we calculate the least squares plane referred to as the local plane and determine its normal vector denoted as $\mathbf{n}_{\phi_p}$. For each of the 4^t local planes, we will quantify how parallel the global and local plane are by the ratio

$$\lambda_p = \frac{1}{\pi}\arccos\left(\frac{\mathbf{n}_\Phi \cdot \mathbf{n}_{\phi_p}}{|\mathbf{n}_\Phi||\mathbf{n}_{\phi_p}|}\right). \tag{8}$$

The angle between the global and local plane, as seen in Fig. 15, follows as $\theta = \pi\lambda_p$. We divide by π in (8) in order that $\lambda_p \in [0, 1]$. When $\lambda_p \approx 0$ the global and local

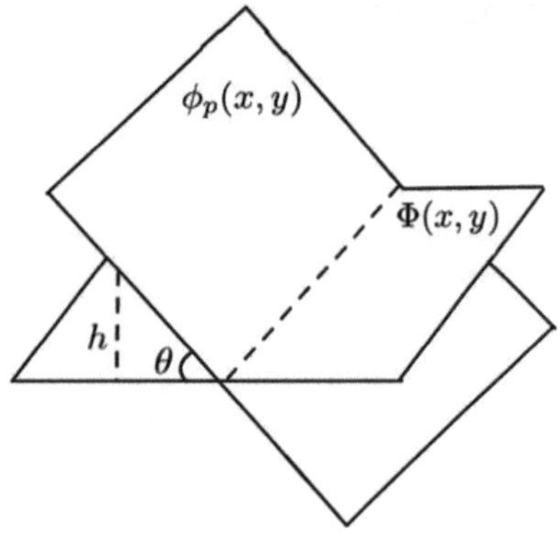

Fig. 15 Intersecting global and local planes

plane are nearly parallel, whereas they are nearly perpendicular when $\lambda_p \approx 1$. After completing n iterations, we compute the degree of global color uniformity Λ of the color distribution as

$$\Lambda = \sum_{t=0}^{n} \sum_{i=0}^{4^t} \lambda_i .$$

We repeat the calculation of Λ for each of the red, blue, and green color distributions. Relatively, small values of Λ are deemed to correspond to moles because their color distribution is nearly uniform, which results in λ_p values near zero. In contrast, the variegated color distribution of a melanoma results in a subdivision of the surface with an increasing number of subsurfaces of larger angle θ, illustrated in Fig. 15. Thus, melanomas with a nonuniform color distribution correspond to high values of Λ.

4 Signature Results

4.1 Data Set

Our data set consists of 60 melanomas and 60 moles, which were acquired from several data bases including Opticom Data Research and MoleMap in New Zealand, [15]. All of the tumors from both databases included an official diagnosis and delineation of the tumor contour. After downloading the images, each image was individually discretized into a set of (x, y) points using active contour segmentation and normalized [9, 10].

Table 1 Intersecting global and local planes

	η_κ	η_{κ_s}	R_κ	R_{κ_s}	ρ_κ	ρ_{κ_s}
Melanoma	16.6	30.63	0.241	0.855	0.0008	0.0057
Moles	3.68	13.08	0.058	0.006	0.0047	0.0007

4.2 *Data and ROC Analysis*

4.2.1 Zero Curvature Points

We include our results of the analysis of the zero curvature points of the signature curve in Table 1. Moles have relatively few zero curvature points because the shape of their contour is elliptical, which also corresponds to a smaller average range. In contrast, melanomas have a greater number of zero curvature points due to their irregularly shaped contour, hence resulting a greater average range.

4.2.2 Global Contour Symmetry

The global contour symmetry algorithm is performed for 37 iterations for each of the $\Delta\theta = \frac{5\pi}{180}$ rotations, where the output is a skewness value. Based on data observation, a *symmetrical axis* λ_1 corresponds to $\delta < 0.01$ and a *very symmetrical axis* λ_2 corresponds to $\delta < 0.001$. The symmetry score $\Lambda = \sum \lambda_1 + \sum \lambda_2$, is calculated for each contour.

The average number of symmetrical and very symmetrical axes for a mole was 27.67 and 13.98, whereas the corresponding averages for a melanoma are 11.39 and 4.45. We calculated an ROC curve, which is a plot of the true positive rate against the false positive rate. The area under the ROC curve indicates the accuracy of our methodology to correctly diagnose melanoma and moles. Our ROC curve is included in Fig. 16, where the area under the curve corresponds to an accuracy of 91.64%.

4.2.3 Local Symmetry

In both the local individual and joint symmetry, a *symmetrical axis* λ_1 corresponds to $\delta < 0.3$, a *very symmetrical axis* λ_2 corresponds to the $\delta < 0.1$. The total symmetry score is calculated by summing the local individual and joint symmetry score using Eq. (8). We have included our results from the local individual and joint symmetry algorithm, which is included in Tables 2 and 3 with each entry listed as the mean $\pm$ standard deviation.

Table 2 Individual symmetry

	Mean λ_1	Mean λ_2
κ-axis		
Moles	3.18 ± 3.25	1.21 ± 1.45
Melanoma	11.83 ± 5.08	4.14 ± 2.21
κ_S-axis		
Moles	1.54 ± 1.10	0.43 ± 0.54
Melanoma	5.03 ± 2.70	1.37 ± 1.08

Table 3 Joint symmetry

	Mean λ_1	Mean λ_2
κ-axis		
Moles	0.78 ± 1.65	0.26 ± 0.67
Melanoma	10.94 ± 9.01	3.42 ± 2.80
κ_S-axis		
Moles	10.47 ± 7.63	3.78 ± 2.75
Melanoma	40.83 ± 26.07	13.23 ± 9.06

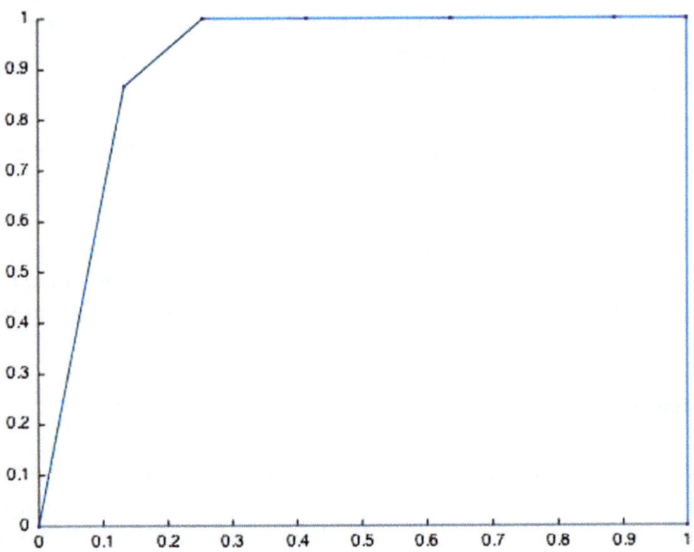

Fig. 16 ROC analysis

Table 4 Global color fractal dimension results

	$\epsilon_r = 0.01$	$\epsilon_r = 0.03$	$\epsilon_g = 0.01$	$\epsilon_g = 0.03$	$\epsilon_b = 0.01$	$\epsilon_b = 0.03$
Moles	0.5438	0.6755	0.4045	0.6115	0.3499	0.5861
Melanomas	0.4043	0.6035	0.3180	0.5376	0.2761	0.5097

5 Color Results

5.1 *Image Processing*

Each image was manually prepared, so that it was suitable to use with our computer aided diagnosis software. We began the image preparation process by removing the background with photoshop so that only the skin lesion is visible. Then magnified the image so that its dimensions were at least 700×700 pixels and cropped the image to minimize the background white space.

5.2 *Data and ROC Analysis*

5.2.1 Global Color Fractal Dimension

As predicted, the global color fractal dimension of moles was significantly higher than that of melanomas. The higher global fractal color dimension correspond to a higher degree of color uniformity because there are more number of pieces of the partition with an average color near the global mean. We have included the average global color fractal dimension of the moles and melanomas in our data set in Table 4.

The columns correspond to our different thresholds when $\epsilon = 0.01$ and $\epsilon = 0.03$. The subscripts r, g, and b correspond to the red, blue, and green color distributions, respectively.

We calculated an ROC curve in order to objectively quantify the accuracy of our algorithm. We combined the global color fractal dimension values calculating for when $\epsilon = 0.01, 0.03$, and 0.05 for the red, green, and blue color distributions. After calculating the ROC curve, we determined that our global color fractal dimension method has an accuracy of 95.71% of diagnosing melanoma in our data set.

5.2.2 Global Plane Method

The measure of global color uniformity using the global plane method is indicated by the value of Λ_p, where $p = r, b$, or g. We have included the average Λ_p value for all of the moles and melanomas in our data set in Table 5, where each entry is the mean $\pm$ the standard deviation.

Table 5 Global color fractal dimension results

	Λ_r	Λ_g	Λ_b
Moles	110.20 ± 27.75	122.38 ± 27.64	123.63 ± 27.53
Melanomas	148.67 ± 24.76	160.70 ± 24.77	165.57 ± 20.90

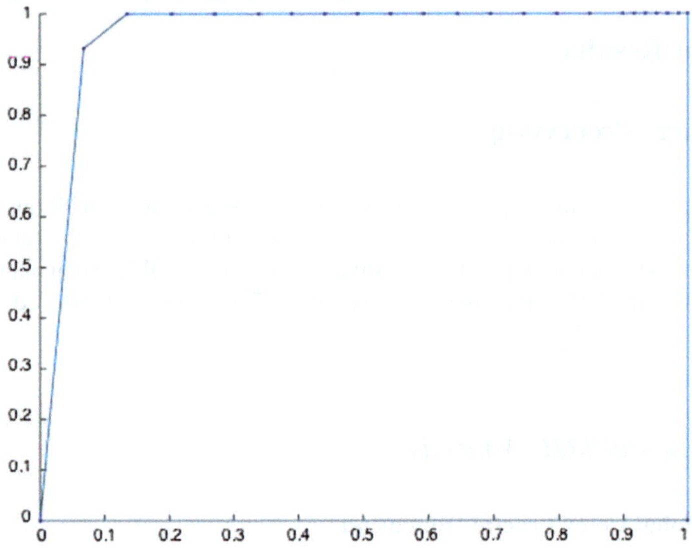

Fig. 17 ROC analysis

A small Λ_p value corresponds to small values of λ_p, which indicates that the global and local plane are nearly parallel. Thus, the color distribution has a higher degree of color uniformity because the local color is nearly the same as the global color. Since the mean Λ_p values are significantly lower than the corresponding melanoma averages, then moles have a higher degree of color uniformity.

We objectively tested that accuracy of the global ball method to diagnose melanoma by calculating an ROC curve. We have included our curve in Fig. 17, where the area under the curve and accuracy of our algorithm is 0.9615.

6 Conclusion

Signature curves capture an object's shape and detect and quantify changes in the curvature of their boundary. They have proven to be effective as a tool detecting global and local symmetry in melanoma and moles. In global contour symmetry, we have shown that they significantly reduce the computational complexity of detecting local symmetry. Further, this algorithm can also be used in a variety of computer vision applications as a means of quantifying global and local symmetry of arbitrary two and three dimensional objects [12, 13]. For example, the signature method has previously been applied to characterize breast tumors by the authors [6] and an adapted version of the symmetry algorithm as a similarity measure has been used to solve spherical jigsaw puzzles [7].

Acknowledgements This work was funded by a CSUMS Grant Number DMS0802959 from the National Science Foundation in collaboration with the University of St. Thomas.

References

1. Mayo Clinic: Melanoma
2. Boutin, M.: Numerically invariant signature curves. Int. J. Comput. Vis. **40** (2014)
3. Calabi, E., Olver, P., Shakiban, C., Tannenbaum, A., Haker, S.: Differential and numerically invariant signature curves applied to object recognition. Int. J. Comput. Vis. **26** (1998)
4. Cartan, É.: La méthode du repère mobile, la théorie des groupes continus, et les espaces généralisés. Exposés de Géométrie **5** (1937)
5. Cartan, É.: Les problêms d'équivalence. In: Oeuvres Complètes, part II, vol. 2, pp. 1311–1334. Gauthiers-Villars, Paris (1953)
6. Grim, A., Shakiban, C.: Applications of signatures in diagnosing breast cancer. Minn. J. Undergrad. Math. **1**(1), 001 (2015)
7. Grim, A., O'Connor, T., Olver, P.J., Shakiban, C., Slechta, R., Thompson, R.: Reassembly of Three-Dimensional Jigsaw Puzzles (2016)
8. Keller, M.: Curvature, geometry and spectral properties of planar graphs. Discret. Comput. Geom. **46**(3), 500–525 (2011)
9. Lankton, S.: Hybrid geodesic region-based curve evolutions for image segmentation. In: Medical Imaging. International Society for Optics and Photonics (2007)
10. Lankton, S., Tannenbaum, A.: Localizing region-based active contours. IEEE Trans. Image Process. **17**(11) (2008)
11. Lloyd, R., Shakiban, C.: Classification of signature curves using latent semantic analysis. Comput. Algebra Geom. Algebra Appl. **3519** (2005)
12. Martinet, A.: Accurate detection of symmetries in 3d shapes. ACM Trans. Graph. **25**(2), 439–464 (2006)
13. Mitra, N., Guibas, L., Pauly, M.: Partial and approximate symmetry detection for 3D geometry. ACM Trans. Graph. **25**(3), 560–568 (2006)
14. Olver, P.J.: Equivalence, Invariants, and Symmetry. Cambridge University Press, Cambridge (1995)
15. MoleMap, New Zealand. http://molemap.co.nz/

Contemporary Interpretation of a Historical Locus Problem with the Use of Computer Algebra

Roman Hašek, Zoltán Kovács and Jan Zahradník

Abstract This paper deals with the joint use of computer algebra and the dynamic geometry features of the mathematics software GeoGebra to solve a locus problem. Through a generally unknown problem from an eighteenth century Latin book of geometry exercises the use of the computer algebra features of GeoGebra will be presented on the one hand as a means of automatic computation of the locus equation and on the other hand as an environment to realize the symbolic step-by-step derivation of the equation. The core principles of the effective implementation of computer algebra functions within the dynamic geometry system will be presented. An enhanced approach to solving the problem, inspired by the findings from the use of the computer to investigate the locus, will cause the appearance of an unexpected and until now not described curve.

Keywords Computer algebra · Locus equation · History of mathematics · GeoGebra

1 Introduction

Issues of the joint use of computer algebra and the dynamic geometry features of the GeoGebra software [14] to solve locus problems will be dealt with in this paper. They will be addressed through the detailed solution of a particular locus problem coming from the eighteenth century Latin book *Exercitationes Geometricae*, authored by Ioannis Holfeld (1747–1814) and published by the Jesuit College of St. Clement in

R. Hašek (✉) · J. Zahradník
University of South Bohemia, Jeronýmova 10, 371 15 České Budějovice, Czech Republic
e-mail: hasek@pf.jcu.cz

J. Zahradník
e-mail: jzahradnik@pf.jcu.cz

Z. Kovács
The Private University College of Education of the Diocese of Linz,
Salesianumweg 3, 4020 Linz, Austria
e-mail: zoltan@geogebra.org

I.S. Kotsireas and E. Martínez-Moro (eds.), *Applications of Computer Algebra*,
Springer Proceedings in Mathematics & Statistics 198,
DOI 10.1007/978-3-319-56932-1_12

Prague in 1773, [15]. From a total of 47 solved geometric problems presented in the book problem number 35 will be dealt with. After the detailed introduction of its original solution, the use of GeoGebra to solve it will be discussed, focusing on the implementation of computer algebra to derive an equation of the respective locus. On the one hand GeoGebra provides a user with the *CAS* environment to derive the equation step by step and on the other hand, it is endowed with powerful tools for its automatic computation on the background of appropriate functions.

The paper reflects two different perspectives; the perspective of a lecturer implementing it in mathematics teacher training courses and the perspective of a developer of the computer algebra features of GeoGebra. Although the authors have proven the benefits of solving the problem with future teachers, in order not to be too extensive the paper does not address this issue. A number of studies have dealt with the importance of the incorporation of the historical perspective into mathematics education. For further reading about the findings of such studies and about the indisputably positive effect of historical issues on the quality of mathematics teaching [4, 9, 10] can be recommended. The use of some other problems from the book by Ioannis Holfeld in mathematics teaching is presented in [12].

1.1 Origin of Problem 35

Problems in *Exercitationes Geometricae* are focused primarily on conic sections. There are, in particular, general problems on conics, locus problems and problems on surface areas and volumes of solids of revolution among the 47 problems in the book. Their assignments, as well as their solutions, illustrate the approach to the geometry of curves and to locus problems typical of mathematics of the seventeenth and eighteenth centuries. They take a reader back to the era before the currently prevalent analytical method based on the Cartesian coordinates was fully established, [16, 22]. These geometry problems were viewed as being more associated with mechanical representations. Curves were drawn by the notional movement of geometric structures. Configurations fitting the assignment of a task but different from its illustrative picture were not considered. Negative results were very rarely dealt with and although variables x and y were used as coordinates to derive an equation of a curve the notion of 'coordinates' was not mentioned. The use of dynamic geometry to solve problem number 35 allows us to replicate the method of the geometric construction of the locus curve typical of that time [9], as well as to examine all possible configurations of the task and subsequently to discover that from the perspective of the contemporary solver the corresponding locus curve can be seen as far more complex than the original solution given in the book.

Despite their date of publication, a number of problems from [15] other than problem 35 are still attractive and are worth resolving with the help of contemporary methods. Assignments of some of them together with a more detailed description of features typical of the method used in the book and some findings about the author's life are provided in [12, 24].

1.2 GeoGebra, the Software Used

GeoGebra is open source software that is free for noncommercial purposes, [14]. Thanks to its unique combination of environments such as *Graphics*, *Algebra*, *Spreadsheet*, *CAS* and *3D Graphics* it allows a user to apply different views on studied objects. In this paper, we are particularly focused on the implementation of computer algebra in GeoGebra and its joint use with the dynamic geometry environment *Graphics*.

GeoGebra has an embedded computer algebra system (CAS), *Giac*, [17, 21]. It is widely used in several types of computations in the background during GeoGebra's internal calculation, including algebraic computations such as expansion and factorization, working with arbitrary big integers (for example, to determine the greatest common divisor or the least common multiplier and prime factorization), solving equations, differential equations and equation systems, computing limits, derivatives and antiderivatives; methods for calculating determinants, function fitting, probability calculations, and statistics are also supported. In addition to this, GeoGebra is capable of outsourcing some special computations to the external CAS *Singular*, [2, 8, 20], including absolute factorization, locus, and envelope equation calculation and solving algebraic equation systems in the frame of its theorem proving subsystem.

2 The Locus Problem and Its Original Solution

Introduced in this section is the assignment of problem 35 together with a detailed presentation of its original solution given in [15].

Problem 35 Given a circle with a diameter *MP* (see Fig. 1); construct a radius AB to this circle and a line segment BO perpendicular to MP so that $MO : AO = r : BC$ (r is the radius of the circle). Find the locus of point C. (*Remark:* Length of the segment BC is the fourth proportional of lengths MO, AO, and r.) ([15], p. 41, Problema 35.).

The original illustration of the problem assignment, in [15] referred to as Fig. 32, is shown in Fig. 1 (a copy of problem 35 original Latin assignment is displayed at http://www.pf.jcu.cz/~hasek/Holfeld). We must not be confused by the points and segments in the picture, which are not mentioned in the assignment. Like certain other figures in the book, this also served to illustrate more than one exercise. For this reason we will use the modified image in Fig. 2, left, containing only given elements and the most currently utilized arrangement of the coordinate axes to illustrate the original solution to the problem.

Ioannis Holfeld (I. H. in the following) begins his solution by labeling lengths of selected segments; $AD = x$, $DC = y$, $AB = r$, $OM = z$. After that, from the similarity of triangles AOB and ADC (without mentioning this concept) he derives the relation $OA : OB = x : y$ and substituting $OA = r - z$ he gets the formula

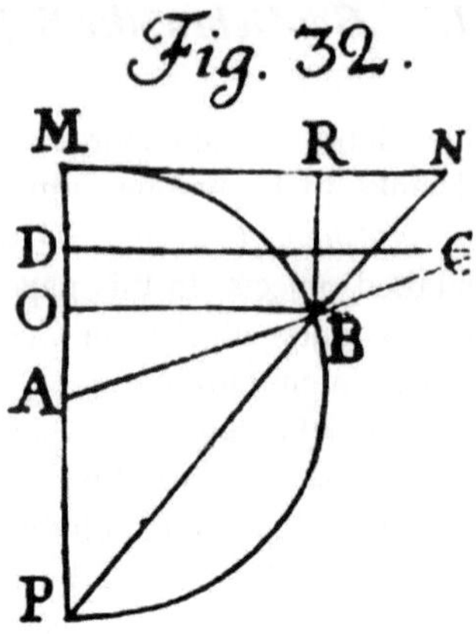

Fig. 1 Original illustration of the assignment of problem 35

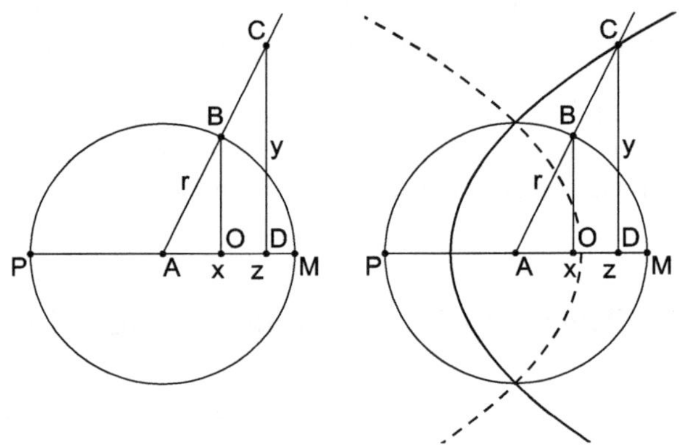

Fig. 2 Problem 35; the given elements and their labeling (*left*), parabola $y^2 = r^2 + 2rx$ (*solid line*) as the solution according to Ioannis Holfeld (*right*)

$OB = \dfrac{ry - zy}{x}$. Then I. H. expresses the length of OB in another way as $OB = \sqrt{2rz - z^2}$. Although he does not mention it, it is evident that he has applied the 'right triangle altitude theorem' (also known as 'geometric mean theorem') on the right triangle MPB, see Fig. 2, where BO is the altitude and $MO = z$ and $OP = 2r - z$ are two corresponding line segments on its hypotenuse MP. Comparing both the expressions of OB I. H. gets

$$\frac{ry - zy}{x} = \sqrt{2rz - z^2}, \tag{1}$$

from which he derives $z = r - \dfrac{rx}{\sqrt{x^2 + y^2}}$ (I. H. does not mention it but it is clear that it is one of the roots of the quadratic equation $(x^2 + y^2)z^2 - 2r(x^2 + y^2)z + r^2y^2 = 0$ in z that is equivalent to (1). Its second root of the form of $z = r + \dfrac{rx}{\sqrt{x^2 + y^2}}$ does not match the considered configuration. Its effect is mentioned later.). Using, this equality I.H. first writes $AO = r - z = \dfrac{rx}{\sqrt{x^2+y^2}}$. Then, substituting for MO, AO and BC into the relation of the fourth proportional $MO : AO = r : BC$, which is used in the assignment of the problem, he gets $\left(r - \dfrac{rx}{\sqrt{x^2+y^2}}\right) : \dfrac{rx}{\sqrt{x^2+y^2}} = r : \left(\sqrt{x^2+y^2} - r\right)$ that he simplifies through $\dfrac{\sqrt{x^2+y^2}}{x} = \dfrac{\sqrt{x^2+y^2}}{\sqrt{x^2+y^2}-r}$ into the form of $x + r = \sqrt{x^2+y^2}$. Finally, squaring both sides of the latter equality and simplifying the resulting one he derives the equation of the studied locus

$$y^2 = r^2 + 2rx \qquad (2)$$

that he immediately interprets as the equation of the parabola with the parameter $2r$ equal to the length of the diameter PM of the given circle. Its plot for the particular value of $r = 1$ is shown in Fig. 2, right, as the solid line parabola (the dashed line parabola corresponds to the above mentioned second root of the quadratic equation equivalent to (1), which was not considered by I.H.). From the perspective of today's solver let us add that the resulting parabola (2) has focus $F = [0, 0]$, directrix $d : x = -r$ and the focal parameter $p = r$.

3 Solution Using Computer Algebra in GeoGebra

GeoGebra features of the dynamic geometry and computer algebra will be used to model the assignment of the problem, to find the shape of the resulting locus curve and to realize the automatic derivation of its algebraic equation in this section.

3.1 The Issue of Computer Algebra Implementation

To create the dynamic geometric model of problem 35 we use the *Graphics* view of GeoGebra. There is more than one way to do so, but not all of them suit the demands of the subsequent utilization of the symbolic algebra tools of GeoGebra for the purpose of determining the locus curve and its equation. Namely, we are referring to the commands *Locus* and *LocusEquation*; the former numerically collects significant points of the curve to support high quality visualization, while the latter uses a pure symbolic algebraic method to compute the equation of a curve.

Partly due to the algebraic theory and partly due to its implementation in the dynamic geometry domain, there are some restrictions in the background which may need to be kept in mind when modeling problems by utilizing computer algebra methods, including GeoGebra. We will mention them successively herein.

The first issue is that the *LocusEquation* command can only process algebraic problems. This means that the construction steps must be described by polynomial equations of the corresponding geometrical objects, which are typically the Cartesian coordinates of the points. Fortunately, Euclidean constructions, that is, constructions by means of a compass and a straightedge ruler, have proven to be algebraic problems (see [5, 7, 19]). Let us note that the converse of this statement is not true: a well-known counterexample is the trisection of an angle which is algebraic but not constructible by Euclidean steps, [5].

3.2 *Euclidean Construction of Problem 35*

As this section will show, the assignment to problem 35 can be realized as a pure Euclidean construction, but not every such a construction is optimal for the application of computer algebra. See Fig. 3. First, using the *Circle with Centre through Point* tool we draw the circle c with the center $A = [0, 0]$ passing through point M on the x-axis and therefore, the radius of c is $r = |AM|$. The point M can be placed on the x-axis arbitrarily, but for simplicity we choose $M = [1, 0]$. The remaining intersection of c with the x-axis is denoted P. Having constructed the circle c we put a movable point B on it and using the *Line* tool construct a line AB. Finally, applying the tool *Perpendicular Line* we construct a line through B perpendicular to PM, which coincides with the x-axis, and mark its foot with O.

Now, we have to construct the point C so that the condition of the fourth proportional $MO : AO = r : BC$ is fulfilled. With GeoGebra we can do it in two ways, either compute the appropriate length of the segment BC and then place the point C, or determine the position of C strictly geometrically. In the following, we use the latter method demonstrating the Euclidean constructibility of the locus' points.

However, this "Euclidization process" is in general not always straightforward and unique for a given locus problem—multiple constructions can be found to describe

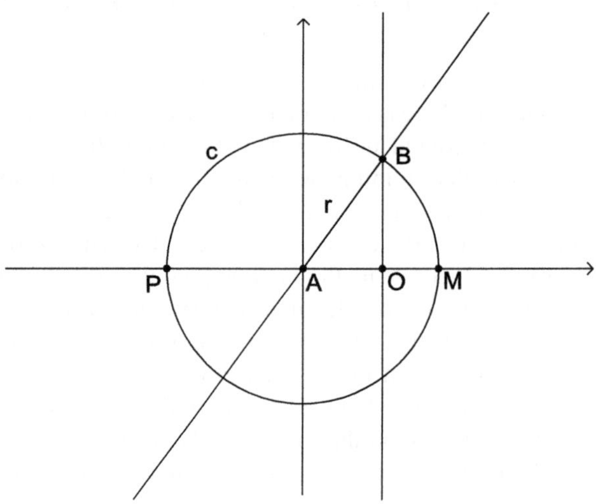

Fig. 3 Startup geometric construction before Euclidization

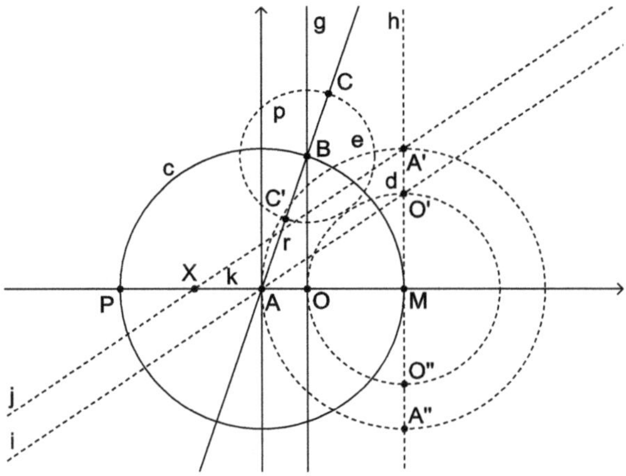

Fig. 4 Geometric construction of point C by using the "naive" approach

a problem when the steps are restricted to being pure Euclidean ones. One "naive" approach can be as described in Fig. 4, by introducing several helper objects to get the required ratios. To make it possible to use the intercept theorem to construct the length of BC, circles d and e have been created. Their intersections with line h (the perpendicular to the x-axis at M) create points O' and A', respectively. Actually, other intersection points O'' and A'' are created silently, because a circle and a line usually have *two* intersection points. After joining A and O' (line i) and creating a parallel line (j) to it which goes through A', it is clear that the intersection point X of line j and the x-axis creates a segment XA of the length k complying with the property $MO' : A'O' = AM : k$. That is, $k = BC$. Now creating another circle p with center B and radius k, one of the intersection points of the circle and line AB must be the sought point C.

By silently introducing points O'', A'' and C' (the last one is the other intersection point of AB and p) we actually define a larger than expected set of points (see Sect. 3.3). To filter out the unwanted points we may want to consider another way for Euclidization.

Our second approach is described in Fig. 5. Compared to Fig. 2, left, the following elements are added: First, we draw a ray MB and denote E its intersection with y-axis. Then, we lead a line parallel to MP through E and label its intersections with the lines OB and AB as Q and C, respectively. Now, we are to prove that the latter point C satisfies the condition $MO : AO = r : BC$, i.e., that it belongs to the investigated locus. To do so, we use two pairs of similar triangles, $\triangle MOB \sim \triangle EQB$ and $\triangle AOB \sim \triangle CQB$. Their similarities give rise to the equalities $MO : BO = EQ : BQ$ and $AB : BO = CB : BQ$, respectively, from the comparison of which we obtain $MO : EQ = AB : CB$. Then, due to the identities $EQ = AO$ (consequence of the construction in Fig. 5) and $AB = r$ we get the final relation $MO : AO = r : BC$. It means that the point C in Fig. 5 satisfies the assignment of problem 35. Justification that the same conclusion applies to the positions of B in other quadrants is analogous to the above procedure and we leave it to the reader. It

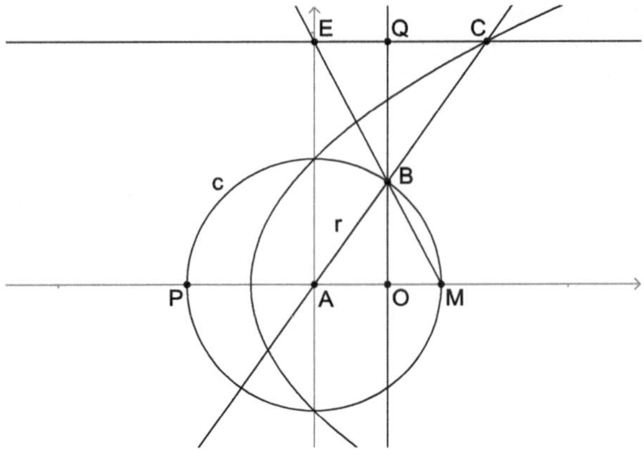

Fig. 5 Geometric construction of point C, point B being in the 1st quadrant

should be noted that a rather special case is the position of B on the y-axis, that, due to zero AO, satisfies the fourth proportional condition only in the modified form of $MO \cdot BC = r \cdot AO$.

3.3 *The Locus Curve*

Having established the geometric construction of the locus' points we use GeoGebra's built-in functions *Locus* and *LocusEquation* to explore the locus curve, namely its shape and equation. To find the "geometric" curve we use the *Locus* tool of the *Graphics* view or enter the command *Locus[C,B]* either into the input line or within *CAS*. To let GeoGebra find the "algebraic" equation of the locus curve we invoke the command *LocusEquation[C,B]*, either through the input line or within *CAS* again.

In the "naive" approach the result is seen in Fig. 6. The plotted "geometric" locus is a smaller set than the "algebraic" one which is shown as dashed. For comparison,

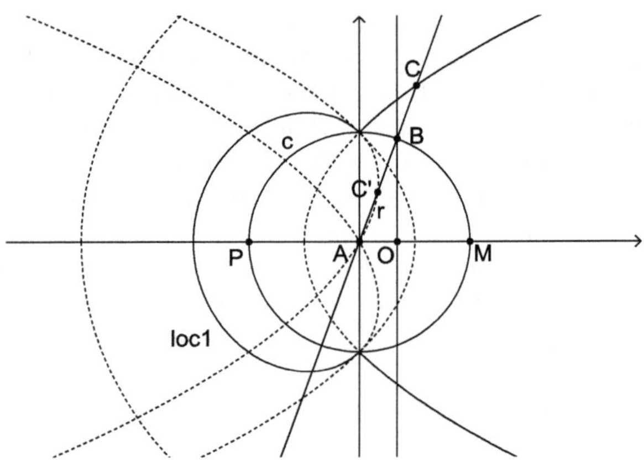

Fig. 6 Locus of point C using the "naive" approach

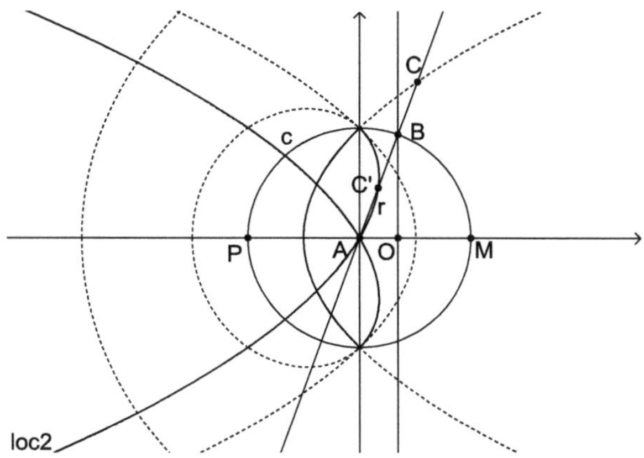

Fig. 7 Locus of point C' using the "naive" approach

another Fig. 7 is demonstrated to support the idea that the loci of points C and C' together represent *certain* parts of the "algebraic" locus. What is more, in some sense the points C and C' cannot be distinguished "algebraically' in the "naive" approach. This confirms that in the "algebraic" approach *both* intersection points of a line and a circle must be handled equally.

The dashed algebraic set is the product of four components, namely $2\,x - y^2 + 1, 2\,x + y^2 - 1, 2\,x^3 - x^2\,y^2 + 5\,x^2 + 2\,x\,y^2 - y^4 + 9\,y^2$ and $2\,x^3 + x^2\,y^2 + 3\,x^2 + 2\,x\,y^2 + y^4 - y^2$. Here the first two factors are parabolas, and one of the last two quartic factors seem to be present also in the geometric locus, at least partially.

In our second approach, Fig. 5 shows the solution only as the parabola. It results in the algebraic equation of the third degree, the polynomial of which can be factored into the product $y(2x - y^2 + 1)$ where along with the expected result $2x - y^2 + 1$ corresponding to Holfeld's parabola we have got an additional factor y, which corresponds to the x-axis. It is clear that the latter is not a geometric solution to problem 35. From the algebraic point of view, factor y arises from the solution of the system of algebraic equations corresponding to the configurations so that $B = M$ or $B = P$.

Dynamic construction of the problem allows us the ease to extend our second inquiry to other possible cases that are not captured in the illustrative picture. Specifically, we take into account the position of point C on the opposite ray with respect to point B which we have not hitherto considered. By using this trick, we can eventually distinguish between loci of C and C'.

Let us therefore use the tool *Reflect Object in Point* to construct point C' symmetric with point C with respect to B. We are interested in the locus of C' if B is being moved along the circle c. To find the locus curve we again use the *Locus* tool or enter the $Locus[C', B]$ command. As a result we get a curve of a surprising shape, looking like a pretzel, see Fig. 8 and [11]. To ask GeoGebra about the equation of this curve we apply the command $LocusEquation[C', B]$. It gives us a 5th degree polynomial, the factorization of which is $y(2x^3 + x^2y^2 + 3x^2 + 2xy^2 + y^4 - y^2) = 0$ where the

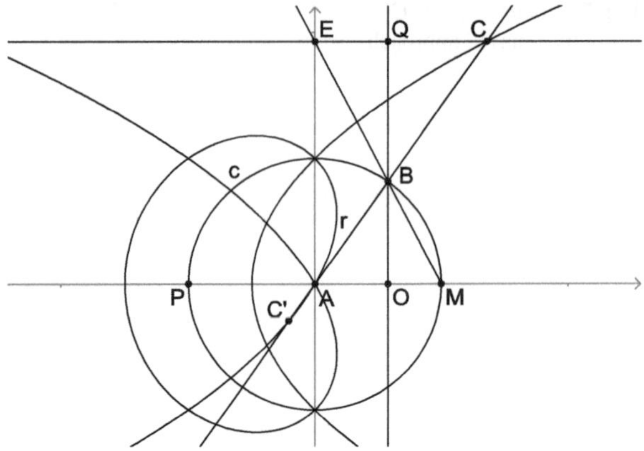

Fig. 8 Geometric construction of point C considering both its possible positions on line AB

first factor y has the same origin as in the case of the result from the command *LocusEquation[C,B]* applied above and the second factor relates to the locus curve we are interested in. Actually, this factor already occurred in the "naive" approach.

A remarkable curve, let us call it 'pretzel curve', of equation $y^4 + x^2y^2 + 2xy^2 + 2x^3 + 3x^2 - y^2 = 0$ thus appeared as an unexpected result of applying the current approach to problem 35 (it was first published in [13]). As far as the authors know only a similar but not identical curve of the fourth degree is mentioned in [6, 23].

In order to apply the commands *Locus[T,M]* and *LocusEquation[T,M]* algebraic problems should be formalized in such a way as to make it possible to eliminate all variables but x and y of the *mover M* (input) point to express the algebraic equation of the *tracer T* (output) point. GeoGebra's underlying CAS requires the description of the problem as a polynomial equation system with rational coefficients—this is the usual way of calculating the locus equation in automatized computation. As we mentioned in Sect. 1.2, GeoGebra uses two possible CAS in the background, but both use Gröbner basis computations to eliminate the intermediate variables.

GeoGebra comes with a subsystem which automatically translates the DGS steps into polynomial equations. This work was recently elaborated on through the extensive work of Recio, Botana, Abanades, Escribano, and Arbeo (see [1]) in the early 2010s, and further refined by others including the authors. Some typical Euclidean steps are, however, as yet not fully implemented or not implemented at all. One such step, for example, is mirroring a line to another line. These missing features in GeoGebra may require additional work to describe the non-implemented step in a substitute way. For example, we realized in Sect. 3.2 that such work was necessary to describe the algebraic expression $MO : AO = r : BC$ in another way, namely, to use similar triangles and the ratios between the appropriate sides.

As an example of the most recent result of the intensive development of the computer algebra tools of GeoGebra we are now able to introduce the pretzel curve by means of an even more natural approach. A new feature in GeoGebra is to compute an implicit locus for a given condition and a mover point. When considering the same

Euclidean construction, one can investigate the output of the command *LocusEquation[MO/AO==r/BC,C]* (see [18]). In this case, we get both the parabola and the pretzel curve at the same time with no additional components.

4 Algebraic Solution

In addition to the automatic derivation of the locus equation based on the geometric construction we can use the computer algebra features of GeoGebra in a less sophisticated way, as it is introduced in this section *CAS*, the CAS of GeoGebra, is used as a suitable environment to the stepwise solution of problem 35 here. As mentioned, there are more ways to derive the equation of the locus curve. One of them, to reproduce the steps of I.H. described in Sect. 2, we leave to the reader. In the following, we will base our procedure on the second approach of the Euclidean construction given in Sect. 3.2.

First, we express the task as the system of nonlinear equations. Then, solving it by elimination we arrive at the algebraic equation of the locus in x and y.

Let us follow Fig. 9. Coordinates of the decisive points are $A = [0, 0]$, $M = [r, 0]$, $B = [r\cos\varphi, r\sin\varphi]$, $O = [r\cos\varphi]$ and $C = [x, y]$, where φ is the angular coordinate of points B and C. Let us mark u and v the directed distance from A to O and from B to C, respectively, so that $u \geq 0$ for O falling on the ray AM and $u \leq 0$ for O falling on the opposite ray AP, and, in the same way, $v \geq 0$ for C laying on the ray opposite to the ray BA and $v \leq 0$ for C laying on the ray BA. Using v, we can write $\dfrac{x}{r+v}$ and $\dfrac{y}{r+v}$, where $r + v \neq 0$, instead of $\cos\varphi$ and $\sin\varphi$, respectively, to get $B = \left[\dfrac{rx}{r+v}, \dfrac{ry}{r+v}\right]$ and $O = \left[\dfrac{rx}{r+v}, 0\right]$. Consequently, the undirected distances used in the formula $\dfrac{MO}{AO} = \dfrac{r}{BC}$ for the fourth geometric proportional are $AO =$

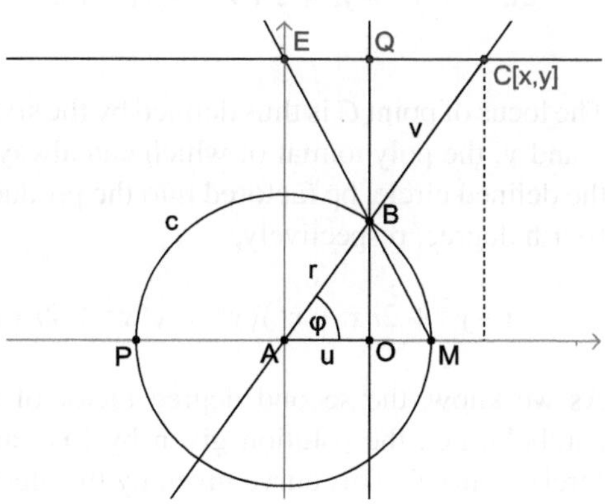

Fig. 9 Problem 35; location of the problem assignment in the coordinate system

$\left|\frac{rx}{r+v}\right|$, $MO = |r-u| = \left|r - \frac{rx}{r+v}\right| = \left|\frac{r^2+rv-rx}{r+v}\right|$, $BC = |v|$ and, allowing $AO = 0$, it can be rewritten into the form $|rv + v^2 - vx| = |rx|$ equivalent to

$$(rv + v^2 - vx)^2 - (rx)^2 = 0, \tag{3}$$

the first equation in x and y defining the locus of point C. The second and third such equations are expressions of conditions for the length of the radius vector of point C and the nonzero value of $r + v$, respectively;

$$x^2 + y^2 - (r+v)^2 = 0, \tag{4}$$

$$k(r+v) + 1 = 0. \tag{5}$$

Now we are to solve the system of algebraic equations (3)–(5). Possible realization of this task by elimination of the parameters v and k in the *CAS* of GeoGebra is as follows:

c1:=(r*v+v^2-v*x)^2-(r*x)^2 → $\mathbf{c1 := -(r\,x)^2 + (v^2 + r\,v - v\,x)^2}$
c2:=x^2+y^2-(r+v)^2 → $\mathbf{c2 := x^2 + y^2 - (r+v)^2}$
c3:=(r+v)*k+1 → $\mathbf{c3 := k\ (r+v) + 1}$
L:=Eliminate[{c1,c2,c3},{k,v}] → $\mathbf{L := \{y^6 + x^2 y^4 - 2 r^2 y^4 - 2 r^2 x^2 y^2 - 4 r^2 x^4 - 8 r^3 x^3 + r^4 y^2 - 3 r^4 x^2\}}$
LE:=Factorise[Element[L,1]]=0 → $\mathbf{LE : -(r^2 - y^2 + 2 r x)\ (y^4 + 2 r x^3 + 3 r^2 x^2 - r^2 y^2 + x^2 y^2 + 2 r x y^2) = 0}$

The locus of point C is thus defined by the sixth degree algebraic equation in variables x and y, the polynomial of which can always, i.e., independently on the radius r of the defined circle, be factored into the product of two polynomials of the second and fourth degree, respectively,

$$(-y^2 + 2rx + r^2)(y^4 + x^2y^2 + 2rxy^2 + 2rx^3 + 3r^2x^2 - r^2y^2) = 0. \tag{6}$$

As we know, the second degree factor of the polynomial (6) corresponds to the parabola, i.e., the solution given by I.H., and the fourth degree factor defines the 'pretzel curve'. This curve given by the algebraic equation

$$y^4 + x^2y^2 + 2rxy^2 + 2rx^3 + 3r^2x^2 - r^2y^2 = 0$$

is worth further exploration and GeoGebra is the appropriate instrument with which to do so. For example, we quite easily reveal the equation in the polar coordinates

$$R = \frac{1 - 2\cos\Phi}{1 - \cos\Phi} \tag{7}$$

or rational parameterization

$$x = \frac{-t^4 + 4t^2 - 3}{2t^2 + 2}, \quad y = \frac{t^3 - 3t}{t^2 + 1}; \quad t \in R, \tag{8}$$

of this curve, both (7) and (8) belonging to $r = 1$.

5 Conclusion

Through the solution of the locus problem, which is not a well-known problem, the effectiveness of the symbolic computer algebra means of GeoGebra and of their joint use with the dynamic geometric feature within this software has been presented in this paper. GeoGebra is endowed with functions that are designed for the automatic computation of a locus curve equation as well as with the environment (*CAS*) enabling the symbolic step by step computation of the equation. This variability of the use of symbolic computer algebra tools determines GeoGebra for utilization in mathematics teaching. It allows a teacher to choose different approaches to the solution of such complex problems, depending on the level of mathematical abstraction of pupils or students.

Utilization of the symbolic algebra feature to investigate a locus which has been presented in the paper is only a part of a much broader base of symbolic algebra tools implemented in GeoGebra. As other examples we can mention the prover subsystem [3] or computation of envelopes [1]. Implementation of the tools of symbolic algebra in GeoGebra is the subject of continuous development. Currently, intensive work is being performed to increase the speed of solving equation systems with a large number of variables and to broaden its availability on several computer platforms including mobile phones and tablets. This work is being performed by the GeoGebra Team under the supervision of Bernard Parisse's, inventor of Giac.

Acknowledgements We are grateful to Francisco Botana for his valuable advice and contribution to the effective use of computer algebra in GeoGebra.

References

1. Botana, F., Kovács, Z.: Teaching loci and envelopes in GeoGebra, GeoGebraBook. http://tube.geogebra.org/material/simple/id/128631# (2014)
2. Botana, F., Kovács, Z.: A singular web service for geometric computations. Ann. Math. Artif. Intell. **74**(3), 359–370 (2015)
3. Botana, F., Hohenwarter, M., Janičić, P., Kovács, Z., Petrović, I., Recio, T., Weitzhofer, S.: Automated theorem proving in GeoGebra: current achievements. J. Autom. Reason. **55**(1), 39–59 (2015)
4. Clark, K.M.: History of mathematics: illuminating understanding of school mathematics concepts for prospective mathematics teachers. Educ. Stud. Math. **81**(1), 67–84 (2012)
5. Courant, R., Robbins, H.: What is Mathematics? An Elementary Approach to Ideas and Methods, 2nd edn. Oxford University Pre, Oxford (1996)
6. Cundy, H.M.: Mathematical Models, 2nd edn. Oxford University Press, Oxford (1961)
7. Czédli, G., Szendrei, Á.: Geometriai szerkeszthetőség (in Hungarian). Polygon, Szeged (1997)
8. Decker, W., Greuel, G.-M., Pfister, G., Schönemann, H.: SINGULAR 4-0-2—a computer algebra system for polynomial computations. http://www.singular.uni-kl.de (2015)
9. Dennis, D.: The role of historical studies in mathematics and science educational research. In: Kelly, R., Lesh, A. (eds.) Research Design in Mathematics and Science Education. Lawrence Erlbaum, Mahwah (2000)
10. Furinghetti, F.: Teacher education through the history of mathematics. Educ. Stud. Math. **66**(2), 131–143 (2007)
11. Hašek , R., Kovács, Z.: The pretzel curve in Ioannis Holfeld's 35th problem, GeoGebra worksheet. http://tube.geogebra.org/m/A1OFuOWt (2015)
12. Hašek, R., Zahradník, J.: Study of historical geometric problems by means of CAS and DGS. The International Journal for Technology in Mathematics Education, pp. 53–58. Research Information Ltd., Burnham (2015)
13. Hašek, R., Zahradník, J.: Současná interpretace vybraných historických úloh na množiny bodů dané vlastnosti (in Czech). In *Sborník příspěvků 33. konference o geometrii a grafice. Horní Lomná, 9. - 12. září 2013*. Ostrava: Vysoká škola báňská – Technická univerzita Ostrava, pp. 115–120 (2013)
14. Hohenwarter, M., Borcherds, M., Ancsin, G., Bencze, B., Blossier, M., Bogner, S., Denizet, C., Éliás, J., Gál, L., Konečný, Z., Kovács, Z., Krismayer, T., Küllinger, W., Lizelfelner, S., Parisse, B., Rathgeb, P., Sólyom-Gecse, C.S., Stadlbauer, C., Tomaschko, M.: GeoGebra 5.0.178.0, free mathematics software for learning and teaching. http://www.geogebra.org (2015)
15. Holfeld, I.: Exercitationes Geometricae. Charactere Collegii Clementini Societas Jesu, Praha (1773)
16. Katz, V.J.: A History of Mathematics: An Introduction, 2nd edn. Adison-Wesley, Reading (1998)
17. Kovács, Z., Parisse, B.: Giac and GeoGebra—improved Gröbner basis computations. In: Gutierrez, J., Schicho, J., Weimann, M. (eds.) Computer Algebra and Polynomials. Lecture Notes in Computer Science 8942, pp. 126–138. Springer, Heidelberg (2015)
18. Kovács, Z.: *Holfeld's 35th Problem as An Implicit Locus*. https://www.geogebra.org/m/Gj6RgKJk (2016)
19. Moise, E.: Elementary Geometry from An Advanced Standpoint, 2nd edn. Addison-Wesley Publishing Company, New York (1990)
20. Montes, A., Wibmer, M.: Groebner bases for polynomial systems with parameters. J. Symb. Comput. **45**, 1391–1425 (2010)
21. Parisse, B.: Giac/Xcas, a free computer algebra system. https://www-fourier.ujf-grenoble.fr/~parisse/giac.html (2015)
22. Struik, D.J.: A Concise History of Mathematics, 4th edn. Dover Publications, New York (1987)
23. Weisstein, E.W.: Knot curve. From MathWorld–a wolfram web resource. http://mathworld.wolfram.com/KnotCurve.html (2015)

24. Zahradník, J.: Problémy z geometrie ve sbírce Ioannise Holfelda Exercitationes geometricae (in Czech). In *Sborník 34. mezinárodní konference Historie matematiky, Poděbrady, 23. - 27. srpna 2013*, Matfyzpress, Praha, p. 191 (2013)

Computing the Chern–Schwartz–MacPherson Class of Complete Simplical Toric Varieties

Martin Helmer

Abstract Topological invariants such as characteristic classes are an important tool to aid in understanding and categorizing the structure and properties of algebraic varieties. In this note, we consider the problem of computing a particular characteristic class, the Chern–Schwartz–MacPherson class, of a complete simplicial toric variety X_Σ defined by a fan Σ from the combinatorial data contained in the fan Σ. Specifically, we give an effective combinatorial algorithm to compute the Chern–Schwartz–MacPherson class of X_Σ, in the Chow ring (or rational Chow ring) of X_Σ. This method is formulated by combining, and when necessary modifying, several known results from the literature and is implemented in Macaulay2 for test purposes.

Keywords Chern–Schwartz–MacPherson class · Chern class · Toric varieties · Computer algebra · Computational intersection theory

1 Introduction and Background

The Chern–Schwartz–MacPherson (c_{SM}) class is a generalization of the total Chern class, that is the Chern class of the tangent bundle, to singular varieties. Unlike other generalizations of the Chern class to the singular setting the c_{SM} class maintains many of the functorial properties of the total Chern class, and in particular maintains the relation to the Euler characteristic. This means, explicitly, that as with the Chern class the c_{SM} class contains the Euler characteristic as the degree of its zero dimensional component, this relationship is discussed in more detail in Sect. 1.2.

Historically, the existence of a functorial theory of Chern classes for singular varieties in terms of a natural transformation from the functor of constructible functions

M. Helmer (✉)
Department of Mathematics, University of California Berkeley,
966 Evans Hall, Berkeley, CA 94720-3840, USA
e-mail: martin.helmer@berkeley.edu

I.S. Kotsireas and E. Martínez-Moro (eds.), *Applications of Computer Algebra*,
Springer Proceedings in Mathematics & Statistics 198,
DOI 10.1007/978-3-319-56932-1_13

to some nice homology theory, and its relation to the Euler characteristic, was conjectured by Deligne and Grothendieck in the 1960s. In the 1974, article [11], MacPherson proved the existence of such a transformation, introducing a new notion of Chern classes for singular algebraic varieties. Independently in the 1960s Schwartz [13] defined a theory of Chern classes for singular varieties in relative cohomology. It was later shown in a paper of Brasselet and Schwartz [3] that these two different notions were in fact equivalent. This construction is now commonly referred to as the Chern–Schwartz–MacPherson class.

In this note, we present Algorithm 1 which computes the Chern–Schwartz–MacPherson class and/or the Euler characteristic of a complete simplicial toric variety X_Σ defined by a fan Σ. The algorithm is based on a result of Barthel et al. [2] which gives an expression for the c_{SM} class of a toric variety in terms of torus orbit closures. Note that, for simplicity, we will only consider toric varieties X_Σ over $\mathbb{C}$.

From a computational point of view the problem of calculating the c_{SM} class for subschemes V of $\mathbb{P}^n$ has been considered by Aluffi in [1], by Jost in [10] and by the author of this note in [8, 9]; this problem has also been considered for subschemes of some smooth complete toric varieties by the author in [7]. All of these algorithms have at their core the need to solve polynomial systems of varying difficulty; for example by means of Gröbner bases calculations or polynomial homotopy continuation. As such the running times of all such algorithms are dependent on the algebraic degrees of the defining equations of V and on other algebraic properties of the defining equations. Given, the often substantial computational cost of solving polynomial systems we believe that an approach to computing c_{SM} classes which is strictly combinatorial in nature is desirable in settings where this is possible, such as the toric setting considered here.

We note that the restriction to complete simplicial toric varieties is not required in the statement of the result of Barthel et al. [2] on which our algorithm is based, indeed these restrictions are present on the algorithm only for the purpose of simplifying the construction of the Chow ring of the toric variety. If one was able to construct the Chow ring in a simple manner with the restrictions removed the algorithm could be applied unchanged in this more general setting.

The Macaulay2 [6] implementation of our algorithm for computing the c_{SM} class and Euler characteristic of a complete simplicial toric variety presented in this note can be found at https://github.com/Martin-Helmer/char-class-calc. This implementation is accessed via the "CharToric" package. Note that this implementation is also available in the github version of the "CharacteristicClasses" Macaulay2 package, see https://github.com/Macaulay2/M2/blob/master/M2/Macaulay2/packages/CharacteristicClasses.m2, and will be included in the next release of Macaulay2.

Example 1.1 Let $\mathcal{H}_r$ denote the rth Hirzebruch surface (see, for example, Cox et al. [4, Example 3.1.16]). Taking $r = 5$ and letting $R = \mathbb{C}[x_0, x_1, x_2, x_3, x_4]$ be the total coordinate ring of the toric variety $\mathcal{H}_r$ we have that

$$c_{SM}(\mathcal{H}_r) = 4x_1x_2 + 2x_1 + 7x_2 + 1 \in A^*(\mathcal{H}_r), \tag{1}$$

where $A^*(\mathcal{H}_r)$ is the Chow ring of $\mathcal{H}_r$. We may write this as

$$A^*(\mathcal{H}_r) \cong \mathbb{Z}[x_0, x_1, x_2, x_3, x_4]/(x_0x_2, x_1x_3, x_0 - x_2, -x_3 + x_1 + 5x_2). \quad (2)$$

From this we deduce that the Euler characteristic is

$$\chi(\mathcal{H}_r) = \int c_{SM}(\mathcal{H}_r) = 4,$$

where $\int \alpha$ denotes the degree of the zero-dimensional part of the cycle class α in some Chow ring (which is the coefficient of x_1x_2 in this case). Note that the Euler characteristic could also be obtained directly as the number of 2-dimensional cones in the fan corresponding to the toric variety $\mathcal{H}_r$ via Theorem 12.3.9 of Cox et al. [4].

The content of this note will be organized as follows. In Sect. 1.1 we will establish the setting for this work and review the construction of the rational Chow ring of a complete and simplicial toric variety. In Sect. 1.2 we will state the problem and briefly review the definition of the c_{SM} class. We then review relevant related results in Sect. 1.3. In Sect. 2 we detail the construction of our algorithm for computing the c_{SM} class in the setting considered here. The problem of computing the multiplicity of a cone in an explicit manner is considered in Sect. 2.1. Our algorithm for computing c_{SM} classes (Algorithm 1), along with the results of some performance testing of Algorithm 1 is given in Sect. 2.2.

1.1 Setting and Notation

Let X_Σ be an n-dimensional complete and simplicial toric variety defined by a fan Σ. Similar to the construction of the Chow ring in the smooth case we may construct the Chow ring of X_Σ from the Chow groups, that is the groups $A^j(X_\Sigma)$ of codimension j-cycles on X_Σ modulo rational equivalence. The only difference in this case will be that we work over the rational number field $\mathbb{Q}$ rather than the integers.

Using, the definition of the intersection product on rational cycles (see Sect. 12.5 of [4]) we have that the rational Chow ring of X_Σ is given by the graded ring

$$A^*(X_\Sigma)_\mathbb{Q} = A^*(X_\Sigma) \otimes_\mathbb{Z} \mathbb{Q} = \bigoplus_{j=0}^{n} A^j(X_\Sigma) \otimes_\mathbb{Z} \mathbb{Q}. \quad (3)$$

For each cone σ in the fan Σ the orbit closure $V(\sigma)$ is a subvariety of codimension $\dim(\sigma)$. We will write $[V(\sigma)]$ for the rational equivalence class of $V(\sigma)$ in $A^{\dim(\sigma)}(X_\Sigma)$.

For convenience of notation we will also write $A_\ell(X_\Sigma)$ for the dimension ℓ-cycles on X_Σ modulo rational equivalence. For a more in depth discussion of rational equivalence, Chow groups, and Chow rings see Fulton [5].

Proposition 1.2 (Lemma 12.5.1 of [4]) *The collections* $[V(\sigma)] \in A_j(X_\Sigma)$ *for* $\sigma \in \Sigma$ *having dimension* $n - j$ *generate* $A_j(X_\Sigma)$, *the Chow group of dimension* j. *Further, the collection* $[V(\sigma)]$ *for all* $\sigma \in \Sigma$ *generates* $A^*(X_\Sigma)$ *as an abelian group.*

The following proposition gives us a simple method to compute the rational Chow ring of a complete, simplicial toric variety X_Σ. We will use this result to compute the rational Chow ring $A^*(X_\Sigma)_\mathbb{Q}$ in Algorithm 1, our algorithm to compute the c_{SM} class of a complete, simplicial toric variety.

Proposition 1.3 (Theorem 12.5.3 of Cox et al. [4]) *Let* N *be an integer lattice with dual* M. *Let* X_Σ *be a complete and simplicial toric variety with generating rays* $\Sigma(1) = \rho_1, \dots, \rho_r$ *where* $\rho_j = \langle v_j \rangle$ *for* $v_j \in N$. *Then, we have that*

$$\mathbb{Q}[x_1, \dots, x_r]/(\mathcal{I} + \mathcal{J}) \cong A^*(X_\Sigma)_\mathbb{Q}, \tag{4}$$

with the isomorphism map specified by $[x_i] \mapsto [V(\rho_i)]$. *Here* $\mathcal{I}$ *denotes the Stanley–Reisner ideal of the fan* Σ, *that is the ideal in* $\mathbb{Q}[x_1, \dots, x_r]$ *specified by*

$$\mathcal{I} = (x_{i_1} \cdots x_{i_s} \mid i_{i_j} \text{ distinct and } \rho_{i_1} + \cdots + \rho_{i_s} \text{ is not a cone of } \Sigma) \tag{5}$$

and $\mathcal{J}$ *denotes the ideal of* $\mathbb{Q}[x_1, \dots, x_r]$ *generated by linear relations of the rays, that is* $\mathcal{J}$ *is generated by linear forms*

$$\sum_{j=1}^{r} m(v_j) x_j \tag{6}$$

for m *ranging over some basis of* M.

1.2 Problem

The main problem considered in this note is the following: given a complete simplical toric variety X_Σ how do does one efficiently compute the class $c_{SM}(X_\Sigma)$ in the Chow ring $A^*(X_\Sigma)_\mathbb{Q}$? We will give a method to solve this problem in Algorithm 1. To further establish the context for this problem, however, we will briefly discuss the definition of the Chern–Schwartz–MacPherson class.

The total Chern class of a j-dimensional nonsingular variety V is defined as the Chern class of the tangent bundle T_V, we write this as $c(V) = c(T_V) \cdot [V]$ in the Chow ring of V, $A_*(V)$. See Fulton [5, Sect. 3.2] for a definition of the Chern class of a vector bundle. As a consequence of the Gauss–Bonnet–Chern theorem (or the Grothendieck–Riemann–Roch theorem, see for example Schürmann and Yokura [12]), we have that the degree of the zero-dimensional component of the total Chern class of a projective variety is equal to the Euler characteristic, that is

$$\int c(T_V) \cdot [V] = \chi(V). \tag{7}$$

Here $\int \alpha$ denotes the degree of the zero-dimensional component of the class $\alpha \in A_*(V)$, i.e., the degree of the part of α in $A_0(V)$.

There are several known generalizations of the total Chern class to singular varieties. All of these notions agree with $c(T_V) \cdot [V]$ for nonsingular V, however the Chern–Schwartz–MacPherson class is the only one of these that satisfies a property analogous to (7) for any V, i.e.,

$$\int c_{SM}(V) = \chi(V). \tag{8}$$

We review here the construction of the c_{SM} classes, given in the manner considered by MacPherson [11]. For a scheme V, let $\mathcal{C}(V)$ denote the abelian group of finite linear combinations $\sum_W m_W \mathbf{1}_W$, where W are (closed) subvarieties of V, $m_W \in \mathbb{Z}$, and $\mathbf{1}_W$ denotes the function that is 1 in W, and 0 outside of W. Elements $f \in \mathcal{C}(V)$ are known as constructible functions and the group $\mathcal{C}(V)$ is referred to as the group of constructible functions on V. To make $\mathcal{C}$ into a functor we let $\mathcal{C}$ map a scheme V to the group of constructible functions on V and a proper morphism $f : V_1 \to V_2$ is mapped by $\mathcal{C}$ to

$$\mathcal{C}(f)(\mathbf{1}_W)(p) = \chi(f^{-1}(p) \cap W), \quad W \subset V_1, \ p \in V_2 \text{ a closed point.}$$

Another functor from algebraic varieties to abelian groups is the Chow group functor $\mathcal{A}_*$. The c_{SM} class may be realized as a natural transformation between these two functors.

Definition 1.4 The Chern–Schwartz–MacPherson class is the unique natural transformation between the constructible function functor and the Chow group functor, that is $c_{SM} : \mathcal{C} \to \mathcal{A}_*$ is the unique natural transformation satisfying:

- (*Normalization*) $c_{SM}(\mathbf{1}_V) = c(T_V) \cdot [V]$ for V nonsingular and complete.
- (*Naturality*) $f_*(c_{SM}(\phi)) = c_{SM}(\mathcal{C}(f)(\phi))$, for $f : X \to Y$ a proper transformation of projective varieties, ϕ a constructible function on X.

For a scheme V let V_{red} denote the support of V, the notation $c_{SM}(V)$ is taken to mean $c_{SM}(\mathbf{1}_V)$ and hence, since $\mathbf{1}_V = \mathbf{1}_{V_{red}}$, we denote $c_{SM}(V) = c_{SM}(V_{red})$.

Note that the c_{SM} classes (and constructible functions) also satisfy the same inclusion/exclusion relation as the Euler characteristic, i.e., for V_1, V_2 subschemes of a scheme W we have

$$c_{SM}(V_1 \cup V_2) = c_{SM}(V_1) + c_{SM}(V_2) - c_{SM}(V_1 \cap V_2).$$

We note that in some settings, such as subschemes of projective spaces or subschemes of some toric varieties, computing the c_{SM} class seems to provide a quite effective means, relative to other available techniques, to compute the Euler characteristic. For a discussion of this see, for example, [7, 8]. For toric varieties themselves, however, this is not the case as there is in fact an explicit formula for the Euler characteristic of a toric variety, see Theorem 12.3.9 of Cox et al. [4].

1.3 Review of Results

In this section, we review the results which will provide the basis for Algorithm 1 below. The main ingredient in this algorithm is the following result of Barthel et al. [2].

Proposition 1.5 (Main Theorem of Barthel et al. [2]). *Let X_Σ be an n-dimensional complex toric variety specified by a fan Σ. We have that the Chern–Schwartz–MacPherson class of X_Σ can be written in terms of orbit closures as*

$$c_{SM}(X_\Sigma) = \sum_{\sigma \in \Sigma} [V(\sigma)] \quad \in A^*(X_\Sigma)_{\mathbb{Q}} \tag{9}$$

where $V(\sigma)$ is the closure of the torus orbit corresponding to σ.

We now recall the definition of the multiplicity of a simplicial cone, for more details see §6.4 of Cox et al. [4]. Let N be an integer lattice with dual lattice M, let $\sigma = \langle v_1, \ldots, v_d \rangle$ be a simplicial cone and let

$$N_\sigma = \mathrm{Span}(\sigma) \cap N, \tag{10}$$

recall that $\mathrm{Span}(\sigma) \subset N_{\mathbb{R}}$ is the smallest subspace of the vector space $N_{\mathbb{R}}$ which contains σ. We note that the index of the subgroup $\mathbb{Z}v_1 + \cdots + \mathbb{Z}v_d \subset N_\sigma$ in N_σ is finite. We define the multiplicity of σ as

$$\mathrm{mult}(\sigma) = [N_\sigma : \mathbb{Z}v_1 + \cdots + \mathbb{Z}v_d \subset N_\sigma] \tag{11}$$

where $[G : H]$ denotes the index of a subgroup H in a group G. In practice, we shall employ Lemma 2.1 to compute $\mathrm{mult}(\sigma)$. Specifically Lemma 2.1 will allow us to compute the multiplicity of a simplicial cone. Since we only consider complete simplicial toric varieties in Algorithm 1 this lemma may be used to compute the multiplicity in all cases considered here.

To compute the classes $[V(\sigma)]$ appearing in (9) we will employ the following proposition combined with Proposition 1.3.

Proposition 1.6 (Theorem 12.5.2. of Cox et al. [4]) *Assume that X_Σ is complete and simplicial. If $\rho_1, \ldots, \rho_d \in \Sigma(1)$ are distinct and if $\sigma = \rho_1 + \cdots + \rho_d \in \Sigma$ then in $A^*(X_\Sigma)$ we have the following:*

$$[V(\sigma)] = \mathrm{mult}(\sigma)[V(\rho_1)] \cdot [V(\rho_2)] \cdots [V(\rho_d)]. \tag{12}$$

Here, $\mathrm{mult}(\sigma)$ *will be calculated using Lemma 2.1.*

2 Algorithm and Performance

In this section, we describe the process by which we turn the Main Theorem of Barthel et al. [2] (Proposition 1.5) into a computational method to find c_{SM} classes of complete simplicial toric varieties.

2.1 Computing Multiplicitiy

One of the main computational steps in Algorithm 1 below, for singular cases, is the computation of the multiplicity of a cone $\sigma \in \Sigma$. In practice this computation will be accomplished using Lemma 2.1. This lemma is a modified version of Proposition 11.1.8. of Cox et al. [4]. We have altered the statement of the result to explicitly show how we will compute these multiplicities in practice. The main point here is to show how the definition of the multiplicity of a cone given in (11) can be phrased in terms of straightforward linear algebra computations in the cases considered in this note.

Lemma 2.1 (Modified version of Proposition 11.1.8. of Cox et al. [4]). *Let $N = \mathbb{Z}^n$ be an integer lattice. For a simplicial cone $\sigma = \rho_1 + \cdots + \rho_d \subset N$ let $\mathfrak{M}_\sigma$ be the matrix with columns specified by the generating vectors of the rays $\rho_1, \ldots, \rho_d$ which define the cone σ; we have*

$$\text{mult}(\sigma) = |\det(\text{Herm}(\mathfrak{M}_\sigma))| \tag{13}$$

where Herm$(\mathfrak{M}_\sigma)$ *denotes the Hermite normal form of matrix $\mathfrak{M}_\sigma$ with all zero rows and/or zero columns removed. Further* mult$(\sigma) = 1$ *if and only if U_σ is smooth.*

Proof Suppose $\rho_1 = \langle u_1 \rangle, \ldots, \rho_d = \langle u_d \rangle$ so that we can write $\sigma = \langle u_1, \ldots, u_d \rangle$. In Proposition 11.1.8. of Cox et al. [4] it is shown that if $e_1, \ldots, e_d$ is a basis for N_σ (see (10)) and $u_i = \sum_{j=1}^d a_{i,j} e_j = E[a_{i,j}]$ (where E is the $n \times d$ matrix with columns $e_1, \ldots, e_d$) then we have that

$$\text{mult}(\sigma) = \left|\det\left([a_{i,j}]\right)\right|. \tag{14}$$

The matrix $\mathfrak{M}_\sigma$ defined by the rays $\rho_1, \ldots, \rho_d$ is the $n \times d$ matrix with columns given by the vectors $u_1, \ldots, u_d$. Note that $\mathfrak{M}_\sigma$ has rank d. Choose $e_1, \ldots, e_d$ to be a basis of N_σ so that the matrix E with columns $e_1, \ldots, e_d$ has the form

$$E = \begin{bmatrix} \tilde{E} \\ \mathbf{0} \end{bmatrix}$$

with $\det(\tilde{E}) = 1$. Now since $\mathfrak{M}_\sigma$ has rank d we may write

$$\mathfrak{M}_\sigma = \begin{bmatrix} \mathrm{Herm}(\mathfrak{M}_\sigma) \\ \mathbf{0} \end{bmatrix} T$$

for $\mathrm{Herm}(\mathfrak{M}_\sigma)$ the $d \times d$ matrix obtained from the Hermite normal form of $\mathfrak{M}_\sigma$ with the zero rows removed and T a $d \times d$ unimodular matrix. Then we have that

$$\begin{bmatrix} \tilde{E} \\ \mathbf{0} \end{bmatrix} [a_{i,j}] = \begin{bmatrix} \mathrm{Herm}(\mathfrak{M}_\sigma) \\ \mathbf{0} \end{bmatrix} T,$$

and hence $\tilde{E}[a_{i,j}] = \mathrm{Herm}(\mathfrak{M}_\sigma)T$. Note that $\det(\tilde{E}) = \det(T) = 1$, this gives that $\det([a_{i,j}]) = \det(\mathrm{Herm}(\mathfrak{M}_\sigma))$ as claimed.

The Hermite normal form of $\mathfrak{M}_\sigma$ is obtained by performing unimodular column operations on $\mathfrak{M}_\sigma$ and thus represents a change of basis of N_σ, we may call this new basis for N_σ $e_1, \dots, e_d$. Since $\mathfrak{M}_\sigma$ has rank d then removing the zero rows we obtain the matrix $\mathrm{Herm}(\mathfrak{M}_\sigma)$ and we may then take this matrix to be the matrix $[a_{i,j}]$ in (14) since the matrix $\mathrm{Herm}(\mathfrak{M}_\sigma)$ specifies the change of basis for N_σ from $u_1, \dots, u_d$ to $e_1, \dots, e_d$. The matrix $\mathfrak{M}_\sigma$ defined by the rays $\rho_1, \dots, \rho_d$ is the matrix with columns given by the vectors $u_1, \dots, u_d$, that is $\mathfrak{M}_\sigma = [u_1, \dots, u_d]$, further $\mathfrak{M}_\sigma$ has rank d. $N_\sigma = \mathrm{Span}(\sigma) \cap N$ is the lattice generated by the columns of the matrix $\mathfrak{M}_\sigma$, that is $N_\sigma = \left\{ y \mid y = \mathfrak{M}_\sigma x,\ x \in \mathbb{R}^d \right\} \cap N$. From the definition of the Hermite form we have that

$$\mathfrak{M}_\sigma T = \begin{bmatrix} \mathrm{Herm}(\mathfrak{M}_\sigma) \\ \mathbf{0} \end{bmatrix}$$

for some unimodular matrix T. Thus, we have

$$N_\sigma = \left\{ y \mid y = \begin{bmatrix} \mathrm{Herm}(\mathfrak{M}_\sigma) \\ \mathbf{0} \end{bmatrix} x,\ x \in \mathbb{R}^d \right\} \cap N,$$

meaning we may take the matrix $[a_{i,j}] = \mathrm{Herm}(\mathfrak{M}_\sigma)$ in (14) and the conclusion follows.

The remaining statements are given in the form stated above in Proposition 11.1.8. of Cox et al. [4]. □

2.2 Algorithm

In Algorithm 1, we present an algorithm to compute $c_{SM}(X_\Sigma)$ for a complete, simplicial toric variety X_Σ defined by a fan Σ. Note that we represent $[V(\rho_j)]$ as x_j via the isomorphism given in Proposition 1.3.

We note that Algorithm 1 is strictly combinatorial; hence the runtime depends only on the combinatorics of the fan Σ defining the toric variety.

Algorithm 1. Input: A complete, simplicial toric variety X_Σ defined by a fan Σ with $\Sigma(1) = \{\rho_1, \dots, \rho_r\}$ and a boolean, Euler_only, indicating if only the Euler characteristic is desired. We assume $\dim(X_\Sigma) \geq 1$.
Output: $c_{SM}(X_\Sigma)$ in $A^*(X_\Sigma)_\mathbb{Q} \cong \mathbb{Q}[x_1, \dots, x_r]/(\mathcal{I} + \mathcal{J})$ and/or the Euler characteristic $\chi(X_\Sigma)$, if Euler_only=true then only $\chi(X_\Sigma)$ will be computed.

- Compute the rational Chow ring $A^*(X_\Sigma)_\mathbb{Q} \cong \mathbb{Q}[x_1, \dots, x_r]/(\mathcal{I} + \mathcal{J})$ using Proposition 1.3.
- csm $= 0$.
- **For** i **from** $\dim(X_\Sigma)$ **to** 1:
  - orbits = all subsets of $\Sigma(1) = \{\rho_1, \dots, \rho_r\}$ containing i elements.
  - total $= 0$.
  - **For** $\rho_{j_1}, \dots, \rho_{j_s}$ **in** orbits:
    - $\sigma = \rho_{j_1} + \cdots + \rho_{j_s}$.
    - Find $w = \text{mult}(\sigma)$ using Lemma 2.1.
    - $[V(\sigma)] = \text{mult}(\sigma)[V(\rho_{i_1})] \cdots [V(\rho_{i_s})] = w \cdot x_{i_1} \cdots x_{i_s}$.
    - total = total + $[V(\sigma)]$.
  - csm = csm + total.
  - **If** $i == \dim(X_\Sigma)$:
    - Set $(c_{SM}(X_\Sigma))_0 =$ csm.
    - Set $\chi(X_\Sigma) =$ sum of the coefficients of the monomials in $(c_{SM}(X_\Sigma))_0$.
    - **If Euler_only==true**:
      - **Return** $\chi(X_\Sigma)$.
- Set $c_{SM}(X_\Sigma) =$ csm.
- **Return** $c_{SM}(X_\Sigma)$ **and/or** $\chi(X_\Sigma)$.

In this subsection, we give the run times for Algorithm 1 applied to a variety of examples. Consider, a complete simplicial toric variety X_Σ. We give two alternate implementations of Algorithm 1 to reflect what we can expect the timings to be in both the smooth cases and singular cases.

Specifically, the running times in Table 1 for Algorithm 1 marked with a † check the input to see if the given fan Σ defines a smooth toric variety, if it does these implementations use the fact that $\text{mult}(\sigma) = 1$ for all $\sigma \in \Sigma$ and hence do not compute the Hermite normal forms and their determinates in Lemma 2.1. However to show how the algorithm would perform on a singular input of a similar size and complexity, we also give running times for an implementation which always computes the Hermite forms and their determinates in Lemma 2.1.

In this way, we see in a precise manner what the extra cost associated to computing the c_{SM} class and Euler characteristic of a singular toric variety would be in comparison to the cost of computing a smooth toric variety defined by a fan having similar combinatorial structure. Hence, the running time for a given example would be very similar to that of a singular toric variety with a similar number and dimension of cones to those considered in the examples in Table 1.

By default the implementation of Algorithm 1 in our "CharToric" package checks if the input defines a smooth toric variety, i.e., performs the procedure of the implementations marked with †. As such the performance of the package methods on smooth cases can be expected to be that of Algorithm 1 † in Table 1.

Table 1 In the table, we present the time to compute the Chow ring separately from the time required for the other computations, as such the total run time for each algorithm will be the time listed in its column plus the time to compute the Chow ring if the Chow ring is not already known. Computations were performed using Macaulay2 [6] on a computer with a 2.9 GHz Intel Core i7-3520M CPU and 8 GB of RAM. The Fano sixfolds are those built by the smoothFanoToricVariety method in the "NormalToricVarieties" Macaulay2 [6] package. $\mathbb{P}^n$ denotes a projective space of dimension n

Input	Algorithm 1 † (s)	Algorithm 1 (Euler only) † (s)	Algorithm 1 (s)	Algorithm 1 (Euler only) (s)	Chow ring (Proposition 1.3) (s)
$\mathbb{P}^6$	0.0	0.0	0.0	0.0	0.1
$\mathbb{P}^{16}$	5.3	0.0	85.4	0.0	0.7
$\mathbb{P}^5 \times \mathbb{P}^6$	0.3	0.0	3.7	0.0	1.2
$\mathbb{P}^5 \times \mathbb{P}^8$	1.1	0.0	16.8	0.1	2.1
$\mathbb{P}^8 \times \mathbb{P}^8$	12.0	0.1	168.5	0.1	4.5
$\mathbb{P}^5 \times \mathbb{P}^5 \times \mathbb{P}^5$	12.8	0.2	156.7	0.6	11.8
$\mathbb{P}^5 \times \mathbb{P}^5 \times \mathbb{P}^6$	28.4	0.3	387.1	0.8	17.0
Fano sixfold 123	0.3	0.0	1.0	0.4	1.1
Fano sixfold 1007	0.4	0.1	1.0	0.1	1.8

We also remark that the extra cost in the singular case (or in the case, where we do not check the input) comes entirely from performing linear algebra with integer matrices. As such the running times in these cases could perhaps be somewhat reduced by using a specialized integer linear algebra package. To give a rough quantification of what performance improvement one might expect from this we performed some testing using LinBox [15] and PARI [16] via Sage [14] on linear systems of similar size and structure to those arising in the examples in Table 1. In this testing, we found that the specialized algorithms seemed to be around two to three times faster than the linear algebra methods used by our implementation in the "CharToric" package, however, this testing is by no means conclusive.

In any case, it seems reasonable to conclude that some performance increase could be expected, for singular examples, if one used a specialized, fast integer linear algebra package to compute the Hermite forms and determinates arising in Algorithms 1. Finally, we note that additional efficiencies in implementation might also be found by a more careful implementation of the combinatorial procedures in a compiled language such as C or C++ rather than the Macaulay2 [6] language used here, which is an interpreted language.

References

1. Aluffi, P.: Computing characteristic classes of projective schemes. J. Symb. Comput. **35**(1), 3–19 (2003)
2. Barthel, G., Brasselet, J.-P., Fieseler, K.-H.: Classes de Chern de variétés toriques singulières. CR Acad. Sci. Paris Sér. I Math. **315**(2), 187–192 (1992)
3. Brasselet, J.-P., Schwartz, M.-H.: Sur les classes de Chern d'un ensemble analytique complexe. Astérisque **82**(83), 93–147 (1981)
4. Cox, D.A., John, B., Schenck, H.K.: Toric varieties. Am. Math. Soc. **124**, 575 (2011)
5. Fulton, W.: Intersection Theory, 2nd edn. Springer, Berlin (1998)
6. Grayson, D.R., Stillman, M.E.: Macaulay2, a software system for research in algebraic geometry. Biometrika **66**(2), 339–344 (2013)
7. Helmer, M.: An algorithm to compute the topological Euler characteristic, the Chern–Schwartz–Macpherson class and the Segre class of subschemes of some smooth complete toric varieties. arXiv:1508.03785 (2015)
8. Helmer, M.: Algorithms to compute the topological Euler characteristic, Chern–Schwartz–Macpherson class and Segre class of projective varieties. J. Symb. Comput. **73**, 120–138 (2015)
9. Helmer, M.: A direct algorithm to compute the topological Euler characteristic and Chern–Schwartz–Macpherson class of projective complete intersection varieties. Submitted to a Special Issue of the Journal of Theoretical Computer Science for SNC-2014. Available on the, arXiv.org/abs/1410.4113 (2015)
10. Jost, C.: An algorithm for computing the topological Euler characteristic of complex projective varieties. arXiv:1301.4128 (2013)
11. Robert, D.: MacPherson. Chern classes for singular algebraic varieties. Ann. Math. **100**(2), 423–432 (1974)
12. Schürmann, J., Yokura, S.: A Survey of Characteristic Classes of Singular Spaces, pp. 865–952. World Scientific, Singapore (2007)
13. Schwartz, M.-H.: Classes caractéristiques définies par une stratification d'une variété analytique complexe. Comptes Rendus de l'Académie des Sciences Paris **260**, 3262–3264 (1965)
14. Stein, W.A et al.: Sage Mathematics Software (Version 5.11). The Sage Development Team, http://www.sagemath.org (2013)
15. The LinBox Group.: LinBox–Exact Linear Algebra Over the Integers and Finite Rings, Version 1.1.6 (2008)
16. The PARI Group, Bordeaux.: *PARI/GP version 2.7.0*. Available from http://pari.math.u-bordeaux.fr/ (2014)

The Generalized Rabinowitsch Trick

Deepak Kapur, Yao Sun, Dingkang Wang and Jie Zhou

Abstract The famous Rabinowitsch trick for Hilbert's Nullstellensatz is generalized and used to analyze various properties of a polynomial with respect to an ideal. These properties include, among others, (i) checking whether the polynomial is a zero divisor in the residue class ring defined by the associated ideal and (ii) checking whether the polynomial is invertible in the residue class ring defined by the associated ideal. Just like using the classical Rabinowitsch's trick, its generalization can also be used to decide whether the polynomial is in the radical of the ideal. Some of the byproducts of this construction are that it is possible to be more discriminatory in determining whether the polynomial is a zero divisor (invertible, respectively) in the quotient ring defined by the ideal, or the quotient ideal constructed by localization using the polynomial. This method also computes the smallest integer which gives the saturation ideal of the ideal with respect to a polynomial. The construction uses only a single Gröbner basis computation to achieve all these results.

Keywords Rabinowitsch trick · Zero divisor · Invertible · Radical membership

1 Introduction

The classical Rabinowitsch trick was first proposed by J.L. Rabinowitsch in his 1-page paper *Zum Hilbertschen Nullstellensatz* in 1929 [9]. This ingenious trick was used to prove the famous Hilbert's Nullstellensatz theorem. Based on this proof, the

The authors were supported by the NSF DMS-1217054 and NSFC 11371356.

D. Kapur
Department of Computer Science, University of New Mexico, Albuquerque, NM, USA

Y. Sun
SKLOIS, Institute of Information Engineering, CAS, Beijing, China

D. Wang · J. Zhou (✉)
KLMM, Academy of Mathematics and Systems Science, CAS, Beijing, China
e-mail: jiezhou@amss.ac.cn

I.S. Kotsireas and E. Martínez-Moro (eds.), *Applications of Computer Algebra*, Springer Proceedings in Mathematics & Statistics 198,
DOI 10.1007/978-3-319-56932-1_14

radical membership problem can be solved. Let $k[X]$ be a polynomial ring over a field k, f be a polynomial and I be an ideal in $k[X]$, where $X = [x_1, \ldots, x_n]$ is a set of variables. The classical Rabinowitsch trick involves adding $fy - 1$ for performing radical membership test of f in I, where y is a new indeterminate different from X. In 2009, Sato and Suzuki [12] used this trick to compute the inverse of a polynomial f in the residue class ring $k[X]/(I : f^\infty)$.

A general construction to determine whether a given polynomial f is a zero divisor or invertible in the quotient ring $k[X]/I$, is proposed. It is proved that all this can be done using a single Gröbner basis construction of I augmented with a generalization of the classical Rabinowitsch trick, $fy - z$, where y, z are new indeterminates not appearing in X. It is also possible to perform radical membership test on f in I using the generalized construction. The generalized construction can be also used to compute the Gröbner bases of a family of related ideals–$I, I : f, I : f^2, \ldots, I : f^\infty$, $I + \langle f \rangle, I : f + \langle f \rangle, I : f^2 + \langle f \rangle, \ldots$, or $I : f^\infty + \langle f \rangle$ simultaneously, where $I : f^s = \{h \mid hf^s \in I\}$.

These results provide a necessary and sufficient condition for deciding whether f is invertible in $k[X]/(I : f^i)$ or whether f is a zero divisor in $k[X]/(I : f^i)$, where i is a nonnegative integer.

This paper is organized as follows. We review the properties of the classical Rabinowitsch trick in Sect. 2; we also relate it to Spear's trick of introducing a tag variable for studying properties of polynomial ideals; Bayer's further exploited the tag variable construction. In Sect. 3, we give two main results about the structure of the Gröbner basis of $I \cup \{fy - z\}$ and discuss how to check invertibility of f, radical membership of f, or f being a zero divisor in the residue class ring defined by I. An application of the generalized Rabinowitsch trick is presented in Sect. 4. Section 5 includes concluding remarks; as said there, constructions proposed in this paper generalize in a natural way to parameterized system using the comprehensive Gröbner system construction [7, 8].

2 Rabinowitsch Trick and Tag Variables

2.1 The Classical Rabinowitsch Trick

The classical Rabinowitsch trick was proposed to prove the famous Hilbert's Nullstellensatz theorem. Given polynomials $f, f_1, \ldots, f_s$ in $k[X]$, if f vanishes on the common zeros of $f_1, \ldots, f_s$, then there exists polynomials $a_0, a_1, \ldots, a_s$ in $k[X, y]$, such that

$$a_0(fy - 1) + a_1 f_1 + \cdots + a_s f_s = 1,$$

where y is an extra variable different from X. Substituting y by $1/f$, there exists an integer m such that f^m in the ideal generated by $f_1, \ldots, f_s$. For details, the reader can refer to [4]. The classical Rabinowitsch's trick can be used to solve the radical membership problem of an ideal by the following proposition (page 176, [3]).

Proposition 1 *Let k be an arbitrary field and let $I = \langle f_1, \ldots, f_s \rangle \subset k[X]$ be an ideal. Then $f \in \sqrt{I}$ if and only if the constant polynomial* 1 *belongs to the ideal $I + \langle fy - 1 \rangle$.*

Sato and Suzuki [12] used the classical Rabinowitsch trick to compute the inverse of a polynomial f in residue class ring $k[X]/(I : f^{\infty})$.

Proposition 2 *Let I be an ideal and f be a polynomial in $k[X]$. If G is a Gröbner basis of the ideal $I + \langle fy - 1 \rangle$ in $k[X, y]$ w.r.t. a term order such that $y >> X$, then f is invertible in $k[X]/(I : f^{\infty})$ if and only if G has a form $G = \{y - h, g_1, \ldots, g_l\}$. Further, h is an inverse of f in $k[X]/(I : f^{\infty})$ and $I : f^{\infty} = \langle g_1, \ldots, g_l \rangle$.*

Proposition 2 can only be used to decide whether f is invertible in $k[X]/(I : f^{\infty})$ directly. To decide whether f is invertible in $k[X]/I$, however, the equality of the two ideals I and $I : f^{\infty}$ needs to be checked.

2.2 Tag Variable

Spear [14] introduced the concept of a *tag* variable and showed how various ideal theoretic operations can be performed with Gröbner basis computations using lexicographic ordering and the associated elimination ideals; please refer to [10] for many interesting comments about Spear's contributions to Gröbner basis theory. In [13], Shannon, and Sweedler used tag variables to test if a given polynomial g of $k[x_1, \ldots, x_n]$ lay in $k[f_1, \ldots, f_s]$.

In [10], Mora credited Bayer [1] for using a tag variable and reverse lexicographic ordering to analyze the properties of a polynomial f with respect to a polynomial ideal $I = \langle f_1, \ldots, f_s \rangle$.

If a Gröbner basis $G = \langle g_1, \ldots, g_t \rangle$ of ideal $I + \langle f - z \rangle$ over $k[X, z]$ is computed w.r.t. a reverse lexicographical ordering such that $X >> z$, then each g_i can be uniquely expressed as

$$g_i = z^{d_i} h_i, \qquad z \nmid h_i, \ h_i \in k[X, z],$$

where d_i is a nonnegative integer. If z divides g_i, let $a_i(X, z) = g_i/z$; otherwise, $a_i = g_i$. Substitute $z = f$ into a_i and h_i, and let

$$A_i(X) = a_i(X, f), \qquad H_i(X) = h_i(X, f).$$

Proposition 3 [10] *Using the above definitions of A_i's and H_j's,*

1. *$\{A_1, \ldots, A_t\}$ is a basis of $I : f$, and*
2. *$\{H_1, \ldots, H_t\}$ is a basis of $I : f^{\infty}$.*

Since the reverse lexicographical (rev-lex) ordering is not a well-ordering, the procedure of computing a Gröbner basis of an ideal w.r.t. the rev-lex ordering may not terminate as illustrated by the following example.

Example 1 Consider $I = \langle x_1, x_2^2 + x_2 \rangle$; let $f = x_1 - x_2$ be a polynomial.

Bayer's method advocates computing a Gröbner basis of $\langle x_1, x_2 + x_2^2, x_1 - x_2 - z \rangle = \langle f_1, f_2, f_3 \rangle$ w.r.t. the rev-lex ordering $x_1 > x_2 > z$. Assuming that the Buchberger's algorithm [2] is used, let $\overline{f}^F$ be the remainder on division of f by the ordered tuple F, and the $S - polynomial$ of f and g is

$$S(f, g) = \frac{x^r}{\mathrm{lt}(f)} f - \frac{x^r}{\mathrm{lt}(g)} g,$$

where $\mathrm{lt}(f)$ is the leading term of polynomial f w.r.t. the rev-lex ordering $x_1 > x_2 > z$, and x^r is the least common multiple of $\mathrm{lt}(f)$ and $\mathrm{lt}(g)$.

Initial: $F = (f_1, f_2, f_3)$;

Step1: $S(f_1, f_2) = x_2 \cdot f_1 - x_1 \cdot f_2 = -x_1 x_2^2 := f_4, \quad \overline{f_4}^F = 0$;

Step2: $S(f_1, f_3) = f_1 - f_3 = x_2 + z := f_5$.

In F, only the leading term of f_2 can divide $\mathrm{lt}(f_5)$. Let $f_5 - f_2 = -x_2^2 + z$, which is still only reduced by f_2. Sequentially, it gives an infinite sequence

$$x_2 + z, -x_2^2 + z, x_2^3 + z, \ldots, (-1)^{k+1} x_2^k + z, \ldots.$$

The procedure of computing a Gröbner basis of $\langle x_1, x_2 + x_2^2, x_1 - x_2 - z \rangle$ w.r.t. the rev-lex ordering $x_1 > x_2 > z$ does not terminate. So Bayer's method can not be used directly in this case.

Mora claimed a way to overcome this problem by homogenizing an ideal. For homogeneous ideals, the Gröbner basis of an ideal w.r.t. rev-lex ordering exists. A nonhomogeneous ideal can thus first be homogenized; use then Proposition 3 on the homogenized ideal basis and then dehomogenize the result. It should be noted however that the dehomogenization does not produce a Gröbner basis of the nonhomogeneous ideal. Moreover, we want to emphasize that Proposition 3 only guarantees as its output, a basis of $I : f$ or $I : f^\infty$, not a Gröbner basis.

Example 2 Let the ideal $I = \langle x_2^2, x_1 x_2 + x_3^2 \rangle$, the polynomial $f = x_1 x_2$.

The Gröbner basis of $I + \langle f - z \rangle$ w.r.t. the rev-lex ordering $x_1 > x_2 > x_3 > z$ is $G = \langle z^2, x_2 z, x_3^2 + z, x_2^2, x_1 x_2 - z \rangle$. By the Proposition 3, $I_1 = \{x_1 x_2, x_2, x_1 x_2 + x_3^2, x_2^2\}$ is a basis of $I : f$. It is easy to check x_3^2 is in $I : f$, but $\mathrm{lt}(x_3^2) = x_3^2$ is not divided by any leading term of polynomials in G. So I_1 is not a Gröbner basis.

3 The Generalized Rabinowitsch Trick

In this section, we generalize the Rabinowitsch trick and discuss properties of f in a quotient ring such as $k[X]/I$, $k[X]/(I : f)$. Specifically, we provide necessary and sufficient conditions to check whether f is invertible or a zero divisor in $k[X]/I$, $k[X]/(I : f)$,..., $k[X]/(I : f^s)$,..., and $k[X]/(I : f^\infty)$. We can also check whether

f is in $\sqrt{I}$, the radical of ideal I, as well as find the smallest integer m such that $I : f^m = I : f^\infty$.

A polynomial f is **invertible** in $k[X]/I$, if $f \notin I$ and there exists g in $k[X]$ such that $fg - 1 \in I$. Moreover, such g is called an inverse of f in $k[X]/I$. A polynomial f is a **zero divisor** in $k[X]/I$, if $f \notin I$ and there exists h in $k[X]$ such that $h \notin I$ and $fh \in I$.

The generalized Rabinoswitsch's trick can be interpreted as integration of Rabinowitsch's trick with that of tag variable as illustrated below. Consider, the following ideal

$$J = I + \langle fy - z\rangle \subset k[X, y, z],$$

associated with I and f, where y and z are two new variables different from X.

Firstly, we analyze some special polynomials in J, which can be expressed as $g = p_t yz^t + p_{t-1}yz^{t-1} + \cdots + p_0 y + q_r z^r + q_{r-1}z^{r-1} + \cdots + q_1 z + q_0$, where $p_0, \ldots, p_t, q_0, \ldots, q_r$ are polynomials in $k[X]$.

Lemma 1 *Let $I = \langle f_1, \ldots, f_s\rangle$ be an ideal, f be a polynomial in $k[X]$, and $J = I + \langle fy - z\rangle$ be an ideal in $k[X, y, z]$. Given a polynomial $g = p_t yz^t + \cdots + p_0 y + q_r z^r + \cdots + q_1 z + q_0$ in J, where $p_0, \ldots, p_t, q_0, \ldots, q_r \in k[X]$, then*

$$p_{i-1} f^{i-1} + q_i f^i \in I,$$

where i is a nonnegative number, $p_j = 0$ when $j > t$, and $q_k = 0$ when $k > r$. Moreover, $p_{i-1} \in I : f^{i-1} + \langle f\rangle$, and when $p_{i-1} = 0$, $q_i \in I : f^i$.

Proof Since g is a polynomial in J, there exists $a_1, \ldots, a_s, a_{s+1} \in k[X, y, z]$, such that

$$p_t yz^t + \cdots + p_0 y + q_r z^r + \cdots + q_1 z + q_0 = a_1 f_1 + \cdots + a_s f_s + a_{s+1}(fy - z). \quad (1)$$

Now setting $z = fy$ in the above Eq. (1) gives

$$p_t (fy)^t y + \cdots + p_0 y + q_r (fy)^r + \cdots + q_1 (fy) + q_0 = a_1' f_1 + \cdots + a_s' f_s,$$

where $a_j' \in k[X, y]$ for $j = 1, \ldots, s$. Viewing the right side of the above equation as a polynomial in $k[X][y]$, it is possible to reformulate it as $a_1' f_1 + \cdots + a_s' f_s = b_k y^k + \cdots + b_1 y + b_0$, where $b_0, \ldots, b_k \in k[X]$. Note that each b_j can also be arranged as an expression of the form $b_j = c_1 f_1 + \cdots + c_t f_t$ for some $c_1, \ldots, c_t \in k[X]$, so $b_0, \ldots, b_k \in I$. Thus,

$$p_t (fy)^t y + \cdots + p_0 y + q_r (fy)^r + \cdots + q_1 (fy) + q_0 = b_k y^k + \cdots + b_1 y + b_0.$$

Comparing each coefficient of y^i, $b_i = p_{i-1} f^{i-1} + q_i f^i$. So $p_{i-i} f^{i-1} + q_i f^i \in I$, i.e. $p_{i-1} + q_i f \in I : f^{i-1}$. It is obvious that p_{i-1} in $I : f^{i-1} + \langle f\rangle$, and $q_i \in I : f^i$ when $p_{i-1} = 0$. □

Lemma 2 *Let I, J be defined as in Lemma 1. For a polynomial h in $k[X]$, $hf^s \in I$ if and only if $hz^s \in J$, where s is any nonnegative integer.*

Proof ($\Rightarrow$) : If $hf^s \in I$, then $hz^s = h(fy - (fy - z))^s = hf^s y^s + hp(fy - z) \in J$, where $p \in k[X, y, z]$. ($\Leftarrow$) : It is obvious from Lemma 1.

□

We analyze the ideal J by studying its Gröbner basis using a block ordering in which $y \gg z \gg X$. Using the structure of this Gröbner basis, we give below the main theoretical result.

Let g be a polynomial in $k[X, y, z]$ and "$\prec$" be an admissible monomial ordering on the set of power products of $X \cup \{y, z\}$. We use notations $\text{lpp}(g)$ and $\text{lc}(g)$ to represent the leading power product and leading coefficient of g with respect to "$\prec$," respectively. The notation "$\prec_{y,z}$" is a restriction of "$\prec$" on the set of power products of $\{y, z\}$. We use the notations $\text{lpp}_{y,z}(g)$ and $\text{lc}_{y,z}(g)$ to represent the leading power product and leading coefficient of g with respect to "$\prec_{y,z}$" respectively. The notation $\text{tail}(g)$ represents the part of $g - \text{lc}(g)\text{lpp}(g)$, i.e., g can be expressed as $g = \text{lc}(g)\text{lpp}(g) + \text{tail}(g)$. For example, let $g = 2x^2yz + x^3z$, and "$\prec$" be the lexicographic ordering w.r.t. $z > y > x$, $\text{lpp}(g) = x^2yz$, $\text{lc}(g) = 2$, $\text{lpp}_{y,z}(g) = yz$, $\text{lc}_{y,z}(g) = 2x^2$ and $\text{tail}(g) = x^3z$. And $\text{lc}_{y,z}(g)$ is in $k[X]$.

Theorem 4 *Let I be an ideal and f be a polynomial in $k[X]$. Let G be a Gröbner basis of ideal $J = I + \langle fy - z\rangle \subset k[X, y, z]$ with respect to a block ordering "$\prec$" such that $y \gg z \gg X$.*

1. *Let $P_s = \{\text{lc}_{y,z}(g) \mid g \in G \cap k[X][z], \text{lpp}_{y,z}(g) = z^k$ and $0 \le k \le s\} \subset k[X]$. For any integer $s \ge 0$, P_s is a Gröbner basis of $I : f^s$.*
2. *Let $Q_s = P_s \cup \{\text{lc}_{y,z}(g) \mid g \in G, \text{lpp}_{y,z}(g) = yz^t$, and $0 \le t \le s\} \subset k[X]$. For any integer $s \ge 0$, Q_s is a Gröbner basis of $I : f^s + \langle f\rangle$.*

Proof (1) First, we prove $P_s \subset I : f^s$. For any $q \in P_s$, by the construction of P_s, there exists a polynomial $g \in G$, such that $g = qz^k + \text{tail}(g)$, where $0 \le k \le s$. From Lemma 1, we know $qf^k \in I$. So $q \in I : f^k \subset I : f^s$. Therefore, we have proved $P_s \subset I : f^s$.

Second, we prove P_s is a Gröbner basis of $I : f^s$, or equivalently, we need to prove that for any $h \in I : f^s$, there exists $q \in P_s$, such that $\text{lpp}(q)$ divides $\text{lpp}(h)$. Let h be any polynomial in $I : f^s$, we have $hf^s \in I$. Hence, we have $hz^s \in J$ by Lemma 2. Since G is a Gröbner basis of J, there exists a polynomial $g \in G$, such that $\text{lpp}(g)$ divides $\text{lpp}(hz^s)$. So g must have the form of $g = qz^k + \text{tail}(g)$, where $q \in k[X]$ and $0 \le k \le s$. Thus, $\text{lpp}(g) \mid \text{lpp}(hz^s)$ means $\text{lpp}(q) \mid \text{lpp}(h)$, and we also have $q \in P_s$ by the construction of P_s.

(2) First, we prove $Q_s \subset I : f^s + \langle f\rangle$. For any $p \in Q_s \subset k[X]$, if $p \in P_s$, then $p \in I : f^s \subset I : f^s + \langle f\rangle$ by (1). Otherwise, if $p \notin P_s$, then there exist a polynomial $g \in G$ having the form of $g = pyz^t + \text{tail}(g)$, where $0 \le t \le s$. By Lemma 1, we have $p \in I : f^t + \langle f\rangle \subset I : f^s + \langle f\rangle$. So we have proved $Q_s \subset I : f^s + \langle f\rangle$.

Second, we show Q_s is a Gröbner basis of $I : f^s + \langle f \rangle$. For any $h \in I : f^s + \langle f \rangle$, there exists $q \in I : f^s$ and $a_1, a_2 \in k[X]$ such that $h = a_1 q + a_2 f$ by the definition of $I : f^s + \langle f \rangle$. Since $q \in I : f^s$, we have $qf^s \in I$, and hence, $qz^s \in J$ by Lemma 2. Next, we construct the polynomial $T = hyz^s - a_2 z^{s+1} = (a_1 q + a_2 f)yz^s - a_2 z^{s+1} = a_1 qyz^s + a_2(fy - z)z^s \in J$. Since G is a Gröbner basis of J and $\text{lpp}(T) = \text{lpp}(h)yz^s$, there exists a polynomial $g \in G$, such that $\text{lpp}(g)$ divides $\text{lpp}(h)yz^s$. This g must have the form of $g = py^k z^t + \text{tail}(g)$, where $0 \le k \le 1$ and $0 \le t \le s$. So we have $\text{lpp}(p) \mid \text{lpp}(h)$. Due to the form of g we also have $p \in Q_s$. This shows that for any $h \in I : f^s + \langle f \rangle$ there exists $p \in Q_s$ such that $\text{lpp}(p) \mid \text{lpp}(h)$. □

If G is a minimal Gröbner basis[1] of J, it is easy to see that $I : f^{i-1} \subsetneq I : f^i$ if and only if $P_{i-1} \subsetneq P_i$, and $I : f^{i-1} + \langle f \rangle \subsetneq I : f^i + \langle f \rangle$ if and only if $Q_{i-1} \subsetneq Q_i$.

The following result serves as the basis for checking if a polynomial is invertible or a zero divisor in a residue class ring as well as for checking its membership in the radical of an ideal.

Theorem 5 *Let I be an ideal and f be a polynomial in $k[X]$. Let G be a minimal Gröbner basis of ideal $J = I + \langle fy - z \rangle \subset k[X, y, z]$ with respect to a block ordering "$\prec$" such that $y \gg z \gg X$, and P_s, Q_s are constructed from G as stated in Theorem 4. The following properties hold:*

1. *f is **invertible** in $k[X]/(I : f^s)$ if and only if $1 \in Q_s$ and $1 \notin P_{s+1}$, i.e., $I : f^s + \langle f \rangle = \langle 1 \rangle$ and $f \notin I : f^s$. The inverse of f in $k[X]/(I : f^s)$ can be obtained from G.*
2. *f is a **zero divisor** in $k[X]/(I : f^s)$ if and only if $P_s \subsetneq P_{s+1}$ and $1 \notin P_{s+1}$, i.e. $I : f^s \subsetneq I : f^{s+1}$ and $f \notin I : f^s$.*
3. *f is **in the radical ideal** $\sqrt{I}$ if and only if there exists an integer s such that $1 \in P_s$, i.e. $I : f^s = \langle 1 \rangle$.*
4. *m is the **smallest** integer such that $I : f^\infty = I : f^m$, if and only if $P_{m-1} \subsetneq P_m = P_s$ for all $s > m$. Further, P_m is a Gröbner basis of $I : f^\infty$.*

Proof (1). ($\Rightarrow$) : If f is invertible in $k[X]/(I : f^s)$, then $f \notin I : f^s$ and there exists h such that $fh - 1 \in I : f^s$. So $1 \notin I : f^{s+1}$ and $1 \in I : f^s + \langle f \rangle$. By Theorem 4 (1) and (2), we have $1 \in Q_s$ and $1 \notin P_{s+1}$.

($\Leftarrow$) : If $1 \notin P_{s+1}$ and $1 \in Q_s$, then $f \notin I : f^s$ and there exists $g \in G$ having the form of $g = yz^t + p_{t-1}yz^{t-1} + \cdots + p_0 y + q_r z^r + \cdots + q_1 z + q_0$, where $p_0, \ldots, p_{t-1}, q_0, \ldots, q_r \in k[X]$ and $0 \le t \le s$. By Lemma 1, $1 + q_{t+1} f \in I : f^t \subset I : f^s$, so f is invertible in $k[X]/(I : f^s)$ and $-q_{t+1}$ is its inverse.

(2). ($\Rightarrow$) : If f is a zero divisor in $k[X]/(I : f^s)$, then $f \notin I : f^s$ and there exists $h \notin I : f^s$ such that $fh \in I : f^s$. So $1 \notin I : f^{s+1}$ and $h \in (I : f^{s+1}) \setminus (I : f^s)$. Then $I : f^s \subsetneq I : f^{s+1}$. By Theorem 4 (1), P_s, P_{s+1} are Gröbner bases of $I : f^s$ and $I : f^{s+1}$ respectively. So $P_s \subsetneq P_{s+1}$ and $1 \notin P_{s+1}$.

---

[1] A set G is a minimal Gröbner basis of I if (1) G is a Gröbner basis of I, and (2) for each $g \in G$, $\text{lpp}(g)$ is not divisible by any leading power products of $G \setminus \{g\}$.

$(\Leftarrow)$: If $1 \notin P_{s+1}$ and $P_s \subsetneqq P_{s+1}$, then $f \notin I : f^s$ and there exists $h \in P_{s+1}$ and $h \notin P_s$. From Theorem 4 (1), there exists $g = hz^{s+1} + \text{tail}(g) \in G$. Then $hf^{s+1} \in I$ by Lemma 1. So $hf \in I : f^s$, and f is a zero divisor in $k[X]/(I : f^s)$.

(3). $(\Rightarrow)$: If $f \in \sqrt{I}$, then there exists an integer t such that $f^t \in I$. So $z^t \in J$ from Lemma 2. Since G is a minimal Gröbner basis of J, there exists $g \in G$, such that $\text{lpp}(g) \mid z^s$. So g must have the form of $g = z^s + \text{tail}(g)$, where $0 \le s \le t$. By Theorem 4 (1), $1 \in P_s$.

$(\Leftarrow)$: If there exists an integer s such that $1 \in P_s$, then there exists a polynomial $g = z^k + \text{tail}(g)$, where $0 \le k \le s$. By Lemma 1, $f^k \in I$, and hence, $f \in \sqrt{I}$.

(4). Since G is a minimal Gröbner basis of J, by Theorem 4 (1), $I : f^{m-1} \subsetneqq I : f^m = I : f^\infty$ if and only if $P_{m-1} \subsetneqq P_m = P_s$, for all $s > m$. Since P_m is a Gröbner basis of $I : f^m$ by Theorem 4 (1), P_m is also a Gröbner basis of $I : f^\infty$.

□

In case f is invertible in $k[X]/(I : f^s)$, the above proof shows how to construct the inverse of f. In particular, f is invertible in $k[X]/I$ if and only if $1 \in Q_0$, implying that G contains a polynomial of the form $y - h$, where $h \in k[X]$. In that case, h is an inverse of f in $k[X]/I$. Similarly, f is a zero divisor in $k[X]/I$ if and only if $P_0 \subsetneqq P_1$ and $1 \notin P_1$.

The following example illustrates Theorems 4 and 5.

Example 3 Let $I = \langle x_1^2(x_1x_2 - 1)\rangle \subset \mathbb{Q}[x_1, x_2]$, and $f = x_1$. Decide the properties of f in $\mathbb{Q}[x_1, x_2]/I$, $\mathbb{Q}[x_1, x_2]/(I : f)$, …, and $\mathbb{Q}[x_1, x_2]/(I : f^\infty)$.

A minimal Gröbner basis of $I + \langle fy - z\rangle \subset \mathbb{Q}[x_1, x_2, y, z]$ using a lexicographic ordering with $(y > z > x_1 > x_2)$ is

$$G = \{x_1^3x_2 - x_1^2, (x_1^2x_2 - x_1)z, (x_1x_2 - 1)z^2, x_1y - z, yz^2 - x_2z^3\}.$$

As per Theorem 4, we construct the following sets:

$$P_0 = \{x_1^3x_2 - x_1^2\}, Q_0 = P_0 \cup \{x_1\},$$

$$P_1 = \{x_1^3x_2 - x_1^2, x_1^2x_2 - x_1\}, Q_1 = P_1 \cup \{x_1\},$$

$$P_2 = \{x_1^3x_2 - x_1^2, x_1^2x_2 - x_1, x_1x_2 - 1\}, Q_2 = P_2 \cup \{x_1, 1\}.$$

From Theorems 4 and 5, we have:

1. P_0 is a Gröbner basis of I; P_1 is a Gröbner basis of $I : f$; P_2 is a Gröbner basis of $I : f^2$.
2. Q_0 is a Gröbner basis of $I + \langle f\rangle$; Q_1 is a Gröbner basis of $I : f + \langle f\rangle$; Q_2 is a Gröbner basis of $I : f^2 + \langle f\rangle$.
3. f is invertible in $\mathbb{Q}[x_1, x_2]/(I : f^2)$, and x_2 is its inverse.
4. f is a zero divisor in $\mathbb{Q}[x_1, x_2]/I$ and $\mathbb{Q}[x_1, x_2]/(I : f)$.
5. The integer 2 is the smallest integer m such that $I : f^\infty = I : f^m$, and P_2 is a Gröbner basis of $I : f^\infty$.

4 Application in Dynamic Evaluation

It is well known that an ideal I can be decomposed using a polynomial f as follows:

$$I = (I : f^{\infty}) \cap (I + \langle f^m \rangle),$$

where m is the smallest number such that $I : f^{\infty} = I : f^m$. From Theorem 4, the smallest m and a Gröbner basis of $I : f^{\infty} = I : f^m$ can be derived from a Gröbner basis of ideal $I + \langle fy - z \rangle$. This means we get a decomposition of I from G. Particularly, this decomposition is not trivial if f is a zero divisor in $k[X]/I$.

In [11], Noro gave a modular method of decomposing a radical and zero-dimensional ideal I into $I : f$ and $I + \langle f \rangle$ to do dynamic evaluation a la Duval [5], where f is a zero divisor in $k[X]/I$. Note that, Noro considered only the case when m is 1 since I is radical. His method needs to compute Gröbner basis for $I : f$ and $I + \langle f \rangle$ separately. In contrast, our approach can produce these two Gröbner bases simultaneously. The following example is taken from [5].

Example 4 Let $\mathbb{Q}(a, b, c, d)$ be ring defined by a, b, c, d, which are the roots of $x^2 - 2$, $x^2 + 3$, $x^2 + 6$, and $x^2 + 1 - 2c$, respectively. Check whether $a + b - d$ is invertible in $\mathbb{Q}(a, b, c, d)$, and compute an inverse if it exists.

The ring $\mathbb{Q}(a, b, c, d)$ is isomorphic to the quotient ring $\mathbb{Q}[X]/I$ where $X = \{x_1, x_2, x_3, x_4\}$ and $I = \langle x_1^2 - 2, x_2^2 + 3, x_3^2 + 6, x_4^2 - 2x_3 + 1 \rangle$. Note that $\mathbb{Q}(a, b, c, d)$ is not a field since I is not maximal, which means $a + b - d$ may not be invertible in $\mathbb{Q}(a, b, c, d)$.

Let $f = x_1 + x_2 - x_4$. Compute a minimal Gröbner bases G of $J = I + \langle fy - z \rangle$ in $\mathbb{Q}[x_1, x_2, x_3, x_4, y, z]$ using a lexicographic ordering with $y > z > x_4 > x_3 > x_2 > x_1$. We get $G = \{x_1^2 - 2, x_2^2 + 3, x_3^2 + 6, x_4^2 - 2x_3 - 1, (x_3x_4 + x_1x_2x_4 + x_2x_3 + x_1x_3 + 2x_2 - 3x_1)z, (x_3 - x_1x_2)y + (1/2)(x_4 + x_2 + x_1)z, (x_4 - x_2 - x_1)y + z, zy + (1/120)(5x_1x_2x_4 + 2x_2x_3 + 3x_1x_3 + 16x_2 - 21x_1)z^2\}$.

As Theorem 4, we construct the following sets:
$Q_0 := \{x_1^2 - 2, x_2^2 + 3, x_3^2 + 6, x_4^2 - 2x_3 - 1\}$,
$P_0 := Q_0 \cup \{x_3 - x_1x_2, x_4 - x_2 - x_1\}$,
$Q_1 := Q_0 \cup \{x_3x_4 + x_1x_2x_4 + x_2x_3 + x_1x_3 + 2x_2 - 3x_1\}$,
$P_1 := Q_1 \cup \{1\}$.
By Theorem 5, f is a zero divisor in $\mathbb{Q}[X]/I$ and hence, not invertible in $\mathbb{Q}[X]/I$. Further, $I : f^{\infty} = I : f$. A nontrivial decomposition of I is thus $I = (I : f) \cap (I + \langle f \rangle) = \langle Q_1 \rangle \cap \langle P_0 \rangle$.

Again using Theorem 5 (1), f is in fact invertible in $\mathbb{Q}[X]/(I : f)$, and an inverse can be obtained from the polynomial $zy + (1/120)(5x_1x_2x_4 + 2x_2x_3 + 3x_1x_3 + 16x_2 - 21x_1)z^2$, i.e. an inverse of f in $\mathbb{Q}[X]/(I : f)$ is $-(1/120)(5x_1x_2x_4 + 2x_2x_3 + 3x_1x_3 + 16x_2 - 21x_1)$.

5 Conclusions

Using a generalization of the classical Rabinowitsch trick, we have proposed a method for checking whether a given polynomial f is invertible or a zero divisor in a residue class ring $k[X]/I$, where I is a polynomial ideal. This check is performed by computing a Gröbner basis of $I \cup \{fy - z\}$ by using a block ordering in which $y \gg z \gg X$, where y, z are new variables different from the variables in X. If f is not invertible in $k[X]/I$, it can be determined using the same Gröbner basis construction whether there is an s such that f is invertible in the residue class ring defined by the colon ideal $I : f^s$ on $k[X]$. As a byproduct, the smallest number s can be computed such that $I : f^s = I : f^\infty$, the saturation ideal of I with respect to f. The method can also be used to determine whether f is invertible or a zero divisor in $k[X]/(I : f)$, $k[X]/(I : f^2)$, $k[X]/(I : f^3)$, etc.

A nice aspect of the proposed construction is that it naturally generalizes to parametric systems using a comprehensive Gröbner system by an algorithm such as in [7, 8]. A paper on this generalization is under preparation; preliminary results on the findings were presented as an invited talk at *the International Workshop on Automated Deduction in Geometry (ADG),* Coimbra, Portugal, in July 2014.

Acknowledgements The authors are grateful to Professors Teo Mora and Yosuke Sato for their helpful discussions and suggestions.

References

1. Bayer, D.: The Division Algorithm and the Hilbert Scheme. Ph.D. thesis, Harvard (1981)
2. Buchberger, B.: Groebner bases: an algorithmic method in polynomial ideal theory. In: Bose, N.K. (ed.) Multidimensional Systems Theory, pp. 184–232. D. Reidel Publishing, Dordrecht (1985)
3. Cox, D., Little, J., O'Shea, D.: Ideals, Varieties, and Algorithms, 3rd edn. Springer, New York (2007)
4. Brownawell, W.D.: Rabinowitsch trick. In: Encyclopedia of Mathematics. Springer, Berlin (2001)
5. Duval, D.: Algebraic numbers: an example of dynamic evaluation. J. Symb. Comput. **18**, 429–445 (1994)
6. Kapur, D.: Geometry theorem proving using Hilbert's Nullstellensatz. In: Proceedings of the ISSAC 1986, pp. 202–208. ACM Press, New York (1986)
7. Kapur, D., Sun, Y., Wang, D.: An efficient algorithm for computing comprehensive Gröbner system for a parametric polynomial system. J. Symb. Comput. **49**, 27–44 (2013)
8. Kapur, D., Sun, Y., Wang, D.: An efficient method for computing comprehensive Gröbner bases. J. Symb. Comput. **52**, 124–142 (2013)
9. Rabinowitsch, J.L.: Zum Hilbertschen Nullstellensatz. Mathematische Annalen **102**(1), 520 (1929)
10. Mora, T.: Solving Polynomial Equation Systems II. Cambridge University Press, New York (2005)
11. Noro, M.: Modular dynamic evaluation. In: Proceedings of the ISSAC 2006, pp. 262–268. ACM Press, New York (2006)

12. Sato, Y., Suzuki, A.: Computation of inverses in residue class rings of parametric polynomial ideal. In: Proceedings of the ISSAC 2009, pp. 311–316. ACM Press, New York (2009)
13. Shannon, D., Sweedler, M.: Using Gröbner bases to determine algebra membership, splitting surjective algebra homomorphisms and determine birational equivalence. J. Symb. Comput. **6**, 267–273 (1988)
14. Spear, D.A.: A constructive approach to commutative ring theory. In: Proceedings of the 1977 MACSYMA Users Conference, pp. 369–376 (1977)

A Web-Based Quantum Computer Simulator with Symbolic Extensions

O.G. Karamitrou, C. Tsimpouris, P. Mavridi and K.N. Sgarbas

Abstract This paper presents a quantum computer simulator with a web interface, based on the circuit model of quantum computation. This is the standard model for which most quantum algorithms have been developed. According to this model, quantum algorithms are expressed as circuits of quantum registers (series of qubits) and quantum gates operating on them. The paper also proposes another version of the existing simulator using symbolic computation in Python programming language, in order to perform quantum calculations.

Keywords Quantum computation · Simulator · Circuit model · Quantum gates · Python language

1 Introduction

Quantum computation and quantum information is the study of the information processing tasks that can be accomplished using quantum mechanical systems. Quantum computing combines quantum mechanics, information theory and aspects of computer science. Quantum computation exploits quantum effects in order to perform calculations more efficiently than ordinary computers do. Superposition and entanglement are two key-phenomena in the quantum field that provide a much more efficient way to perform computations than classical algorithms.

O.G. Karamitrou · C. Tsimpouris · P. Mavridi · K.N. Sgarbas (✉)
University of Patras, Patras, Greece
e-mail: sgarbas@upatras.gr

O.G. Karamitrou
e-mail: okaramitrou@upatras.gr

C. Tsimpouris
e-mail: xtsimpouris@upatras.gr

P. Mavridi
e-mail: petramauridi@gmail.com

I.S. Kotsireas and E. Martínez-Moro (eds.), *Applications of Computer Algebra*, Springer Proceedings in Mathematics & Statistics 198,
DOI 10.1007/978-3-319-56932-1_15

The bit is the fundamental concept of classical computation and information. The quantum analog of a bit is called quantum bit or qubit. A classical bit has a state, which is either 0 or 1. A qubit can be in a composition of its basis states (denoted in Dirac notation as $|0\rangle$ and $|1\rangle$), called superposition:

$$|x\rangle = a|0\rangle + b|1\rangle$$

where a, b are complex numbers called probability amplitudes. When we measure a qubit we get 0 with probability a^2 or 1 with probability b^2. So $a^2 + b^2 = 1$. Quantum gates are the basic components for quantum computation. In contrast to classical gates, quantum gates are not circuits with input and output, but operators over a quantum register that consists of several qubits. Consequently, quantum algorithms are implemented as a series of applications of quantum gates over the contents of a quantum register. The most popular quantum algorithms are Grover's algorithm, which is able to search an unsorted database in time $O(\sqrt{N})$ [2, 3], quantum Fourier transform (QFT) [1] and Shor's algorithm [5].

The objective of this paper is to present a quantum computer simulator. This is a useful tool for students and researchers in quantum computation and quantum information. The proposed simulator is based on the circuit model [4] of quantum computation. This is the standard model in which most quantum algorithms have been developed. The usage of this tool is focused in studying and understanding quantum circuits, quantum computations and well known quantum algorithms, such as Grover's algorithm [1, 5] and Quantum Fourier Transform. It may also be very useful for the development of new quantum algorithms and the construction of new quantum gates. A demo of this tool with a web interface has been developed and it is available from this URL: http://www.wcl.ece.upatras.gr/en/ai/resources/demo-quantum-simulation.

The algorithm of the simulator and an example of its usage are presented in Sect. 2. Section 3 proposes another version of the quantum computer simulator using symbolic processing and Sect. 4 concludes the paper with an overview and some final remarks.

2 Quantum Computer Simulator

The quantum computer simulator presented here can simulate the operation of quantum circuits. The inputs of the simulator are the number of qubits, the number of computation steps, the initial state of the quantum register, and the gates that are applied at each step. The outputs of the simulator are the quantum register state at each step (the probability of measuring each one of the possible states and the phases of each state).

The pseudocode of the simulator is shown in Table 1. The simulation starts with specific choices from the end-user. The user has to select the number of qubits of

Table 1 The pseudocode of the quantum computer simulator

Pseudocode
1. Start
2. Read the number of qubits
3. Read the number of computations steps
4. Read the initial state of quantum register
5. Calculate the tensor product of the initial state of quantum register
6. $i = 1$
7. Read the matrix of quantum gate at ith step
8. Calculate the tensor product of the matrices of quantum gates at the ith step
9. Calculate the new state of the quantum register
10. If i is less than the number of computations steps, $i = i + 1$, go to step 7,
11. Produce the output (measure and phase of quantum register)
12. End

the input quantum register and the number of computation steps. After this, the user selects the initial quantum state of the quantum register and the appropriate quantum gates that are applied at eash step, as shown in Fig. 1. The available quantum gates are:

1. Identity
2. Hadamard
3. Controlled Not
4. Toffoli gate, also known as CCNOT
5. Phase Shift
6. Controlled Phase Shift
7. Fredkin

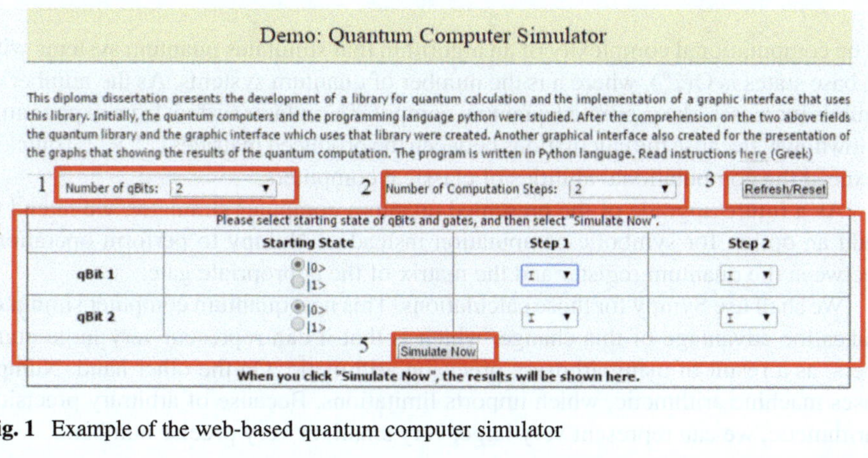

Fig. 1 Example of the web-based quantum computer simulator

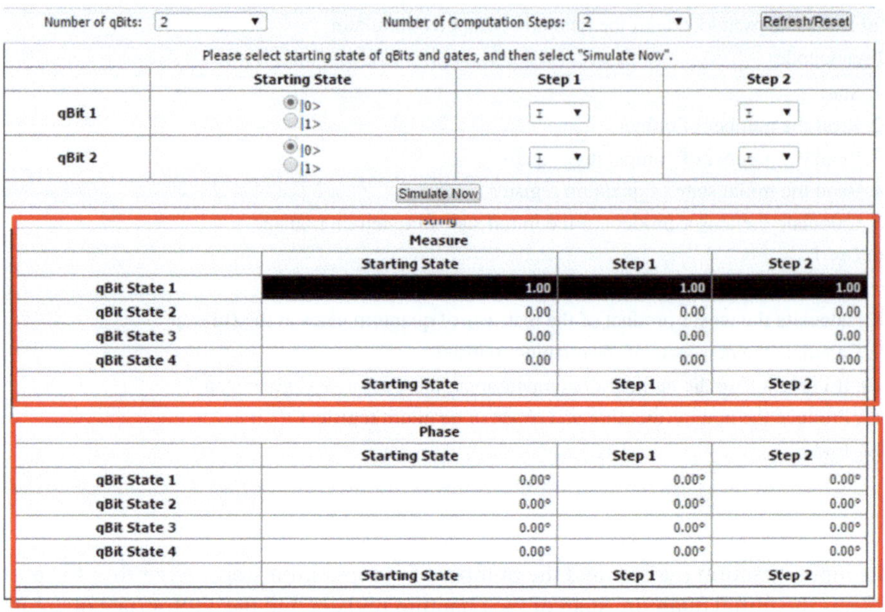

Fig. 2 Output of the web-based quantum computer simulator

The output of the simulator is shown in Fig. 2.

The quantum computer simulator has been developed in Python, using some extra libraries for our purposes. The fundamental library that is used is Numpy: the package for scientific computing in Python.

3 Future Work

The computational complexity of an algorithm that simulates quantum systems with 2 base states is $O(2^n)$, where n is the number of quantum systems. As the number of qubits increases, the presented quantum computer simulator suffers from exponential slowdown, because the calculations between the produced matrices ($2^n \times 2^n$) quickly exceed the computational abilities of classical computers.

As a future upgrade of the presented quantum computer simulator, we intend to add an option for symbolic computation instead of Numpy to perform operations between the quantum register and the matrix of the appropriate gate.

We shall use Sympy for those calculations. This new quantum computer simulator takes the advantage of this change, which is that it can represent very large numbers, as a result of using arbitrary precision arithmetic. On the other hand, Numpy uses machine arithmetic, which imports limitations. Because of arbitrary precision arithmetic, we can represent very large, very small, or very precise numbers.

The basic idea of this approach is:

- to use Sympy library of Python to do symbolic manipulation of quantum computations.
- to use mpmath library of Sympy for the numerical computations at the final step that we get the final quantum register (output), in order to compute the measure and phase of each state of quantum register.

4 Conclusion

A web-based quantum computer simulator has been developed and used for quantum computations. Using this quantum computer simulator, a significant number of quantum computations can be performed in short time. The symbolic extension which we described may be helpful for future development of a simulator that can perform quantum computations with larger input.

Acknowledgements The publication for this article was covered by the University of Patras Research Committee (ELKE/TSMEDE fund).

References

1. Coppersmith, D.: An Approximate Fourier Transform Useful in Quantum Factoring, IBM Research Report RC 19642 (1994)
2. Grover, L.K.: Quantum mechanics helps in searching for a needle in a haystack. Phys. Rev. Lett. **79**(2), 325–328 (1997)
3. Grover, L.K.: Quantum computers can search rapidly by using almost any transformation. Phys. Rev. Lett. **80**(19), 4329–4332 (1998)
4. Nielsen, M.A., Chuang, I.L.: Quantum Computation and Quantum Information. Cambridge University Press, Cambridge, UK (2000)
5. Shor, P.: Algorithms for quantum computation: discrete logarithms and factoring. In: Proceedings of the 35th Annual Symposium on Foundations of Computer Science, pp. 124–134 (1994)

Dixon-EDF: The Premier Method for Solution of Parametric Polynomial Systems

Robert H. Lewis

Abstract Using examples of interest from real problems, we will discuss the Dixon-EDF resultant as a method of solving parametric polynomial systems. We will briefly describe the method itself, then discuss problems arising in geometric computing, flexibility of structures, pose estimation, robotics, image analysis, physics, differential equations, and others. We will compare Dixon-EDF to several respected implementations of Gröbner bases algorithms on several systems. We find that Dixon-EDF is greatly superior, usually by orders of magnitude, in both CPU usage and RAM usage.

Keywords Polynomial system · Resultant · Dixon · Determinant · Symbolic computing · Gröbner basis

1 Polynomial Systems

"Solve a system of polynomial equations" means different things to different people. Everyone will agree that we take a collection of multivariate polynomials, set each to 0, and search for the common roots. For us in this paper, we have a ground ring K, variables $x_1, x_2, \ldots, x_n$, and parameters $a_1, a_2, \ldots, a_m$, so we are working over $K[x_1, x_2, \ldots, x_n, a_1, \ldots, a_m]$. K is primarily **Z** or **Q**; secondarily, **Z/p** for p "large," $40000 - 2^{31}$; possibly another finite field. We are not interested in $K = \mathbf{Z/2}$ or cryptology. We are not interested in purely numerical solution. We want an exact symbolic solution, at least to the point where we have an equation in one variable we could turn over to numerical solvers (after choosing numerical values for the parameters.) We typically have n equations in n variables $x_1, x_2, \ldots, x_n$ and some parameters. Ideally, the system is neither over- nor under-determined, though it could be. Always, $n \geq 2$; usually $3 \leq n \leq 15$. There are always parameters. Usually there are as many parameters as variables.

R.H. Lewis (✉)
Fordham University, New York, NY 10458, USA
e-mail: rlewis@fordham.edu

I.S. Kotsireas and E. Martínez-Moro (eds.), *Applications of Computer Algebra*,
Springer Proceedings in Mathematics & Statistics 198,
DOI 10.1007/978-3-319-56932-1_16

What does "solve the system" mean? In this work, it means to eliminate all but one of the variables. We are then left with one equation in one variable and the parameters—the *resultant* [6, 7, 24]. If desired, numerical values for the parameters can then be substituted, and the variable obtained numerically. If desired, to get numerical values for all the variables we could run this method in parallel on different machines and then test all combinations of values for each variable. For most problems this last step is no problem. But for many problems, the resultant is really the desired solution already.

The Bezout–Dixon method produces a matrix whose determinant is a multiple of the resultant. Dixon-EDF [13] is a way to compute the resultant without finding the entire determinant. Often the determinant is too large to compute, but it has many factors and so the resultant is much smaller than the determinant. Often the resultant occurs with multiplicity. We detect these polynomial factors "early," hence EDF = Early Detection of Factors. The output of the algorithm is a list of polynomials whose product is the determinant. Interesting problems tend to have many factors. There is no guarantee that this will work better than a standard determinant method. However, on many real problems from interesting applications, it does very well [14, 16, 17].

Gröbner Bases are well known [6, 24]. They have many applications, among which is solving systems of polynomial equations.

Let us emphasize a key difference between resultant solutions and Gröbner bases solutions. The resultant of a system of n polynomials in n variables is a single polynomial in one variable (and, usually, parameters). A Gröbner basis used this way yields a number polynomials in triangular form. That is, the first polynomial contains only one variable (and the parameters), the second has two variables, etc. It seems reasonable that the Gröbner basis is more "complete," and may well take longer to compute. However, often one or two resultants are not only sufficient, they are exactly what is desired—the other variables are mere artifacts. In any event, one may compute the resultants for different desired variables in parallel on different machines. This is a huge advantage.

Over the last 15 years we have noticed again and again that when engineers, scientists, and most mathematicians want to solve a polynomial system, they want a symbolic solution. They try Gröbner bases, usually in either Maple, Mathematica, or Magma. Many times, the program crashes or the user gives up after many hours. Almost always these systems would be enormously easier to solve with Dixon-EDF. I do not know of any examples of the type of problem described here where Gröbner bases are better than Dixon-EDF.

There is a further advantage to Dixon-EDF. Since we are computing the determinant of a matrix (or factors thereof), there are many ways to do that. Basically Dixon-EDF is a modified and adaptive row and column reduction. But one can easily examine the state of the computation and interrupt it part way to switch to another method. As we will see below, it is often useful to switch to the Gentleman Johnson idea of expansion by minors with storage of minors [10].

Computations in this paper were run on an Intel Imac at 2.3 GHz with 16 gigabytes of RAM, and on faster Linux servers with 130 gig. Dixon-EDF was run in Fermat [11]. Some Fermat code for Dixon-EDF is at [12]. The actual commands used in

Magma and Maple are in an appendix. Maple has some built-in triangularize and Gröbner bases routines, as well as the FGb package of Faugere [8]. FGb is usually superior and the others were not often used here. The Maple FGb commands were explained to us by Faugere [9]. Two ways to use Magma are given in the appendix. One was recommended by an experienced user of Magma, the other was specified by Magma programmer Steel [23]. These are referred to as Magma 1 and Magma 2. Magma 1 is usually inferior so it was not used on some examples.

This is a paper in applied experimental mathematics. It is not a paper in pure mathematics. We do not have a theorem saying that under certain conditions Dixon-EDF is many times more efficient than Gröbner bases. Perhaps someone will be inspired to produce such a theorem.

2 Brief Explanation of Dixon-EDF

This has been published [13, 15].

- The resultant is a factor of the determinant of the "second Dixon matrix," M. M contains polynomials in the parameters and usually one remaining variable.
- We wish to avoid simply computing the determinant of M. Instead we begin to column normalize the matrix M.
- To avoid creating large messy denominators (rational functions) we pull out denominators from each row as soon as they arise. Then later we factor out gcds whenever possible from the numerators in each row and column.
- We keep track of all denominators and gcds so discovered. We check often to see if some polynomial in the denominator list has a common gcd with some polynomial in the numerator list; if so we divide it out. In the end, the denominator list must be all 1. The product of the numerator list is Det[M].
- This can work efficiently because the determinant of the second Dixon matrix M usually has many factors. This is a bad way of computing the determinant of a "random" matrix. But random matrices are seldom of interest.

There are subtleties that can make an enormous difference in execution time. First, the strategy of picking the pivot on each reduction step. Second, one may run the algorithm first over **Z/p**. Quite often this gives the actual answer over **Z** if p is fairly large, say close to 2^{31}. Third, the second Dixon matrix is often really the third, because if the second is not square, one extracts a maximal minor. Usually there are many maximal minors. One tries to select the "best" one by some heuristic, such as sparseness.

Here is a simple example. Given initially

$$M_0 = \begin{pmatrix} 9 & 2 \\ 4 & 4 \end{pmatrix} \quad \text{numerators:} \qquad \text{denominators:}$$

We factor a 2 out of the second column, then a 2 from the second row. Thus:

$$M_0 = \begin{pmatrix} 9 & 1 \\ 2 & 1 \end{pmatrix} \quad \text{numerators: } 2, 2 \quad \text{denominators:}$$

We change the second row by subtracting 2/9 of the first:

$$M_0 = \begin{pmatrix} 9 & 1 \\ 0 & 7/9 \end{pmatrix}$$

numerators: 2, 2 denominators:

We pull out the denominator 9 from the second row, and factor out 9 from the first column:

$$M_0 = \begin{pmatrix} 1 & 1 \\ 0 & 7 \end{pmatrix}$$

numerators: 2, 2, 9 denominators: 9

We "clean up" by dividing out the common factor of 9 from the numerator and denominator lists; any 1 that occurs may be erased and the list compacted. Since the first column is canonically simple, we are finished with one step of the algorithm, and have produced a one-smaller M_1 for the next step.

$$M_1 = (7)$$

numerators: 2, 2 denominators: 1

The algorithm terminates by pulling out the 7:

numerators: 2, 2, 7 denominators: 1

At end: three numerators, one denominator (= 1).

As expected (since the original matrix contained all integers) the denominator list is empty. The product of all the entries in the numerator list is the determinant, but we never needed to deal with any number larger than 9.

We now illustrate the power of Dixon-EDF with a series of examples. The ground ring $K = \mathbf{Z}$ in all cases.

3 A Motion Controller

In 2009, an engineer named Nachtwey [18] was dealing with a motion controller leading to the following system:

$x0 + v0(a1 - a0)/j + a0/2(a1 - a0)^2/j^2 + j/6(a1 - a0)^3/j^3 - x1,$
$v0 + a0(a1 - a0)/j + j/2(a1 - a0)^2/j^2 - v1,$
$x1 + v1(a1 - a5)/j + a1/2(a1 - a5)^2/j^2 - j/6(a1 - a5)^3/j^3 - x5,$
$v1 + a1(a1 - a5)/j - j/2(a1 - a5)^2/j^2 - v5,$
$x5 + v5(-a5/j) + a5/2(-a5/j)^2 + j/6(-a5/j)^3 - x7,$
$v5 + a5(-a5/j) + j/2(-a5/j)^2$

Set each expression to 0, multiply out the denominators. There are six variables $(x1, x5, v1, v5, a1, a5)$ and five parameters. Results vary depending on the variable being solved for:

for $v5$, the answer has 95 terms:
- Dixon-EDF: takes 0.04 s, 9.0 meg RAM
- Maple's FGb: takes 0.36 s, 358 meg RAM
- Maple's Basis cmd: killed after 4h, 235 meg
- Maple's Triangular: takes 1.13 s, 124 meg
- Magma 2: takes 0.16 s, 32 meg

for $x1$, the answer has 244 terms:
- Dixon-EDF: takes 0.12 s, 9.1 meg.
- Maple's FGb: takes 1.56 s, 371 meg
- Maple's Basis cmd: killed after 3 h, 227 meg
- Maple's Triangular: takes 16 mins, 520 meg
- Magma 2: takes 0.46 s, 31 meg.

4 Physics: A Spinning Double Pendulum

J. Tot has studied the motion of a spinning double pendulum [26]. He derived three equations in two variables s, t and three parameters q, mu, ld.

$q\,ld\,s^3\,t^2 - s^3\,t^2 - q\,ld\,s\,t^2 + 2\,q\,s\,t^2 - s\,t^2 - q\,mu\,ld\,s^4\,t + q\,mu\,ld\,t + q\,ld\,s^3 - s^3 - q\,ld\,s + 2\,q\,s - s,$

$q\,ld\,s\,t^4 - q\,s\,t^4 - q\,ld\,s^2\,t^3 + q\,s^2\,t^3 - s^2\,t^3 - q\,ld\,t^3 + q\,t^3 - t^3 + q\,ld\,s^2\,t + q\,s^2\,t - s^2\,t + q\,ld\,t + q\,t - t - q\,ld\,s + q\,s,$

$10\,q^2\,ld^2\,s^3\,t^5 - 10\,q^2\,ld\,s^3\,t^5 - 10\,q\,ld\,s^3\,t^5 + 10\,q\,s^3\,t^5 - 2\,q^2\,ld^2\,s\,t^5 + 6\,q^2\,ld\,s\,t^5 - 2\,q\,ld\,s\,t^5 - 4\,q^2\,s\,t^5 + 2\,q\,s\,t^5 - 15\,q^2\,mu\,ld^2\,s^4\,t^4 - 5\,q^2\,ld^2\,s^4\,t^4 + 15\,q^2\,mu\,ld\,s^4\,t^4 + 5\,q^2\,ld\,s^4\,t^4 - 5\,q\,s^4\,t^4 + 5\,s^4\,t^4 - 10\,q^2\,ld^2\,s^2\,t^4 + 12\,q^2\,ld\,s^2\,t^4 - 2\,q\,ld\,s^2\,t^4 - 2\,q^2\,s^2\,t^4 - 6\,q\,s^2\,t^4 + 8\,s^2\,t^4 - q^2\,mu\,ld^2\,t^4 + 3\,q^2\,ld^2\,t^4 + q^2\,mu\,ld\,t^4 - 9\,q^2\,ld\,t^4 + 6\,q\,ld\,t^4 + 6\,q^2\,t^4 - 9\,q\,t^4 + 3\,t^4 + 10\,q^2\,mu\,ld^2\,s^5\,t^3 - 10\,q^2\,mu\,ld\,s^5\,t^3 + 10\,q\,mu\,ld\,s^5\,t^3 + 12\,q^2\,mu\,ld^2\,s^3\,t^3 + 12\,q^2\,ld^2\,s^3\,t^3 - 12\,q^2\,mu\,ld\,s^3\,t^3 + 12\,q\,mu\,ld\,s^3\,t^3 - 12\,q^2\,ld\,s^3\,t^3 - 12\,q\,ld\,s^3\,t^3 + 12\,q\,s^3\,t^3 + 2\,q^2\,mu\,ld^2\,s\,t^3 - 4\,q^2\,ld^2\,s\,t^3 - 2\,q^2\,mu\,ld\,s\,t^3 + 2\,q\,mu\,ld\,s\,t^3 + 12\,q^2\,ld\,s\,t^3 - 4\,q\,ld\,s\,t^3 - 8\,q^2\,s\,t^3 + 4\,q\,s\,t^3 - 10\,q^2\,ld^2\,s^4\,t^2 + 8\,q^2\,ld\,s^4\,t^2 + 2\,q\,ld\,s^4\,t^2 - 8\,q\,s^4\,t^2 + 8\,s^4\,t^2 - 12\,q\,s^2\,t^2 + 12\,s^2\,t^2 + 2\,q^2\,ld^2\,t^2 - 8\,q^2\,ld\,t^2 + 6\,q\,ld\,t^2 + 8\,q^2\,t^2 - 12\,q\,t^2 + 4\,t^2 - 2\,q^2\,mu\,ld^2\,s^5\,t - 2\,q^2\,mu\,ld\,s^5\,t + 2\,q\,mu\,ld\,s^5\,t - 4\,q^2\,mu\,ld^2\,s^3\,t + 2\,q^2\,ld^2\,s^3\,t - 4\,q^2\,mu\,ld\,s^3\,t + 4\,q\,mu\,ld\,s^3\,t - 2\,q^2\,ld\,s^3\,t - 2\,q\,ld\,s^3\,t + 2\,q\,s^3\,t$

$-2q^2 mu\, ld^2 s\, t - 2q^2 ld^2 s\, t - 2q^2 mu\, ld\, s\, t + 2q\, mu\, ld\, s\, t + 6q^2 ld\, s\, t - 2q\, ld\, s\, t - 4q^2 s\, t + 2q\, s\, t - q^2 mu\, ld^2 s^4 + 3q^2 ld^2 s^4 + q^2 mu\, ld\, s^4 + 3q^2 ld\, s^4 - 6q\, ld\, s^4 - 3q\, s^4 + 3s^4 + 2q^2 ld^2 s^2 + 4q^2 ld\, s^2 - 6q\, ld\, s^2 + 2q^2 s^2 - 6q\, s^2 + 4s^2 + q^2 mu\, ld^2 - q^2 ld^2 - q^2 mu\, ld + q^2 ld + 2q^2 - 3q + 1$

The third equation is the Jacobian determinant of the first two. He wants the resultant found by eliminating the two variables.

Dixon-EDF solves this is 8.4 s, 51 meg.

Maple-FGb succeeds in 340 s, 2 gig.

Magma 1 succeeds in 5280 s and 520 meg.

Magma 2 succeeds after 5910 s and 294 meg.

5 Computational Geometry: Heron's Formula

The familiar Heron's formula is the relation between the area A and sides a, b, c of a triangle:

$$A^2 = s(s-a)(s-b)(s-c)$$

($s = (a+b+c)/2$.) This can be easily derived as the resultant of a polynomial system: Place $(0, a)$ on x-axis, point (x, y) in first quadrant, lengths b and c, yielding obvious equations:

$x^2 + y^2 - c^2,$

$(x-a)^2 + y^2 - b^2,$

$2A - a\,y$

We eliminate x and y, getting the answer of A in terms of a, b, c. This is a classic example of where the resultant for one variable is exactly what we want.

Now, we easily generalize this idea to 3, 4, 5, ..., dimensions. The figure below shows the set up for three dimensions, a tetrahedron (Fig. 1).

Here is the system for five dimensions, after some routine simplifications, there are 15 equations:

$y^2 + x^2 - cs^2, \quad -2as\,x + cs^2 - bs^2 + as^2,$

$z_1^2 + y_1^2 + x_1^2 - fs^2, \quad -2as\,x_1 + fs^2 - ds^2 + as^2,$

$-2y\,y_1 - 2x\,x_1 + fs^2 - es^2 + cs^2,$

$w_2^2 + z_2^2 + y_2^2 + x_2^2 - gs^2, \quad -2as\,x_2 - hs^2 + gs^2 + as^2,$

$-2y\,y_2 - 2x\,x_2 - is^2 + gs^2 + cs^2,$

$-2z_1\,z_2 - 2y_1\,y_2 - 2x_1\,x_2 - js^2 + gs^2 + fs^2,$

$u_3^2 + w_3^2 + z_3^2 + y_3^2 + x_3^2 - ks^2, \quad -2as\,x_3 - ls^2 + ks^2 + as^2,$

$-2y\,y_3 - 2x\,x_3 - ms^2 + ks^2 + cs^2,$

$-2z_1\,z_3 - 2y_1\,y_3 - 2x_1\,x_3 - ns^2 + ks^2 + fs^2,$

$-2w_2\,w_3 - 2z_2\,z_3 - 2y_2\,y_3 - 2x_2\,x_3 - os^2 + ks^2 + gs^2,$

$-as\,y\,z_1\,w_2\,u_3 + 120\,V$

There are fourteen coordinate variables to eliminate. The (second) Dixon matrix is 313×313, very sparse. EDF finishes completely in 6.5 min, 844 meg RAM. But the answer appears before that. As often happens with EDF, the answer appears

Fig. 1 Tetrahedron in 3D

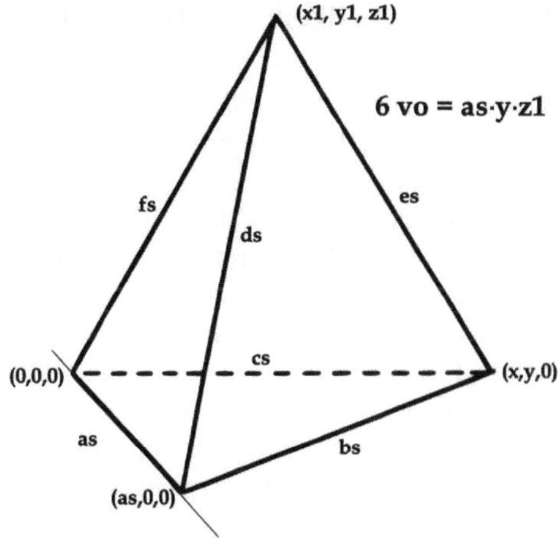

on the list of factors early, at row 266, after 1.8 min, 50 meg. It has 823 terms. At completion, there are hundreds of numerators. Most are monomials, plus the following list showing the number of terms in each polynomial:
6 6 6 6 6 6 6 6 6 6 6 6 6 6 6 6 6 6 6 6 6 6 6 6 6 6 4161 130 6 823 130 6 823 6 22 22 22 6 22 6 130 6 10000 6 130 6 10000 22 6 6 22 6 22 10000 57147 57147

After canceling all common factors (which is fast) we have finally:
1 823 130 6 22

Gröbner basis Maple and Mathematica, fail:

Maple FGb: killed after 9.5 h, 53 gig.

Magma 1: killed after 9.5 h, 2.6 gig. Magma 2: killed after 33 h, 4 gig.
The answer has 823 terms, $230400vo^2 - bs^2 fs^4 os^4 - cs^2 es^2 fs^2 os^4 + \cdots + cs^2 ds^2 es^2 gs^2 hs^2 - bs^2 ds^2 es^2 gs^4$.
In four dimensions, we have the same idea, with ten equations. Now Dixon-EDF finishes completely in 0.944 s, 24 meg:

131 1 22 1 6

Maple FGb: succeeds 31 min, 13 gig.

Magma 1: killed after 360 min, 4 gig. Magma 2: succeeds 1.90 s, 32 meg.

This problem is not just a good example for Dixon-EDF. These resultants come up in a proof of the Bellows Conjecture [3].

6 A Differential Equation for a Circle Rolling on an Ellipse

There is a subject called differential algebra in which symbolic algebra helps to solve differential equations [2, 22]. Here we have a geometric problem that results in a similar idea.

Problem: Describe a circle rolling on an ellipse (like an epicycloid, which is circle on circle). I can find no reference to this problem anywhere.

Let the ellipse have semi-axes a and b. Let the circle have radius r.

Solution: Set up parameters that are certain angles. Label important points like the point of intersection of the circle and ellipse. Get equations from the geometry of the circle and ellipse. As with an epicycloid, the key equation says that the distance rolled along the circle equals that along the ellipse.

The problem: arclength on an ellipse is an elliptic integral! How do we get polynomials? Answer: the derivative of the arclength can be expressed in trig functions. In this way, we eventually get nine polynomial equations in eleven variables.

$v^2 - 2\,k\,v + k^2 + h^2 - 2\,u\,h + u^2 - r^2,$

$t^4\,k^2 + 2\,t^2\,k^2 + k^2 - 4\,b\,t^4\,k - 4\,b\,t^2\,k + t^4\,h^2 + 2\,t^2\,h^2 + h^2 - 4\,a\,t^3\,h - 4\,a\,t\,h + 4\,b^2\,t^4 - r^2\,t^4 + 4\,a^2\,t^2 - 2\,r^2\,t^2 - r^2,$

$-2\,b\,t^3\,k - 2\,b\,t\,k + a\,t^4\,h - a\,h + 4\,b^2\,t^3 - 2\,a^2\,t^3 + 2\,a^2\,t,$

$-r^2\,t^2\,cg - r^2\,cg - t^2\,k\,v - k\,v + 2\,b\,t^2\,v + t^2\,k^2 + k^2 - 2\,b\,t^2\,k + t^2\,h^2 + h^2 - t^2\,u\,h - u\,h - 2\,a\,t\,h + 2\,a\,t\,u,$

$-r^2\,t^4\,dcg^2 - 2\,r^2\,t^2\,dcg^2 - r^2\,dcg^2 - a^2\,t^4\,cg^2 - 4\,b^2\,t^2\,cg^2 + 2\,a^2\,t^2\,cg^2 - a^2\,cg^2 + a^2\,t^4 + 4\,b^2\,t^2 - 2\,a^2\,t^2 + a^2,$

$t^4\,k\,dk + 2\,t^2\,k\,dk + k\,dk - 2\,b\,t^4\,dk - 2\,b\,t^2\,dk + t^4\,h\,dh + 2\,t^2\,h\,dh + h\,dh - 2\,a\,t^3\,dh - 2\,a\,t\,dh - 2\,b\,t^3\,k - 2\,b\,t\,k + a\,t^4\,h - a\,h + 4\,b^2\,t^3 - 2\,a^2\,t^3 + 2\,a^2\,t,$

$v\,dv - k\,dv - v\,dk + k\,dk + h\,dh - u\,dh - du\,h + du\,u,$

$-2\,b\,t^3\,dk - 2\,b\,t\,dk + a\,t^4\,dh - a\,dh + b\,t^4\,k - b\,k + 2\,a\,t^3\,h + 2\,a\,t\,h - 2\,b^2\,t^4 + a^2\,t^4 + 6\,b^2\,t^2 - 6\,a^2\,t^2 + a^2,$

$-r^2\,t^2\,dcg - r^2\,dcg - t^2\,k\,dv - k\,dv + 2\,b\,t^2\,dv - t^2\,v\,dk - v\,dk + 2\,t^2\,k\,dk + 2\,k\,dk - 2\,b\,t^2\,dk + 2\,t^2\,h\,dh + 2\,h\,dh - t^2\,u\,dh - udh - 2\,a\,tdh + 2\,b\,t\,v - 2\,b\,t\,k - t^2\,du\,h - du\,h + a\,t^2\,h - a\,h - a\,t^2\,u + a\,u + 2\,a\,t\,du.$

Note du = derivative of u, dh = derivative of h, t = time, $u = x-$coordinate of the contact point, $v = y-$coordinate. Eliminate all variables except u, du, t. Leave parameters a, b, r.

This is a challenging problem. The second matrix is 42×42 (for u and du) or 45×45 (for v and dv). The time and space for Dixon-EDF depends on how one adjusts the algorithm. First, if one spends 15 minutes of real time, one can find much better (sparser) maximal minors when one extracts the square matrix from the second Dixon matrix. Then a straight forward EDF all the way to step 42 (for u and du) takes about 8 h. However, we can run EDF through step 34, leaving an 8×8 matrix with number of terms:

826 301 1085 159 472 1162 157 119
0 20 0 7 19 0 7 14
2657 1096 3449 896 1676 3545 598 613
552 357 738 247 428 852 235 257
2086 831 2490 667 1283 2450 373 374
196 23 316 24 88 351 0 0
653 299 862 235 448 989 162 185
276 141 306 107 199 382 77 100

Now use Gentleman–Johnson [10] to expand the determinant. In total, this takes 4.5 h, 8 gig. An even better choice of maximal minor run to step 37 leaves a 5×5 matrix.

332 608 1526 2715 1923
1808 2583 5453 8160 6512
1323 1926 4212 6482 5117
700 1145 2579 4244 3119
76 122 440 847 672

At this point, there are no denominators (!) and the list of numerators has 219 polynomials. Most have fewer than three terms, but there are also:

3 8 6 8 9 10 6 6 9 7 9 8 5 8 8 6 8 9 8 8 8 9 6 7 8 7 8 8 8 6 7 6 6 8 6 8 7 6 7 8 3 4 3 3 3 4 3 3 4 8 20 4 3 4 4 4 34

Finishing the 5×5 determinant, in total this takes less than 90 min and 1.44 gig. After dividing common factors, the final list of numerator terms is:

1 3 1 2 2 4 10 45623 2 102

The resultant (45623 terms) factors into two pieces, corresponding to rolling inside and outside. The pieces have 2264 and 2146 terms and are degree 4 in u and du. Similar for v and dv. The other six variables are artifacts of no real interest (Fig. 2).

If we plug in $a = 2, b = 3, r = 1$, we may use standard numerical software (Maple) to solve the two differential equations (the resultants just computed) numerically. The result is in Fig. 2.

Attempting this system on Maple and Magma yielded:

Maple FGb: killed after 30 h, 22.8 gig.

Magma 1: killed after 18 h CPU, 43 gig RAM, 40 h real time (disk thrashing).

Magma 2: killed after 48 h, 5.8 gig.

7 Flexibility of Structures

This is joint work with Evangelos A. Coutsias, UNM, and Stony Brook.

As part of his research into flexible octahedra, Bricard [1] described this planar system of seven rigid rods (Fig. 3), which he said was equivalent to the octahedra problem. The structure is described by a system of six equations in six variables (the sine and cosine of the angles α, β, γ) and 11 parameters (the sides). The resultant res, computable only by EDF (in 4 minutes), has 190181 terms. The determinant of the 29×29 matrix contains res^2 and many other factors. res appears first at step 26.

Fig. 2 Circle rolls on ellipse

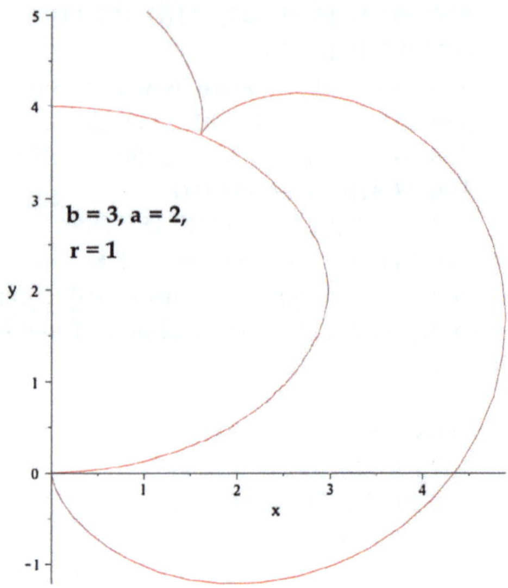

The problem: the structure is generically rigid. Find all ways that it becomes flexible. One obvious way is if all the quadrilaterals are parallelograms (Fig. 4).

Coutsias and colleagues have developed complex numerical algorithms for the general subject of flexible structures [4, 5].

I developed an algorithm that analyses the resultant to find modes of flexibility. Several surprising new modes were found in 2014 [16] in this way, which we call “exotic configurations.” These could not have been found numerically.

This has ramifications for computational chemistry [25].

Fig. 3 Bricard’s quadrilaterals

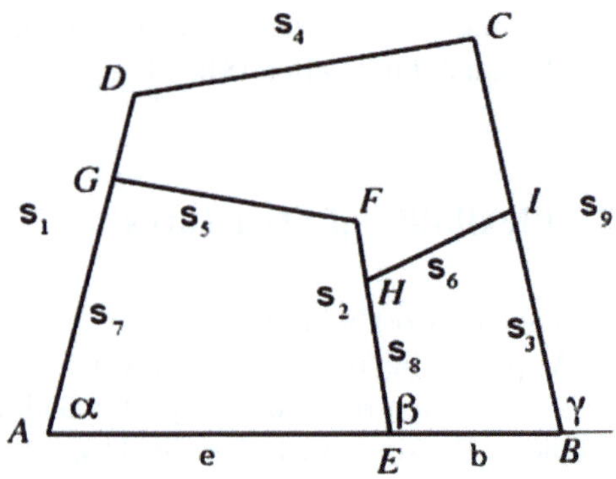

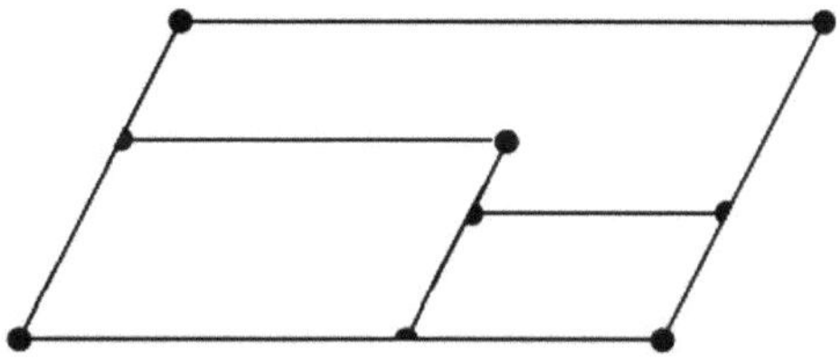

Fig. 4 Flexible because parallelograms

There is a way to reduce the system of six equations to three using the well-known half-angle tangent substitution. With $t_1 = \tan(\alpha/2)$, $t_2 = \tan(\beta/2)$, $t_3 = \tan(\gamma/2)$ we have equations

$a_1\, t_2^2\, t_1^2 + b_1\, t_1^2 + 2\, c_1\, t_2\, t_1 + d_1\, t_2^2 + e_1,$
$a_2\, t_3^2\, t_2^2 + b_2\, t_2^2 + 2\, c_2\, t_3\, t_2 + d_2\, t_3^2 + e_2,$
$a_3\, t_3^2\, t_1^2 + b_3\, t_1^2 + 2\, c_3\, t_3\, t_1 + d_3\, t_3^2 + e_3,$

where the 15 parameters are certain combinations of the eleven sides. This system also describes the molecule cyclo-hexane [5].

Dixon-EDF computes the resultant of this system (5685 terms) in 2.58 s using 60 meg RAM. Maple-FGb was killed after 20 min, using 28 gig RAM. Magma 1 was killed after 1800 min, 1.58 gig RAM. Magma 2 crashed after 7 min and 480 meg RAM.

8 Robotics: A Writing System

Two robotics students posted a question concerning the device in Fig. 5 [21]. There are four arms *AB, BC, CD, DO* connected in pivoting joints *O, A, B, C, D*. There is a pen at (x, y) on a rigid plate (shaded). $\beta_1, \beta_2, \theta_1, \theta_2, x, y$ vary with the motion. *O* and *A* do not move. $\alpha, l_1, l_2, l_3, aa$ are parameters. The problem: find expressions for θ_1, θ_2 in terms of x, y and the parameters.

It is not difficult to write equations for (x, y) in terms of sine and cosine of the various angles. As in the previous section, we then use the half-angle tangent identities to form four equations. $t_1 = \tan(\theta_1/2)$, $b_1 = \tan(\beta_1/2)$, $al = \tan(\alpha/2)$, etc.

$aa\; t_2^2\; t_1^2\; b_2^2\; b_1^2 + 2\; l_1\; t_1^2\; b_2^2\; b_1^2 + aa\; t_1^2\; b_2^2\; b_1^2 - 2\; l_1\; t_2^2\; b_2^2\; b_1^2 + aa\; t_2^2\; b_2^2\; b_1^2 + aa\, b_2^2\, b_1^2 + 2\, l_2\, t_2^2\, t_1^2\, b_1^2 + aa\, t_2^2\, t_1^2\, b_1^2 + 2\, l_2\, t_1^2\, b_1^2 + 2\, l_1\, t_1^2\, b_1^2 + aa\, t_1^2\, b_1^2 + 2\, l_2\, t_2^2\, b_1^2 - 2\; l_1\; t_2^2\; b_1^2 + aa\; t_2^2\; b_1^2 + 2\; l_2\; b_1^2 + aa\; b_1^2 - 2\; l_2\; t_2^2\; t_1^2\; b_2^2 + aa\; t_2^2\; t_1^2\; b_2^2 - 2\; l_2\; t_1^2\; b_2^2 + 2\, l_1\, t_1^2\, b_2^2 + aa\, t_1^2\, b_2^2 - 2\, l_2\, t_2^2\, b_2^2 - 2\, l_1\, t_2^2\, b_2^2 + aa\, t_2^2\, b_2^2 - 2\, l_2\, b_2^2 + aa\, b_2^2 + aa\, t_2^2\, t_1^2 + 2\; l_1\; t_1^2 + aa\; t_1^2 - 2\; l_1\; t_2^2 + aa\; t_2^2 + aa,$

$2\, l_1\, t_2\, t_1^2\, b_2^2\, b_1^2 - 2\, l_1\, t_2^2\, t_1\, b_2^2\, b_1^2 - 2\, l_1\, t_1\, b_2^2\, b_1^2 + 2\, l_1\, t_2\, b_2^2\, b_1^2 + 2\, l_2\, t_2^2\, t_1^2\, b_2\, b_1^2 + 2\; l_2\; t_1^2\; b_2\; b_1^2 + 2\; l_2\; t_2^2\; b_2\; b_1^2 + 2\; l_2\; b_2\; b_1^2 + 2\; l_1\; t_2\; t_1^2\; b_1^2 - 2\; l_1\; t_2^2\; t_1\; b_1^2 - 2\; l_1\; t_1\; b_1^2 + 2\, l_1\, t_2\, b_1^2 - 2\, l_2\, t_2^2\, t_1^2\, b_2^2\, b_1 - 2\, l_2\, t_1^2\, b_2^2\, b_1 - 2\, l_2\, t_2^2\, b_2^2\, b_1 - 2\, l_2\, b_2^2\, b_1 - 2\, l_2\, t_2^2\, t_1^2\, b_1 -$

Fig. 5 Robotic arms

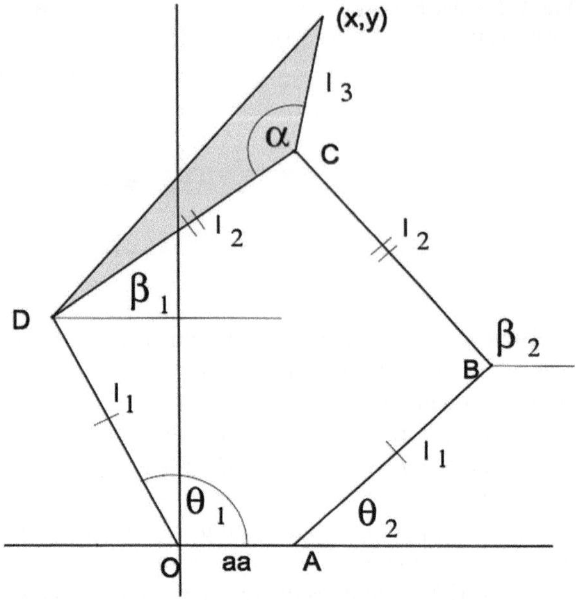

$2\,l_2\,t_1^2\,b_1 - 2\,l_2\,t_2^2\,b_1 - 2\,l_2\,b_1 + 2\,l_1\,t_2\,t_1^2\,b_2^2 - 2\,l_1\,t_2^2\,t_1\,b_2^2 - 2\,l_1\,t_1\,b_2^2 + 2\,l_1\,t_2\,b_2^2 +$
$2\,l_2\,t_2^2\,t_1^2\,b_2 + 2\,l_2\,t_1^2\,b_2 + 2\,l_2\,t_2^2\,b_2 + 2\,l_2\,b_2 + 2\,l_1\,t_2\,t_1^2 - 2\,l_1\,t_2^2\,t_1 - 2\,l_1\,t_1 + 2\,l_1\,t_2,$
$-al^2\,x\,t_1^2\,b_1^2 - x\,t_1^2\,b_1^2 - l_3\,al^2\,t_1^2\,b_1^2 - l_2\,al^2\,t_1^2\,b_1^2 - l_1\,al^2\,t_1^2\,b_1^2 + l_3\,t_1^2\,b_1^2 -$
$l_2\,t_1^2\,b_1^2 - l_1\,t_1^2\,b_1^2 - al^2\,x\,b_1^2 - x\,b_1^2 - l_3\,al^2\,b_1^2 - l_2\,al^2\,b_1^2 + l_1\,al^2\,b_1^2 + l_3\,b_1^2 -$
$l_2\,b_1^2 + l_1\,b_1^2 - 4\,l_3\,al\,t_1^2\,b_1 - 4\,l_3\,al\,b_1 - al^2\,x\,t_1^2 - x\,t_1^2 + l_3\,al^2\,t_1^2 + l_2\,al^2\,t_1^2 -$
$l_1\,al^2\,t_1^2 - l_3\,t_1^2 + l_2\,t_1^2 - l_1\,t_1^2 - al^2\,x - x + l_3\,al^2 + l_2\,al^2 + l_1\,al^2 - l_3 + l_2 + l_1,$
$-al^2\,y\,t_1^2\,b_1^2 - y\,t_1^2\,b_1^2 - 2\,l_3\,al\,t_1^2\,b_1^2 + 2\,l_1\,al^2\,t_1\,b_1^2 + 2\,l_1\,t_1\,b_1^2 - al^2\,y\,b_1^2 -$
$y\,b_1^2 - 2\,l_3\,al\,b_1^2 + 2\,l_3\,al^2\,t_1^2\,b_1 + 2\,l_2\,al^2\,t_1^2\,b_1 - 2\,l_3\,t_1^2\,b_1 + 2\,l_2\,t_1^2\,b_1 + 2\,l_3\,al^2\,b_1 +$
$2\,l_2\,al^2\,b_1 - 2\,l_3\,b_1 + 2\,l_2\,b_1 - al^2\,y\,t_1^2 - y\,t_1^2 + 2\,l_3\,al\,t_1^2 + 2\,l_1\,al^2\,t_1 + 2\,l_1\,t_1 -$
$al^2\,y - y + 2\,l_3\,al$

We eliminate t_2, b_1, b_2 to form the resultant for t_1, then eliminate t_1, b_1, b_2 to form the resultant for t_2. The first is easier because α on the rigid plate is more easily related to β_1, θ_1 and the origin.

For the t_1 resultant: Dixon-EDF takes 40 s, 132 meg. Maple FGb was killed after 200 min and 20 gig. Magma 2 was killed after 24 h and 6 gig.

For the t_2 resultant: Dixon-EDF takes between 60 and 150 min, depending on whether Gentleman–Johnson [10] is used, and between 1.6 and 2.6 gig. Maple FGb was killed after 25 h and 68 gig. Magma 2 was killed after 25 h and 9 gig.

9 Pose Estimation

Suppose we have a quadrilateral $ABCE$; it does not have to be planar. The distances between each pair of vertices are known. The object moves. We observe it from point P, noting the angles spanned by each pair of vertices. The classic four point pose problem is to deduce the distances X_1, X_2, X_3, X_4.

There are four variables X_1, X_2, X_3, X_4. The parameters are AB, BC, CE, AE, AC, BE. An overdetermined system results from applying the law of cosines to each triangle having vertex P.

Using four equations including the diagonals AC and BE gives an easy system of equations, solvable by many means. But suppose the object could be flexible! Then we have to use only the outside edges; diagonal distances might change (Figs. 6 and 7).

Maple and Magma both fail on this. FGb was killed after 25 h, exhausting 62 gig of RAM. Magma 1 was killed after 40 h. Magma 2 crashed after 308 min, 8.2 gig.

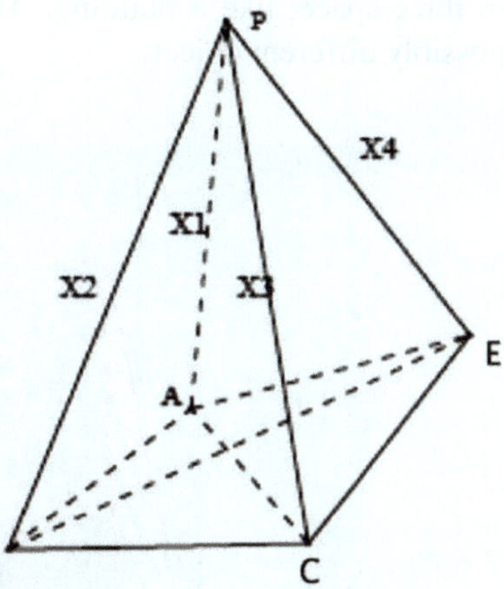

Fig. 6 Viewing four points on an object

Fig. 7 Equations from law of cosines, angles p, q, r, s, t, u

$$
\begin{aligned}
X_1^2 + X_2^2 - X_1X_2r - |AB|^2 &= 0\\
X_1^2 + X_3^2 - X_1X_3q - |AC|^2 &= 0\\
X_2^2 + X_3^2 - X_2X_3p - |BC|^2 &= 0\\
X_1^2 + X_4^2 - X_1X_4s - |AE|^2 &= 0\\
X_4^2 + X_3^2 - X_3X_4t - |CE|^2 &= 0\\
X_2^2 + X_4^2 - X_2X_4u - |BE|^2 &= 0
\end{aligned}
$$

Dixon-EDF finishes in 36 s, 275 meg RAM. However, the resultant can be computed in less than 1 s by the variation of EDF in which we run EDF to a certain point (in this case after the third row) and then use Gentleman–Johnson [10] to compute the determinant of the remaining 9×9 matrix. This is a good example of the enormous flexibility of Dixon-EDF, in which the user is in control throughout the process. The resultant for X_1 has 24068 terms.

The analogous problem with a five-sided figure, using only the outside edges, is also solvable by Dixon by a two step process. We now have variables X_1, X_2, X_3, X_4, X_5 and five equations. Use four of them to eliminate all variables but X_1, X_2. That takes 144 s. Then take that resultant (47295 terms) and the remaining fifth equation and eliminate X_2. That takes 4 h. The final answer has 37291784 terms.

10 The Six-Line Problem

This was a problem of great interest around 1996–2000. Imagine a man-made object in three-space, like a building. Abstract six lines. Imagine later photographing a possibly different object.

Fig. 8 Six lines on a building

Problem: Can we decide from the two-dimensional representation (photograph) that this is the original object? Can we at least show that it is not the original object? That is, develop an algorithm to reject incorrect objects from the 2D data. This is a problem in automated image recognition (Fig. 8).

People who worked on this besides the author: Peter Stiller, Texas A & M; Robert M. Williams, Naval Research Center; George Nakos, U.S. Naval Academy; Frank Grosshans, Westchester University (Pennsylvania); Ronald Gleeson, The College of New Jersey; Michael Hirsch, student.

Using algebraic geometry, Peter Stiller produced a transformation: object $\Rightarrow$ six lines $\Rightarrow$ nine three-dimensional ("3D") invariants [17]. Later, we get a two-dimensional photograph of some possibly different object. From the photograph $\Rightarrow$ four two-dimensional ("2D") invariants. There are four variables representing a transformation matrix, but one can be eliminated. This yields four polynomial equations in three variables involving the $9 + 4 = 13$ invariants as parameters. The goal is to eliminate the three remaining variables (which are a_{12}, a_{21}, and a_{22}) producing a resultant in the 13 parameters. This is exactly what one wants to test the data for any object.

A full symbolic solution seemed impossible. Everyone at the time acknowledged that a Gröbner bases attack was hopeless. By experimenting with substituting numerical values for most of the parameters on known objects, it became clear that the resultant was actually quite small—less than 500 terms. Yet a symbolic solution seemed hopeless.

The second Dixon matrix is 24 × 24:

416 0 0 0 352 8 838 507 0 1035 132 0 636 782 0 48 88 320 0 1006 242 8 679 488
880 494 57 22 532 93 0 0 238 1091 0 0 278 0 380 0 0 164 34 1273 0 88 489 922
2942 1382 68 104 2068 32 2898 1548 222 5110 0 88 2987 3738 917 12 484 1706
56 4550 ...
684 356 168 74 204 0 57 0 158 572 0 0 154 288 416 0 0 108 0 455 0 0 0 498
3250 2460 328 196 2127 202 3548 1630 644 5774 20 88 3847 4801 1957 12 508
2029 98 ...
1670 805 242 126 802 136 0 0 446 1494 0 0 734 0 797 0 0 453 48 1580 0 132 612
1500
70 14 22 8 12 0 0 0 16 38 0 0 4 0 32 0 0 4 0 30 0 0 0 48
96 0 0 0 160 0 680 248 0 950 0 0 452 620 0 0 0 278 0 660 52 0 160 84
792 0 0 0 376 83 140 116 0 960 184 52 674 612 0 60 234 312 26 1569 342 85 1002
872
28 0 0 0 0 0 178 72 0 148 0 0 54 90 0 0 0 36 0 92 0 0 16 24
562 476 16 16 352 0 1024 125 76 1645 0 0 866 1158 298 0 0 462 0 1355 120 12 401
372
2 0 0 0 2 0 8 0 0 8 0 0 0 8 0 0 0 2 0 0 0 0 0 0
0 0 0 0 24 0 146 48 0 116 0 0 34 82 0 0 0 30 0 56 0 0 0 0
386 208 17 6 224 29 0 0 78 425 0 0 202 0 140 0 0 104 10 509 0 24 201 384
4665 2389 636 264 2913 500 1118 768 1016 4818 445 204 2770 2933 2042 150 668
... 5390 ...

614 612 24 0 480 0 2005 774 88 2330 0 0 1318 1512 374 0 0 674 0 2056 242 24 753 566
1026 0 0 0 638 92 510 360 0 1334 264 48 1180 1244 0 98 341 510 32 1494 641 96 504 1099
688 600 28 16 480 0 1855 652 96 2312 0 0 1256 1640 392 0 0 668 0 2052 226 24 725 526
6156 2986 727 354 3826 511 2709 1733 832 6376 220 196 4786 5640 2500 98 960 ... 6336 ...
1490 922 114 52 938 166 124 100 474 2039 152 36 540 518 738 48 188 290 66 2333 290 ...
3085 1448 430 186 1793 298 576 230 656 2596 0 0 1693 1875 1298 0 0 888 104 2814 458 ...
474 0 0 0 220 40 88 76 0 542 140 29 384 348 0 48 168 178 14 902 228 36 658 488
1854 1106 32 98 1311 88 1924 668 190 3376 0 0 1538 2802 744 0 0 982 2672 ... 1048 1518
1196 0 0 0 474 56 1220 446 0 2556 12 32 1853 1742 0 0 184 885 0 2554 442 82 1024 1048

Again, these are the numbers of terms in the polynomial at each spot in the matrix. For example, one of the smallest polynomials (70 terms) is

$$-l_1 l_2 n_{11} p_{22} q_2 q_4 + l_2 n_{11} p_{22} q_2 q_4 + l_1 n_{11} p_{22} q_2 q_4 - n_{11} p_{22} q_2 q_4 - l_2^2 g p_{21} q_2 q_4 + l_1 l_2 g p_{21} q_2 q_4 + l_2 g p_{21} q_2 q_4 - l_1 g p_{21} q_2 q_4 + l_2^2 g p_{12} q_2 q_4 - l_1 l_2 g p_{12} q_2 q_4 - l_2 g p_{12} q_2 q_4 + l_1 g p_{12} q_2 q_4 + l_1 l_2 n_{22} p_{11} q_2 q_4 - l_2 n_{22} p_{11} q_2 q_4 - l_1 n_{22} p_{11} q_2 q_4 + n_{22} p_{11} q_2 q_4 + l_1 l_2 n_{22} p_{22} q_4 - l_2 n_{22} p_{22} q_4 - l_2^2 n_{11} p_{22} q_4 + l_1 l_2 n_{11} p_{22} q_4 + l_2 n_{11} p_{22} q_4 - l_1 n_{11} p_{22} q_4 + l_2^2 p_{22} q_4 - 2 l_1 l_2 p_{22} q_4 + l_1 p_{22} q_4 + l_2^2 g p_{21} q_4 - l_2 g p_{21} q_4 - l_2^2 g p_{12} q_4 + 2 l_1 l_2 g p_{12} q_4 + l_2 g p_{12} q_4 - 2 l_1 g p_{12} q_4 - l_1 l_2 n_{22} p_{11} q_4 + l_2 n_{22} p_{11} q_4 + l_1 l_2 p_{11} q_4 - l_2 p_{11} q_4 - l_1 l_2 n_{22} p_{22} q_2 + l_2 n_{22} p_{22} q_2 + l_1 l_2 n_{11} p_{22} q_2 - l_2 n_{11} p_{22} q_2 + l_2^2 g p_{21} q_2 - 2 l_1 l_2 g p_{21} q_2 - l_2 g p_{21} q_2 + 2 l_1 g p_{21} q_2 - l_2^2 g p_{12} q_2 + l_2 g p_{12} q_2 + l_2^2 n_{22} p_{11} q_2 - l_1 l_2 n_{22} p_{11} q_2 - l_2 n_{22} p_{11} q_2 + l_1 n_{22} p_{11} q_2 - l_2^2 n_{22} q_2 + 2 l_1 l_2 n_{22} q_2 - l_1 n_{22} q_2 - l_1 l_2 n_{11} q_2 + l_2 n_{11} q_2 + l_2^2 n_{11} p_{22} - l_1 l_2 n_{11} p_{22} - l_2^2 p_{22} + l_1 l_2 p_{22} - l_2^2 g p_{21} + l_1 l_2 g p_{21} + l_2^2 g p_{12} - l_1 l_2 g p_{12} - l_2^2 n_{22} p_{11} + l_1 l_2 n_{22} p_{11} + l_2^2 p_{11} - l_1 l_2 p_{11} + l_2^2 n_{22} - l_1 l_2 n_{22} - l_2^2 n_{11} + l_1 l_2 n_{11}$$

This matrix is hopelessly large for Dixon-EDF running on any now-conceivable computer system—unless the answer were to emerge very early in the process. It does not.

We solved the problem then [17] by a long series of tricks: interpolation, modding out by certain polynomials, etc. It was very difficult. The answer has only 239 terms!

Already in 1998 we had the idea to do Dixon in stages, as in the previous section on pose estimation.

- Take three of the equations and eliminate two variables. The resultant would have one variable and 13 parameters.
- Take another set of three, do it again, obtaining a second resultant in the same variable.
- Finally take those two resultants and eliminate the last variable.

This helps a bit. Suppose we eliminate the two variables a_{12} and a_{21}. The second Dixon matrix is 10×10:

96 0 44 0 82 38 12 24 4 94
24 8 0 0 0 6 18 6 0 12
610 0 356 0 707 313 64 130 68 719
210 0 0 8 0 98 76 78 24 130
34 0 84 0 60 6 37 2 0 33
168 10 60 22 44 96 44 96 24 126
1168 32 510 44 472 644 360 484 160 978
1043 8 986 32 999 414 444 363 84 917
1673 24 990 46 1713 937 294 662 232 1538
980 12 513 24 918 472 162 325 110 998

This is still hopeless. The resultant, in a_{22}, would have hundreds of millions of terms. No significant factor is found early. We would need to then feed two such massive polynomials to Dixon again. This approach was therefore abandoned in 1998.

However, in July 2015 I decided to try to eliminate instead the two variables a_{22} and a_{21}. Astonishingly, the second Dixon matrix is

104 0 0 0 72
1232 96 594 180 1489
1195 72 581 0 0
642 0 0 104 545
148 0 96 0 0

Dixon-EDF finishes in only 33 s. The numerator list is

1 1 1 1 1 104 2 96 216340

But the last polynomial is easily found to have contents (simple factors):

1 1 104 2 96 192 130 48 2

Only the factor of 192 terms involves all the parameters. It is the resultant in a_{12}.

Repeat with another set of three equations. The same thing happens. Then feed the two 192 term polynomials to Dixon-EDF to eliminate a_{12}. In 10 s we have the answer of 1086153 terms. But this has an easily discovered content of 22726 terms, yielding the answer of 239 terms. The total elapsed time for Dixon-EDF is 96 s, using 610 meg RAM.

Try the first stage with Gröbner bases:

- Maple FGb: killed after 6.4 h, 48 gig RAM.
- Magma 1: killed after 1.1 h, 48 gig RAM.
- Magma 2: killed after 6.7 h, 31 gig RAM.

Try the second stage with Gröbner bases:

- Maple FGb: crashed after 6.2 h, 12.5 gig RAM.
- Magma 1: killed after 14 h, 5 gig RAM.
- Magma 2: killed after 15 h, 28 gig RAM.

11 Conclusions

We found great success in applying Dixon-EDF to polynomial systems arising in important applications. Dixon-EDF succeeds on many other systems [14, 19, 20]. Maple failed repeatedly with several implementations of Gröbner bases, as did two methods in Magma. Some of the systems above were also tried in Mathematica and Singular and failed.

- Dixon-EDF is a powerful tool for symbolic solution of systems of multivariate equations.
- Dixon-EDF succeeds where other methods fail. It is usually orders of magnitude more effective, at least on systems with parameters.
- Dixon-EDF challenges the user's creativity. There are many variations and options.

Appendix

The Maple-FGb commands for the pose example:

```
    |\^/|     Maple 2015 (X86 64 LINUX)
._|\|   |/|_. Copyright (c) Maplesoft, a division of Waterloo Maple Inc. 2015
 \  MAPLE  /  All rights reserved. Maple is a trademark of
 <____ ____>  Waterloo Maple Inc.
      |       Type ? for help.

> with(FGb):
p := 0;
v1 := [ x2, x3, x4 ];
v2 := [ x1, b1,b2,b3,b4,c12,c23,c34,c41 ];
sys := [x1^2 + x2^2 - c12*x1*x2 - b1, x2^2 + x3^2 - c23*x2*x3 - b2,
           x3^2 + x4^2 - c34*x3*x4 - b3, x4^2 + x1^2 - c41*x4*x1 - b4];
> lll:=fgb_gbasis_elim(sys, p,v1,v2,{"step"=8,"verb"=3,"index"=40000000});
```

Magma 1 commands for the pose example:

```
Magma V2.21-8   Thu Dec 10 2015 13:26:28 on ace-math01 [Seed = 2343837211]
Type ? for help.  Type <Ctrl>-D to quit.
Q := RationalField();
A<x1,x2,x3,x4,b1,b2,b3,b4,c12,c23,c34,c41> := AffineSpace(Q,12);
X := Scheme(A, [x1^2 + x2^2 - c12*x1*x2 - b1, x2^2 + x3^2 - c23*x2*x3 - b2,
       x3^2 + x4^2 - c34*x3*x4 - b3, x4^2 + x1^2 - c41*x4*x1 - b4]);
I := Ideal(X);
time J := EliminationIdeal(I, { x1,b1,b2,b3,b4,c12,c23,c34,c41 });
```

Magma 2 commands for the pose example:

```
Magma V2.21-8   Thu Dec 10 2015 13:26:28 on ace-math01 [Seed = 2343837211]
Type ? for help.  Type <Ctrl>-D to quit.
```

```
Q:=RationalField();
F<b1,b2,b3,b4,c12,c23,c34,c41> := FunctionField(Q,8);
R<x1,x2,x3,x4> := PolynomialRing(F,4,"elim", [2,3,4]);
I := Ideal ([x1^2 + x2^2 - c12*x1*x2 - b1, x2^2 + x3^2 - c23*x2*x3 - b2,
      x3^2 + x4^2 - c34*x3*x4 - b3, x4^2 + x1^2 - c41*x4*x1 - b4]);
time G := GroebnerBasis(I);
```

References

1. Bricard, R.: Mémoire sur la théorie de l'octaèdre articulé. J. Math. Pures Appl. 3, pp. 113–150 (1897). (English translation: http://www.math.unm.edu/~vageli/papers/bricard.pdf)
2. Carrà-Ferro, G.: A resultant theory for the systems of two ordinary algebraic differential equations. Appl. Algebra Eng. Commun. Comput. **8**, 539–560 (1997)
3. Connelly, R., Sabitov, I., Walz, A.: The bellows conjecture. Contrib. Algebra Geom. **38**, 1–10 (1997)
4. Coutsias, E., Seok, C., Jacobson, M.P., Dill, K.A.: A kinematic view of loop closure. J. Comput. Chem. **25**(4), 510–528 (2004)
5. Coutsias, E., Seok, C., Wester, M.J., Dill, K.A.: Resultants and loop closure. Int. J. Quantum Chem. **106**(1), 176–189 (2005)
6. Cox, D., Little, J., O'Shea, D.: Using Algebraic Geometry. Graduate Texts in Mathematics, vol. 185. Springer, New York (1998)
7. Dixon, A.L.: The eliminant of three quantics in two independent variables. Proc. Lond. Math. Soc. **6**, 468–478 (1908)
8. Faugere, J.-C.: A new efficient algorithm for computing Gröbner bases (F4). J. Pure Appl. Algebra **139**, 61–88 (1999)
9. Faugere, J.-C.: Personal communication, July 8 (2014)
10. Gentleman, W., Johnson, S.: The evaluation of determinants by expansion by minors and the general problem of substitution. Math. Comput. **28**(126), 543–548 (1974)
11. Lewis, R.H.: Computer algebra system Fermat. http://home.bway.net/lewis/
12. Lewis, R.H.: Fermat code for Dixon-EDF. http://home.bway.net/lewis/dixon
13. Lewis, R.H.: Heuristics to accelerate the Dixon resultant. Math. Comput. Simul. **77**(4), 400–407 (2008)
14. Lewis, R., Bridgett, S.: Conic tangency equations arising from Apollonius problems in biochemistry. Math. Comput. Simul. **61**(2), 101–114 (2003)
15. Lewis, R.H., Coutsias,E.A.: Algorithmic search for flexibility using resultants of polynomial systems. In: Automated Deduction in Geometry, 6th International Workshop, ADG, 2006. LNCS. Springer. **4869**, 68–79 (2007)
16. Lewis, R.H., Coutsias, E.A.: Flexibility of Bricard's Linkages and Other Structures via Resultants and Computer Algebra. Math. Comput. Simul. (2014). http://arxiv.org/abs/1408.6247
17. Lewis, R.H., Stiller, Peter: Solving the recognition problem for six lines using the Dixon resultant. Math. Comput. Simul. **49**, 203–219 (1999)
18. Nachtwey, P.: Delta Computer Systems, personal communication (2009)
19. Palancz, B., Lewis, R.H., Zaletnyik, P., Awange, J.: Computational study of the 3D affine transformation part I. 3-point problem (2008). http://library.wolfram.com/infocenter/MathSource/7090/
20. Palancz, B., Awange, J., Zaletnyik, P., Lewis, R.H.: Linear homotopy solution of nonlinear systems of equations in geodesy. J. Geod. (2009). http://www.springerlink.com/content/78qh80606j224341/
21. Post to reddit (2014). https://www.reddit.com/r/math/comments/1z7j3x/geometry_me_and_my_friends_are_working_on_a/

22. Ritt, J.F.: Differential Equations from the Algebraic Standpoint, College Publications 14. American Mathematical Society, New York (1932)
23. Steel, A.: Personal communications, September 2–7 (2015)
24. Sturmfels, B.: Solving systems of polynomial equations. In: CBMS Regional Conference Series in Mathematics 97 (2003). American Mathematical Society, Providence
25. Thorpe, M., Lei, M., Rader, A.J., Jacobs, D.J., Kuhn, L.: Protein flexibility and dynamics using constraint theory. J. Mol. Gr. Model. **19**, 60–69 (2001)
26. Tot, J.: Spinning double pendulum: equilibria and bifurcations. In: ACA 2014 Conference, New York, In Session: Innovative Applications and Emerging Challenges

Visualization of Orthonormal Triads in Cylindrical and Spherical Coordinates

J. López-García, J.J. Jiménez Zamudio and M.E. Canut Díaz Velarde

Abstract According to Committee on Programs for Advanced Study of Mathematics and Science in American High Schools [1] (Gollub et al. (eds.) in Learning and Understanding. Improving Advanced Study of Mathematics and Science in U.S. Hight Schools. National Academic Press, Washington, 2002), "the primary goal of advanced study in any discipline should be for students to achieve a deep conceptual understanding of the disciplines content". It is undoubted that abstraction is one of the skills that teachers wish to improve in their students, but, how can teachers take advantage of technological resources, such as CAS or DGS, as help in their classes in undergraduate courses? One concept, whose importance is both theoretical and practical, corresponds to the coordinate transformation, in particular orthogonal coordinate systems. We can use trigonometric constructions to find the transformation equations, namely, the algorithm for transforming a Cartesian system into other coordinate system, as cylindrical or spherical coordinates. Not only, if we add the knowledge and some techniques from Linear Algebra, we can motivate new mathematical properties, but also we will increase considerably the abstract reasoning and symbolic calculation. We know that visualization helps intuitive understanding. Therefore, we propose using CAS and DGS to show how a triad, of basis vectors, is continuously changing direction, keeping the norm vector without change, and how this match visualization with the reasoning from theories of linear algebra.

Keywords Visualization in Mathematics · Orthonormal triads · Basis vectors · Cylindrical coordinates · Spherical coordinates · CAS

J. López-García (✉) · J.J. Jiménez Zamudio (✉) · M.E. Canut Díaz Velarde (✉)
FES-Acatlán, UNAM, Av. Alcanfores y San Juan Totoltepec s/n,
Santa Cruz Acatlán, 53150 Naucalpan, Estado de México, Mexico
e-mail: jeanettlg@hotmail.com

J.J. Jiménez Zamudio
e-mail: jzamudio02@yahoo.com

M.E. Canut Díaz Velarde
e-mail: maru@gmail.com

I.S. Kotsireas and E. Martínez-Moro (eds.), *Applications of Computer Algebra*,
Springer Proceedings in Mathematics & Statistics 198,
DOI 10.1007/978-3-319-56932-1_17

1 Introduction

In recent years, new technologies have enhanced the theories using different representations in teaching math concepts, to make possible new ways of representing mathematical objects, through visualization by the students. It is for this reason that technological innovations have changed the technical and didactic models that were handled in the teaching of mathematics. Currently, this technological development allows more often to symbolize some mathematical concepts through different representations in arithmetic, numerical, graphical, algebraic, verbal, and symbolic more quickly and easily with calculators and computers. Thus, the display is related to the operation of cognitive structures, establishing the variety of representations with a mathematical object. Visualization is simply a means by which a student could improve mathematical understanding. When we refer to visualizing a concept, we are talking about understanding a concept through a visual image [2] which claims that: Computers have a direct and specific role in this renewal of the display due to the ways in which computers can generate math graphs. A particular case of study is when students work in coordinate systems different to Cartesian coordinates, they tend to make just algorithmic tasks, and sometimes cannot answer questions like: In spherical coordinates, do unit vectors dependent on the local point?

2 Visualization in Mathematical Education

In the field of mathematical education, in recent years some approaches have arisen in order to understand mathematical concepts during teaching of mathematics. Various aspects that are important for learning fundamental cognitive activities such as representation, conceptualization, reasoning, visualization, comprehension, problem solving, etc., require math teacher of various systems of semiotic representation, simultaneous and articulated [3]. Within these cognitive approaches, there are basic concepts with the same meaning, even though, terminology used is comparable; this happens with notions such as visualization, spatial ability, geometric reasoning, spatial thinking, and spatial vision. This paper aims to define some concepts used to characterize the cognitive processes involved and developed when geometry problems are solved, taking a fundamental basis proposed by Duval [4]. There are different views on the meaning of the display, which researchers [5] point out that the notion of visualization or visual thinking is strongly linked with the ability to form mental images. What characterizes a mental image is to enable the evocation of an object, without the right to be present [5]. Following this idea makes a very general description of the display as "the act by which an individual establishes a strong connection between internal construction and something to which access is gained through the senses" [6].

The spatial visualization has received much attention as a research topic in Mathematical Education [7]. It is evaluating the processes and capabilities of individuals to perform certain tasks that require "seeing" or "imagining" mentally, mathematical or geometric objects. Zimmermann and Cunningham, describe it as "the process of production or use of geometric or graphical representations of mathematical concepts, principles or problems, either hand drawn or computer generated" [2]. Also, they mention that a diagram display simply means to form a mental picture of the diagram, but a problem is displayed to understand it in terms of a chart or a visual figure. In general, it is considered that the term display is used in reference to pictorial figures representations, which can be internal or external. Therefore, it must be graphs, diagrams, shapes that are built manually or computer-generated internal representations, and they strengthen the cognitive process that leads to learning. When speaking of internal representations, it refers to mental images, visual images, and conceptual images. Gilbert [8] emphasizes that any visualization produces models, which play a central role in learning science. Therefore, we understand visual images, like mental schemes that represent visual information. Specifically, we are interested in linking algebraic concepts with geometric concepts for improving the students' understanding as the main purpose of any advanced study [1].

3 Orthonormal Triads in Cylindrical and Spherical Coordinates

As Haaser et al. [9] state, the geometric idea behind an analytical model is the concept of coordinate system. In geometry, a coordinate system is a system that uses one or more numbers (coordinates) to determine the position of a point or other geometric object. A coordinate system is a continuous one-one correspondence between the points of a space and the n-tuples of real numbers or points in space.

A topic that is commonly taught in undergraduate courses is transformation of coordinates; basically, teachers used to work with cylindrical and spherical coordinates instead use Cartesian coordinates. For example, they would teach how to operate these formulas

$$x = r \sin\theta \ \cos\phi \tag{1}$$

$$y = r \sin\theta \ \sin\phi \tag{2}$$

$$z = r \cos\theta \tag{3}$$

They used to introduce some of the main uses of the coordinates, for example to solve some kind of integrals

$$\int_a^b \int_{\phi_1(\theta)}^{\phi_2(\theta)} \int_{r_1(\theta,\phi)}^{r_2(\theta,\phi)} f(r,\phi,\theta) r^2 \sin\phi dr d\phi d\theta \tag{4}$$

or to find different expressions for some points, by example: the way of transforming (1, 1, 1) given in Cartesian Coordinates to $(\sqrt{3}, \arctan\sqrt{2}, \pi/4)$ in Spherical Coordinates.

3.1 Teaching New Mathematical Objects

If teachers want to go farther into linking algebraic concepts with geometric concepts for improving the students understanding, our proposal is to teach what the rotation matrices makes of over a triad in the process of coordinate transformations, analytically and geometrically with help of DGS or CAS. Therefore, we started asking students if they are able to figure out how the scalars of the transformation matrix are linked with the rotate picture of one given basis, such as, the triads shown (Figs. 1, 2 and 3).

It is possible to verify that three vectors of a basis of a coordinate system are related to another basis, through the linear transformation given by Eq. (5).

$$\begin{pmatrix} \hat{\mathbf{e}}'_1 \\ \hat{\mathbf{e}}'_2 \\ \hat{\mathbf{e}}'_3 \end{pmatrix} = \begin{pmatrix} \hat{\mathbf{e}}_1 \cdot \hat{\mathbf{e}}'_1 & \hat{\mathbf{e}}_2 \cdot \hat{\mathbf{e}}'_1 & \hat{\mathbf{e}}_3 \cdot \hat{\mathbf{e}}'_1 \\ \hat{\mathbf{e}}_1 \cdot \hat{\mathbf{e}}'_2 & \hat{\mathbf{e}}_2 \cdot \hat{\mathbf{e}}'_2 & \hat{\mathbf{e}}_3 \cdot \hat{\mathbf{e}}'_2 \\ \hat{\mathbf{e}}_1 \cdot \hat{\mathbf{e}}'_3 & \hat{\mathbf{e}}_2 \cdot \hat{\mathbf{e}}'_3 & \hat{\mathbf{e}}_3 \cdot \hat{\mathbf{e}}'_3 \end{pmatrix} \begin{pmatrix} \hat{\mathbf{e}}_1 \\ \hat{\mathbf{e}}_2 \\ \hat{\mathbf{e}}_3 \end{pmatrix} \tag{5}$$

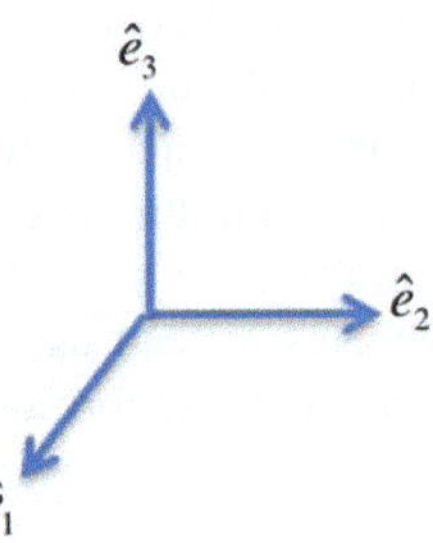

Fig. 1 Cartesian basis

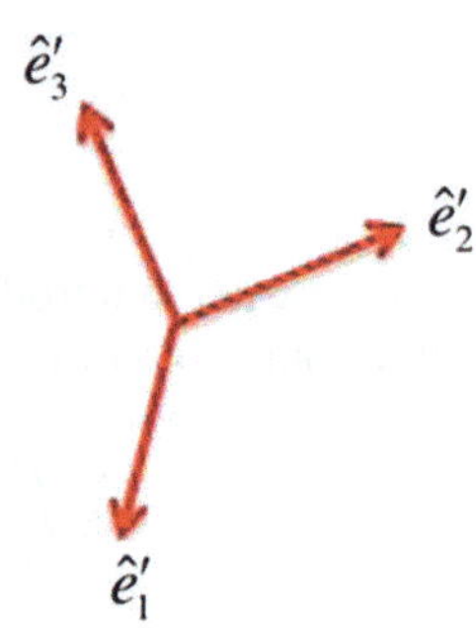

Fig. 2 Rotated orthogonal basis

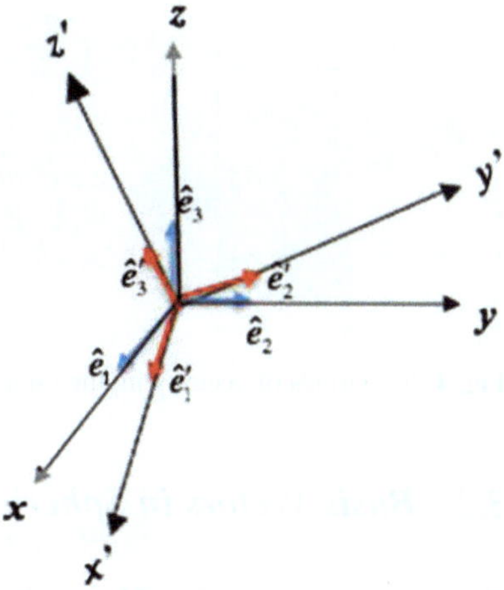

Fig. 3 Former and rotated orthogonal basis

In particular, we establish the notation as $(\hat{\mathbf{i}}, \hat{\mathbf{j}}, \hat{\mathbf{k}}) = (\hat{\mathbf{e}}_1, \hat{\mathbf{e}}_2, \hat{\mathbf{e}}_3)$ (Fig. 1) for Cartesian coordinates, and $(\hat{\mathbf{e}_1'}, \hat{\mathbf{e}_2'}, \hat{\mathbf{e}_3'})$ (Fig. 2) for spherical or cylindrical coordinates. Consequently, for the cases mentioned, the transformation respectively are the Eqs. (6) and (7)

$$\begin{pmatrix} \hat{\mathbf{e}}_r \\ \hat{\mathbf{e}}_\theta \\ \hat{\mathbf{e}}_\phi \end{pmatrix} = \begin{pmatrix} \sin\theta\ \cos\phi & \sin\theta\ \sin\phi & \cos\theta \\ \cos\theta\ \cos\phi & cos\theta\ \sin\phi & -\sin\theta \\ -\sin\phi & \cos\phi & 0 \end{pmatrix} \begin{pmatrix} \hat{\mathbf{i}} \\ \hat{\mathbf{j}} \\ \hat{\mathbf{k}} \end{pmatrix} \tag{6}$$

$$\begin{pmatrix} \hat{\mathbf{e}}_\rho \\ \hat{\mathbf{e}}_\theta \\ \hat{\mathbf{e}}_z \end{pmatrix} = \begin{pmatrix} \cos\theta & \sin\theta & 0 \\ -\sin\theta & \cos\theta & 0 \\ 0 & 0 & 1 \end{pmatrix} \begin{pmatrix} \hat{\mathbf{i}} \\ \hat{\mathbf{j}} \\ \hat{\mathbf{k}} \end{pmatrix} \tag{7}$$

After this explanation, some students cannot visualize what is happening yet, mathematically speaking.

Students must understand how a basis is produced. When the topic is chosen in three dimensions, three mutually perpendicular planes, $x =$constant, $y =$constant and $z =$constant are selected, then a trihedral formed by three orthogonal dihedrals is obtained. This means, it has gotten rectangular coordinates as the result of the intersection of three orthogonal surfaces.

In rectangular coordinates when we translate the point (x, y, z) to $(x + \Delta x, y + \Delta y, z + \Delta z)$, the triad of vectors of the basis associated to both points is parallel (Fig. 4).

Cylindrical coordinates or spherical coordinates come from the intersection of two planes and one curved surface (a cylinder), and the intersection of two curved surfaces and one plane, respectively, which will be explained in the next section.

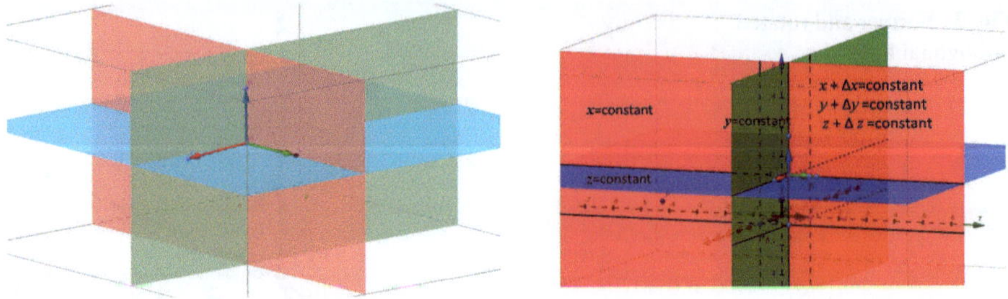

Fig. 4 Two triads of vector with the same orientation

3.2 *Basis Vectors in Spherical and Cylindrical Coordinates*

Kurmyshev and Sánchez-Yañez [10] point out that often mathematical physicists or engineers are used to fitting the best system of coordinates for facing an specific problem, instead of Cartesian coordinates. The main reason is to pose and solve a problem, which could be easier if they choose a natural system for the problem.

According to Hauser [11], Arfken and Weber [12], cylindrical, spherical, parabolic coordinates, hyperbolic, just for naming a few, are only some particular cases of the generalized coordinates q_1, q_2, q_3, depending on which it can analyze the motion of a particle.

The equations which relate the Cartesian coordinates x, y, z with generalized coordinates q_1, q_2, q_3, are $x = q_1$, $y = q_2$, $z = q_3$. Then any vector $\mathbf{r}$ can be written as $\mathbf{r} = r_1\hat{\mathbf{e}}_1 + r_2\hat{\mathbf{e}}_2 + r_3\hat{\mathbf{e}}_3$ where $(\hat{\mathbf{e}}_1, \hat{\mathbf{e}}_2, \hat{\mathbf{e}}_3)$ are the unit basis vectors.

In spherical coordinates, as we mentioned, one of the surfaces is a plane. The others two surfaces are a cone and a sphere (Fig. 5). Of course, their basis vectors are normal to these surfaces (Fig. 6). In this case, the generalized coordinates are $q_1 = r, q_2 = \theta, q_3 = \phi$.

When the variable theta is incremented, with ϕ = constant, and r = constant, we can realize how these vectors are parallels no more (Fig. 7). Moreover, the use of DGS shows how is changing the normal vector as theta angle is changing (Fig. 6). Students have to analyze that the basis depends on each point and in that sense the coordinate system is local. In other words, students have to be aware, that the basis is dependent on each point $P(r, \theta, \phi)$, and in that sense the coordinate system $(\hat{\mathbf{e}}_r, \hat{\mathbf{e}}_\theta, \hat{\mathbf{e}}_\phi)$ is local. It is the challenge to resolve using visualization theories.

Similar things happen when we are working in cylindrical coordinates. Their normal vectors are associated at two planes and a cylinder. One of the planes is parallel to the plane xy and the other is perpendicular at the first plane.

If the radius ρ (circular cylinder no change) and the plane z are maintained constant, and the angle ϕ is continuously varied, we can notice that the triads of vectors are changing constantly. Analogously, the variations happen if we change two or three coordinates instead of just one.

Fig. 5 *Plane*, *cone*, and *sphere*

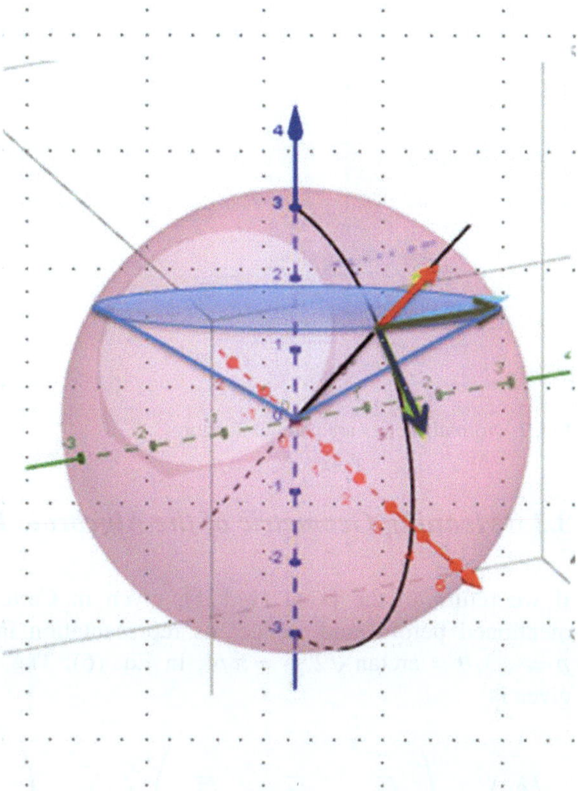

Fig. 6 Normal basis vectors

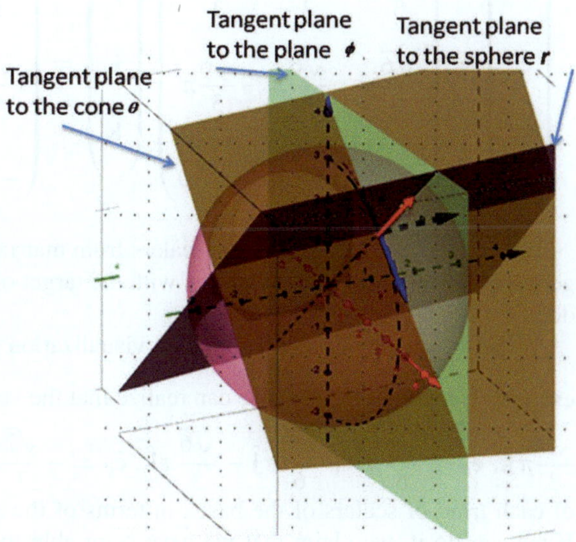

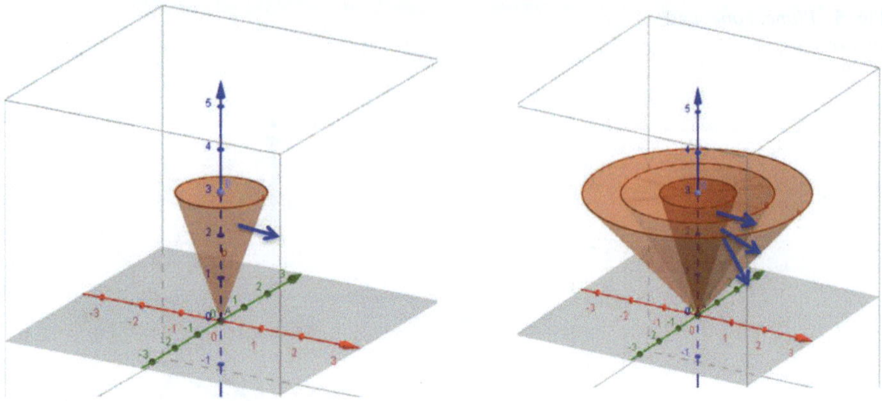

Fig. 7 Normal vectors are parallel no more

3.3 *Meaning Geometric of the Algebraic Results*

If we return to the point (1, 1, 1), given in Cartesian coordinates, which was mentioned before, and we get its representation in spherical coordinates using $\rho = \sqrt{3}, \theta = \arctan \sqrt{2}, \phi = \pi/4$, in Eq. (6). The basis associated to the point given is

$$\begin{pmatrix} \hat{\mathbf{e}}_r \\ \hat{\mathbf{e}}_\theta \\ \hat{\mathbf{e}}_\phi \end{pmatrix} = \begin{pmatrix} \frac{\sqrt{3}}{3}\pi & \frac{\sqrt{3}}{3}\pi & \frac{\sqrt{3}}{3}\pi \\ \frac{\sqrt{6}}{6}\pi & \frac{\sqrt{6}}{6}\pi & -\frac{\sqrt{6}}{3}\pi \\ -\frac{\sqrt{2}}{2}\pi & \frac{\sqrt{2}}{2}\pi & 0 \end{pmatrix} \begin{pmatrix} \hat{\mathbf{i}} \\ \hat{\mathbf{j}} \\ \hat{\mathbf{k}} \end{pmatrix} = \begin{pmatrix} \frac{\sqrt{3}}{3}\pi\hat{\mathbf{i}} + \frac{\sqrt{3}}{3}\pi\hat{\mathbf{j}} + \frac{\sqrt{3}}{3}\pi\hat{\mathbf{k}} \\ \frac{\sqrt{6}}{6}\pi\hat{\mathbf{i}} + \frac{\sqrt{6}}{6}\pi\hat{\mathbf{j}} - \frac{\sqrt{6}}{3}\pi\hat{\mathbf{k}} \\ -\frac{\sqrt{2}}{2}\pi\hat{\mathbf{i}} + \frac{\sqrt{2}}{2}\pi\hat{\mathbf{j}} \end{pmatrix} \quad (8)$$

But, what is the meaning of the scalars from matrix in Eq. (8)? We wish to give a geometric meaning at these numbers with the target of understanding what we have done.

To get it, we might help us of some visualization tools. If we use any DGS, by example GeoGebra (Fig. 8), we can realize that the vectors $\hat{\mathbf{e}}_r = \frac{\sqrt{3}}{3}\pi\hat{\mathbf{i}} + \frac{\sqrt{3}}{3}\pi\hat{\mathbf{j}} + \frac{\sqrt{3}}{3}\pi\hat{\mathbf{k}}$, $\hat{\mathbf{e}}_\theta = \frac{\sqrt{6}}{6}\pi\hat{\mathbf{i}} + \frac{\sqrt{6}}{6}\pi\hat{\mathbf{j}} - \frac{\sqrt{6}}{3}\pi\hat{\mathbf{k}}$, $\hat{\mathbf{e}}_\phi = -\frac{\sqrt{2}}{2}\pi\hat{\mathbf{i}} + \frac{\sqrt{2}}{2}\pi\hat{\mathbf{j}}$ are the resultants of each triad of scalars of the basis, in terms of the standard unit vectors $(\hat{\mathbf{i}}, \hat{\mathbf{j}}, \hat{\mathbf{k}})$. If we can do it, we claim that we have been able to understand the link between algebraic concepts and geometric representations, of course with the help of some

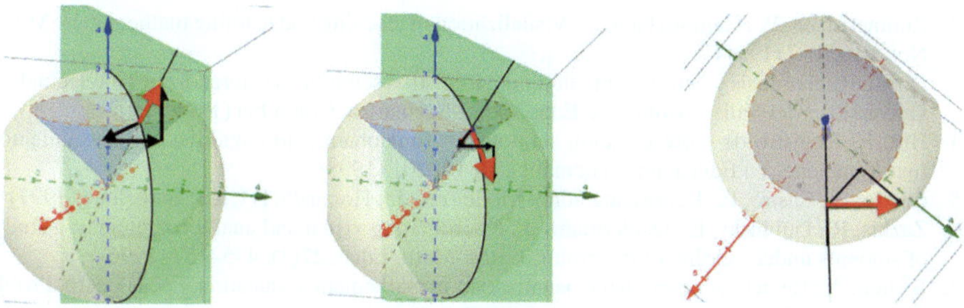

Fig. 8 Basis vectors $\hat{e}_1, \hat{e}_2, \hat{e}_3$ in spherical coordinates

DGS or CAS (they are the unitary vectors). Therefore, our students could deepen their knowledge as we wish in learning math.

4 Conclusions and Reflections

With some CAS (by example MAPLE or GeoGebra) is easy to show how a triad, of basis vectors, is continuously changing its direction when we are working in spherical coordinates or cylindrical coordinates.

The learning–teaching process is short and more dynamic in order to visualize that the change of triad depends on the point of analysis and it gives meaning to algebraic construction.

It is possible to switch to several representations, such as, geometrical and algebraic when you need it. We think that great transformation in the fieldwork in math could be supported by a lot of micro-changes in everyday duties in the classroom, gathering and sharing research on mathematical education to improve our function as teachers.

We can help more and more students in their tasks if we were able to allow ourselves a little change and incorporate the technological advance into our classrooms, of course keeping a high degree of mathematical abstraction.

We are interested in proceeding to use of visualization. Therefore, the next step in the subject of coordinates transformation could be to go further and try to understand what happen in other kind of coordinates, by example hyperbolic coordinates, whose basis vectors are no longer mutually perpendicular to each other.

References

1. Gollub, J., Bertenthal, W., Labov, J., Curtis, P. (eds.): Learning and Understanding. Improving Advanced Study of Mathematics and Science in U.S. Hight Schools. National Academic Press, Washington D.C. (2002)

2. Zimmermann, W., Cunningham, S.: Visualization in teaching and learning mathematics. MAA Notes **19**, 1–7 (1991)
3. Duval, R.: Semiosis y pensamiento humano. Registros semióticos y aprendizajes intelectuales. Universidad del Valle, Instituto de Educación Matemática, Colombia (1999)
4. Hitt, F.: Registro de representación semiótica y funcionamiento cognitivo del pensamiento. Investigaciones en Educación Matemática **II**, 173201 (1998)
5. Castro, E., Castro, E.: Representación y Modelización. Horsiri/ICE UAB, Barcelona (1997)
6. Zazkis, R., Dubinsky, E., Dautermann, J.: Coordinating visual and analytic strategies: a study of students understanding of the group. J. Res. Math. Educ. **27**(4), 435–457 (1996)
7. Bishop, A.: Review of research on visualization in mathematics education. Focus Learn. Probl. Math. **11**(1), 716 (1989)
8. Gilbert, E.: Models and Modeling in Science Education. Visualization in Science Education. Springer, Dordrecht (2005)
9. Haaser, N., LaSalle, J., Sullivan, J.: Análisis Matemático, vol. 1. Trillas, México (1970)
10. Kurmyshev, E., Sánchez-Yañez, R.: Fundamentos de Métodos Matemáticos para Física e Ingeniería. LIMUSA, México (2003)
11. Hauser, W.: Introducción a los Principios de Mecánica. UTEHA, México (1969)
12. Arfken, G., Weber, H.: Mathematical Methods for Physicists. Elsevier, Boston (2005)

Geometric and Computational Approach to Classical and Quantum Secret Sharing

Ryutaroh Matsumoto and Diego Ruano

Abstract Secret sharing is a cryptographic scheme to encode a secret to multiple shares being distributed to participants, so that only qualified (or authorized) sets of participants can reconstruct the original secret from their shares. It is also known that every linear ramp secret sharing can be expressed by a nested pair of linear codes $C_2 \subset C_1 \subset \mathbf{F}_q^n$. On the other hand, a nest code pair $C_2 \subset C_1 \subset \mathbf{F}_q^n$ can also give a quantum secret sharing. Since C_1 and C_2 are linear codes, it is natural to use algebraic geometry codes to construct C_1 and C_2. The purpose of this work is to find sufficient conditions for qualified or forbidden sets by using geometric properties of the set of points.

Keywords Algebraic geometry codes · Quantum secret sharing · Access structure

1 Introduction

Secret sharing (SS) [15] is a cryptographic scheme to encode a secret to multiple shares being distributed to participants, so that only qualified (or authorized) sets of participants can reconstruct the original secret from their shares. Traditionally both secret and shares were classical information (bits). Several authors [5, 7, 16] extended the traditional SS to a quantum one so that a quantum secret can be encoded to quantum shares.

When we require unqualified sets of participants to have zero information of the secret, the size of each share must be larger than or equal to that of the secret. By tolerating partial information leakage to unqualified sets, the size of shares can be smaller

R. Matsumoto (✉)
Department of Communications and Computer Engineering,
Tokyo Institute of Technology, Tokyo, Japan
e-mail: ryutaroh@rmatsumoto.org

D. Ruano
Department of Mathematical Sciences, Aalborg University, Aalborg, Denmark
e-mail: diego@math.aau.dk

I.S. Kotsireas and E. Martínez-Moro (eds.), *Applications of Computer Algebra*,
Springer Proceedings in Mathematics & Statistics 198,
DOI 10.1007/978-3-319-56932-1_18

than that of the secret. Such an SS is called a ramp (or non-perfect) SS [2, 13, 17]. The quantum ramp SS was proposed by Ogawa et al. [14]. In their construction [14] as well as its improvement [18], the size of shares can be L times smaller relative to quantum secret than its previous construction [5, 7, 16], where L is the number of qudits in quantum secret.

Classical secret sharing is said to be linear if a linear combination of shares corresponds to the linear combination of the original secrets [3]. It is also known that every linear ramp secret sharing can be expressed by a nested pair of linear codes $C_2 \subset C_1 \subset \mathbf{F}_q^n$. On the other hand, a nest code pair $C_2 \subset C_1 \subset \mathbf{F}_q^n$ can also give a quantum secret sharing as described in [10]. A share set is said to be forbidden if it has no information about the secret. It is natural to express conditions for qualified and forbidden sets in terms of $C_2 \subset C_1$, and the following is known:

Theorem 1 ([1, 9, 10]) *Let $J \subseteq \{1, \ldots, n\}$, and define $P_J : \mathbf{F}_q^n \to \mathbf{F}_q^{|J|}$, $(x_1, \ldots, x_n) \mapsto (x_j : j \in J)$. We consider classical and quantum secret sharing constructed from $C_2 \subset C_1$. J can be regarded as a share set, and J is qualified in the classical secret sharing if and only if*

$$\dim P_J(C_1)/P_J(C_2) = \dim C_1/C_2, \tag{1}$$

and J is forbidden in the classical secret sharing if and only if

$$P_J(C_1) = P_J(C_2). \tag{2}$$

Let $\overline{J} = \{1, \ldots, n\} \setminus J$. In the quantum secret sharing, J is qualified if and only if

$$\text{both} \begin{cases} \text{(1) is true,} \\ P_{\overline{J}}(C_1) = P_{\overline{J}}(C_2) \end{cases} \text{i.e.,} \begin{cases} J \text{ is classically qualified,} \\ \overline{J} \text{ is classically forbidden} \end{cases} \tag{3}$$

hold, and J is forbidden if and only if $\overline{J}$ is qualified.

Since C_1 and C_2 are linear codes, it is natural to use algebraic geometry codes to construct C_1 and C_2 [4]. Let F be an algebraic function field of one variable with genus $g(F)$, $P_1, \ldots, P_n$ its rational places, $G_1 \geq G_2$ divisors whose support contain none of $P_1, \ldots, P_n$. Define $C(P_1 + \cdots + P_n, G_1) = \{(f(P_1), \ldots, f(P_n)) \mid f \in \mathscr{L}(G_1)\}$. By the Riemann–Roch theorem, for $C_1 = C(P_1 + \cdots + P_n, G_1)$ and $C_2 = C(P_1 + \cdots + P_n, G_2)$, it is straightforward to see

Theorem 2 *Equation (1) holds if*

$$|J| \geq 1 + \deg G_1. \tag{4}$$

Equation (2) holds if

$$|J| \leq \deg G_2 - 2g(F) + 1. \tag{5}$$

Equation (3) holds if

$$|J| \geq \max\{1 + \deg G_1, n - (\deg G_2 - 2g(F) + 1)\}. \tag{6}$$

The purpose of this work is to find sufficient conditions **less** demanding than (4)–(6) by using geometric properties of the set of points $\{P_j \mid j \in J\}$.

2 Geometric and Computational Analysis of Qualified and Forbidden Sets

2.1 Computational Approach

Fix a rational place Q arbitrarily. When $C_1 = C(P_1 + \cdots + P_n,\ G_1)$ and $C_2 = C(P_1 + \cdots + P_n, G_2)$, (1) holds

$$\Leftrightarrow C\Big(\sum_{j\in J} P_j, G_1\Big)/C\Big(\sum_{j\in J} P_j, G_2\Big) \simeq C(P_1 + \cdots + P_n, G_1)/C(P_1 + \cdots + P_n, G_2)$$
$$\Leftrightarrow \ker(P_J) \cap C(P_1 + \cdots + P_n, G_1) = \ker(P_J) \cap C(P_1 + \cdots + P_n, G_2)$$
$$\Leftrightarrow C\Big(\sum_{j\notin J} P_J, G_1 - \sum_{j\in J} P_j\Big) = C\Big(\sum_{j\notin J} P_J, G_2 - \sum_{j\in J} P_j\Big)$$
$$\Leftrightarrow f_1 \in \mathcal{L}\Big(G_1 - \sum_{j\in J} P_j\Big) \Rightarrow \exists f_2 \in \mathcal{L}\Big(G_2 - \sum_{j\in J} P_j\Big) \text{ s.t. } f_1(P_j) = f_2(P_j) \forall j \notin J$$
$$\Leftrightarrow f_1 \in \mathcal{L}\Big(G_1 - \sum_{j\in J} P_j\Big) \Rightarrow \exists f_2 \in \mathcal{L}\Big(G_2 - \sum_{j\in J} P_j\Big) \text{ s.t. } f_1 - f_2 \in \mathcal{L}\Big(G_1 - \sum_{j\notin J} P_j\Big)$$
$$\Leftrightarrow \forall f_1 \in \mathcal{L}\Big(G_1 - \sum_{j\in J} P_j\Big), \exists f_2 \in \mathcal{L}\Big(G_2 - \sum_{j\in J} P_j\Big), \exists f_3 \in \mathcal{L}\Big(G_1 - \sum_{j=1}^{n} P_j\Big) \text{ s.t. } f_1 = f_2 + f_3$$
$$\Leftrightarrow \mathcal{L}\Big(G_1 - \sum_{j\in J} P_j\Big) \subseteq \mathcal{L}\Big(G_1 - \sum_{j=1}^{n} P_j\Big) + \mathcal{L}\Big(G_2 - \sum_{j\in J} P_j\Big)$$
$$\Leftrightarrow v_Q\Big(\mathcal{L}\Big(G_1 - \sum_{j\in J} P_j\Big)\Big) \subseteq v_Q\Big(\mathcal{L}\Big(G_1 - \sum_{j=1}^{n} P_j\Big) + \mathcal{L}\Big(G_2 - \sum_{j\in J} P_j\Big)\Big)$$
$$\Leftarrow v_Q\Big(\mathcal{L}\Big(G_1 - \sum_{j\in J} P_j\Big)\Big) \subseteq v_Q\Big(\mathcal{L}\Big(G_1 - \sum_{j=1}^{n} P_j\Big)\Big) \cup v_Q\Big(\mathcal{L}\Big(G_2 - \sum_{j\in J} P_j\Big)\Big). \tag{7}$$

For any rational place Q and any divisor G of F, $v_Q(\mathcal{L}(G))$ can be computed by **Gröbner bases** and the algorithm in [11], provided that the defining equation of F is in special position with respect to Q [6, 8, 12].

We turn our attention to (2). Equation (2) holds

$$\Leftrightarrow C\Big(\sum_{j\in J} P_j, G_1\Big) = C\Big(\sum_{j\in J} P_j, G_2\Big)$$

$$\Leftrightarrow \forall f_1 \in \mathcal{L}(G_1), \exists f_2 \in \mathcal{L}(G_2) \text{ s.t. } f_1 - f_2 \in \mathcal{L}\Big(-\sum_{j\in J} P_j + G_1\Big)$$

$$\Leftrightarrow \forall f_1 \in \mathcal{L}(G_1), \exists f_2 \in \mathcal{L}(G_2), \exists f_3 \in \mathcal{L}\Big(-\sum_{j\in J} P_j + G_1\Big) \text{ s.t. } f_1 = f_2 + f_3$$

$$\Leftrightarrow \mathcal{L}(G_1) = \mathcal{L}(G_2) + \mathcal{L}\Big(G_1 - \sum_{j\in J} P_j\Big)$$

$$\Leftrightarrow v_Q(\mathcal{L}(G_1)) = v_Q\Big(\mathcal{L}(G_2) + \mathcal{L}\Big(G_1 - \sum_{j\in J} P_j\Big)\Big)$$

$$\Leftarrow v_Q(\mathcal{L}(G_1)) = v_Q\Big(\mathcal{L}(G_2)\Big) \cup v_Q\Big(\mathcal{L}(G_1 - \sum_{j\in J} P_j\Big)\Big). \tag{8}$$

A similar sufficient condition for (3) can be deduced from (4) and (5).

2.2 *Explicit Sufficient Conditions*

We explicitly write sufficient conditions for (7) and (8), and examine if they are easier to hold than (4) and (5) for one point AG codes with $G_1 = m_1 Q$ and $G_2 = m_2 Q$. For any divisor G, let $H_Q(G) = -v_Q(\mathcal{L}(G + \infty Q) \setminus \{0\})$. Observe that $H_Q(0)$ is the Weierstrass semigroup at Q. The conductor of $H_Q(G)$ is defined as $\min\{i \in H_Q(G) \mid i \le j \in \mathbf{N} \Rightarrow j \in H_Q(G)\}$, which generalizes the conductor of the Weierstrass semigroup $H_Q(0)$.

Equation (7) holds if

$$v_Q\Big(\mathcal{L}\Big(m_1 Q - \sum_{j\in J} P_j\Big) \setminus \{0\}\Big) = \emptyset$$

$$\Leftrightarrow m_1 \le \min H_Q\Big(-\sum_{j\in J} P_j\Big) - 1 \tag{9}$$

We see that condition (9) is less demanding than (4), because $\min H_Q(-\sum_{j\in J} P_j) \ge |J|$.

Similarly, (8) holds if

$$m_2 \ge \text{the conductor of } H_Q\Big(-\sum_{j\in J} P_j\Big) - 1 \tag{10}$$

We also see that condition (10) is less demanding than (5), because the conductor of $H_Q(-\sum_{j\in J} P_j)$ is $\le 2g(F)$. We can also make a similar improvement over (6):

Condition (6) holds if

$$m_1 \le \min H_Q\Big(-\sum_{j\in J} P_j\Big) - 1 \text{ and } m_2 \ge \text{the conductor of } H_Q\Big(-\sum_{j\notin J} P_j\Big) - 1.$$

In particular, for elliptic function fields ($g(F) = 1$),

$$(9) \Leftrightarrow \begin{cases} m_1 + 1 \le |J| \text{ if } \exists f \in \mathscr{L}(\infty Q), (f)_0 = \sum_{j\in J} P_j, \\ m_1 \le |J| \quad \text{otherwise} \end{cases} \tag{11}$$

$$(10) \Leftrightarrow \begin{cases} |J| \le m_2 - 1 \text{ if } \exists f \in \mathscr{L}(\infty Q), (f)_0 = \sum_{j\in J} P_j, \\ |J| \le m_2 \quad \text{otherwise} \end{cases} \tag{12}$$

Acknowledgements The authors gratefully acknowledge the support from Japan Society for the Promotion of Science (Grant Nos. 23246071 and 26289116), from the Spanish MINECO/FEDER (Grant No. MTM2012-36917-C03-03 and No. MTM2015-65764-C3-2-P), the Danish Council for Independent Research (Grant No. DFF-4002-00367) and from the "Program for Promoting the Enhancement of Research Universities" at Tokyo Institute of Technology.

References

1. Bains, T.: Generalized Hamming weights and their applications to secret sharing schemes. Master's Thesis, University of Amsterdam (2008). Supervised by R. Cramer, G. van der Geer, and R. de Haan
2. Blakley, G.R., Meadows, C.: Security of ramp schemes. In: Advances in Cryptology—CRYPTO'84. Lecture Notes in Computer Science, vol. 196, pp. 242–269. Springer (1985). doi:10.1007/3-540-39568-7_20
3. Chen, H., Cramer, R., Goldwasser, S., de Haan, R., Vaikuntanathan, V.: Secure computation from random error correccting codes. In: Advances in Cryptology—EUROCRYPT 2007. Lecture Notes in Computer Science, vol. 4515, pp. 291–310. Springer (2007). doi:10.1007/978-3-540-72540-4_17
4. Chen, H., Cramer, R., de Haan, R., Cascudo Pueyo, I.: Strongly multiplicative ramp schemes from high degree rational points on curves. In: Smart, N. (ed.) Advances in Cryptology—EUROCRYPT 2008. Lecture Notes in Computer Science, vol. 4965, pp. 451–470. Springer (2008). doi:10.1007/978-3-540-78967-3_26
5. Cleve, R., Gottesman, D., Lo, H.K.: How to share a quantum secret. Phys. Rev. Lett. **83**(3), 648–651 (1999). doi:10.1103/PhysRevLett.83.648
6. Geil, O., Pellikaan, R.: On the structure of order domains. Finite Fields Appl. **8**, 369–396 (2002)
7. Gottesman, D.: Theory of quantum secret sharing. Phys. Rev. A **61**(4), 042311 (2000). doi:10.1103/PhysRevA.61.042311
8. Heegard, C., Little, J., Saints, K.: Systematic encoding via Gröbner bases for a class of algebraic-geometric Goppa codes. IEEE Trans. Inf. Theory **41**(6), 1752–1761 (1995). doi:10.1109/18.476247
9. Kurihara, J., Uyematsu, T., Matsumoto, R.: Secret sharing schemes based on linear codes can be precisely characterized by the relative generalized Hamming weight. IEICE Trans. Fundam. **E95-A**(11), 2067–2075 (2012). doi:10.1587/transfun.E95.A.2067
10. Matsumoto, R.: Coding theoretic construction of quantum ramp secret sharing, Version 4 or later (2014)

11. Matsumoto, R., Miura, S.: Finding a basis of a linear system with pairwise distinct discrete valuations on an algebraic curve. J. Symb. Comput. **30**(3), 309–323 (2000). doi:10.1006/jsco.2000.0372
12. Matsumoto, R., Miura, S.: On construction and generalization of algebraic geometry codes. In: Katsura, T. et al. (eds.) Proceedings of Algebraic Geometry, Number Theory, Coding Theory, and Cryptography, pp. 3–15. University of Tokyo, Japan (2000). http://www.rmatsumoto.org/repository/weight-construct.pdf
13. Ogata, W., Kurosawa, K., Tsujii, S.: Nonperfect secret sharing schemes. In: Advances in Cryptology—AUSCRYPT '92. Lecture Notes in Computer Science, vol. 718, pp. 56–66. Springer (1993). doi:10.1007/3-540-57220-1_52
14. Ogawa, T., Sasaki, A., Iwamoto, M., Yamamoto, H.: Quantum secret sharing schemes and reversibility of quantum operations. Phys. Rev. A **72**(3), 032318 (2005). doi:10.1103/PhysRevA.72.032318
15. Shamir, A.: How to share a secret. Commun. ACM **22**(11), 612–613 (1979). doi:10.1145/359168.359176
16. Smith, A.D.: Quantum secret sharing for general access structures. arXiv:quant-ph/0001087 (2000)
17. Yamamoto, H.: Secret sharing system using (k, l, n) threshold scheme. Electron. Commun. Jpn. (Part I: Communications) **69**(9), 46–54 (1986). doi:10.1002/ecja.4410690906 (The original Japanese version published in 1985)
18. Zhang, P., Matsumoto, R.: Quantum strongly secure ramp secret sharing. Quantum Inf. Process. **14**(2), 715–729 (2015). doi:10.1007/s11128-014-0863-2

Computing the Dixon Resultant with the Maple Package DR

Manfred Minimair

Abstract The Maple package DR provides functions for computing the Dixon resultant of a system of parametric multi-variate polynomials. The Dixon resultant constitutes a necessary condition for the polynomials to have a common root after specializing their parameters. The newest version 2 of the package DR includes the new heuristic pivot row detection of factors for extracting the Dixon resultant from the Dixon matrix. It is shown to be efficient on systems of benchmark polynomials, outperforming other heuristics for a majority of systems.

Keywords Resultant · Dixon matrix · Polynomial system

1 Introduction

The Maple package DR [19] implements algorithms for computing the Dixon resultant of a list $f_0, \ldots, f_n$ of parametric polynomials, with parameters p_j, variables x_k, and integer coefficients, in $\mathbb{Z}[p_1, \ldots, p_k, x_1, \ldots, x_n]$, where $\mathbb{Z}$ stands for the ring of integers [6, 11]. The Dixon resultant of the f_i's is a polynomial in the parameters p_i contained in the ring $\mathbb{Z}[p_1, \ldots, p_k]$ and vanishes whenever the f_i's have a common root in an appropriate space, explicated in Sect. 2. Because of these properties, applications [4, 5, 9, 10, 17, 18, 21, 22, 24] commonly use the Dixon resultant to eliminate variables from systems of equations [20].

The objective of this paper is to explain the usage and design of the Maple package DR [19] for computing the Dixon resultant and to introduce the new heuristic pivot row detection of factors (PRDF), applied in the package for efficiently extracting the Dixon resultant from a maximal-rank submatrix of the Dixon matrix. The Maple package DR has been created through merging Manfred Minimair's and Arthur Chtcherba's packages for Dixon Resultant computation in 2006, also including some

M. Minimair (✉)
Department of Mathematics and Computer Science,
Seton Hall University, South Orange, NJ 07079, USA
e-mail: manfred.minimair@shu.edu

I.S. Kotsireas and E. Martínez-Moro (eds.), *Applications of Computer Algebra*,
Springer Proceedings in Mathematics & Statistics 198,
DOI 10.1007/978-3-319-56932-1_19

code contributed by Hoon Hong. The package is being expanded and maintained by Manfred Minimair who has presented it at the conference applications of computer algebra (ACA) 2015, Kalamata, Greece, for the first time. The new updated release 2 with expanded features has been made available since ACA 2015. The package has been designed for ease of use, minimizing the need for user interventions, and efficiency.

As the number of bibliographic references on the Dixon resultant indicates, Dixon Resultant computation has been implemented by several authors. However, only the package DR [19] and a package [15] implemented in the computer algebra system Fermat seem to be currently available on the Internet.

Subsequently, Sect. 2 fixes some notation, defines Dixon resultant, and presents some properties of the Dixon resultant and Sect. 3 describes the usage, design, and implementation of the Maple package DR and introduces the new heuristic PRDF for Dixon resultant extraction. Furthermore, Sect. 4 addresses the efficiency of Dixon Resultant computation and experimentally demonstrates the efficiency of PRDF outperforming other heuristics and Sect. 5 concludes the paper with a discussion of the package and the computational results.

2 Preliminaries on the Dixon Resultant

This section introduces the basic notation used throughout this paper, defines Dixon resultant and gives some fundamental properties of the Dixon resultant. In this paper, we work with polynomials over the integers because the Maple package DR processes polynomial systems containing such polynomials as inputs. All definitions and statements in this section can naturally be generalized to polynomial systems over algebraically closed fields [1].

2.1 Definition of Dixon Resultant

Let $f_0, \ldots, f_n$ be a list of parametric polynomials $f_i \in \mathbb{Z}[p_1, \ldots, p_k, x_1, \ldots, x_n]$ with parameters p_j, variables x_k, and integer coefficients. Furthermore, let $\overline{\mathbb{Q}}$ denote the algebraic closure of the rational numbers $\mathbb{Q}$. Kapur et al. [11] define the Dixon resultant $\text{DRes}(f_0, f_1, \ldots, f_n)$ of the f_i's to be a necessary condition for the existence, in $\overline{\mathbb{Q}}^n$, of a common root of the f_i's obtained through Dixon's construction of [6]. They also give the RSC condition (rank submatrix construction) for which they show that Dixon's construction yields a necessary condition for the existence of a common root. Furthermore, Chtcherba's and Kapur's later Generalized RSC condition [3] implies that the appropriately generalized Dixon's construction yields a necessary condition for the existence of a common root if $\overline{\mathbb{Q}}^n$ is replaced with a different space. Consequently, the subsequently presented definition of Dixon Resultant generalizes [11], referring to any expression obtained through Dixon's construction as a Dixon Resultant, not necessarily an expression yielding a necessary condition. Moreover, Sect. 2.2 explicates according to [3] what space for the common roots may be chosen such that Dixon's construction yields a necessary condition for the existence of common roots in that space.

In the following, the Dixon resultant $\mathrm{DRes}(f_0, f_1, \ldots, f_n)$ of the f_i's with respect to the variables $x_1, \ldots, x_n$ is defined up to a rational factor in $\mathbb{Q}(p_1, \ldots, p_k)$. The definition constructively proceeds through computing the determinant of a maximal-rank submatrix of the Dixon matrix, which is determined from the Dixon polynomial of the f_i's.

2.1.1 Compute the Dixon Polynomial

Let $\overline{x}_1, \ldots, \overline{x}_n$ denote additional variable symbols distinct from the x_i's. Then the Dixon polynomial of the f_i's is

$$\frac{\det\begin{pmatrix} f_0(x_1, x_2, \ldots, x_n) & \ldots & f_n(x_1, x_2, \ldots, x_n) \\ f_0(\overline{x}_1, x_2, \ldots, x_n) & \ldots & f_n(\overline{x}_1, x_2, \ldots, x_n) \\ \vdots & & \vdots \\ f_0(\overline{x}_1, \overline{x}_2, \ldots, \overline{x}_n) & \ldots & f_n(\overline{x}_1, \overline{x}_2, \ldots, \overline{x}_n) \end{pmatrix}}{(x_1 - \overline{x}_1)(x_2 - \overline{x}_2) \ldots (x_n - \overline{x}_n)},$$

where the lth row in the matrix is obtained from the first row by replacing $x_1, \ldots, x_{l-1}$ with $\overline{x}_1, \ldots, \overline{x}_{l-1}$. (The matrix in the formula for the Dixon polynomial is usually called the Cancellation Matrix.)

Example 1 Let

$$\begin{aligned} f_0 &= -x_1^3 + 3x_1x_2^2 + 3x_1 - 3p_1, \\ f_1 &= -3x_1^2x_2 + x_2^3 - 3x_2 - 3p_2, \\ f_2 &= x_1^2 - x_2^2 - p_3, \end{aligned}$$

with variables x_1 and x_2 and parameters p_1, p_2 and p_3. Then the Dixon polynomial of f_0, f_1 and f_2 is

$$\frac{\begin{pmatrix} x_1^3 + 3x_1x_2^2 + 3x_1 - 3p_1 & -3x_1^2x_2 + x_2^3 - 3x_2 - 3p_2 & x_1^2 - x_2^2 - p_3 \\ \overline{x}_1^3 + 3\overline{x}_1x_2^2 + 3\overline{x}_1 - 3p_1 & -3\overline{x}_1^2x_2 + x_2^3 - 3x_2 - 3p_2 & \overline{x}_1^2 - x_2^2 - p_3 \\ \overline{x}_1^3 + 3\overline{x}_1\overline{x}_2^2 + 3\overline{x}_1 - 3p_1 & -3\overline{x}_1^2\overline{x}_2 + \overline{x}_2^3 - 3\overline{x}_2 - 3p_2 & \overline{x}_1^2 - \overline{x}_2^2 - p_3 \end{pmatrix}}{(x_1 - \overline{x}_1)(x_2 - \overline{x}_2)} =$$

$$(3x_1^2 - 3x_2^2 - 3p_3)\overline{x}_1^4 + (-x_1^2 + x_2^2 + p_3)\overline{x}_1^2\overline{x}_2^2 + (6x_1x_2^2 + (-3p_3 + 9)x_1 - 9p_1)\overline{x}_1^3 +$$
$$(2x_1^2x_2 - 6x_2^3 + (-8p_3 - 6)x_2 - 6p_2)\overline{x}_1^2\overline{x}_2 + (-2x_1x_2^2 + (p_3 - 3)x_1 + 3p_1)\overline{x}_1\overline{x}_2^2 +$$
$$(-x_1^2x_2^2 + 3x_2^4 + (-3p_3 + 3)x_1^2 + p_3x_2^2 - 9p_1x_1 - 6p_2x_2 + 6p_3)\overline{x}_1^2 +$$
$$(-8p_3x_1x_2 - 6p_2x_1 - 6p_1x_2)\overline{x}_1\overline{x}_2 + (x_1^2x_2^2 - 3x_2^4 + p_3x_1^2 + (-3p_3 - 3)x_2^2 + 3p_1x_1 - 3p_3)\overline{x}_2^2 +$$

$$((-8p_3+6)x_1x_2^2-6p_2x_1x_2-6p_1x_2^2+(-3p_3+9)x_1-9p_1)\overline{x}_1+((p_3+3)x_1^2x_2+$$
$$(-3p_3-9)x_2^3+3p_2x_1^2-6p_1x_1x_2-9p_2x_2^2+(-3p_3-9)x_2-9p_2)\overline{x}_2+$$
$$p_3x_1^2x_2^2-3p_3x_2^4-6p_1x_1x_2^2+3p_2x_1^2x_2-9p_2x_2^3-3p_3x_1^2+6p_3x_2^2-9p_1x_1-9p_2x_2+9p_3.$$

2.1.2 Set Up the Dixon Matrix

The Dixon matrix M is the matrix of coefficients of the Dixon Polynomial D uniquely defined[1] by the equality

$$D = rMc,$$

up to the orderings of the row vector r and column vector c, where r and c contain all the monomials in $\overline{x}_1, \ldots, \overline{x}_n$ and, respectively, $x_1, \ldots, x_n$ of the Dixon polynomial D [11].

Example 2 (cont. Example 1) The Dixon matrix M is implicitly given by

$$\begin{aligned} D &= rMc \\ r &= (1, \overline{x}_2, \overline{x}_1, \overline{x}_2{}^2, \overline{x}_1\overline{x}_2, \overline{x}_1{}^2, \overline{x}_1\overline{x}_2{}^2, \overline{x}_1{}^2\overline{x}_2, \overline{x}_1{}^3, \overline{x}_1{}^2\overline{x}_2{}^2, \overline{x}_1{}^4) \\ c &= (1, x_2, x_1, x_2{}^2, x_1x_2, x_1{}^2, x_2{}^3, x_1x_2{}^2, x_1{}^2x_2, x_2{}^4, x_1{}^2x_2{}^2)^{\mathrm{T}} \end{aligned}$$

and therefore M is

$$\begin{pmatrix} 9p_3 & -9p_2 & -9p_1 & 6p_3 & 0 & -3p_3 & -9p_2 & -6p_1 & 3p_2 & -3p_3 & p_3 \\ -9p_2 & -3p_3-9 & 0 & -9p_2 & -6p_1 & 3p_2 & -3p_3-9 & 0 & p_3+3 & 0 & 0 \\ -9p_1 & 0 & -3p_3+9 & -6p_1 & -6p_2 & 0 & 0 & -8p_3+6 & 0 & 0 & 0 \\ -3p_3 & 0 & 3p_1 & -3p_3-3 & 0 & p_3 & 0 & 0 & 0 & -3 & 1 \\ 0 & -6p_1 & -6p_2 & 0 & -8p_3 & 0 & 0 & 0 & 0 & 0 & 0 \\ 6p_3 & -6p_2 & -9p_1 & p_3 & 0 & -3p_3+3 & 0 & 0 & 0 & 3 & -1 \\ 3p_1 & 0 & p_3-3 & 0 & 0 & 0 & 0 & -2 & 0 & 0 & 0 \\ -6p_2 & -8p_3-6 & 0 & 0 & 0 & 0 & -6 & 0 & 2 & 0 & 0 \\ -9p_1 & 0 & -3p_3+9 & 0 & 0 & 0 & 0 & 6 & 0 & 0 & 0 \\ p_3 & 0 & 0 & 1 & 0 & -1 & 0 & 0 & 0 & 0 & 0 \\ -3p_3 & 0 & 0 & -3 & 0 & 3 & 0 & 0 & 0 & 0 & 0 \end{pmatrix}$$

with respect to r and c.

2.1.3 Choose a Maximal-Rank Submatrix of the Dixon Matrix

View the Dixon matrix M as a matrix with scalar entries in the field of rational expressions $\mathbb{Q}(p_1, \ldots, p_k)$. Then, let S be any (square) submatrix of the Dixon matrix M that has maximal rank.

---

[1] This definition differs from the notion of Dixon matrix used in [7].

Example 3 (cont. Example 2) The matrix

$$\begin{pmatrix} 9p_3 & -9p_2 & -9p_1 & 6p_3 & 0 & -3p_3 & -9p_2 & -6p_1 & -3p_3 \\ -9p_2 & -3p_3-9 & 0 & -9p_2 & -6p_1 & 3p_2 & -3p_3-9 & 0 & 0 \\ -9p_1 & 0 & -3p_3+9 & -6p_1 & -6p_2 & 0 & 0 & -8p_3+6 & 0 \\ -3p_3 & 0 & 3p_1 & -3p_3-3 & 0 & p_3 & 0 & 0 & -3 \\ 0 & -6p_1 & -6p_2 & 0 & -8p_3 & 0 & 0 & 0 & 0 \\ 6p_3 & -6p_2 & -9p_1 & p_3 & 0 & -3p_3+3 & 0 & 0 & 3 \\ 3p_1 & 0 & p_3-3 & 0 & 0 & 0 & 0 & -2 & 0 \\ -6p_2 & -8p_3-6 & 0 & 0 & 0 & 0 & -6 & 0 & 0 \\ p_3 & 0 & 0 & 1 & 0 & -1 & 0 & 0 & 0 \end{pmatrix}$$

is a maximal-rank submatrix S of the Dixon matrix M.

2.1.4 Dixon Resultant Extraction

In the final computational step, the Dixon resultant is extracted from the Dixon matrix by computing the determinant of the maximal-rank submatrix S of Sect. 2.1.3.

Definition 1 (*Dixon resultant*) The determinant of the maximal-rank submatrix S of the Dixon Matrix M is the Dixon resultant $\mathrm{DRes}(f_0, \ldots, f_n)$ of the f_i's with respect to the variables $x_1, \ldots, x_n$.

Since in Sect. 2.1.3 *any* maximal-rank submatrix can be chosen, the Dixon resultant is not uniquely defined.

Example 4 (cont. Example 3) The Dixon resultant $\mathrm{DRes}(f_0, \ldots, f_n)$, i.e., the determinant of the maximal-rank submatrix S, is

$$\begin{gathered} 36864\,p_3^9 - 248832\,p_1^2p_3^6 + 248832\,p_2^2p_3^6 - 699840\,p_1^4p_3^3 - \\ 3639168\,p_1^2p_2^2p_3^3 - 2239488\,p_1^2p_3^5 - 699840\,p_2^4p_3^3 - 2239488\,p_2^2p_3^5 - \\ 663552\,p_3^7 - 419904\,p_1^6 + 1259712\,p_1^4p_2^2 - 2519424\,p_1^4p_3^2 - 1259712\,p_1^2p_2^4 - \\ 3732480\,p_1^2p_3^4 + 419904\,p_2^6 + 2519424\,p_2^4p_3^2 + 3732480\,p_2^2p_3^4 + \\ 419904\,p_1^4p_3 - 839808\,p_1^2p_2^2p_3 + 2239488\,p_1^2p_3^3 + 419904\,p_2^4p_3 + \\ 2239488\,p_2^2p_3^3 + 2985984\,p_3^5. \end{gathered}$$

2.2 *Existence of Common Roots*

Knowing when the Dixon resultant yields a necessary condition for the existence of common roots of the f_i's is crucial for applications and therefore, subsequently, a result relating the vanishing of the Dixon resultant to the existence of common roots is presented. The generalized rank submatrix construction (RSC) Theorem 1 [3] shows some space in which the f_i's have a common root if the Dixon resultant vanishes. Before stating the theorem, some auxiliary notation is introduced.

Let the subset $U_0 \subseteq \overline{\mathbb{Q}}^n$ and $V_0 \subseteq \overline{\mathbb{Q}}^k$ be embedded in the projective set U and, respectively, V and be the domain of the tuples of the variables $(x_1, \ldots, x_n)$ and, respectively, parameters $(p_1, \ldots, p_k)$, over which solutions of the polynomial system $f_0 = \cdots = f_n = 0$ are thought. Furthermore, let m be the number of columns of the Dixon matrix M, that is, the length of the vector c from Sect. 2.1.2. Then [3], sets U_0, V_0, U and V and a map $\phi : U \to \mathbb{P}^{m-1}$ exist such that ϕ restricted to U_0 equals c. According to [3], V can be obtained as a subset of $\mathbb{P}^k$, the k-dimensional projective space over $\overline{\mathbb{Q}}$, whereas U may have to embedded in a projective space of a dimension possibly higher than n to allow the map ϕ.

Next, let q_v be obtained by evaluating q at $v \in V \subseteq \mathbb{P}^k$, as $p_1 = v_1, \ldots, p_k = v_k$, where q is a polynomial or vector of polynomials containing the parameters $p_1, \ldots, p_k$, and let M^{h} be obtained from the Dixon matrix M by homogenizing the matrix entries of each row to the same minimal total degree with respect to the homogenizing variable p_0, however, total degrees are allowed to differ across rows. Now, let q be restricted to ranging over tuples of m homogeneous polynomials, of equal total degrees, in $\mathbb{Z}[p_0, p_1, \ldots, p_k]$, that are the members of the kernel of the matrix M^{h} and let Z_M be the set of all non-vanishing q_v, for $v \in V$, viewed as members of $\mathbb{P}^{m-1}$. Furthermore, let $\phi^{-1}(Z_M)$ denote the inverse set of $\phi(U) \cap Z_M$ under ϕ and $\overline{U - \phi^{-1}(Z_M)}$ be the projective closure of the set $U - \phi^{-1}(Z_M)$.

Theorem 1 (Generalized RSC [3]) *The gcd of the determinants of the maximal-rank submatrices of the Dixon matrix M evaluated at v vanishes if the f_{i_v}'s have a common root in $\overline{U - \phi^{-1}(Z_M)}$, for $v \in V$.*

Consequently, the vanishing of $\mathrm{DRes}(f_0, \ldots, f_n)_v$, for $v \in V$, is a necessary condition for the existence of a common root of the f_i's in $\overline{U - \phi^{-1}(Z_M)}$.

Example 5 (cont. Example 4) The map ϕ is obtained from c by homogenizing the entries of c with respect to x_1 and x_2 individually with respect to two different homogenizing variables s and t. Then $U_0 = \overline{\mathbb{Q}}^2$ which can be embedded into a suitable $U \subseteq \mathbb{P}^4$ [3]. Additionally, the f_i's are homogenized like c to allow evaluating them at U. Furthermore, V_0 and V are chosen to be $\overline{\mathbb{Q}}^k$ and, respectively, $\mathbb{P}^k$. Then, the Dixon resultant $\mathrm{DRes}(f_0, f_1, f_2)$ vanishes if f_0, f_1 and f_2 have a common root in $\overline{U - \phi^{-1}(Z_M)}$. (Section 5.1 of [3] illustrates the computation of $\phi(U) \cap Z_M$ and roots in $\overline{U - \phi^{-1}(Z_M)}$ for similar polynomials f_i.)

3 Maple Package DR

The basic functions of the Maple Package DR are laid out in this section, followed by descriptions of additional features for benchmarking and selecting particular implementations of subalgorithms.[2] The new heuristic PRDF for extracting the Dixon resultant from a maximal-rank submatrix of the Dixon matrix is explained in Sect. 3.1.4.

3.1 Basic Functions and PRDF Heuristic

To simplify the notation, $f_1, f_2, \ldots$ and $x_1, x_2, \ldots$ also stand for the corresponding Maple symbols f1, f2, ..., x1, x2, ... and objects represented by DR in Maple. Similarly, $\overline{x}_1, \overline{x}_2, \ldots$, shown in the subsequent text, are represented as x1_, x2_, ... through appending an underscore to x1, x2, ... by DR. Additional symbols defined through Maple commands will be shown with upright font, such as the subsequently used name DP.

After loading the Maple package DR, Maple computes $\mathrm{DRes}(f_0, \ldots, f_n)$ with the command

$$\text{DR:-DixonResultant}([f_0, \ldots, f_n], [x_1, \ldots, x_n]).$$

This command implements the four-step scheme for computing the Dixon resultant [6, 11] defined in Sect. 2,

(1) compute the Dixon polynomial
  DP := DR:-DixonPolynomia($f_0, \ldots, f_n$)
(2) set up the Dixon matrix
  DM := DR:-DixonMatrix(DP)
(3) determine a maximal-rank submatrix of the Dixon matrix
  RS := DR:-RankSubMatrix(DM)
(4) compute the determinant of the maximal-rank submatrix
  DRES := DR:-DixonExtract(RS)

with the corresponding Maple commands provided by DR. The subsequent paragraphs elaborate on the above steps and the corresponding functions provided by DR.

3.1.1 Compute the Dixon Polynomial

The command

$$\text{DP} := \text{DR:-DixonPolynomial}(f_1, \ldots, f_n)$$

[2]The subsequent function specifications refer to subversion 2.1 of the package DR.

computes the Dixon polynomial and assigns the triple $D, \overline{v}, v$ consisting of the Dixon polynomial D and the lists $\overline{v} = [\overline{x}_1, \ldots, \overline{x}_n]$ and $v = [x_1, \ldots, x_n]$ to the Maple symbol DP. The package DR uses Maple's built-in standard functions for expanding and dividing the determinant in the formula for the Dixon Polynomial.

3.1.2 Set up the Dixon Matrix

The command

$$\text{DM} := \text{DR:-DixonMatrix(DP)}$$

computes the Dixon matrix from the Dixon polynomial DP. It assigns the tuple $M, r, \overline{v}, c, v$, defined in Sect. 2.1.2, to the Maple symbol DM.

3.1.3 Choose a Maximal-Rank Submatrix of the Dixon Matrix

A maximal-rank (square) submatrix of the Dixon matrix M is obtained by specializing the parameters p_j of M with random integers and determining the rows and columns of a maximal-rank submatrix of the specialized M with PLU-decomposition.[3] It is expected that the same rows and columns constitute a maximal-rank submatrix of the original M. This method for determining the maximal-rank submatrix is a randomized scheme of Las Vegas type. That is, it may fail in some rare cases. The package DR is able to detect and report such rare events of failure. The command

$$\text{RS} := \text{DR:-RankSubMatrix(DM)}$$

determines the maximal-rank submatrix S of the Dixon matrix M, which is assigned to the Maple symbol RS. The failure case is detected by the command DR:-DixonExtract presented in Sect. 3.1.4.

3.1.4 Dixon Resultant Extraction

This final step in the computation yields the Dixon resultant of the f_i's by computing the determinant of the maximal-rank submatrix S of the Dixon matrix M. The package DR provides a specialized version of Gaussian elimination for computing this determinant. This version has been optimized to speed-up the computation by taking advantage of factors commonly arising during Gaussian elimination on Dixon

[3]The Fermat implementation [15] computes the PLU-decomposition modulo a random prime number rather than over the rational numbers. The package DR does not use a random prime to reduce the failure probability of this step. Furthermore, the speed-up gained in Maple through computing modulo a random prime has been insignificant as compared to the dominating computational step Sect. 3.1.4 for benchmark systems of Sect. 3.2 and therefore is ignored up by the current version of DR.

Matrices and to provide a representation of the resultant that is compact by computing the determinant in factored form. It usually outperforms the built-in determinant function of Maple which relies on versions of fraction-free Gaussian elimination that ignore the special factors arising from the Dixon matrix. (For remarks on the efficiency of this operation see Sect. 4.)

Computation of the Dixon resultant proceeds by Gaussian elimination with row-pivoting on the matrix S such that the determinant of S is the product of all the pivots determined during elimination. The Gaussian elimination and determinant computation uses the following operations. At the lth elimination step, after row pivoting, let S_{ll} be the pivot, then, for all $u > l$ and $v > l$, let $t = \text{normal}(S_{ul}/S_{ll})$ and update S_{uv} with $\text{normal}(S_{uv} - t \cdot S_{lv})$, where normal is Maple's normalizing function that cancels common factors in the numerators and denominators of rational expressions. When running the computation, the determinant of S is obtained incrementally from the pivots S_{ll}. Initially the partial determinant is set to 1 and after each elimination step the pivot S_{ll} is factored and multiplied with the partial determinant.

Lewis [13] observed that for certain polynomial systems of f_i during Gaussian elimination the row gcds, $\text{gcd}(S_{ul}, S_{u\,(l+1)}, \ldots)$ for $u \geq l$, and the column gcds, $\text{gcd}(S_{lv}, S_{(l+1)\,v}, \ldots)$ for $v \geq l$, are often non-trivial and dividing out these gcds before performing the Gaussian elimination step sometimes dramatically speeds up Dixon Resultant computation. He named this heuristic EDF (Early Detection of Factors) and provided an implementation in the computer algebra system Fermat [15]. This strategy does not perform as well when implemented in Maple because the implementation of polynomial gcds seems to be faster in Fermat [14].

Therefore, I propose a variation of the EDF heuristic, called PRDF (Pivot Row Detection of Factors) which requires fewer gcd computations than EDF and is implemented by the Maple package DR. The timing results of Sect. 4 indicate that PRDF is more efficient than EDF for a large number of polynomial systems.

Definition 2 (*PRDF* (*pivot row detection of factors*)) Let S_{ll} be the pivot, at the lth elimination step after row pivoting. Before computing $\text{normal}(S_{uv} - t \cdot S_{lv})$, with $t = \text{normal}(S_{ul}/S_{ll})$, in Gaussian elimination, divide out the gcd of the pivot row, $\text{gcd}(S_{ll}, S_{l\,(l+1)}, \ldots)$, from the entries u_{lv}, for $v \geq l$, of the pivot row and update the partial determinant with this gcd.

Accordingly, the command

$$\text{DRES} := \text{DR:-DixonExtract(RS)}$$

computes the determinant of the maximal-rank submatrix S using PRDF and assigns the pair e, R to the Maple symbol DRES. The string e is an error message if the maximal-rank submatrix RS does not have the expected full rank, and otherwise e is empty. Furthermore, R is the Dixon resultant, if e is empty.

3.2 Additional Features

The package DR provides various additional features whose documentation can be accessed through the Maple command Describe(DR). Subsequently, a few main items are surveyed.

Timing and memory usage The Dixon resultant command

$$\text{DR:-DixonResultant}([f_0, \ldots, f_n], [x_1, \ldots, x_n], \text{t})$$

allows an optional output parameter, here denoted by t. After execution, the unassigned symbol t is assigned a table with timing and memory usage information of all the steps carried out when computing the Dixon Resultant.

Detection of factors during Gaussian elimination: The submodule DR:-DF implements heuristics for detection of factors during Gaussian elimination (see Sect. 3.1.4) and the submodule DR:-DRes provides front-end functions employing these heuristics for computing the Dixon Resultant. The matrix function DR:-DF:-ColRow implements EDF [13] and DR:-DF:-PivotRow implements PRDF of Sect. 3.1.4. While PRDF is run by the default command DR:-DixonResultant, EDF is used by the Dixon resultant command DR:-DRes:-RankColRow. The DF submodule contains additional heuristics such as extracting factors from all rows and no columns, all columns and no rows, and all columns and the pivot row. During computational test runs, it was found that PRDF usually surpasses all other tried heuristics and therefore this paper will not address these heuristics any further. (See Sect. 4 for comparisons of PRDF and EDF.)

Ignore factor detection: The command DR:-DRes:-MaxMinor computes the Dixon resultant with Gaussian elimination without using any heuristics for factor detection. This is the default in versions of the package DR prior to version 2.

Random polynomials: The submodule DR:-GenPoly contains the functions RandParamTotalDeg and RandParamMultiDeg for creating random parametric multivariate total- and, respectively, multi-degree polynomials.

Sample polynomials: The submodule DR:-Samples provides sample polynomial systems used for benchmarking in Sect. 4. These samples include random systems as well as systems found in the literature on resultants, documented in the source code of the package DR. This repository is accessed via the functions DR:-Samples:-Available and DR:-Samples:-Get. The function Available returns a list of the names of all available polynomial systems and the command Get(s), where s is a string, the name of a polynomial system, returns the corresponding polynomial system, including its degree, variable, and parameter lists.

4 Efficiency of Dixon Resultant Extraction

The dominating step in computing Dixon resultants is Sect. 3.1.4, computing the determinant of a maximal-rank submatrix of the Dixon Matrix. The subsequent timing results illustrate the efficiency of the package DR employing the PRDF heuristic on a set of benchmark polynomial systems, included in the submodule Samples of DR. Some of these systems are sourced from research literature and others are randomly generated as it is indicated in the column Source of Fig. 1. The first column in the table in Fig. 1 assigns a label to each system which is used in the subsequent figures and the second column lists the name of the system used by DR. The following columns list the number of variables and parameters of the systems and the number of rows and columns and the rank of the systems' Dixon Matrices. The rows of the table have been sorted according to the number of rows of the Dixon matrices.

Figure 2 shows the running times for Dixon resultant extraction using Maples' built-in determinant function LinearAlgebra:-Determinant, Gaussian elimination

System	DR Name	Var.	Param.	Rows	Col.	Rank	Source
S01	sparse_2v2_2p3	2	3	5	5	5	random
S02	Bricard	2	16	8	8	8	[14]
S03	sparse_3v2_2p2	2	2	9	9	9	random
S04	cubic	2	3	11	11	11	[10]
S05	Enneper	2	3	11	11	9	[10]
S06	sec11_2_2	2	3	12	12	5	[10]
S07	sphere	2	3	11	12	10	[10]
S08	sparse_3v2_1p2	2	2	12	12	11	random
S09	sparse_3v2_1p3	2	3	12	12	11	random
S10	sparse_3v2_2p3	2	3	12	12	11	random
S11	sparse_4v2_1p3	2	3	17	17	17	random
S12	bicubic	2	0	18	18	18	[10]
S13	sparse_4v2_1p2	2	2	21	21	19	random
S14	sparse_4v2_2p2	2	2	21	22	20	random
S15	sparse_4v2_2p3	2	3	22	21	20	random
S16	ParamElim	3	2	25	16	16	[15]
S17	sparse_5v2_1p2	2	2	25	25	24	random
S18	sparse_5v2_1p3	2	3	27	26	26	random
S19	sparse_5v2_2p2	2	2	34	31	27	random
S20	sparse_6v2_2p2	2	2	35	35	33	random
S21	sparse_6v2_1p2	2	2	44	44	40	random
S22	KK5	5	1	81	81	81	[16]
S23	Cyclic 6-Root	5	1	86	86	78	[16]
S24	SB L1 M5 K1	8	0	111	136	76	[16]
S25	KK6	6	1	193	193	193	[18], [16]
S26	Cyclic 7-Root	6	1	348	349	314	[16]
S27	KK7	7	1	449	449	449	[16]

Fig. 1 Benchmark systems

System	Maple Determinant	DR Prior Versions Gaussian Elimination	DR Version 2 PRDF	Fermat EDF
S08	0.016	0.015	0.014	0.03
S09	0.338	0.547	0.514	1.55
S10	11.061	11.086	11.516	NA
S11	24.087	14.947	14.335	NA
S12	0.0211	0.040883	0.0401	0.01
S13	0.252	0.139	0.137	0.46
S14	1.275	0.475	0.428	12
S15	5000	37.373	65.144	NA
S16	0.016667	0.009117	0.008333	0.00312
S17	1.02	0.769	0.753	13.36
S18	159.285	57.723	83.318	NA
S19	10.129	2.636	2.522	97.9
S20	22.656	8.189	6.839	753.8
S21	28.743	19.623	19.2	NA
S22	0.131767	0.025	0.023183	0.05
S23	0.60755	0.089333	0.081767	0.1
S24	3.139317	1.25	1.2547	3.31
S25	4.297133	0.411467	0.399483	2.05
S26	NA	17.486467	15.674217	NA
S27	153.448183	12.796883	11.147667	269.48

Fig. 2 Running times of Dixon resultant extraction in minutes

implemented by DR in Maple, DR's PRDF run in Maple and EDF run in Fermat [15]. The computations were carried out on a computer with Intel Core i7-3770 CPU at 3.40 GHz with 12 GB RAM running Windows 8.1, Maple 2015 and Fermat 3.9.999. The benchmark systems from Fig. 1 have been chosen to allow computing Dixon resultants within 3 h, however, there were exceptions when some computations did not finish within several hours which is indicated with NA in Fig. 2. Since computing the Dixon resultant of each system among S01-S07 took much less than one second, results from these computations are excluded from the figures.

Figures 3, 4, and 5 compare the running times of A=DR's Gaussian elimination, DR's PRDF and EDF in Fermat to the respective B=Maple's built-in determinant function, DR's Gaussian elimination and DR's PRDF. In each figure the height of the bars is the percentage by which A is faster than B, that is $100 \times (b - a)/a$, where a and b is the running time of A and, respectively, B shown in Fig. 2.

Figure 3 shows that DR's Gaussian elimination, the method employed by default in versions of DR prior to the current version 2, is often much faster than Maple's built-in determinant function when applied to submatrices of the Dixon matrix. The speed-up of Gaussian elimination ranges from 6.6 to 1099.1% for S08 and, respectively, S27. Furthermore, Maple's determinant function was unable to finish for S15 and S26 within 3 h, whereas Gaussian elimination took 37.3 and, respectively, 17.4 min.

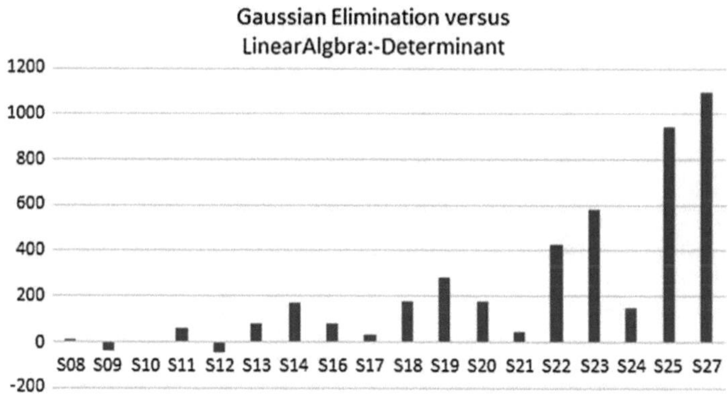

Fig. 3 Percentage by which DR's Gaussian elimination is faster than Maple's built-in determinant function

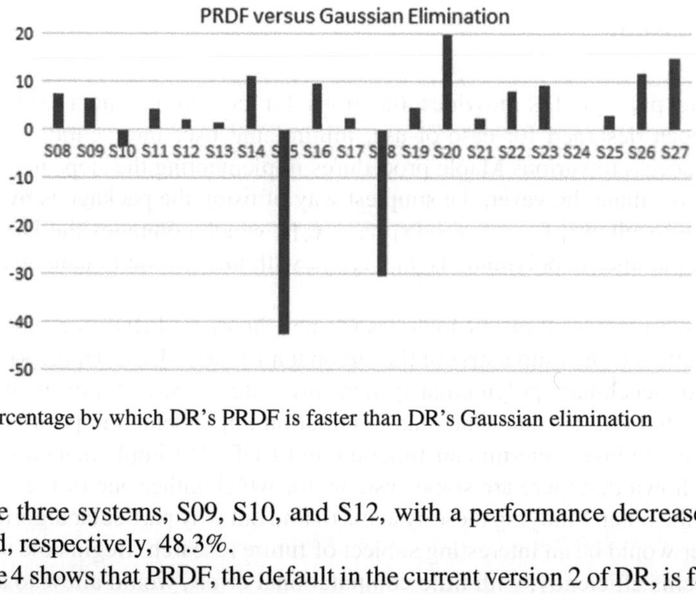

Fig. 4 Percentage by which DR's PRDF is faster than DR's Gaussian elimination

There are three systems, S09, S10, and S12, with a performance decrease of 38.2, 0.2% and, respectively, 48.3%.

Figure 4 shows that PRDF, the default in the current version 2 of DR, is faster than Gaussian elimination for most systems by up to 19.7%. There are two systems, S10 and S24 where PRDF is slightly slower by 3.7% and, respectively, 0.3%. Furthermore, there are two outlier systems, S15 and S18, where PRDF is slower by 42% and, respectively, 30%.

In Fig. 5, the bars below the x-axis show that PRDF is faster than EDF in all but two cases. The speed-up of PRDF ranges from 18.2 to 99% for S23 and, respectively, S20. Furthermore, EDF was unable to finish within 3 h for S10, S11, S15, S18, S21, and S26, whereas PRDF took 11.5, 14.3, 65.1, 83.3, 19.2 and, respectively, 15.6 min. There are only two systems, S12 and S16, where EDF is faster, by 170% and, respectively, 167.3%.

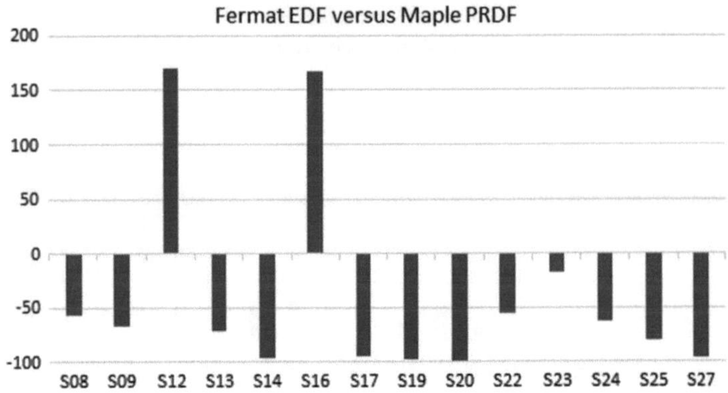

Fig. 5 Percentage by which EDF in Fermat is faster than DR's PRDF

5 Discussion

The Maple package DR provides functions for computing the Dixon resultant and has been designed for ease-of-use, minimizing user intervention. The package gives access to various Maple procedures implementing the steps in computing the Dixon resultant, however, the simplest way of using the package is by invoking DR:-DixonResultant([$f_0, \ldots, f_n$], [$x_1, \ldots, x_n$]), which computes the Dixon resultant of the parametric polynomials $f_0, \ldots, f_n$ with integer coefficients with respect to the variables $x_1, \ldots, x_n$.

The newest release 2 of DR includes the new heuristic PRDF for extracting the Dixon resultant from a submatrix of the Dixon matrix Sect. 3.1.4. Timings of computations with benchmark polynomial systems presented in Sect. 4 indicate that PRDF is expected to be faster than Gaussian elimination implemented in prior versions of DR, Maple's built-in determinant function and EDF [15] implemented in Fermat. It is also shown that there are some systems for which either one of these methods surpasses the other. Studying these systems to find out why particular algorithms perform faster would be an interesting subject of future research. Begin able to classify these systems efficiently, with little computational effort, would allow to automatically choose the most efficient algorithm and therefore help further speed-up Dixon resultant computation.

Algorithms for efficiently constructing the Dixon matrix have been proposed [2, 25]. Since Dixon resultant extraction, Sect. 3.1.4, is the computationally dominating step in computing the Dixon resultant, these algorithms have not yet been implemented in DR. However, future versions of DR may incorporate these algorithms. Furthermore, future work on DR may include versions for other systems and languages, such as Sage [23], and interfacing with the benchmarking framework SDEval [8].

References

1. Bus, L., Elkadi, M., Mourrain, B.: Using projection operators in computer aided geometric design. Comtempor. Math. **334**, 321–342 (2003)
2. Chionh, E.-W., Zhang, M., Goldman, R.N.: Fast computation of the Bezout and Dixon resultant matrices. J. Symb. Comput. **33**(1), 13–29 (2002)
3. Chtcherba, A.D., Kapur, D.: Conditions for determinantal formula for resultant of a polynomial system. In: Proceedings of the 2006 International Symposium on Symbolic and Algebraic Computation, pp. 55–62. ACM (2006)
4. Chtcherba, A., Kapur, D., Minimair, M.: Cayley–Dixon projection operator for multi-univariate composed polynomials. J. Symb. Comput. **44**(8), 972–999 (2009)
5. Coutsias, E.A., Seok, C., Jacobson, M.P., Dill, K.A.: A kinematic view of loop closure. J. Comput. Chem. **25**, 510–528 (2004)
6. Dixon, A.L.: The eliminant of three quantics in two independent variables. Proc. Lond. Math. Soc. **7**(49–69), 473–492 (1908)
7. Emiris, I.Z., Mourrain, B.: Matrices in elimination theory. J. Symb. Comput. **28**(12), 3–44 (1999)
8. Heinle, A., Levandovskyy, V.: The SDEval benchmarking toolkit. ACM Commun. Comput. Algebr. **49**(1), 1–9 (2015)
9. Hu, H.Y., Wang, Z.: Dynamics of Controlled Mechanical Systems with Delayed Feedback. Springer, New York (2002)
10. Kapur, D., Minimair, M.: Multivariate resultants in Bernstein basis. In: Proceedings of the 7th International Conference on Automated Deduction in Geometry. Lecture Notes in Computer Science, vol. 6301, pp. 60–85. Springer, Shanghai (2011)
11. Kapur, D., Saxena, T. Yang, L.: Algebraic and geometric reasoning using dixon resultants. In: Proceedings of the International Symposium on Symbolic and Algebraic Computation (ISSAC '94), pp. 99–107. ACM, New York (1994)
12. Lewis, R.H.: Comparing acceleration techniques for the Dixon and Macaulay resultants. Math. Comput. Simul. (2008)
13. Lewis R.H.: Heuristics to accelerate the Dixon resultant. Math. Comput. Simul. **77**(4), 400–407 (2008)
14. Lewis, R.H.: Comparison of GCD in several systems. https://home.bway.net/lewis/fermat/gcdcomp
15. Lewis, R.H.: Dixon resultant computation in the Fermat system. http://home.bway.net/lewis/. Accessed 29 Sept 2015
16. Lewis, R.H.: Parametric polynomial system motivated by Bricard. http://home.bway.net/lewis/dixon/. Accessed 29 Sept 2015
17. Lewis, R.H., Stiller, P.: Solving the recognition problem for six lines using the Dixon resultant. Math. Comput. Simul. **49**(3), 205–219 (1999)
18. Little, J.B.: Solving the SelesnickBurrus filter design equations using computational algebra and algebraic geometry. Adv. Appl. Math. **31**(2), 463–500 (2003)
19. Minimair, M.: DR: Maple package for Dixon resultant computation (2015). http://minimair.org/dr/
20. Nakos, G., Williams, R.M.: Elimination with the Dixon resultant. Math. Educ. Res. **6**(3), 11–21 (1997)
21. Paláncz, B.: Application of Dixon resultant to satellite trajectory control by pole placement. J. Symb. Comput. **50**, 79–99 (2013)
22. Paláncz, B., Zaletnyik, P., Awange, J.L., Grafarend, E.W.: Dixon resultants solution of systems of geodetic polynomial equations. J. Geodesy **82**(8), 505–511 (2007)
23. Stein, W.: Sage—open-source mathematical software system (2008)
24. Sun, W.K.: Solving 3–6 parallel robots by Dixon resultant. Appl. Mech. Mater. **235**, 158–163 (2012)
25. Zhao, S., Fu, H.: An extended fast algorithm for constructing the Dixon resultant matrix. Sci. China Ser. A Math. **48**(1), 131–143 (2005)

Collaborative Computer Algebra

Manfred Minimair

Abstract A definition of Collaborative Computer Algebra as a field of research is proposed. The significance of this field is examined and theoretical frameworks that have the potential to form its foundation are surveyed. Furthermore, the state of the art and open questions of Collaborative Computer Algebra are discussed.

Keywords Computer algebra · Symbolic computation · Collaborative computing · Social computing

1 Introduction

The paper's objective is to propose a definition of Collaborative Computer Algebra as a field of research, highlight the significance of this field, describe its foundations and discuss its state of the art and potential open questions. Manfred Minimair has informally introduced the concept of Collaborative Computer Algebra, without providing a definition, at the conference Applications of Computer Algebra (ACA) 2014 [53] and additionally elaborated at ACA 2015 [54]. Accordingly, this paper serves to motivate, specify, and elaborate this concept.

Complex human endeavors are often beyond the reach of any single individual and therefore require collaboration. The fields of engineering, natural sciences and mathematics are ripe with examples, including designing some complex machines such as space satellites or industrial robots, conducting scientific experiments and analyzing data, and solving challenging mathematical problems. Collaborative Engineering is a distinct field of academic inquiry studying collaboration processes and designing systems to facilitate cooperative work in engineering [34, 43, 51, 55]. In recent years, scientific workflows of collaborative experimentation, data collec-

Partially funded by Seton Hall University's Office of the Provost.

M. Minimair (✉)
Department of Mathematics and Computer Science, Seton Hall University,
South Orange, NJ 07079, USA
e-mail: Manfred.Minimair@shu.edu

I.S. Kotsireas and E. Martínez-Moro (eds.), *Applications of Computer Algebra*,
Springer Proceedings in Mathematics & Statistics 198,
DOI 10.1007/978-3-319-56932-1_20

tion, computation, and data analysis have been intensely studied and supporting software has been designed [9, 15, 24]. In mathematics, blogs and online wikis support large-scale collaborations, such as the polymath blog [26] and, according to [1], another blog focused on complexity theory [16]. Furthermore, the PlanetMath community has prepared several thousand mathematical encyclopedia articles [63], an accomplishment beyond an individual's capabilities.

This paper suggests that collaboration is as central to the practice of computer algebra as for the cited areas of engineering, natural sciences, and mathematics, however, independently and not only because computer algebra software is applied in these areas. To substantiate this suggestion, Sect. 2 reviews domains of collaboration in computer algebra and Sect. 3 surveys software tools supporting collaboration. Subsequently, Sect. 4 proposes a formal definition of Collaborative Computer Algebra and discusses its significance. Motivated by the definition, Sect. 5 reviews theoretical frameworks from human-centered computing that have the potential for guiding software design to facilitate collaboration in computer algebra. Section 6 concludes the paper with discussing the state of the art and potential open questions in Collaborative Computer Algebra and addressing the wider context of computation in mathematics.

2 Domains of Collaboration

In the introduction, it has been pointed out that collaboration is quite common in areas of engineering, natural sciences, and mathematics. Similarly, computer algebra work is often cooperative. Accordingly, the following sections examine computer algebra research and applications, system development, as well as eduction, and give examples of collaboration.

2.1 Research and Applications

Several journals and conferences cover computer algebra research and applications, such as Journal of Symbolic Computation, International Symposium on Symbolic and Algebraic Computation and Applications of Computer Algebra. As an example, examine the yearly conference Applications of Computer Algebra which has been organized since 1995 [74]. To illustrate the quantitative development of working groups over the course of several recent years, information about the submissions to ACA conferences from 2001 to 2015 have been collected, whenever the conference pages were available and listed the sessions and its submissions.

The findings can be summarized as:

- collaboration is common;
- collaboration seems to grow in prevalence;
- working groups are relatively small;
- and working groups seem to be growing.

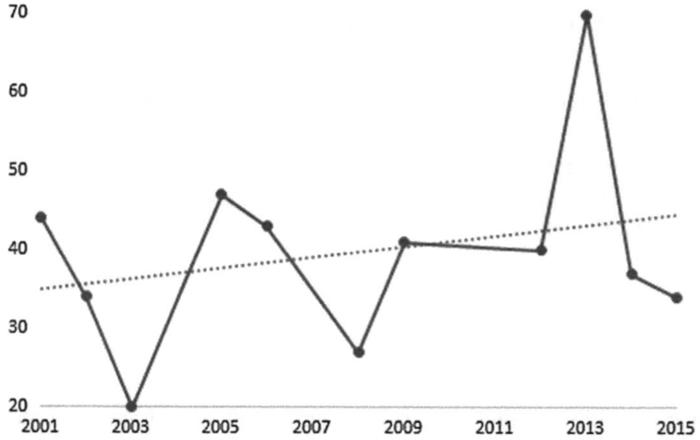

Fig. 1 Percentage of group submissions to ACA

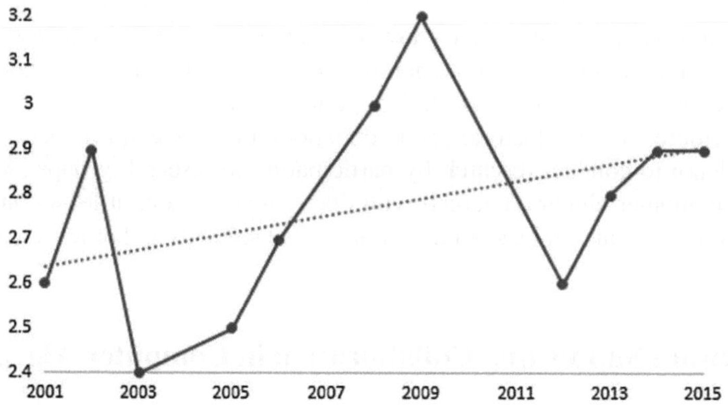

Fig. 2 Average group size of ACA group submissions

To arrive at these conclusions, the number of submissions with published titles and authors, for each conference session, and the number of authors, for each submission, have been counted. Aggregates of the counts are shown in Figs. 1 and 2. Figure 1 shows the percentage of group submissions to ACA, that is, the percentage of submissions with more than one author, and, respectively, Fig. 2 shows the average sizes of groups. Figure 1 indicates that collaboration is common. The percentages of contributions to ACA that are collaborative range between 20 and 70%. Furthermore, it seems that the percentages have an upward trend. Figure 2 shows that average group sizes are in the range 2–3, which is relatively small. Note that the maximal group size has been found to be 11 in 2009. Additionally, in average of the sizes of groups seem to be growing over the past years since 2001.

2.2 Computer Algebra System Development

Computer algebra systems and libraries are often developed by working groups organized as open source software development projects. These are multi-year efforts involving numerous participants which frequently change. The number of contributors can range up to several hundreds, as the web sites of some projects, such as [14, 19, 78, 81], report.

2.3 Education

In secondary and higher education, students are often encouraged to collaborate on solving mathematical problems to prepare them for industry and academia where team work is common. Furthermore, many students naturally tend to collaborate forming study groups to complete assignments or to prepare for exams. During collaboration, students may jointly use a computer algebra system installed on a dedicated computer or rely on collaborative learning environments available online that include access to a computer algebra system [72].

Furthermore, at the doctoral-level, collaboration is the norm because Ph.D. students learn to conduct research by participating in research groups. A research group in computer algebra, charged with doctoral education, at least consists of a Ph.D. adviser and one Ph.D. student, if not several students and other research staff.

3 Software Supporting Collaboration in Computer Algebra

Collaboration in Computer Algebra usually relies on software to exchange ideas, mathematical documents and code implementing algorithms and to coordinate work. This section provides a list of software products that are used by computer algebra teams to support this exchange and coordination. The list is intended to comprehensively cover the major needs that arise through collaboration, however, not necessarily contains all currently available software.

At the software infrastructure level, protocols and mechanisms for communication and interoperability provide foundations for software and its users to cooperate. For example, the Symbolic Computation Software Composability Protocol (SCSCP) [50] allows to connect different mathematical software. For another example, the SymbolicData project provides databases with mathematical information and supports benchmarking by employing Python to connect with different computer algebra systems [35].

Collaborators need to share ideas, findings, and implementations of algorithms. Worksheets, notebooks or active mathematical documents, user interfaces for computer algebra systems, such as Maple [84], Mathematica [86], and Sage [73], which allow running computations and entering text annotations, support such exchange. These electronic documents may be shared by e-mail or even edited collaboratively [73].

Furthermore, computer algebra work may require interactively executing mathematical computations. Consequently, team members may jointly use command shells for interactive command execution which are often included in computer algebra systems. A group may assemble around a computer display, and a dedicated group member may use a keyboard to enter commands that conform with the decisions made by the group.

Software development is part of computer algebra, as new algorithms are being implemented or systems are being developed. Much computer algebra software is being developed in the framework of open source projects, such as the projects for GAP [77], Mathemagix [82], and Singular [13]. Such work arrangements typically include bug trackers, wikis, blogs, and messaging systems which help the developers to interact and design software collaboratively.

General-purpose features of groupware, including virtual whiteboard, online chat, e-mail messages, wikis and blogs, also support computer algebra work. VirtualMathTeams, an online environment with virtual whiteboard and online chat for collaborative mathematical problem solving provide access to a computer algebra system [72]. Blogs are being used to discuss mathematical and computational problems [26, 52].

4 Proposed Definition and Significance

Work in computer algebra uniquely involves mathematics and computer science by being concerned "with the development, implementation, and application of algorithms that manipulate and analyze mathematical expressions" [11]. (See also for an equivalent statement and proposed variants in the German-language article [28]). That is, the goal of activities in computer algebra is to compute representations of mathematical objects and to extract information from them [28]. Consequently, collaboration in computer algebra has the same unique focus. Due to this particular aim, collaborators rely on a diverse combination of particular tools which have been surveyed in Sect. 3, besides basic objects such as paper and pencil. Humans and their tools form systems as they engage in activities (solving problems of computer algebra, conducting research, running computations, etc.) Communication in such systems may occur through personal and direct physical interactions, however, is often facilitated by computer networks. Emphasizing tools based on computing technology connected through computer networks, such systems are called cyber-human systems (CHS) [58].

CHS, in general, are commonly studied in the field of Human–Computer Interaction, which is represented by well-known professional societies such as ACM SIGCHI [71] and IEEE Computer Society [40]. CHS potentially amplify the capabilities of individuals by allowing people to work together and by allowing humans to offload challenging cognitive tasks onto software. These benefits motivate individuals to join CHS because they enable them to transcend their individual capabilities. This feature also motivates CHS as a subject of research.

Hence, I propose the following definition of Collaborative Computer Algebra.

Definition: Collaborative Computer Algebra is concerned with designing, evaluating, and applying CHS for collaboratively conducting computer algebra. Such CHS are called Collaborative Computer Algebra Systems.

Therefore, Collaborative Computer Algebra can be viewed as descending from "symbolic and algebraic manipulation" and from "collaborative and social computing." The latter is a sub-category of "Human-centered computing" in the 2012 ACM Computing Classification System [76] and includes theory, concepts, and paradigms, such as social networks and computer supported cooperative work, as well as design, evaluation methods, systems and tools.

5 Human-Centered Computing

Various theoretical frameworks have been developed to study CHS, such as in [2, 6, 21, 29, 61, 62, 65, 85], considered in areas of human-centered computing, including human–computer interaction. Accordingly, these frameworks may also be applicable to studying and designing Collaborative Computer Algebra Systems, and they include Activity Theory [22, 49, 83], Actor-Network Theory [46, 56], Distributed Cognition [37], and, recently, Connectivism [18, 70]. (The term connectivism has been used in earlier literature [25] which refers to a different notion than [18, 70]).

The following sections survey the basic principles of Activity Theory, Actor-Network Theory, Distributed Cognition and Connectivism. The surveys are brief because these frameworks have been extensively reviewed in other works such as [4, 5, 37, 45, 48, 57].

5.1 Activity Theory

Activity Theory uses the activity as the basic unit of analysis [45, 57]. An activity is composed of subject, object, actions, and operations. The subject is a person or group engaged in an activity. The term object represents the objectified motive of an activity. Actions are conscious goal-directed processes to fulfill the object. Operations describe the way actions are carried out and can become routinized and unconscious in the course of an activity.

Activity Theory assumes an asymmetrical relationship between people and things as the activity is mediated by artifacts, including instruments, signs, language, and machines, with a particular culture and history which may persist across activities. The internal context of people engaged in activities and the external context of mediating artifacts are both considered as important and are seen as fused, not to be considered independently. Therefore the activity itself is regarded as the comprehensive context. The constituents of an activity can change dynamically and the participating humans may be transformed through engaging in the activity.

Applications of Activity Theory include studies of Lotus Notes groupware [30], hospital work [3], design of interactive systems [21], e-learning in corporate training [36], activity recognition in computing systems [68], computer-supported collaborative business process modeling [12], and learning analytics in Virtual Math Teams [87].

5.2 *Actor-Network Theory*

The components of a network are called actants, which can be humans and nonhumans, including ideas and concepts [4, 48]. Unlike Activity Theory, human and nonhuman actants are considered symmetrically and not assigned a presupposed role. The notion of network is more general than computer network and current "social networks" on the Internet. It encompasses the connectivity of the actants by arbitrary means and also potential transformations of actants as they pass through the network.

In Actor-Network analyses, actants may be viewed as black boxes, consisting of sub-networks (punctualization) and actants may be combined to sub-networks (compartmentalization). Analyses usually identify important actants, uncover how networks are created, identify how actors are enrolled in (added to) the network, observe how actors move around the networks, trace interactions, associations, and alliances between actors and investigate how individual parts come together to form a whole network.

Applications of Actor-Network Theory include studies of scientific discovery in research labs [47], e-government in developing countries [29], online communities [62], online social networks [42], air traffic control [56], mobile media consumption [44], and sustainable cyberinfrastructure [66].

5.3 *Distributed Cognition*

Following approaches from cognitive psychology, Distributed Cognition focuses on representing and processing of information by social groups, including CHS [37, 57]. It recognizes that, for analyzing cognitive processes, it is not enough to focus on individuals and that social groups and their environments should be included in the analysis. Processes may involve coordination between internal and external (material or environmental) structure (including representation of information), and may be distributed through time in such a way that the products of earlier events can transform the nature of later events. Like Actor-Network Theory, people and things are viewed symmetrically and considered equivalent as information processors.

The tenets of Distributed Cognition are:

Socially distributed cognition: Cognitive phenomena emerge in social interactions as well as interactions between people and structure in their environments.
Embodied cognition: The organization of mind is an emergent property of interactions among internal and external resources.
Culture and cognition: The study of cognition is not separable from the study of culture.
Ethnography of distributed cognitive systems: It is necessary to investigate how people go about using what they know to do what they do.

Applications include studies of airline cockpit automation [39], air traffic control [32], laboratory research and learning [60], end-user software engineering [8], healthcare technology [65], reuse process in collaboration [61], and collective intelligence [31].

5.4 Connectivism

Connectivism [17, 18, 69, 70] aims at providing a theory of learning that is suitable for CHS. It studies how CHS learn, how learning of individuals is impacted by participating in these systems, and how these systems are designed to support learning. The development of connectivism has been motivated by the desire to explain learning in Massive Open Online Courses (MOOCs). Consequently, the Internet and computer networking with their abilities to connect large numbers of individuals, devices, and software systems are seen as the key technologies that motivate connectivism. However, connectivist principles apply to any CHS of any scale and using any type of communication mechanism, not necessarily exclusively relying on computer networking. Accordingly, the main principles of connectivism [70] do not focus on computer networks and rather talk about connecting, by some unspecified mechanisms, nodes which are implied to include humans, devices, or, more generally, any information sources.

The principles of connectivism are:

- Learning and knowledge rest in diversity of opinions.
- Learning is a process of connecting specialized nodes or information sources.
- Learning may reside in nonhuman appliances.
- Capacity to know more is more critical than what is currently known.
- Nurturing and maintaining connections is needed to facilitate continual learning.
- Ability to see connections between fields, ideas, and concepts is a core skill.
- Currency (accurate, up-to-date knowledge) is the intent of all connectivist learning activities.
- Decision-making is itself a learning process. Choosing what to learn and the meaning of incoming information is seen through the lens of a shifting reality. While

there is a right answer now, it may be wrong tomorrow due to alterations in the information climate affecting the decision.

Connective knowledge is one of the theoretical pillars of connectivism which is the kind of knowledge that emerges in a CHS through the connectedness of its participants [17, 18]. In general, this knowledge is distributed across the CHS and, as a whole, it does not necessarily reside in any single participant, whether human or nonhuman. Downes [17] states that connective knowledge is knowledge of the interaction among the participants. Therefore the nature of the connectivity and the types of interactions of the participants, as the system evolves over time, can also contribute to the system's knowledge. One of the open questions relating to connectivism is how precisely connective knowledge emerges in CHS [4], and it has been postulated that learning of CHS is analogous to connectionist learning of neural networks [4, 17, 18].

Connectivism and its principles have been studied and applied in various contexts, such as distributed professional learning communities [67], workplace learning and e-learning [75], use of the WEB in education [5], information literacy [20], and social media [79].

5.5 *Classification and Comparison*

Activity Theory, Actor-Network Theory, Distributed Cognition and Connectivism can be classified according to different properties, namely, date of emergence, status of human participants of CHS and whether they are scientific theories. Activity Theory has the longest history of development going back to the 1920's [57]. The emergence of Actor-Network Theory, Distributed Cognition and Connectivism respectively date to the 1980's [48], 1980's [38] and 2000's [18, 70]. Activity Theory assigns a special role to humans participating in CHS, whereas the other cited frameworks, Distributed Cognition, Actor-Network Theory and Connectivism, do not impose a special significance on human as opposed to nonhuman components of CHS [48, 57, 70]. A scientific theory of a class of phenomena needs to be predictive, that is, allow to formulate testable hypotheses, and not only describe the phenomena [64]. Distributed Cognition is a scientific theory of the organization of cognitive systems [38]. Activity Theory is not a theory, that is, not a "fixed body of accurately defined statements" [45]. Actor-Network Theory is only descriptive and not a scientific theory despite its name [48]. There have been debates whether Connectivism is a scientific theory [5] because it seems to lack testable hypotheses, even though its foundational literature [18, 70] presents it as a theory.

These frameworks can be individually used to study CHS, and some promote that combining some of them may lead to deeper insights for the functioning and design of CHS, such as Activity Theory and Distributed Cognition [57], Actor-Network Theory and Distributed Cognition [56], Activity Theory and Connectivism

[7], or Actor-Network Theory and Connectivism [4]. Additionally, [33] argues that both Activity Theory and Distributed Cognition are useful for, however, individually insufficient as theoretical foundations of computer supported collaborative work.

6 Discussion

The proposed definition of Collaborative Computer Algebra in Sect. 4 naturally leads to the questions: What is the state of the art of this area? What are the open problems? What may be the impact of solutions? Let us consider theory, social networks, systems, tools, design and evaluation methods, sub-fields of the area of "Collaborative and social computing" mentioned in Sect. 4 and, subsequently, address the wider context of computation in mathematics.

Section 5 proposes several frameworks that have the potential to form the theoretical foundations, considering that they are widely employed in human-centered computing. However, their practical applicability in the context of computer algebra still needs to be demonstrated.

The SymbolicData project [27, 35], presented at ACA 2015, is developing a social network for computer algebra, Computer Algebra Social Network (CASN). It incorporates Semantic Web technology [23] to share data and to run benchmarks, addressing important communal needs. Since this project has only emerged over recent years, the CASN still needs to grow, and work is ongoing to expand its features, such as adding more data.

Section 3 reviews some systems and tools commonly used to collaborate in computer algebra. Additionally, the formulae system [80], presented at ACA 2015, is being designed to facilitate collaborative implementations of symbolic computation packages. However, to date, there is no integrated system that supports the whole range of work in computer algebra, when people collaborate, from mathematical discoveries, algorithm design, implementation, evaluation and documentation to applications.

Workflow management systems for natural science have been motivated by the need to connect and coordinate large working groups and disparate groups of collaborators [9, 15, 24] and to document their work. So far, in symbolic computation, workflow management has only been studied for distributed computations, that is, symbolic grid services [9]. However, a system has yet to emerge to manage the workflows of human collaboration in computer algebra, which could potentially make collaborations more efficient or make it easier to scale the number of collaborators significantly beyond single digit numbers, illustrated in Sect. 2.1.

Evaluations of existing systems for collaboration can motivate improvements and encourage and inform the design of new systems. To date, there have not been any systematic evaluation studies of collaboration in computer algebra. Potentially, methods from Computer Supported Collaborative Work [59] could be adopted to evaluate Collaborative Computer Algebra Systems.

Additionally, let us consider the wider context of mathematical computation. Recognizing that other computational areas of mathematics are related to computational algebra, the term Computer Mathematics has emerged to denote a larger field which includes Computer Algebra. In 1995, Grabmeier [28] proposed a definition of Computer Mathematics by adding other computational domains such as numeric, statistical, combinatorial, and graph-theoretical computations. (This definition differs from the usage by International Journal of Computer Mathematics [41], first published in 1964, referring to computer systems theory and computational mathematics and its applications). Furthermore, Conference on Intelligent Computer Mathematics (CICM) which promotes the advancement of "machine-supported reasoning, computation, and knowledge management in Science, Technology, Engineering, and Mathematics" [10] implicitly expands Computer Mathematics by adding machine-supported reasoning and knowledge management. By this extension, the need for collaborative work arises in these areas as in Computer Algebra.

Definition: Basing on CICM's notion of Intelligent Computer Mathematics, I define Collaborative Computer Mathematics analogously to the definition of Collaborative Computer Algebra from Sect. 4. (The adjective Intelligent has not been included in the name for the sake of brevity). Therefore, Collaborative Computer Mathematics relies on human-centered computing like Collaborative Computer Algebra and consequently is expected to be impacted by analogous concerns and research problems.

Acknowledgements I thank Suzy Maddah for pointing out the example of open source development of mathematical software and Elena Varbanova and Hans-Gert Gräbe for discussions on connectivism and respectively Grabmeier's paper [28]. I also thank the audiences of my ACA 2014 and 2015 presentations for sharing their thoughts on collaboration.

References

1. Autexier, S., David, C., Dietrich, D., Kohlhase, M., Zholudev, V.: Workflows for the management of change in science, technologies, engineering and mathematics. In: Davenport, J.H., Farmer, W.M., Urban, J., Rabe, F. (eds.) Intelligent Computer Mathematics. Lecture Notes in Computer Science, pp. 164–179. Springer, Berlin (2011)
2. Ayyad, M.: Using the actor-network theory to interpret e-government implementation barriers. In: Proceedings of the 3rd International Conference on Theory and Practice of Electronic Governance, ICEGOV '09, pp. 183–190. ACM, New York (2009)
3. Bardram, J., Doryab, A.: Activity analysis: applying activity theory to analyze complex work in hospitals. In: Proceedings of the ACM 2011 Conference on Computer Supported Cooperative Work, CSCW '11, pp. 455–464. ACM, New York (2011)
4. Bell, F.: Network theories for technology-enabled learning and social change: connectivism and actor network theory. In: Networked Learning Conference 2010: Seventh International Conference on Networked Learning, Aalborg (2010)
5. Bell, F.: Connectivism: its place in theory-informed research and innovation in technology-enabled learning. Int. Rev. Res. Open Distance Learn. **12**(3), 98–118 (2010)
6. Bodker, S.: A human activity approach to user interfaces. Hum. Comput. Interact. **4**(3), 171–195 (1989)
7. Boitshwarelo, B.: Proposing an integrated research framework for connectivism: utilising theoretical synergies. Int. Rev. Res. Open Distance Learn. **12**(3), 161–179 (2011)

8. Burnett, M., Bogart, C., Cao, J., Grigoreanu, V., Kulesza, T., Lawrance, J.: End-user software engineering and distributed cognition. In: Proceedings of the 2009 ICSE Workshop on Software Engineering Foundations for End User Programming, SEEUP '09, pp. 1–7. IEEE Computer Society, Washington (2009)
9. Carstea, A., Macariu, G., Frincu, M., Petcu, D.: Workflow management for symbolic grid services. In: 10th International Symposium on Symbolic and Numeric Algorithms for Scientific Computing, 2008. SYNASC '08, pp. 373–379. (2008)
10. CICM Steering Committee.: CICM - Conferences on Intelligent Computer Mathematics. http://cicm-conference.org/cicm.php Accessed 11 Sept 2015
11. Cohen, J.S.: Computer Algebra and Symbolic Computation: Mathematical Methods. A K Peters, Wellesley (2003)
12. Coskunay, D.F., Akir, M.P.: Examination of computer supported collaborative business process modeling with activity theory. In: Proceedings of the XV International Conference on Human Computer Interaction, Interacción '14, pp.15:1–15:8. ACM, New York (2014)
13. Decker, W., Greuel, G.-M., Pfister,G., Schnemann, H.: SINGULAR—a computer algebra system for polynomial computations. Free software under the GNU General Public License. Accessed 19 Oct 2014
14. Decker, W., Greuel, G.-M., Pfister, G., Schönemann, H.: SINGULAR 4-0-2—a computer algebra system for polynomial computations. http://www.singular.uni-kl.de (2015)
15. Deelman, E., Gannon, D., Shields, M., Taylor, I.: Workflows and e-science: an overview of workflow system features and capabilities. Future Gener. Comput. Syst. **25**(5), 528–540 (2009)
16. Deolalikar p vs NP paper-polymath1wiki. http://michaelnielsen.org/polymath1/index.php?title=Deolalikar_P_vs_NP_paper. Accessed 26 Aug 2014
17. Downes, S.: An introduction to connective knowledge. http://www.downes.ca/cgi-bin/page.cgi?post=33034. Accessed 25 Sept 2014
18. Downes, S.: Learning networks and connective knowledge. In: Yang, H.H., Yuen, S.C.Y. (eds.) Collective Intelligence and E-Learning 2.0: Implications of Web-Based Communities and Networking. IGI Global, Washington (2010)
19. Dumas, J.-G., Gautier, T., Pernet, C., Saunders, B.D.: LinBox founding scope allocation, parallel building blocks, and separate compilation. In: Mathematical SoftwareICMS 2010. pp. 77–83. Springer (2010)
20. Dunaway, M.K.: Connectivism. Ref. Serv. Rev. **39**(4), 675–685 (2011)
21. Dweling, S., Schmidt, B., Gb, A.: A model for the design of interactive systems based on activity theory. In: Proceedings of the ACM 2012 Conference on Computer Supported Cooperative Work, CSCW '12, pp. 539–548. ACM, New York (2012)
22. Engestrm, Y., Miettinen, R., Punamki-Gitai, R.-L. (eds.): Perspectives on Activity Theory. Cambridge University Press, Cambridge (1999)
23. Feigenbaum, L., Herman, I., Hongsermeier, T., Neumann, E., Stephens, S.: The semantic web in action. Sci. Am. **297**(6), 90–97 (2007)
24. Garijo, D., Alper, P., Belhajjame, K., Corcho, O., Gil, Y., Goble, C.: Common motifs in scientific workflows: an empirical analysis. Future Gener. Comput. Syst. **36**, 338–351 (2014)
25. Geszti, T.: Physical Models of Neural Networks. World Scientific Pub Co Inc, Singapore (1990)
26. Gowers, T.: The polymath blog. http://polymathprojects.org/. Accessed 19 Oct 2014
27. Gräbe, H.-G., Nareike, A., Johanning, S: The SymbolicData ProjectTowards a Computer Algebra Social Network (2014)
28. Grabmeier, J.: Computeralgebra-eine Säule des Wissenschaftlichen Rechnens/Computer-Algebra—a part of the Foundation of Scientific Computing. Inform. Technol. **37**(6), 5–30 (1995)
29. Gunawong, P., Gao, P.: Challenges of egovernment in developing countries: actor-network analysis of Thailand's smart ID card project. In: Proceedings of the 4th ACM/IEEE International Conference on Information and Communication Technologies and Development, ICTD '10, pp. 17:1–17:9. ACM, New York (2010)
30. Halloran, J., Rogers, Y., Scaife, M.: Taking the 'No' out of lotus notes: activity theory, groupware, and student groupwork. In: Proceedings of the Conference on Computer Support for

Collaborative Learning: Foundations for a CSCL Community, CSCL '02, pp. 169–178. International Society of the Learning Sciences, Boulder (2002)
31. Halpin, H.: Does the web extend the mind? In: Proceedings of the 5th Annual ACM Web Science Conference, WebSci '13, pp. 139–147. ACM, New York (2013)
32. Halverson, CA.: Inside the cognitive workplace: new technology and air traffic control. (Doctoral dissertation, University of California, San Diego) (1995)
33. Halverson, C.A.: Activity theory and distributed cognition: or what does CSCW need to DO with theories? Comput. Support. Coop. Work (CSCW) **11**(1–2), 243–267 (2002)
34. Hammond, J., Koubek, R.J., Harvey, C.M.: Distributed collaboration for engineering design: a review and reappraisal. Hum. Factors Ergon. Manuf. **11**(1), 35–52 (2001)
35. Heinle, A., Levandovskyy, V.: The SDEval benchmarking toolkit. ACM Commun. Comput. Algebra **49**(1), 1–9 (2015)
36. Henneke, M., Matthee, M.: The adoption of e-learning in corporate training environments: an activity theory based overview. In: Proceedings of the South African Institute for Computer Scientists and Information Technologists Conference, SAICSIT '12, pp. 178–187. ACM, New York (2012)
37. Hollan, J., Hutchins, E., Kirsh, D.: Distributed cognition: toward a new foundation for human-computer interaction research. ACM Trans. Comput. Hum. Interact. **7**(2), 174–196 (2000)
38. Hutchins, E.: Distributed cognition. In: Internacional Enciclopedia of the Social and Behavioral Sciences (2000)
39. Hutchins, E.: How a cockpit remembers its speeds. Cognit. Sci. **19**(3), 265–288 (1995)
40. IEEE.: IEEE computer society-premier organization of computer professionals. http://www.computer.org/portal/web/guest/home. Accessed 26 Aug 2014
41. Informa UK Limited.: International Journal of Computer Mathematics. ISSN 1029-0265
42. Kaldoudi, E., Dovrolis, N., Dietze, S.: Information organization on the internet based on heterogeneous social networks. In: Proceedings of the 29th ACM International Conference on Design of Communication, SIGDOC '11, pp. 107–114. ACM, New York (2011)
43. Kirschman, J.S., Greenstein, J.S.: The use of groupware for collaboration in distributed student engineering design teams. J. Eng. Educ. **91**(4), 403–407 (2002)
44. Kumar, N., Rangaswamy, N.: The mobile media actor-network in Urban India. In: Proceedings of the SIGCHI Conference on Human Factors in Computing Systems, CHI '13, pp. 1989–1998. ACM, New York (2013)
45. Kuutti, K.: Activity theory as a potential framework for human-computer interaction research. In: Nardi, B.A. (ed.) Context and Consciousness: Activity Theory and Human-Computer Interaction. MIT Press, Cambridge (1996)
46. Latour, B.: Reassembling the Social: An Introduction to Actor-Network-Theory. Oxford University Press, Oxford (2007)
47. Latour, B., Woolgar, S.: Laboratory Life: The Construction of Scientific Knowledge. Princeton University Press, Princeton (1986)
48. Law, J.: Actor-network theory and material semiotics. In: Turner, B.S. (ed.) The New Blackwell Companion to Social Theory, 3rd edn, pp. 141–158. Blackwell, Oxford (2008)
49. Leont'ev, A.: The problem of activity in psychology. J. Russ. East Eur. Psychol. **13**(2), 4–33 (1974)
50. Linton, S., Hammond, K., Konovalov, A., Brown, C., Trinder, P.W., Loidl, H.W., Horn, P., Roozemond, D.: Easy composition of symbolic computation software using SCSCP: a new Lingua Franca for symbolic computation. J. Symb. Comput. **49**, 95–119 (2013)
51. Lu, S.Y., Elmaraghy, W., Schuh, G., Wilhelm, R.: A scientific foundation of collaborative engineering. CIRP Ann. Manuf. Technol. **56**(2), 605–634 (2007)
52. MathOverflow.: MathOverflow site. http://mathoverflow.net/
53. Minimair, M.: Collaborative Computer Algebra Systems. Applications of Computer Algebra (ACA) (2014)
54. Minimair, M.: Collaborative computer algebra: review of foundations. Applications of Computer Algebra (ACA) (2015)

55. Monell, D.W., Piland, W.M.: Aerospace systems design in NASA's collaborative engineering environment. Acta Astronaut. **47**(2), 255–264 (2000)
56. Moran, S. Nakata, K., Inoue, S.: Bridging the analytical gap between distributed cognition and actor network theory using a tool for information trajectory analysis. In: Proceedings of the 30th European Conference on Cognitive Ergonomics, ECCE '12, pp. 72–77, ACM, New York (2012)
57. Nardi, B.A.: Studying context: a comparison of activity theory, situated action models, and distributed cognition. Context Conscious. pp. 69–102. (1996)
58. National Science Foundation.: US NSF-CISE-IIS-Cyber-Human Systems (CHS). http://www.nsf.gov/cise/iis/chs_pgm13.jsp. Accessed 20 Oct 2014
59. Neale, D.C., Carroll, J.M., Rosson, M.B.: Evaluating computer-supported cooperative work: models and frameworks. pp. 2–121. ACM (2004)
60. Newstetter, W., Johri, A., Wulf, V.: Laboratory learning: industry and University Research as site for situated and distributed cognition. In: Proceedings of the 8th International Conference on International Conference for the Learning Sciences-Volume 3, ICLS'08, pp. 290–297. International Society of the Learning Sciences, Utrecht (2008)
61. Nobarany, S., Haraty, M., Fisher, B.: Facilitating the reuse process in distributed collaboration: a distributed cognition approach. In: Proceedings of the ACM 2012 Conference on Computer Supported Cooperative Work, CSCW '12, pp. 1223–1232. ACM, New York (2012)
62. Pelizza, A.: Openness as an asset: a classification system for online communities based on actor-network theory. In: Proceedings of the 6th International Symposium on Wikis and Open Collaboration, WikiSym '10, pp. 8:1–8:10, ACM, New York (2010)
63. planetmath.org | math for the people, by the people. http://planetmath.org/. Accessed 24 Aug 2014
64. Popper, K.R.: Conjectures and Refutations: The Growth of Scientific Knowledge. Psychology Press, New York (2002)
65. Rajkomar A., Blandford, A.: Distributed cognition for evaluating healthcare technology. In: Proceedings of the 25th BCS Conference on Human-Computer Interaction, BCS-HCI '11, pp. 341–350, British Computer Society, Swinton (2011)
66. Randall, D.P., Diamant, E.I., Lee, C.P.: Creating sustainable cyberinfrastructures. In: Proceedings of the 33rd Annual ACM Conference on Human Factors in Computing Systems, CHI '15, pp. 1759–1768, ACM, New York (2015)
67. Rousseau, J.J.: Historical case study of the supernet consortium: investigating change management and the concepts of connectivism and distributed professional learning communities. ProQuest (2007)
68. Saguna, S., Zaslavsky, A., Chakraborty, D.: Complex activity recognition using context-driven activity theory and activity signatures. ACM Trans. Comput. Hum. Interact. **20**(6), 1–32 (2013)
69. Siemens, G.: http://www.knowingknowledge.com (2006)
70. Siemens, G.: Connectivism: a learning theory for the digital age. Int. J. Instruct. Technol. Distance Learn. **2**(1), 3–10 (2005)
71. SIGCHI.: Welcome SIGCHI. http://www.sigchi.org/. Accessed 26 Aug 2014
72. Stahl, G.: Studying Virtual Math Teams. Springer, Berlin (2009)
73. Stein, W.A.: Sage notebook. http://sagenb.org/. Accessed 19 Oct 2014
74. Steinberg, S., Wester, M.: Conferences on Applications of Computer Algebra
75. Strong, K., Hutchins, H.M.: Connectivism: a theory for learning in a world of growing complexity. Impact **1**(1), 53–67 (2009)
76. The 2012 ACM Computing Classification System Association for Computing Machinery
77. The GAP Group.: GAP system for computational discrete algebra. http://www.gap-system.org/. Accessed 19 Oct 2014
78. The GAP Group.: GAP–Groups, Algorithms, and Programming, Version 4.7.8 (2015)
79. Tinmaz, H.: Social networking websites as an innovative framework for connectivism. Contemp. Educ. Technol. **3**(3), 234–245 (2012)
80. Ugalde, L.R.: http://formulae.org (2015)
81. van der Hoeven, J.: GNU TeXmacs. SIGSAM Bull. **38**(1), 24–25 (2004)

82. van der Hoeven, J., Lecerf, G., Mourrain, B.: Mathemagix. Accessed 19 Oct 2014
83. Vygotsky, L.S.: Mind in Society: The Development of Higher Psychological Processes. Harvard University Press, Cambridge (1980)
84. Waterloo Maple, Inc. Maplesoft-technical computing software for engineers, mathematicians, scientists, instructors and students. http://maplesoft.com/. Accessed 19 Oct 2014
85. Waycott, J., Jones, A., Scanlon, E.: PDAs as lifelong learning tools: an activity theory based analysis. Learn. Media Technol. **30**(2), 107–130 (2005)
86. Wolfram Research.: Wolfram: computation meets knowledge. http://www.wolfram.com/. Accessed 19 Oct 2014
87. Xing, W., Wadholm, B., Goggins, S.: Learning analytics in CSCL with a focus on assessment: an exploratory study of activity theory-informed cluster analysis. In: Proceedings of the Fourth International Conference on Learning Analytics And Knowledge, LAK '14, pp. 59–67, ACM, New York (2014)

States and Channels in Quantum Mechanics Without Complex Numbers

J.A. Miszczak

Abstract In the presented work, we aim at exploring the possibility of abandoning complex numbers in the representation of quantum states and operations. We demonstrate a simplified version of quantum mechanics in which the states are represented using real numbers only. The main advantage of this approach is that the simulation of the n-dimensional quantum system requires n^2 real numbers, in contrast to the standard case where n^4 real numbers are required. The main disadvantage is the lack of hermicity in the representation of quantum states. Using *Mathematica* computer algebra system we develop a set of functions for manipulating real-only quantum states. With the help of this tool, we study the properties of the introduced representation and the induced representation of quantum channels.

Keywords Quantum states · Random density matrix · Quantum mathematics

1 Introduction

Quantum information theory aims at harnessing the behavior of quantum mechanical objects to store, transfer and process information. This behavior is, in many cases, very different from the one we observe in the classical world [8]. Quantum algorithms and protocols take advantage of the superposition of states and require the presence of entangled states. Both phenomena arise from the rich structure of the space of quantum states [1]. Hence, to explore the capabilities of quantum information processing, one needs to fully understand this space. Quantum mechanics provides us also with much larger allowed operations than in classical case space. It

J.A. Miszczak (✉)
Institute of Theoretical and Applied Informatics,
Polish Academy of Sciences, Baltycka 5, 44100 Gliwice, Poland
e-mail: jmiszczak@acm.org

J.A. Miszczak
Applied Logic, Philosophy and History of Science Group,
University of Cagliari, Via Is Mirrionis 1, 09123 Cagliari, Italy

I.S. Kotsireas and E. Martínez-Moro (eds.), *Applications of Computer Algebra*,
Springer Proceedings in Mathematics & Statistics 198,
DOI 10.1007/978-3-319-56932-1_21

can be used to manipulate quantum states. However, the exploration of the space of quantum operations is fascinating, but a cumbersome task.

Functional programming is frequently seen as an attractive alternative to the traditional methods used in scientific computing, which are based mainly on the imperative programming paradigm [4]. Among the features of functional languages which make them suitable for the use in this area is the easiness of execution of the functional code in the parallel environments.

During the past few years *Mathematica* computing systems have become very popular in the area of quantum information theory and the foundations of quantum mechanics. The main reason for this is its ability to merge the symbolic and numerical capabilities, both of which are often necessary to understand the theoretical and practical aspects of quantum systems [3, 5, 10, 11].

In this paper, we utilize the ability to merge symbolical and numerical calculations offered by *Mathematica* to investigate the properties of the variant of quantum theory based on the representation of density matrices built using real-numbers only. We start by introducing the said representation, including the *Mathematica* required functions. Next, we test the behavior of selected partial operations in this representation and consider the general case of quantum channels acting on the space of real-only density matrices. In the last part, we provide some insight into the spectral properties of the real-only density matrices. Finally, we provide the summary and the concluding remarks.

1.1 Preliminaries

In quantum mechanics the state is represented by positive semidefinite, normalized matrix. In the following, we focus on this property as it is crucial for the properties of quantum states and channels. To be more specific, we aim at using symbolic matrix which is Hermitian. Using the symbolic capabilities of *Mathematica* they can be expressed as

```
SymbolicDensityMatrix[a_, b_, d_] := Array[
  If[#1 < #2, a#1,#2 + I b#1,#2, If[#1 > #2, a#2,#1 − I b#2,#1, a#1,#2]] &, {d, d}]
```

In the above definition slots a_ and b_ are used to specify the symbols used to denote the real and the imaginary parts of the matrix elements.

Additionally one has to take into account the fact that symbols a_{i, j} and b_{i, j} represent real numbers. This fact is useful during the simplifications in the formulas and can be expressed using the function

```
SymbolicDensityMatrixAssume[a_, b_, d_] :=
  $Assumptions = Map[Element[#, Reals] &,
    Flatten[Join[
        Table[ai,j, {i,1,d}, {j,i,d}], Table[bi,j, {i,1,d}, {j, i+1, d}]
      ]]
  ]
```

It is easy to see that the normalization condition can be easily added to the list of assumptions. However, the conditions for the positivity, *e.g.* in the form of the positivity conditions for the principal minors, are more complicated [2, Chap. 1].

One should note that, in order to utilize the hermicity conditions for a matrix defined using function **SymbolicDensityMatrix**, is it necessary to execute function **SymbolicDensityMatrixAssume** with the same symbolic arguments.

Another function useful for the purpose of analyzing the operation on quantum states is **SymbolicMatrix** function defined as

```
SymbolicMatrix[a_, d1_, d2_] := Array[Subscript[a, #1, #2] &, {d1, d2}]
```

Using **Flatten** function in combination with **Map** we can impose a list of assumptions on the elements of the symbolic matrix. For example, if one needs to ensure that the elements of the matrix mA are real, this can be achieved as

```
mA = SymbolicMatrix[a, 2, 2];
$Assumptions = Map[Element[#, Reals] &, Flatten[mA]]
```

2 Using Real Density Matrices

Clearly, the representation of the density used in Sect. 1.1 is redundant as the off-diagonal element $a_{i,j} + ib_{i,j}$ is conjugate to $a_{j,i} - ib_{j,i}$. Using this observation, we can represent any density matrix as a real matrix with elements defined as

$$\mathcal{R}[\rho]_{ij} = \begin{cases} \mathbf{Re}\rho_{ij} & i \leq j \\ -\mathbf{Im}\rho_{ij} & i > j \end{cases}. \tag{1}$$

The above definition can be translated into *Mathematica* code as

```
ComplexToReal[denMtx_] := Block[{d = Dimensions[denMtx][[1]]},
  Array[If[#1 <= #2, Re[denMtx[[#1, #2]]], - Im[denMtx[[#1, #2]]]] &, {d, d}]]
```

Thus, for a given density matrix, describing d-dimensional system we get a matrix with n^2 real elements, instead of a matrix with n^2 complex (or n^4 real) elements. Note, that these numbers can be reduced during the simulation due to the positivity and normalization conditions, but this requires distinguishing between diagonal and off-diagonal elements.

In the following, we denote the map defined by the **ComplexToReal** function as $\mathcal{R}[\cdot]$. One should note that $\mathcal{R} : \mathbb{M}_n(\mathbb{C}) \mapsto \mathbb{M}_n(\mathbb{R})$. However, we will only consider multiplication by real numbers as it does not affect the hermicity of the density matrix.

The real representation of a density matrix contains the same information as the original matrix. As such it can be used to reconstruct the initial density matrix.

Assuming that realMtx represents a real matrix obtained as a representation of the density matrix one can reconstruct the original density matrix as

```
RealToComplex[realMtx_] := Block[{d = Dimensions[realMtx][[1]]},
  Array[If[#1 < #2, realMtx[[#1, #2]] + I realMtx[[#2, #1]],
     If[#1 > #2, realMtx[[#2, #1]] - I realMtx[[#1, #2]],
      realMtx[[#1, #2]]]] &, {d, d}]
  ]
```

The map defined by the function **RealToComplex** will be denoted as $\mathcal{C}[\cdot]$. It is easy to see that for any ρ we have $\mathcal{R}[\mathcal{C}[\rho]] = \rho$.

One can also see that maps $\mathcal{R}$ and $\mathcal{C}$ are linear if one considers the multiplication by real numbers only. Thus, it can be represented as a matrix on the Hilbert–Schmidt space of density matrices. Using this representation one gets

$$\mathcal{R}[\rho] = \mathbf{res}^{-1}\left(M_{\mathcal{R}}\,\mathbf{res}(\rho)\right) \tag{2}$$

where **res** is the operation of reordering elements of the matrix into a vector [6].

The introduced representation can be utilized to reduce the amount of memory required during the simulation. For the purpose of modelling the discrete time evolution of quantum system, one needs to transform the form of quantum maps into the real representation. For a map Φ given as a matrix M_{Φ} one obtains its real representation as

$$M_{\mathcal{R}[\Phi]} = M_{\mathcal{R}} M_{\Phi} M_{\mathcal{C}} \tag{3}$$

One can see that this allows the reduction of the number of multiplication operations required to simulate the evolution.

3 Examples

Let us now consider some examples utilizing maps $\mathcal{R}$ and $\mathcal{C}$. We will focus on the computation involving symbolic manipulation of states and operations. Only in the last example, we use the statistical properties of density matrices which have to be calculated numerically.

3.1 One-Qubit Case

In the simplest case of two-dimensional quantum system, the symbolic density matrix can be obtained as

```
SymbolicDensityMatrix[a, b, 2]
```

which results in

$$\begin{pmatrix} a_{1,1} & a_{1,2} + ib_{1,2} \\ a_{1,2} - ib_{1,2} & a_{2,2} \end{pmatrix}. \tag{4}$$

The list of assumptions required to force *Mathematica* to simplify the expressions involving the above matrix can be obtained as

SymbolicDensityMatrixAssume[a, b, 2]

which results in storing the following list

$\{a_{1,1} \in \textbf{Reals},\ a_{1,2} \in \textbf{Reals},\ a_{2,2} \in \textbf{Reals},\ b_{1,2} \in \textbf{Reals}\}$

in the global variable $**Assumptions**.

In *Mathematica* the application of map $\mathcal{R}$ on the above matrix results in

$$\begin{pmatrix} \textbf{Re}\left(a_{1,1}\right) & \textbf{Re}\left(a_{1,2}\right) - \textbf{Im}\left(b_{1,2}\right) \\ \textbf{Re}\left(b_{1,2}\right) - \textbf{Im}\left(a_{1,2}\right) & \textbf{Re}\left(a_{2,2}\right) \end{pmatrix}, \tag{5}$$

where **Re** and **Im** are the functions for taking the real and the imaginary parts of the number. Only after using function **FullSimplify** one gets the expected form of the output

$$\begin{pmatrix} a_{1,1} & a_{1,2} \\ b_{1,2} & a_{2,2} \end{pmatrix}. \tag{6}$$

In the one-qubit case, it is also easy to check that map $\mathcal{R}$ is represented by the matrix

$$M_{\mathcal{R}}^{(2)} = \frac{1}{2}\begin{pmatrix} 2 & 0 & 0 & 0 \\ 0 & 1 & 1 & 0 \\ 0 & -i & i & 0 \\ 0 & 0 & 0 & 2 \end{pmatrix}. \tag{7}$$

The matrix representation of the map $\mathcal{C}$ reads

$$M_{\mathcal{C}}^{(2)} = (M_{\mathcal{R}}^{(2)})^{-1} = \begin{pmatrix} 1 & 0 & 0 & 0 \\ 0 & 1 & i & 0 \\ 0 & 1 & -i & 0 \\ 0 & 0 & 0 & 1 \end{pmatrix}. \tag{8}$$

The above consideration can be repeated and in the case of three-dimensional quantum system the matrix representation of the $\mathcal{R}$ map reads

$$M_{\mathcal{R}}^{(3)} = \frac{1}{2}\begin{pmatrix} 2 & 0 & 0 & 0 & 0 & 0 & 0 & 0 & 0 \\ 0 & 1 & 0 & 1 & 0 & 0 & 0 & 0 & 0 \\ 0 & 0 & 1 & 0 & 0 & 0 & 1 & 0 & 0 \\ 0 & -i & 0 & i & 0 & 0 & 0 & 0 & 0 \\ 0 & 0 & 0 & 0 & 2 & 0 & 0 & 0 & 0 \\ 0 & 0 & 0 & 0 & 0 & 1 & 0 & 1 & 0 \\ 0 & 0 & -i & 0 & 0 & 0 & i & 0 & 0 \\ 0 & 0 & 0 & 0 & 0 & -i & 0 & i & 0 \\ 0 & 0 & 0 & 0 & 0 & 0 & 0 & 0 & 2 \end{pmatrix}. \tag{9}$$

3.2 One-Qubit Channels

The main benefit of the real representation of density matrices is the smaller number of multiplications required to describe the evolution of the quantum system.

To illustrate this, let us consider a bit-flip channel defined by Kraus operators

$$\left\{\begin{pmatrix} \sqrt{1-p} & 0 \\ 0 & \sqrt{1-p} \end{pmatrix}, \begin{pmatrix} 0 & \sqrt{p} \\ \sqrt{p} & 0 \end{pmatrix}\right\}, \tag{10}$$

or equivalently as a matrix

$$M_{BF}^{(2)} = \begin{pmatrix} 1-p & 0 & 0 & p \\ 0 & 1-p & p & 0 \\ 0 & p & 1-p & 0 \\ p & 0 & 0 & 1-p \end{pmatrix}. \tag{11}$$

The form of this channel on the real density matrices is given by

$$M_{\mathcal{R}}^{(2)} M_{BF}^{(2)} M_{\mathcal{C}}^{(2)} = \begin{pmatrix} 1-p & 0 & 0 & p \\ 0 & 1 & 0 & 0 \\ 0 & 0 & 1-2p & 0 \\ p & 0 & 0 & 1-p \end{pmatrix}. \tag{12}$$

This map acts on the real density matrix as

$$\begin{pmatrix} pa_{2,2} - (p-1)a_{1,1} & a_{1,2} \\ (1-2p)b_{1,2} & pa_{1,1} - (p-1)a_{2,2} \end{pmatrix}. \tag{13}$$

One should note that in *Mathematica* the direct application of the map $\mathcal{R}$ on the output of the channel, *ie.* $M_R M_{BF}$ **res** ρ, results in

$$\begin{pmatrix} \mathbf{Re}\left(pa_{2,2} - (p-1)a_{1,1}\right) & a_{1,2} + 2\mathbf{Im}(p)b_{1,2} \\ (1-2\mathbf{Re}(p))b_{1,2} & \mathbf{Re}\left(pa_{1,1} - (p-1)a_{2,2}\right) \end{pmatrix} \tag{14}$$

In order to get the simplified result, one needs to explicitly specify assumptions p ∈ **Reals**. This is important if one aims at testing the validity of the symbolic computation, as without these assumptions *Mathematica* will not be able to evaluate the result.

3.3 Werner States

As the first example of the quantum states of the composite system, let us use the Werner states defined for two-qubit systems as

$$W(a) = \begin{pmatrix} \frac{a+1}{4} & 0 & 0 & \frac{a}{2} \\ 0 & \frac{1-a}{4} & 0 & 0 \\ 0 & 0 & \frac{1-a}{4} & 0 \\ \frac{a}{2} & 0 & 0 & \frac{a+1}{4} \end{pmatrix}. \tag{15}$$

The partial transposition transforms $W(a)$ as

$$W(a)^{T_A} = \begin{pmatrix} \frac{a+1}{4} & 0 & 0 & 0 \\ 0 & \frac{1-a}{4} & \frac{a}{2} & 0 \\ 0 & \frac{a}{2} & \frac{1-a}{4} & 0 \\ 0 & 0 & 0 & \frac{a+1}{4} \end{pmatrix} \tag{16}$$

and this matrix has one negative eigenvalue for $a > 1/3$, which indicates a presence of quantum entanglement.

In this case, the real representation of quantum states reduces one element from the $W(a)$ matrix and we get

$$\mathcal{R}[W(a)] = \begin{pmatrix} \frac{a+1}{4} & 0 & 0 & \frac{a}{2} \\ 0 & \frac{1-a}{4} & 0 & 0 \\ 0 & 0 & \frac{1-a}{4} & 0 \\ 0 & 0 & 0 & \frac{a+1}{4} \end{pmatrix}. \tag{17}$$

This matrix has eigenvalues

$$\left\{ \frac{1-a}{4}, \frac{1-a}{4}, \frac{a+1}{4}, \frac{a+1}{4} \right\} \tag{18}$$

and we have that the sum of smaller eigenvalues is greater than the larger eigenvalue for $a > 1/3$.

3.4 *Partial Transposition*

Another important example related to the composite quantum systems is the case of partial quantum operations. Such operations arise in the situation when one needs to distinguish between the evolution of the system and the evolution of the same system treated as a part of a bigger subsystem.

Let us consider the partial transposition of the two-qubit density matrix

```
ρ = SymbolicDensityMatrix[x, y, 4]
```

which is given by

$$\rho^{T_A} = \begin{pmatrix} x_{1,1} & x_{1,2}+iy_{1,2} & x_{1,3}-iy_{1,3} & x_{2,3}-iy_{2,3} \\ x_{1,2}-iy_{1,2} & x_{2,2} & x_{1,4}-iy_{1,4} & x_{2,4}-iy_{2,4} \\ x_{1,3}+iy_{1,3} & x_{1,4}+iy_{1,4} & x_{3,3} & x_{3,4}+iy_{3,4} \\ x_{2,3}+iy_{2,3} & x_{2,4}+iy_{2,4} & x_{3,4}-iy_{3,4} & x_{4,4} \end{pmatrix} \tag{19}$$

One can easily check that in this case

$$\mathcal{R}[\rho^{T_A}] = \begin{pmatrix} x_{1,1} & x_{1,2} & x_{1,3} & x_{2,3} \\ y_{1,2} & x_{2,2} & x_{1,4} & x_{2,4} \\ -y_{1,3} & -y_{1,4} & x_{3,3} & x_{3,4} \\ -y_{2,3} & -y_{2,4} & y_{3,4} & x_{4,4} \end{pmatrix} \tag{20}$$

and

$$(\mathcal{R}[\rho])^{T_A} = \begin{pmatrix} x_{1,1} & x_{1,2} & y_{1,3} & y_{2,3} \\ y_{1,2} & x_{2,2} & y_{1,4} & y_{2,4} \\ x_{1,3} & x_{1,4} & x_{3,3} & x_{3,4} \\ x_{2,3} & x_{2,4} & y_{3,4} & x_{4,4} \end{pmatrix} \tag{21}$$

and thus

$$\mathcal{R}[\rho^{T_A}] \neq (\mathcal{R}[\rho])^{T_A}. \tag{22}$$

For this reason one cannot change the order of operations. However, the explicit form of the partial transposition on the real density matrices can be found by representing operation of partial transposition as a matrix [6],

ChannelToMatrix[**PartialTranspose**[#, {2, 2}, {1}] &, 4]

and using Eq. (3).

One should note that this method can be used to obtain an explicit form of any operation of the form $\Phi \otimes \mathbb{1}$, where $\mathbb{1}$ denotes the identity operation of the subsystem.

3.5 *Partial Trace*

The second important example of a partial operation is the partial trace. This operation allows obtaining the state of the subsystem.

For two-qubit density matrix we have

$$\mathrm{tr}_A\rho = \begin{pmatrix} x_{1,1} + x_{3,3} & x_{1,2} + x_{3,4} + i\left(y_{1,2} + y_{3,4}\right) \\ x_{1,2} + x_{3,4} - i\left(y_{1,2} + y_{3,4}\right) & x_{2,2} + x_{4,4} \end{pmatrix}. \tag{23}$$

One can verify that the operation of tracing-out the subsystem commutes with the map $\mathcal{R}$ and in this case we have

$$\mathcal{C}[\mathrm{tr}_A\mathcal{R}[\rho]] = \mathrm{tr}_A\rho. \tag{24}$$

Thus, one can calculate the reduced state of the subsystem using the real value representation.

3.6 *Random Real States*

In this section, we focus on the statistical properties of the matrices representing real quantum states. The main difficulty here is that, in contrast to the random density matrices, real representations can have complex eigenvalues.

Random density matrices play an important role in quantum information theory and they are useful in order to obtain information about the average behavior of quantum protocols. Unlike the case of pure states, mixed states can be drawn uniformly using different methods, depending on the used probability measure [1, 7, 9].

One of the methods is motivated by the physical procedure of tracing-out a subsystem. In a general case, one can seek a source of randomness in a given system, by studying the interaction of the n-dimensional system in question with the environment. In such situation, the random states to model the behaviour of the system should be generated by reducing a pure state in $N \times K$-dimensional space. In what follows we denote the resulting probability measure by $\mu_{N,K}$.

Using Wolfram language, the procedure for generating random density matrices with $\mu_{N,K}$ can be implemented as

```
RandomState[n_, k_] := Block[{gM},
  gM = GinibreMatrix[n, k];
  Chop[#/Tr[#]] & @(gM.ConjugateTranspose[gM])
]
```

where function **GinibreMatrix** is defined as

```
GinibreMatrix[n_, k_] := Block[{dist},
    dist = NormalDistribution[0,1];
    RandomReal[dist,{n,k}] + I RandomReal[dist,{n,k}]
]
```

3.7 *Spectral Properties*

In the special case of $K = N$ we obtain the Hilbert–Schmidt ensemble. The distribution of eigenvalues for $K = N = 4$ (i.e. Hilbert–Schmidt ensemble for ququart) is presented in Fig. 1.

The real representation for the Hilbert–Schmidt ensemble for one ququart consists of matrices having four eigenvalues. Two of these values are complex and mutually conjugate (see Fig. 2).

3.7.1 Form of the Resulting Matrix Elements

Using **SymbolicMatrix** function one can easily analyze the dependency of the elements of the resulting matrix on the element of the Ginibre matrix.

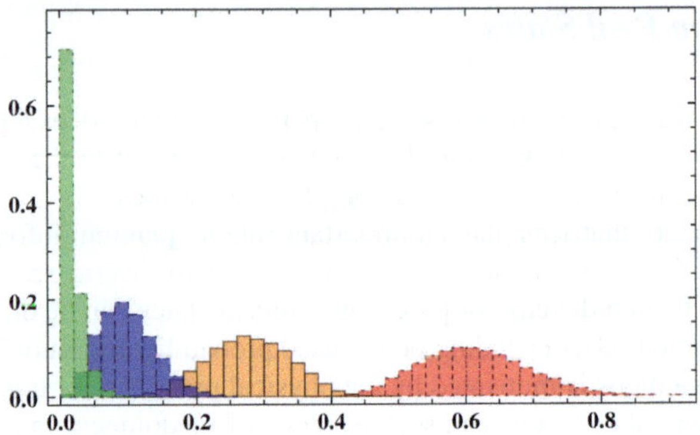

Fig. 1 Distribution of eigenvalues for 4-dimensional random density matrices distributed uniformly with Hilbert–Schmidt measure for the sample of size 10^4. Each color (and contour style) correspond to the subsequent eigenvalue, ordered by their magnitude

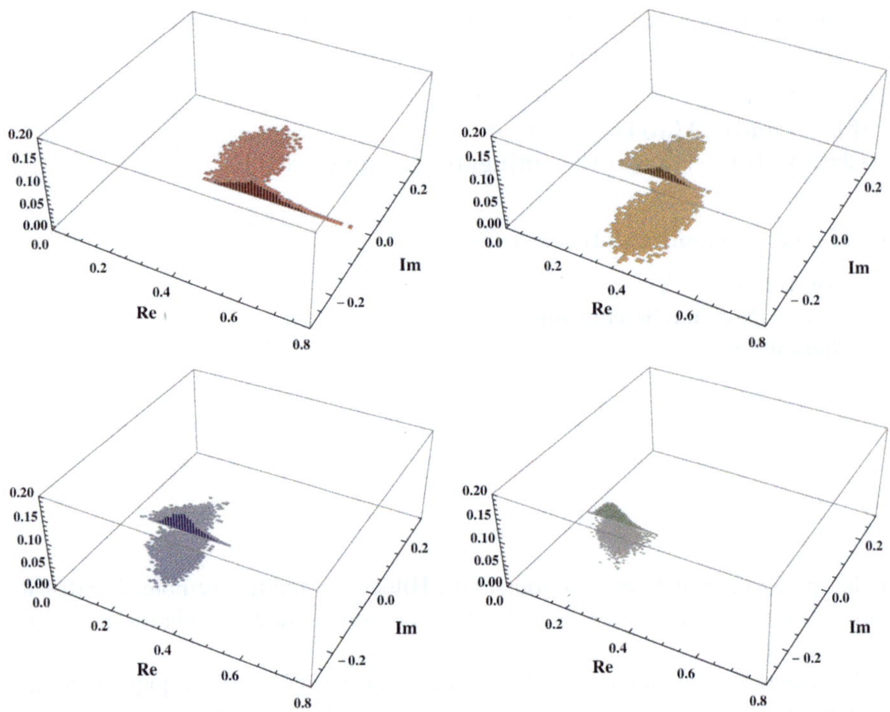

Fig. 2 Distribution of eigenvalues for 4-dimensional random density matrices distributed uniformly with Hilbert–Schmidt measure for the sample of size 10^4. Eigenvalues were ordered according to their absolute value

For the sake of simplicity we demonstrate this on one-qubit states from the Hilbert–Schmidt ensemble. In this case, the Ginibre matrix can be represented as

```
mA = SymbolicMatrix[a, 2, 2];
mB = SymbolicMatrix[b, 2, 2];
m2 = mA + I mB
```

The resulting density matrix has (up to the normalization) elements given by the matrix

```
m2.ConjugateTranspose[m2].
```

In this case, the real representation is given by

$$\begin{pmatrix} q_{1,1} & q_{1,2} \\ q_{2,1} & q_{2,2} \end{pmatrix}, \tag{25}$$

with

$$\begin{aligned} q_{1,1} &= a_{1,1}^2 + a_{1,2}^2 + b_{1,1}^2 + b_{1,2}^2, \\ q_{1,2} &= a_{1,1}a_{2,1} + a_{1,2}a_{2,2} + b_{1,1}b_{2,1} + b_{1,2}b_{2,2}, \\ q_{2,1} &= a_{2,1}b_{1,1} + a_{2,2}b_{1,2} - a_{1,1}b_{2,1} - a_{1,2}b_{2,2}, \\ q_{2,2} &= a_{2,1}^2 + a_{2,2}^2 + b_{2,1}^2 + b_{2,2}^2. \end{aligned} \tag{26}$$

Here $a_{i,j}$ and $b_{i,j}$ are independent random variables used in the definition of the Ginibre matrix.

From the above, one can see that the elements of the density matrix resulting from the procedure for generating random quantum states are obtained as a product and a sum of the elements of real and imaginary parts of the Ginibre matrix. In the case of density matrices, the normalization imposes the condition $q_{1,1} = 1 - q_{2,2}$. Thus, one can also see that the elements are not independent.

4 Final Remarks

In this work, we have introduced a simplified version of quantum states' representation using the redundancy of information in the standard representation of density matrices. Our aim was to the find out if such representation can be beneficial from the point of view of the symbolic manipulation of quantum states and operations.

To achieve this goal we have used *Mathematica* computing system to implement the functions required to operate on real quantum states and demonstrated some examples where this representation can be useful from the computational point of view. Its main advantage is that it can be used to reduce the memory requirements for the representation of quantum states. Moreover, in some particular cases where the density matrix contains only real numbers, the real representation reduces to the upper triangular matrix.

The real representation can be also beneficial for the purpose of modelling quantum channels. Here, its main advantage is that it can be used to reduce the number of multiplications required during the simulation of the discrete quantum evolution. As a particular example, we have studied the form of partial quantum operations in the introduced representation. In the case of the partial trace for the bi-bipartite system, the introduced representation allows the calculation of the reduced dynamics using the real representation only.

Unfortunately, the introduced representation poses some disadvantages. The main drawback of the introduced representation is the lack of hermicity of real density matrices. This makes the analysis of the spectral properties of real quantum states much more complicated.

Acknowledgements This work has been supported by Polish National Science Centre project number 2011/03/D/ST6/00413 and RAS project on: "Modeling the uncertainty: quantum theory and imaging processing", LR 7/8/2007. The author would like to thank G. Sergioli for motivating discussions.

References

1. Bengtsson, I., Zyczkowski, K.: Geometry of Quantum States: An Introduction to Quantum Entanglement. Cambridge University Press, Cambridge, UK (2006)
2. Bhatia, R.: Positive Definite Matrices. Princton University Press, Princeton (2007)
3. Gerdt, V., Kragler. R., Prokopenya, A.: A Mathematica package for simulation of quantum computation. In: Gerdt, V., Mayr, E., Vorozhtsov, E. (eds.) Computer Algebra in Scientific Computing/CASC2009. LNCS, vol. 5743, pp. 106–117. Springer, Berlin (2009)
4. Hinsen, K.: The promises of functional programming. Comput. Sci. Eng. **11**(4), 86–90 (2009). doi:10.1109/MCSE.2009.129
5. Juliá-Díaz, B., Burdis, J., Tabakin, F.: QDENSITY—a Mathematica quantum computer simulation. Comput. Phys. Commun. **174**, 914–934 (2006). doi:10.1016/j.cpc.2005.12.021. arXiv:quant-ph/0508101
6. Miszczak, J.: Singular value decomposition and matrix reorderings in quantum information theory. Int. J. Mod. Phys. C **22**(9), 897–918 (2011). arXiv:1011.1585
7. Miszczak, J.: Generating and using truly random quantum states in mathematica. Comput. Phys. Commun. **183**(1), 118–124 (2012a). doi:10.1016/j.cpc.2011.08.002. arXiv:1102.4598
8. Miszczak, J.: High-Level Structures for Quantum Computing. Synthesis Lectures on Quantum Computing, vol. #6. Morgan and Claypol Publishers (2012b). doi:10.2200/S00422ED1V01Y201205QMC006
9. Miszczak, J.: Employing online quantum random number generators for generating truly random quantum states in mathematica. Comput. Phys. Commun. **184**(1), 257258 (2013a). doi:10.1016/j.cpc.2012.08.012. arXiv:1208.3970
10. Miszczak, J.: Functional framework for representing and transforming quantum channels. In: Galan Garcia, J., Aguilera Venegas, G., Rodriguez Cielos, P. (eds.) Proceedings of the Applications of Computer Algebra (ACA2013), Malaga, 2–6 July 2013 (2013b). arXiv:1307.4906
11. Tabakin, F., Juliá-Díaz, B.: QCWAVE—a Mathematica quantum computer simulation update. Comput. Phys. Commun. **182**(8), 1693–1707 (2011). doi:10.1016/j.cpc.2011.04.010. arXiv:1101.1785

Double Hough Transform for Estimating the Position of the Mandibular Canal in Dental Radiographs

Darian Onchis-Moaca, Simone Zappalá, Smaranda Laura Goţia and Pedro Real

Abstract In this work, a multiple generalized anisotropic Hough transform (AGHT) is used to detect the mandibular canal in dental panoramic radiographs. The proposed method relies on a sequential application of the Hough transform that we call double Hough transform. The recognition of the mandibular canal is based on a double template matching compared with the clinical detection using the fact that the shape of the mandibular canal is usually the same and it is situated inside the mandibular bone. The experiments performed on real orthopantomographic images shown that the risk of false detection is significantly decreased, while the recognition is not affected by occlusion and by the presence of additional structures, e.g., teeth, projection errors.

Keywords Pattern recognition · Hough transform · Mandibular canal

---

D. Onchis-Moaca · S. Zappalá (✉)
Faculty of Mathematics, University of Vienna, Oskar-Morgenstern-Platz 1, Vienna, Austria
e-mail: simone.zappala@univie.ac.at

D. Onchis-Moaca
Faculty of Mathematics and Computer Science, West University of Timisoara, Vasile Pârvan Street 4, Timişoara, Romania
e-mail: darian.onchis@univie.ac.at

S.L. Goţia
Department of Physiology, Victor Babes University of Medicine and Pharmacy, Eftimie Murgu Street 2, Timisoara, Romania
e-mail: lauragotia@yahoo.com

P. Real
Department of Applied Mathematics I, University of Seville, Av. Reina Mercedes, Seville, Spain
e-mail: real@us.es

I.S. Kotsireas and E. Martínez-Moro (eds.), *Applications of Computer Algebra*, Springer Proceedings in Mathematics & Statistics 198,
DOI 10.1007/978-3-319-56932-1_22

1 Introduction

From a clinical point of view, the marking of the mandibular canal is useful in detecting the nerve for inferior teeth called inferior dental nerve, which is found inside it. While there are many research studies trying to visually identify the mandibular canal, e.g., [1, 7, 9] or to mark the canal by searching the whole image, like in [11], in this paper, we propose a double application of the generalized anisotropic Hough transform, first used to detect a part of the mandibular bone and to restrict the search area, followed by second application in the detection for marking the mandibular canal. This procedure decreases the risk of false detection of the mandibular canal by focalizing on the exact area where the canal is situated. The method is based on template matching of a shape which can be found in other zones of the dental radiography, representing other anatomical features.

The **the Hough transform** is a popular technique to extract features from an image. The method was patented in 1962 [5] for the detection of lines in photographs. The functioning of the algorithm lies in a proper choice of the parameters space for the set of lines on the plane [6].

In order to develop a method for the recognition of a generic template in an image, [2] used the following parameters for a shape:

$$\boldsymbol{a} = \{\boldsymbol{y}, \boldsymbol{S}, \theta\} \tag{1}$$

where $\boldsymbol{y} = (x_r, y_r)$ is a reference point to represent the translations, $\boldsymbol{S} = \left(S_x, S_y\right)$ are scale values for the orthogonal shearing deformations, and θ is an angle that represents the rotations.

The reference point $\boldsymbol{y}$ is described in terms of a table, called the **R-table** of the template, of possible edge pixel orientations. The other parameters are described in terms of transformations of the aforementioned table.

The key for generalizing the Hough transform is to use the directional information. Given a template, i.e., a set of boundary points $\{\boldsymbol{x}_B\}$, a reference point $\boldsymbol{y}$ is chosen. After the discretization of the straight angle through a uniform partition $\{0, \Delta t, 2\Delta t, \ldots, N\Delta t\}$, for every boundary point the tangent direction $\phi(\boldsymbol{x}_B)$ is computed, then $\boldsymbol{r} = \boldsymbol{y} - \boldsymbol{x}_B$ is stored in the nth bin of the R-Table if $\mod(\phi(\boldsymbol{x}_B), \pi) \in [(n-1)\Delta t, n\Delta t)$ as in [11]. Given this simple structure, with the use of rotation and shearing operators, i.e. rot_θ and def_S, we can build the following procedure to detect the template in the set of edge points of any image.

Data: `IMG`: Image
`R-T`: R-Table of a template
$\{(S, \theta)\}$: Set of scale and rotation parameters
`WHERE`:subregion of `IMG`
Result: Pattern Localization

Compute $(\boldsymbol{x}_e, \phi(\boldsymbol{x}_e))$ edge points and their gradient;
for *each* (S, θ) **do**
  for *each* $\boldsymbol{x}_e$ **do**
    Find $n\Delta t$ s.t. $defs_S(rot_\theta(\phi(\boldsymbol{x}_e))) \in n\Delta t$;
    for *each* $\boldsymbol{r} \in n\Delta$ **do**
      Compute $\boldsymbol{y}'_{(S,\theta)} = \boldsymbol{x}_e + defs_S(rot_\theta(\boldsymbol{r}))$;
      if $\boldsymbol{y}' \in$ *WHERE* **then**
        Report an occurrence of $\boldsymbol{y}'_{(S,\theta)}$.
      end
    end
  end
  Select $\boldsymbol{y}_{(S,\theta)}$ with higher occurrence ;
end
return $\boldsymbol{y}$ with higher occurrence;

Algorithm 1: `Hough_Recognition`

In this way, we find the edge point that satisfy the non-analytic version of

$$f(\boldsymbol{x}, \boldsymbol{a}) = 0 \tag{2}$$

and

$$\frac{\partial f}{\partial \boldsymbol{x}}(\boldsymbol{x}, \boldsymbol{a}) = 0 \tag{3}$$

The space of occurrences for the reference point is called **Accumulator Space**. Regarding complexity concerns of the algorithm, we point out that all of the 3 nested **for** loops could be parallelised.

2 The Recognition Procedure

Problems with AGTH recognition may arise when we have to deal with real images which could be corrupted by noise. That is the case of radiography where the structure of a bone is not well defined and where some part of the bone which can be disguised with unwanted details.

The position of the mandibular canal is described by medical indications as follows: the canal starts at the mandibular foramen in the middle part of the vertical ramus. It continues through the mandibular bone and ends in the menton foramen between apexes of the two inferior premolars.

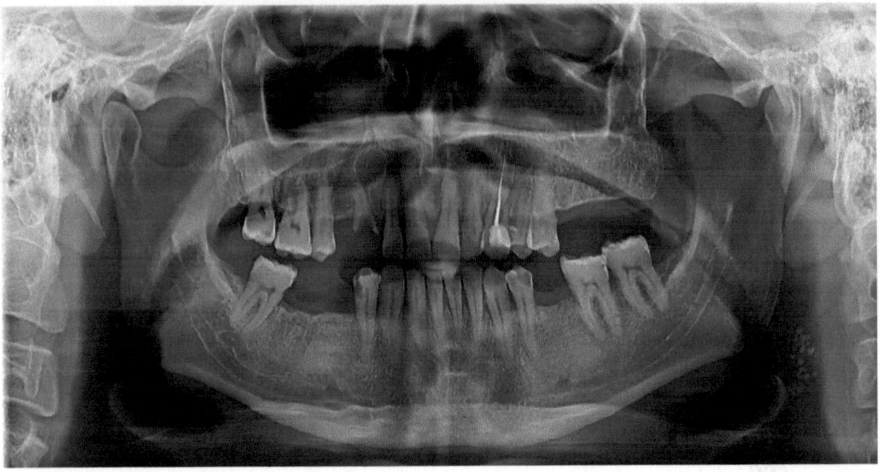

Fig. 1 Typical panoramic radiography

With this indicaton we can roughly compare the position of the canal against the barycenter of the mandible. This is how the doctor's mind work, by focusing on the interest area representing the horizontal ramus of the mandibular bone (first Hough transform) and recognizing a pattern which represents the mandibular canal (second Hough recognition). This is the algorithm that we want to mimic.

As shown below, the shape of the canal can become misleading if we analyze radiograph of a patient who has lost some teeth.

Any surgical intervention in the mandibular area must prevent any nerve injury. The injury of the nerve would result in prolonged local and lower lip anesthesia for a minimum period of six weeks. Estimating the position of the mandibular canal means knowing the position of the nerve and by this the surgeon can estimate the risks and to adapt the surgical procedure to the individual case.

In our test, we used Fig. 1 to extract the template of the canal. The recognition performs well on the same panoramic radiography as it can be seen in Fig. 4, but the aforementioned image is in a critical situation: the patient has only one molar on the right part of the mandible.

When the mandible is edentated, without any further restriction, the canal could not be recognized anymore because the AGHT algorithm matches the template in Fig. 2 with the top part of the alveolar process, the part of the bone where the teeth should be. It happens because the process is detected with a thicker edge than the canal, but has the same gradient direction and shape, so in the recognition process it will have more importance. This undesired, unavoidable, matching has been soften by using the modulus-π direction of the gradient: the canal, as modeled in Fig. 2, looks like the empty space in an edentated mouth; by using the modulus we remove every information about the inside-outside of the model. This seems the best choice

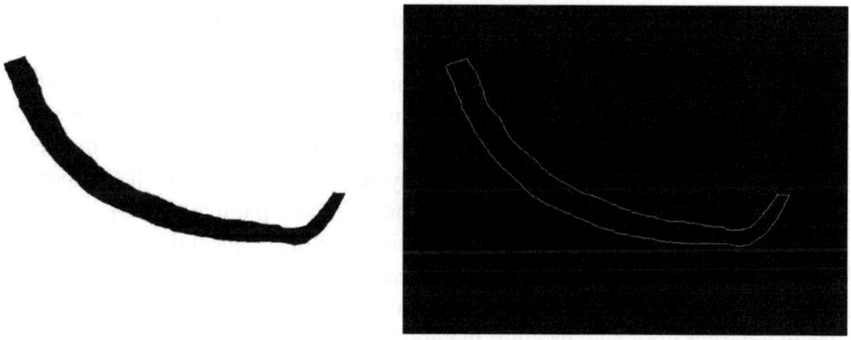

Fig. 2 Canal and its contour

for the case of the mandibular canal, a poorly defined region of the mandible enclosed by a slightly brighter contour.

This problem is related to how the AGHT is implemented: one of the weakness of this transform is that the accumulation space does not carry any information about the position in space of the template nor the mutual relation of different shape in the image. We identify two ways to overcome this problem:

- Manual solution: the user should restrict the area to be investigated manually through anatomical information given beforehand.
- Double-automatic solution: after a first, coarse and less accurate search for the mandible template through AGHT, the area to be investigated for the mandibular canal is automatically restricted.

We used the second method as described in the sequel.

3 The Double Hough Transform Method

The canal template described in Sect. 2 has the following characteristics: it is a connected, compact and simply connected region of the plane. Therefore, after the binarization of the template, we can run a contour-following algorithm to detect the boundary points. This way, one could reach the first purpose for the proposed pattern recognition: to obtain an easy manipulable set of data samples.

The sorted array of boundary point that we obtain through a contour-following algorithm is well suited for the double Hough transform. As mentioned before, we do not need an accurate detection of the mandible, because this pattern belongs to the high level set of structure in the hierarchy of the image. In this way we can easily

- Compute the gradient of a boundary point knowing its neighborhood
- Subsample the high level templates in a set of equally spaced point

The same technique cannot be applied directly to a panoramic radiograph. Edge points of are the result of an high pass filter applied to the image.

We find out that common edge detector filters such as Canny, Sobel, etc., fail. Radiographs are spurious images which contain a great amount of unwanted details. So, we concentrated our work on the choice of the good parameters for the detection of the edges of the teeth (for a good survey see [4]). This means the proper choice of the variance in the Gaussian filter and the threshold parameter [10].

The processes of low-pass filtering and thresholding, cancel the mandibular canal from the image; so we have to create an *ad hoc* method for detecting the edges. It can be seen from Fig. 1 that mandibular canals are drawn by two bright gray curved lines in a darker gray background; the good point is that the orientation of the canal is steady. Therefore, the natural idea is to use a high-pass anisotropic filter mask adjusted on the shape of the mandibular canal.

To calculate the gradient needed in the implementation of the AGHT, we used a Sobel mask for both the boundary and the edge points to have consistency in the calculation.

After the calculation of the barycenter of the template, for the construction of the R-table we chose to store $\boldsymbol{r} = \boldsymbol{y} - \boldsymbol{x}_B$ in Cartesian coordinates to follow the natural discretization introduced by an image. This also helps us to understand the worst-case scenario: after the recognition process the accumulator space will be a matrix with the same dimension of the image so we could print it on screen in grayscale to understand how the error spread and which other shape can be disguised as a mandibular canal.

We increment the accumulator space as in [11] and we use the Gabor transfrom [3, 8] to rank it.

The last remark should be about the parameters expressed in (1). We have not used the rotational parameter θ inasmuch as every panoramic radiography is taken with the patient's head fixed. The important parameter is $\boldsymbol{S} = \left(S_x, S_y\right)$ which helps us to reconstruct the anatomical difference among human beings. In this way, it is possible to find the best deformed version of the template in Fig. 2 that matches the canal in the panoramic radiography under analysis.

We sum up in Algorithm 2 all the remarks we expressed in the previous section.

Data: `IMG`: Image
`TMP_C`: Mandibular Canal
`TMP_M`: Mandible
$\{(S_x^C, S_y^C)\}$ Possible Canal Scales
$\{(S_x^M, S_y^M)\}$ Possible Mandible Scales
Result: Template Localization

$\{\boldsymbol{x}_B^M\}$=`Contour-Following(TMP_M)`;
`R-T_M = R-Table_Build(subsample(`$\{\boldsymbol{x}_B^M\}$`))`;
$\boldsymbol{y}^M$=`Hough_Recognition(IMG,R-T_M,`$\{(S_x^M, S_y^M)\}$`,IMG)`;

Select `A_M` an Area around $\boldsymbol{y}^M$;

$\{\boldsymbol{x}_B^C\}$=`Contour-Following(TMP_C)`;
`R-Table_C = R-Table_Build(`$\{\boldsymbol{x}_B^C\}$`)`;
$\boldsymbol{y}^C$=`Hough_Recognition(IMG,R-T_C,`$\{(S_x^C, S_y^C)\}$`,A_M)`;
return $\boldsymbol{y}^C$

4 Results

In this section, we present the experimental results of the proposed algorithm.

The first test is to search for the template extracted from Fig. 1 in the same image. The overlapping is perfect even without the double technique. The same argument can be brought forward for the flipped template, since there is no significant anatomical difference between the left and the right side of the same patient.

Fig. 3 Accumulator space for canal recognition

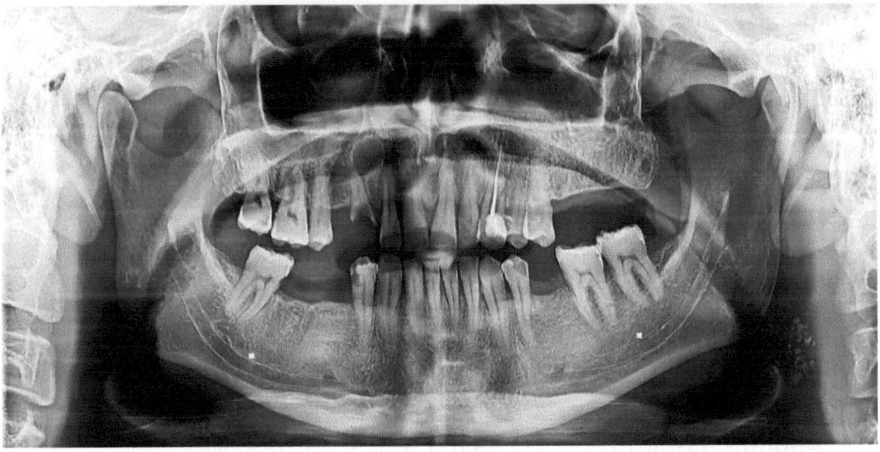

Fig. 4 Perfect matching of the template on the original image. Optimal matching of the *flipped-left* template

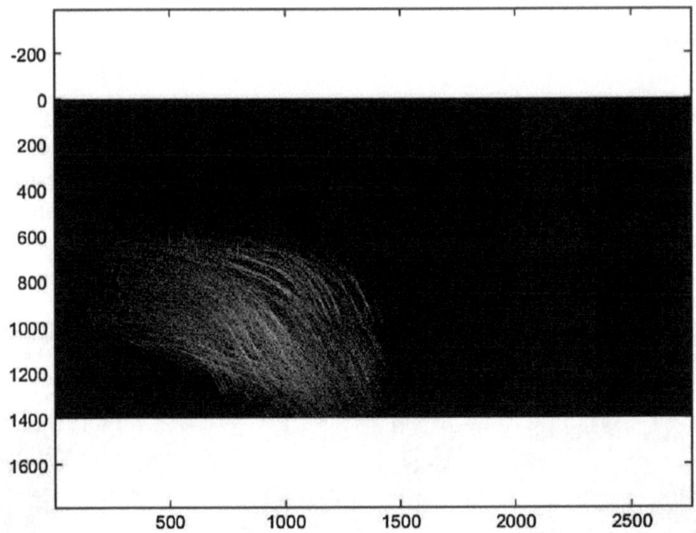

Fig. 5 Accumulator space for mandibular canal. Restriction to *bottom-left* part of picture 6

The approach proposed in this paper is based on a *Double Hough* transform, named in this way because it searches for the whole mandible to find the area of interest, then the AGHT recognition process for the canal is performed.

This process is possible because the canal lies in the center of the mandible, so their reference points (their barycenters) are really close. After the detection of reference points for the mandible $\mathbf{y}_M = (x_M, y_M)$, we force the reference points of

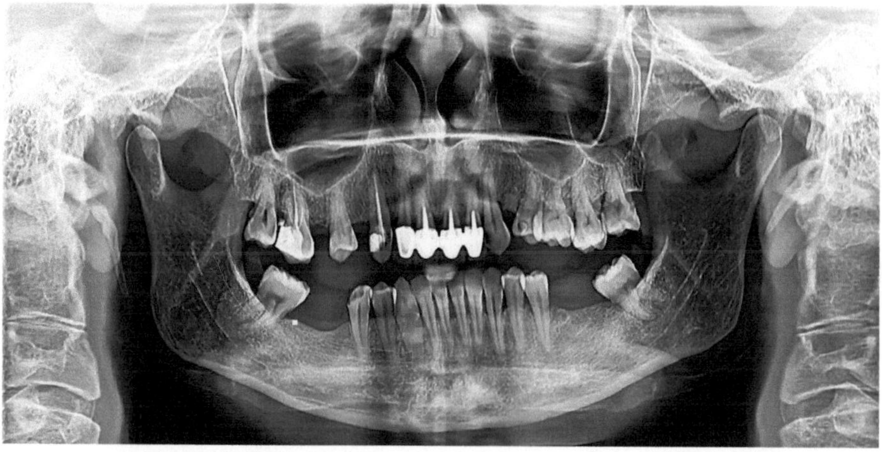

Fig. 6 Hough recognition without area restriction

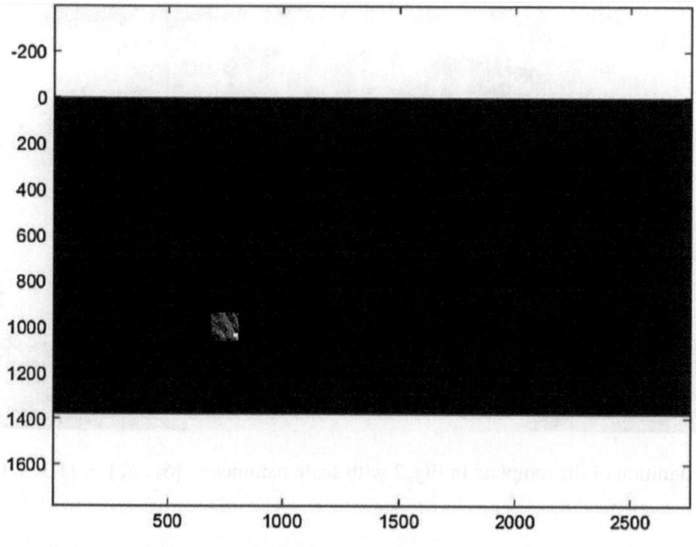

Fig. 7 Accumulator space for mandibular canal. Double Hough restriction

the canal to be in the square window $[x_M - 50, x_M + 50] \times [x_M - 50, x_M + 50]$, with underlining unit of measure the pixel (Figs. 6, 8).

It is easily observed in the Fig. 9 that if the patient has all his teeth, the area-restriction process is not mandatory for the canal recognition.

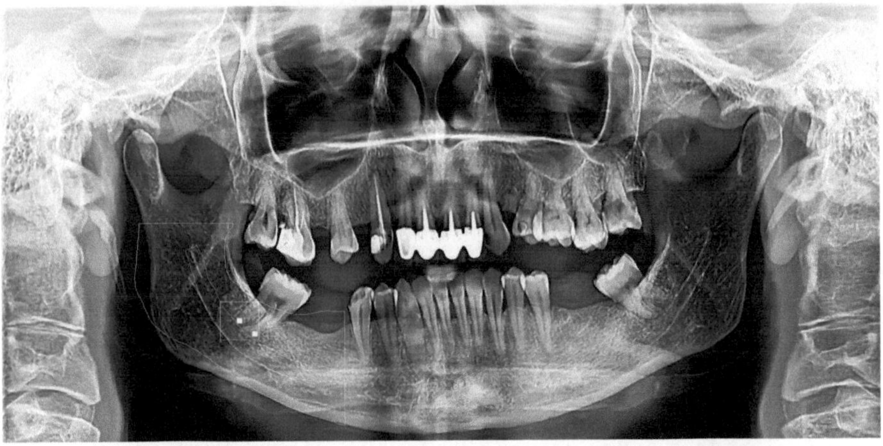

Fig. 8 Double Hough solution through area restriction

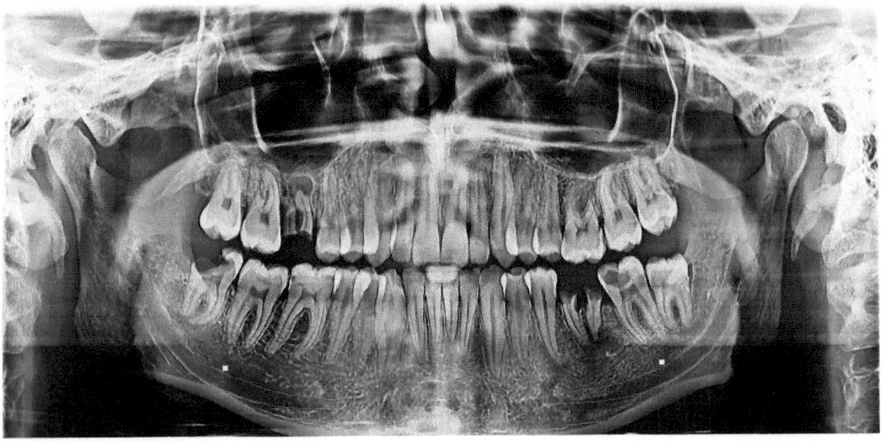

Fig. 9 Recognition of the template in Fig. 2 with scale parameters $(S_x, S_y) = (1.1, 1.15)$

5 Conclusions

In this work, we used the double AGHT for the detection of the mandibular canal in a panoramic radiograph. We had to deal with different shapes which can be mistaken with each other, such as the top of the mandible and the mandibular canal (Figs. 3, 4, 5, 6, 7, 8, 9).

We wanted to restrict the area to be analyzed using medical information with an automatic strategy.

Therefore, we used an automatic selection based on the mutual relation of different patterns in an image.

Acknowledgements The first two co-authors gratefully acknowledge the support of the Austrian Science Fund (FWF): project number P27516.

References

1. Atieh, M.A.: Diagnostic accuracy of panoramic radiography in determining relationship between inferior alveolar nerve and mandibular third molar. J. Oral Maxillofac. Surg. **68**, 74–82 (2010)
2. Ballard, D.H.: Generalizing the Hough transform to detect arbitrary shapes. Pattern Recognit. **13**, 111–122 (1981)
3. Feichtinger, H.G., Strohmer, T.: Gabor Analysis and Algorithms: Theory and Applications. Birkhäuser, Boston (1998)
4. Gráfová, L., Kašparová, M., Kakawand, S., Procházka, A., Dostálová, T.: Study of edge detection task in dental panoramic radiographs. Dentomaxillofac. Radiol. **42**, 20120391 (2013)
5. Hough Paul, V.C.: Method and means for recognizing complex patterns. US Patent 3069654 (1962)
6. Illingworth, J., Kittler, J.: A survey of the Hough transform. Comput. Vis. Gr. Image Process. **44**, 87–116 (1988)
7. Jhamb, A., Dolas, R.S., Pandilwar, P.K., Mohanty, S.: Comparative efficacy of spiral computed tomography and orthopantomography in preoperative detection of relation of inferior alveolar neurovascular bundle to the impacted mandibular third molar. J. Oral Maxillofac. Surg. **67**, 58–66 (2009)
8. Karlheinz, G.: Foundations of Time-Frequency Analysis. Birkhäuser, Boston (2001)
9. Mehra, A., Pai, K.M.: Evaluation of dimensional accuracy of panoramic cross-sectional tomography, its ability to identify the inferior alveolar canal, and its impact on estimation of appropriate implant dimensions in the mandibular posterior region. Clin. Implant Dent. Relat. Res. **14**, 100–111 (2012)
10. Onchis, D.M., Real, P., Gillich, G.R.: Gabor frames and topology-based strategies for astronomical images. In: Proceedings of CTIC 2010 Computational Topology in Image Context, pp. 159–166. Chipiona, Spain (2010)
11. Onchis, D.M., Zappala, S., Gotia, S.L., Real, P., Pricop, M.: Detection of the mandibular canal in orthopantomography using a Gabor-filtered anisotropic generalized Hough transform. Pattern Recognit. Lett. (In press) (2015)

Simple and Nearly Optimal Polynomial Root-Finding by Means of Root Radii Approximation

Victor Y. Pan

Abstract We propose a new simple but nearly optimal algorithm for the approximation of all sufficiently well isolated complex roots and root clusters of a univariate polynomial. Quite typically the known root-finders at first compute some crude but reasonably good approximations to well-conditioned roots (that is, those well isolated from the other roots) and then refine the approximations very fast, by using Boolean time which is nearly optimal, up to a polylogarithmic factor. By combining and extending some old root-finding techniques, the geometry of the complex plane, and randomized parametrization, we accelerate the initial stage of obtaining crude approximations to all well-conditioned simple and multiple roots as well as to all isolated root clusters. Our algorithm performs this stage at a Boolean cost dominated by the nearly optimal cost of the subsequent refinement of these approximations, which we can perform concurrently, with minimum processor communication and synchronization. Our techniques are quite simple and elementary; their power and application range may increase in their combination with the known efficient root-finding methods.

Keywords Polynomials · Root-finding · Root isolation · Root radii

1 Introduction

1.1 The Problem and Our Progress

We seek the roots $x_1, \ldots, x_n$ of a univariate polynomial of degree n with real or complex coefficients,

V.Y. Pan (✉)
Departments of Mathematics and Computer Science, Lehman College
and the Graduate Center of the City University of New York, Bronx, NY 10468, USA
email: victor.pan@lehman.cuny.edu
URL: http://comet.lehman.cuny.edu/vpan/

V.Y. Pan
Ph.D. Programs in Mathematics and Computer Science, The Graduate Center
of the City University of New York, New York, NY 10036, USA

I.S. Kotsireas and E. Martínez-Moro (eds.), *Applications of Computer Algebra*,
Springer Proceedings in Mathematics & Statistics 198,
DOI 10.1007/978-3-319-56932-1_23

$$p(x) = \sum_{i=0}^{n} p_i x^i = p_n \prod_{j=1}^{n} (x - x_j), \quad p_n \neq 0. \tag{1}$$

This classical problem is four millennia old, but is still important, e.g., for Geometric Modeling, Control, Robotics, Global Optimization, Financial Mathematics, and Signal and Image Processing (cf. [8, Preface]).

Quite typically a fast root-finder consists of two stages. At first one computes some crude but reasonably good approximations to well-conditioned roots (that is, those well isolated from the other roots) and then refines the approximations very fast, within nearly optimal Boolean time. Here and hereafter "nearly optimal" means "optimal up to a polylogarithmic factor", and we measure the isolation of two roots x_g and x_h from one another by the ratio $|x_g - x_h|/\max_{1 \le i,j \le n} |x_i - x_j|$.

We obtain substantial progress the initial stage of computing crude but reasonably close initial approximations to all well-conditioned and possibly multiple roots. The Boolean cost of performing our algorithm can be complemented by the nearly optimal Boolean cost of refining our initial approximation by means of the algorithms of [16–19]. By combining them with our present algorithm, we approximate all well-conditioned roots of a polynomial by matching the record and nearly optimal Boolean complexity bound of [11] and [14]. Our present algorithm, however, is much less involved, more transparent and more accessible for the implementation (see the next subsection).

Approximation of the well-conditioned roots is already an important sub-problem of the root-finding problem, but can also facilitate the subsequent approximation of the ill-conditioned roots. E.g., having approximated the well-conditioned roots $1, -1, \sqrt{-1}$, and $-\sqrt{-1}$ of the polynomial $p(x) = (x^4 - 1)(x^4 - 10^{-200})$, we can deflate it explicitly or implicitly (cf. [15]) and more readily approximate its ill-conditioned roots $10^{-50}, -10^{-50}, 10^{-50}\sqrt{-1}$, and $-10^{-50}\sqrt{-1}$ as the well-conditioned roots of the deflated polynomial $x^4 - 10^{-200}$ (cf. the first paragraph of Sect. 5).

Moreover, our algorithm can be readily extended to computing crude approximations of small discs covering all isolated root clusters (cf. Remark 3). Then again the Boolean cost of this computation is dominated by the cost of the refinement of the computed approximations to the clusters.

Clearly, the refinement of well-conditioned roots and root clusters can be performed concurrently, with minimum communication and synchronization among the processors. The existence of non-isolated roots and root clusters little affects our algorithm; our cost estimate does not depend on the minimal distance between the roots and includes no terms like $\log(\mathrm{Discr}(p)^{-1})$.

1.2 Our Technical Means

We achieve our progress by means of exploiting the geometry of the complex plane, randomized parametrization, and an old algorithm of [21], which approximates all root radii of a polynomial, that is, the distances from all roots to the origin, at a low

Boolean cost. We refer the reader to [1, 3, 6], and [10] on the preceding works related to that algorithm and cited in [21, Section 14] and [12, Section 4], and one can also compare the relevant techniques of the power geometry in [1] and [2], developed for the study of algebraic and differential equations. By combining the root-radii algorithm with a shift of the origin, one readily extends it to fast approximation of the distances of all roots to any selected point of the complex plane (see Corollary 1).

Computations of our algorithm amount essentially to approximation of the root radii of three polynomials obtained from a polynomial $p(x)$ by means of three shifts of the variable x, versus many more computations of this kind and application of many other nontrivial techniques in the algorithm of [11] and [14]. This makes it much harder to implement and even to comprehend than our present algorithm.

Schönhage in [21] used only a small part of the potential power of his root-radii algorithm by applying it to the rather modest auxiliary task of the isolation of a single complex root, and we restricted ourselves to similar auxiliary applications in [11] and [14]. The algorithm and its basic concept of Newton's polygon, however, deserve greater attention of the researchers in univariate polynomial root-finding.

In the next two subsections we outline our algorithms of [20] and the present paper, which should demonstrate the power of our approach.

1.3 Approximation of Well-Conditioned Real Roots: An Outline

It is instructive to recall the algorithm of [20], which approximates all simple and well-conditioned real roots in nearly optimal Boolean time. At first it approximates all the n root radii. They define n narrow annuli at the complex plane, all of them centered at the origin and each of them containing a root of the polynomial $p(x)$. Multiple roots define multiple annuli. Clusters of roots define clusters of overlapping annuli. The intersections of at most n narrow annuli with the real axis define at most $2n$ small intervals, which contain all real roots. By applying to these intervals a known efficient real root-refiner, e.g., that of [17, 19], we readily approximate all well-conditioned real roots within a desired precision at a nearly optimal Boolean cost.

In [20] the resulting real root-finder was tested for some benchmark polynomials, each having a small number of real roots. The tests, performed numerically, with the IEEE standard double precision, gave encouraging results; in particular the number of the auxiliary root-squaring iterations (3) grew very slowly as the degree of input polynomials increased from 64 to 1024.

1.4 Approximation of Well-Conditioned Complex Roots: An Outline

Next we outline our main algorithm, which we specify in some detail and analyze in Sect. 4. The algorithm approximates the well-conditioned complex roots of a polynomial by means of incorporating the root-radii algorithm into a rather sophisticated construction on the complex plane.

At first we compute a sufficiently large positive value r_1^+ such that the disc $D = \{x : |x| \leq r_1^+\}$ on the complex plane contains all roots of $p(x)$. Then we approximate the distances to the roots from the two points, ηr_1^+ on the real axis and $\eta r_1^+ \sqrt{-1}$ on the imaginary axis, for a reasonably large value of η, so that these two points lie reasonably far from the disc D.

Having the distances approximated, we obtain two families of narrow annuli centered at the latter pair of points. Each family is made up of n annuli that contain the n roots of a polynomial $p(x)$, all lying in the disc D. Its intersections with the annuli are closely approximated by n narrow vertical and n narrow horizontal rectangles on the complex plane. Every root lies in the intersection of two rectangles of the vertical and horizontal families, and there are $N \leq n^2$ intersections overall, each approximated by a square.

At most n squares contain all $n_- \leq n$ well-conditioned roots of a polynomial $p(x)$, and we can identify these squares by evaluating $p(x)$ or applying proximity tests at the centers of N candidate squares (and then we would discard the other $N - n_-$ squares). The cost of these computations would be prohibitively large, however, and so instead we identify the desired n_- squares probabilistically, by applying the root-radii algorithm once again.

This time we approximate the distances to all the n roots from a complex point $\eta r_1^+ \exp(\frac{\gamma\sqrt{-1}}{2\pi})$ where we choose the angle γ at random in a fixed range. Having the distances approximated, we obtain at most n_- narrow rectangles that contain all the n roots. The long sides of the rectangles are directed at the angle γ to the real axis. We choose the range for γ such that with a probability close to 1 each rectangle intersects a single square. Then we readily compute the centers of all these squares in a nearly optimal randomized Boolean time and notice that all the well-conditioned roots are expected to be closely approximated by some of these centers. We can refine these approximations readily by applying the efficient algorithms of [7] or [19].

1.5 Organization of Our Paper

We organize our presentation as follows. In the next two sections we recall some auxiliary results. In Sect. 4 we describe our main algorithm. In Sect. 5 we briefly comment on some directions to its strengthening and extension.

2 Some Definitions and Auxiliary Results

Hereafter "flop" stands for "arithmetic operation", and "lg" stands for "$\log_2$".

$O_B(\cdot)$ and $\tilde{O}_B(\cdot)$ denote the Boolean complexity up to some constant and polylogarithmic factors, respectively.

$||p(x)|| = \max_{i=0}^{n} |p_i|$ and $\tau = \lg\left(||p(x)|| + \frac{1}{||p(x)||}\right)$ for a polynomial $p(x)$ of (1).

Definition 1 $D(z, \rho) = \{x : |x - z| \leq \rho\}$ denotes the closed disc with a complex center z and a radius ρ. Such a disc is γ*-isolated*, for $\gamma > 1$, if the disc $D(z, \gamma\rho)$ contains no other roots of a polynomial $p(x)$ of Eq. (1). Its root x_j is γ*-isolated* if no other roots of the polynomial $p(x)$ lie in the disc $D(x_j, (\gamma + 1)|x_j|)$.

Suppose that crude but reasonably close approximations to the set of well-isolated roots of a polynomial are available. Then, by applying the algorithms of [7] or [17], [19], one can refine these approximations to these roots at a low Boolean cost. In the rest of our paper we present and analyze our new algorithm for computing such crude initial approximations to the well-isolated roots.

3 Approximation of Root Radii and Distances to Roots

Definition 2 List the absolute values of the roots of $p(x)$ in non-increasing order, denote them $r_j = |x_j|$ for $j = 1, \dots, n$, $r_1 \geq r_2 \geq \cdots \geq r_n$, and call them the *root radii* of the polynomial $p(x)$.

The following theorem bounds the largest root radius r_1, and then we bound the Boolean cost of the approximation of all root radii.

Theorem 1 *(See* [22].*) For a polynomial $p(x)$ of* (1) *and $r_1 = \max_{j=1}^{n} |x_j|$, it holds that*

$$0.5r_1^+/n \leq r_1 \leq r_1^+ \quad \text{for } r_1^+ = 2\max_{i=1}^{n} |p_{n-i}/p_n|. \tag{2}$$

Theorem 2 *Assume that we are given a polynomial $p = p(x)$ of* (1) *and $\theta > 1$. Then, within the Boolean cost bound $\tilde{O}_B(\tau n + n^2)$, one can compute approximations $\tilde{r}_j$ to all root radii r_j such that $1/\theta \leq \tilde{r}_j/r_j \leq \theta$ for $j = 1, \dots, n$, provided that $\lg(\frac{1}{\theta-1}) = O(\lg(n))$, that is, $|\tilde{r}_j/r_j - 1| \geq c/n^d$ for a fixed pair of constants $c > 0$ and d.*

Proof This is [21, Corollary 14.3].

Let us sketch this proof and the supporting algorithm. At first approximate the n root radii at a dominated cost in the case where $\theta = 2n$ (see [21, Corollary 14.3] or [12, Section 4]). In order to extend the approximation to the case where $\theta = (2n)^{1/2^k}$ for any positive integer k, apply k Dandelin's root-squaring iterations to the monic polynomial $q_0(x) = p(x)/p_n$ (cf. [5]), that is, compute recursively the polynomials

$$q_i(x) = (-1)^n q_{i-1}(\sqrt{x}\,)q_{i-1}(-\sqrt{x}\,) = \prod_{j=1}^{n}\left(x - x_j^{2^i}\right), \quad \text{for } i = 1, 2, \dots \tag{3}$$

Then approximate the root radii $r_j^{(k)}$ of the polynomial $q_k(x)$ by applying Theorem 2 for $\theta = 2n$ and for $p(x)$ replaced by $q_k(x)$. Finally approximate the root radii r_j of the polynomial $p(x)$ as $r_j = (r_j^{(k)})^{1/2^k}$.

Having isolation ratio $2n$ for $q_k(x)$ is equivalent to having isolation ratio $(2n)^{1/2^k}$ for $p(x)$, which is $1 + c/n^d = 1 + 2^{O(\lg(n))}$ for $k = O(\lg(n))$ and any fixed pair of constants $c > 0$ and d. Each Dandelin's iteration amounts to convolution, and Schönhage in [21] estimates that the Boolean cost of performing $k = O(\lg(n))$ iterations is within the cost bound of Theorem 2.

Corollary 1 *Assume that we are given a polynomial $p(x)$ of (1) and a complex z. Then, within the Boolean cost bound $\tilde{O}_B((\tau + n(1 + \beta))n)$, for $\beta = \lg(2 + |z|)$, one can compute approximations $\tilde{r}_j \approx \bar{r}_j$ to the distances $\bar{r}_j = |z - x_j|$ from the point z to all roots x_j of the polynomial $p(x)$ such that $1/\theta \le \tilde{r}_j/\bar{r}_j \le 1 < \theta$, for $j = 1, \ldots, n$, provided that $\lg(\frac{1}{\theta - 1}) = O(\lg(n))$.*

Proof The root radii of the polynomial $q(x)$ for a complex scalar z are equal to the distances $|x_j - z|$ from the point z to the roots x_j of $p(x)$. Let $\bar{r}_j$ for $j = 1, \ldots, n$ denote these root radii listed in the non-increasing order. Then, clearly, $\bar{r}_j \le r_j + |z|$ for $j = 1, \ldots, n$.

Furthermore, the coefficients of the polynomial $q(x) = p(x - z) = \sum_{i=0}^{n} q_i x^i$ have bit-size $\tilde{O}(\tau + n(1 + \beta))$ for $\beta = \lg(2 + |z|)$. By applying Theorem 2 to the polynomial $q(x)$, extend the cost bounds from the root radii to the distances.

To complete the proof, recall that, for a polynomial $p(x)$ of (1) and a complex scalar z, one can compute the coefficients of the polynomial $q(x) = p(x + z)$ by using $O(n \lg(n))$ flops (cf. [13]) and at a dominated Boolean cost (cf. [4]).

4 Approximation of Well-Conditioned Complex Roots by Using Root-Radii Algorithm

4.1 Approximation of Well-Conditioned Roots: An Algorithm

Let us specify our new algorithm.

Algorithm 1 *Approximation of Well-Conditioned Complex Roots.*
INPUT: two positive numbers ρ and ϵ and the coefficients of a polynomial $p(x)$ of (1).
OUTPUT: A set of approximations to the roots of the polynomial $p(x)$ within $\rho/\sqrt{2}$ such that with a probability at least $1 - \epsilon$ this set approximates all roots that have δ-neighborhoods containing no other roots of the polynomial $p(x)$ for

$$\delta = n^2(n^2 - 1)\frac{4\rho}{\pi\epsilon}.$$

INITIALIZATION: Fix a reasonably large scalar η, say, $\eta = 100$. Generate a random value ϕ under the uniform probability distribution in the range $[\pi/8, 3\pi/8]$.
COMPUTATIONS:

1. (Three Long Shifts of the Variable.) Compute the value $r_1^+ = 2\max_{i=1}^n |\frac{p_{n-i}}{p_n}|$ of (2). Then compute the coefficients of the three polynomials

$$q(x) = p(x - \eta r_1^+),$$

$$q_-(x) = p(x - \eta r_1^+\sqrt{-1}),$$

$$q_\phi(x) = p\left(x - \eta r_1^+ \exp\left(\frac{\phi\sqrt{-1}}{2\pi}\right)\right).$$

2. Compute approximations to all the n root radii of each of these three polynomials within the error bound $\rho/2$.
   This defines three families of large thin annuli having width at most ρ. Each family consists of n annuli, and each annulus contains a root of $p(x)$. Multiple roots define multiple annuli. Clusters of roots define clusters of overlapping annuli.
   At most $2n$ coordinates on the real and imaginary axis define the intersections of all pairs of the annuli from the first two families and of the disc $D = D(0, r_1^+)$. We only care about the roots of $p(x)$, and all of them lie in the disc D.
   We have assumed that the value η is large enough and now observe the following properties.
   - The intersection of each annulus with the disc D is close to a vertical or horizontal rectangle on the complex plane.
   - Every rectangle has width about ρ or less because every annulus has width at most ρ.
   - The intersection of any pair of annuli from the two families is close to a square having vertical and horizontal edges of length about ρ or less. We call such a square a *node*.
   - The disc D contains a *grid* made up of N such nodes, for $N \le n^2$.
   - The center of the (i, j)th node has real part $r_i^{(1)} - \eta r_1^+$ and imaginary part $r_j^{(2)} - \eta r_1^+$, for $i, j = 0, 1, \ldots, n$. Here $r_i^{(1)}$ and $r_j^{(2)}$ denote the distances of the roots x_i and x_j, respectively, from the real point ηr_1^+ and the complex point $\eta r_1^+\sqrt{-1}$, respectively.
3. For each annulus of the third family, determine whether it intersects only a single node of the grid. If so, output the center of this node.

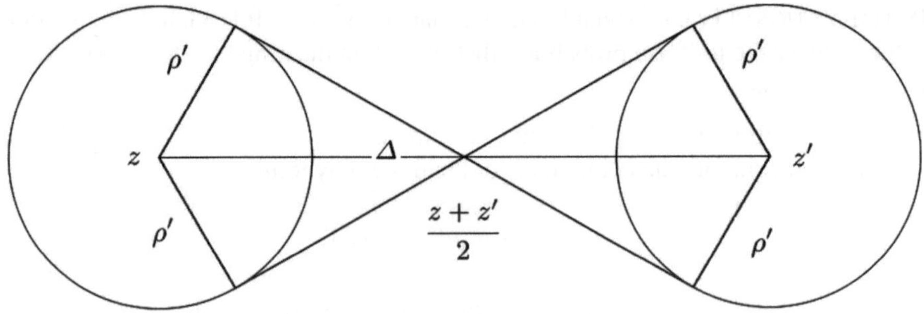

Fig. 1 The discs $D(z, \rho)$ and $D(z, \rho)$ are from the proof of Lemma 1. The parameter $\Delta = |z - z|$ is used in Theorems 3 and 4

4.2 *Approximation of Complex Roots: Correctness of Algorithm 1*

Let us prove *correctness* of Algorithm 1.

For simplicity assume that the annuli computed by it and the nodes of a grid are replaced by their approximating rectangles and squares, respectively.

At first readily verify the following lemma.

Lemma 1 *Suppose that z and z' are two complex numbers, ρ' is a positive number, and a straight line passes through a disc $D(z, \rho')$ under an angle β with the real axis where we choose β at random under the uniform probability distribution in the range $[\alpha, \alpha + \gamma]$ for $0 < \gamma \leq 2\pi$ and $0 < \alpha \leq 2\pi$. Then the line intersects a disc $D(z', \rho')$ with a probability at most $P = \frac{2}{\gamma} \sin(\frac{2\rho'}{|z-z'|})$.*

Proof (See Fig. 1.) Consider the two tangent lines common for the discs $D(z, \rho')$ and $D(z', \rho')$ and both passing through the complex point $\frac{z+z'}{2}$. Then any straight line intersects both discs if and only if it lies in the angle $2\sin(\frac{2\rho'}{|z-z'|})$, formed by these two lines. This implies the lemma.

Next apply it to a pair of nodes of the grid having centers z and z', $\alpha = \frac{\pi}{8}$, and $\gamma = \frac{\pi}{4}$. Let the two nodes lie in the two discs $D(z, \rho')$ and $D(z', \rho')$ for $\rho' = \rho/\sqrt{2}$ and $|z - z'| > 2\rho'$, and obtain

$$\frac{\rho'}{|z-z'|} - \frac{1}{6}\left(\frac{\rho'}{|z-z'|}\right)^3 < \sin\left(\frac{\rho'}{|z-z'|}\right) < \frac{\rho'}{|z-z'|}.$$

Then the lemma implies the strict upper bound $\frac{4}{\pi}\frac{\rho'}{|z-z'|}$ on the probability P. Substitute $\rho' = \rho\sqrt{2}$ and obtain

$$P < \frac{4\rho\sqrt{2}}{\pi\,|z-z'|}. \tag{4}$$

Theorem 3 *Let the grid of Algorithm 1 have N_ρ nodes overall, $N_\rho \le n^2$. Define the smallest superscribing disc for every node of a grid. Fix $\Delta > 2\rho$ and call a node of the grid Δ-isolated if the Δ-neighborhood of its center contains no centers of any other node of the grid. Suppose that a rectangle of the third family intersects a fixed Δ-isolated node. Then*

(i) this rectangle intersects the smallest superscribing disc of another node of the grid with a probability less than $\frac{4\rho\sqrt{2}}{\pi\Delta}(N_p - 1)$,
(ii) the probability that any fixed rectangle of the third family intersects the smallest superscribing discs at least two Δ-isolated nodes is less than $\frac{2\rho\sqrt{2}}{\pi\Delta}(N_p - 1)N_p$, and
(iii) if

$$\frac{2\rho\sqrt{2}}{\pi\Delta}(N_p - 1)N_p \le \epsilon, \tag{5}$$

then Algorithm 1 outputs the claimed set of the roots of a polynomial $p(x)$ with a probability more than $1 - \epsilon$.

Proof Apply bound (4) to the fixed node and obtain part (i). Apply bound (4) to all $(N_p - 1)N_p/2$ pairs of distinct nodes of the grid and obtain part (ii). Substitute bound (5) and obtain part (iii).

Now correctness of the algorithm follows because every root of the polynomial $p(x)$ lies in some annulus of each of the three families.

4.3 Approximation of Complex Roots: Complexity of Performing Algorithm 1 and Further Comments

Remark 1 The estimates of Theorem 3 are pessimistic because, for any integer $k > 1$, every k-tuple of nodes intersected by a single straight line contributes to the probability count of Theorem 3 just as much as a single pair of nodes, but we count the contribution of such a k-tuple as that of $(k-1)k/2$ pairs of nodes.

Theorem 4 *Suppose that we are given the coefficients of a polynomial $p(x)$ of equation (1) and two constants ρ and Δ such that $0 < \rho < \Delta$. Then (i) with a probability of success estimated in Theorem 3, the algorithm approximates all roots in Δ-isolated nodes within the error bound ρ, and (ii) the algorithm performs at the Boolean cost within the randomized cost bound of Corollary 1.*

Proof Both claims of the theorem are readily verified as soon as we ensure that the Boolean complexity of Stage 3 of Algorithm 1 is within the claimed bound. To achieve this, apply a bisection process as follows. At first, for a fixed rectangle of the third family, determine whether it intersects any node of the grid below or any node of the grid above its mean node. If the answer is "yes" in both cases, then the rectangle must intersect more than one node of the grid. Otherwise discard about one half of the nodes of the grid and apply similar bisection process to the remaining nodes. Repeat such computation recursively.

Every recursive step either determines that the fixed rectangle intersects more than one node of the grid or discards about 50% of the remaining nodes of the grid. So in $O(\lg(n))$ recursive applications we determine whether the rectangle intersects only a single node of the grid or not.

Application of this process to every rectangle of the third family (made up of n rectangles) requires only $O(n\lg(n))$ tests of the intersections of rectangles with a mean node in the set of the remaining nodes of the grid. Clearly the overall cost of these tests is dominated.

Remark 2 Suppose that we modify Algorithm 1 by collapsing every chain of m pairwise overlapping or coinciding root radii intervals, for $m \leq n$, into a single interval that has a width at most $m\rho$ and by assigning multiplicity m to this interval. Such extended root radius defines an annulus having multiplicity $m \geq 1$ and width in the range from ρ to $m\rho$. Suppose that a pair of such new annuli of a vertical and horizontal families and the disc $D = D(0, r_1^+)$ has multiplicity m_1 and m_2, respectively. Then the intersection of these two annuli defines a node of a new grid, to which we assign multiplicity $\min\{m_1, m_2\}$, and then at Stage 3 of the modified algorithm an output node of multiplicity m contains at most m roots of the polynomial $p(x)$, each counted according to its multiplicity. The probability of success of the algorithm does not decrease and typically increases a little, although the estimation of the increase would be quite involved.

Remark 3 Suppose that Algorithm 1 modified according to the previous remark outputs a node that covers an isolated root or an isolated cluster of the roots of an input polynomial $p(x)$. Then the algorithms of [7, 17, 19] would compress the superscribing disc of this node at a nearly optimal Boolean cost.

Remark 4 We can decrease a little the precision of computing by applying the algorithm with a smaller value of η, although in that case our proof of Theorem 3 would be invalid, and the algorithm would become heuristic.

Remark 5 Suppose that we apply our algorithm as before, but fix an angle ϕ instead of choosing it at random in the range $[\pi/8, 3\pi/8]$. Then for almost all such choices, the algorithm (at its Stage 4) would correctly determine at most n nodes of the grid intersected by the rectangles-annuli of the third family, but finding any specific angle ϕ with this property deterministically would be costly because we would have to ensure that the angle of the real axis with neither of up to $(n^2 - 1)n^2/2$ straight lines passing through the $(n^2 - 1)n^2/2$ pairs of the nodes of the grid approximates ϕ closely. Clearly, this could require us to perform up to $(n^2 - 1)n^2/2$ flops.

5 Conclusions

Algorithm 1 approximates all the isolated single and multiple roots of a polynomial, and its modification of Remark 3 enables us to approximate also all the isolated root clusters. Having specified a modified node containing such a cluster, we can split out a factor $f(x)$ of the polynomial $p(x)$ whose root set is precisely this cluster. Based on the algorithms [7] or [19], we can approximate the factor $f(x)$ at a nearly optimal Boolean cost. Then we can work on root-finding separately for this factor and for the complementary factor $\frac{p(x)}{f(x)}$, both having degrees smaller than n and possibly having better isolated roots.

We plan to work on enhancing the power of this algorithm by means of its combination with various efficient techniques known for root-finding. In particular, the coefficient size of an input polynomial grows very fast in Dandelin's root-squaring iterations, thus requiring high precision computations. We can avoid this growth by applying the algorithm of [9], which uses the tangential representation of the coefficients, but then the Boolean cost bound would increase by a factor of n. So we are challenged to explore alternative techniques for root-radii approximations. We would be interested even in heuristic algorithms, as long as they produce correct outputs for a large input class and perform the computations by using a small number of flops and bounded precision.

Acknowledgements Our work has been supported by NSF Grants CCF 1116736 and CCF 1563942 and PSC CUNY Awards 67699-00 45 and 68862–00 46. We are also grateful to the reviewer for helpful comments.

References

1. Bruno, A.D.: The Local Method of Nonlinear Analysis of Differential Equations. Nauka, Moscow (1979) (in Russian). English Translation: Soviet Mathematics. Springer, Berlin (1989)
2. Bruno, A.D.: Power Geometry in Algebraic and Differential Equations. Fizmatlit, Moscow (in Russian). English Translation: North-Holland Mathematical Library, vol. 57, Elsevier, Amsterdam (2000), also reprinted in 2005, ISBN: 0-444-50297
3. Graham, R.I.: An efficient algorithm for determining the convex hull of a finite planar set. Inf. Process. Lett. **1**, 132–133 (1972)
4. von zur Gathen, J., Gerhard, J.: Fast algorithms for Taylor shifts and certain difference equations. In: Blochinger, W., Küchlin, W. (eds.) Proceedings of the 1997 International Symposium on Symbolic and Algebraic Computation (ISSAC '97), pp. 40–47. ACM, New York (1997)
5. Householder, A.S.: Dandelin, Lobachevskii, or Graeffe. Am. Math. Mon. **66**, 464–466 (1959)
6. Henrici, P.: Applied and Computational Complex Analysis. Wiley, New York (1974)
7. Kirrinnis, P.: Partial fraction decomposition in $C(z)$ and simultaneous Newton iteration for factorization in $C[z]$. J. Complex. **14**, 378–444 (1998)
8. McNamee, J.M., Pan, V.Y.: Numerical Methods for Roots of Polynomials. Part 2 (XXII + 718 pages). Elsevier, Amsterdam (2013)
9. Malajovich, G., Zubelli, J.P.: Tangent Graeffe iteration. Numer. Math. **89**(4), 749–782 (2001)
10. Ostrowski, A.M.: Recherches sur la méthode de Graeffe et les zéros des polynomes et des series de Laurent. Acta Math. **72**, 99–257 (1940)

11. Pan, V.Y.: Optimal (up to polylog factors) sequential and parallel algorithms for approximating complex polynomial zeros. In: Proceedings of the 27th Annual ACM Symposium on Theory of Computing, pp. 741–750. ACM Press, New York (1995)
12. Pan, V.Y.: Approximating complex polynomial zeros: modified quadtree (Weyl's) construction and improved Newton's iteration. J. Complex. **16**(1), 213–264 (2000)
13. Pan, V.Y.: Structured Matrices and Polynomials: Unified Superfast Algorithms. Birkhäuser/Springer, Boston/New York (2001)
14. Pan, V.Y.: Univariate polynomials: nearly optimal algorithms for factorization and root-finding. J. Symb. Comput. **33**(5), 701–733 (2002). Proceedings version in ISSAC' 2001, pp. 253–267. ACM Press, New York (2001)
15. Pan, V.Y.: Completion of Newton's Iterations for Univariate Polynomial Root-finding Initialized at a Quasi-Universal Set. arXiv:1606.01396, math.NA (June 2016)
16. Pan, V.Y., Tsigaridas, E.P.: On the Boolean complexity of the real root refinement. In: Kauers, M. (ed) Proceedings of the International Symposium on Symbolic and Algebraic Computation (ISSAC 2013), pp. 299–306, Boston, June 2013. ACM Press, New York (2013)
17. Pan, V.Y., Tsigaridas, E.P.: Accelerated approximation of the complex roots of a univariate polynomial. In: Proceedings the International Conference on Symbolic and Numeric Computation (SNC'2014), Shanghai, China, July 2014, pp. 132–134. ACM Press, New York (2014) Also April 18, 2014. arXiv:1404.4775 [math.NA]
18. Pan, V.Y., Tsigaridas, E.P.: Nearly optimal refinement of real roots of a univariate polynomial. J. Symb. Comput. **74**, 181–204 (2016). doi:10.1016/j.jsc.2015.06.009
19. Pan, V.Y., Tsigaridas, E.P.: Accelerated approximation of the complex roots and factors of a univariate polynomial. In: Watt, S., Verschelde, J., Zhi, L. (eds.) Theoretical Computer Science, Special Issue on Symbolic—Numerical Algorithms. http://dx.doi.org/10.1016/j.tcs.2017.03.030
20. Pan, V.Y., Zhao, L.: Real root isolation by means of root radii approximation. In: Gerdt, V.P., Koepf, V., Vorozhtsov, E.V. (eds.) Proceedings of the 17th International Workshop on Computer Algebra in Scientific Computing (CASC' 2015), Lecture Notes in Computer Science, vol. 9301, pp. 347–358. Springer, Heidelberg (2015)
21. Schönhage, A.: The fundamental theorem of algebra in terms of computational complexity. Mathematics Department, University of Tübingen (1982)
22. Van der Sluis, A.: Upper bounds on the roots of polynomials. Numer. Math. **15**, 250–262 (1970)

A Fast Schur–Euclid-Type Algorithm for Quasiseparable Polynomials

Sirani M. Perera and Vadim Olshevsky

Abstract In this paper, a fast $\mathcal{O}(n^2)$ algorithm is presented for computing recursive triangular factorization of a Bezoutian matrix associated with quasiseparable polynomials via a displacement equation. The new algorithm applies to a fairly general class of quasiseparable polynomials that includes real orthogonal, Szegö polynomials, and several other important classes of polynomials, e.g., those defined by banded Hessenberg matrices. While the algorithm can be seen as a Schur-type for the Bezoutian matrix it can also be seen as a Euclid-type for quasiseparable polynomials via factorization of a displacement equation. The process, i.e., fast Euclid-type algorithm for quasiseparable polynomials or Schur-type algorithm for Bezoutian associated with quasiseparable polynomials, is carried out with the help of a displacement equation satisfied by Bezoutian and Confederate matrices leading to $\mathcal{O}(n^2)$ complexity.

Keywords Quasiseparable matrices · Bezoutians · Euclid algorithm · Schur algorithm · Fast algorithms · Displacement structure

1 Introduction

It is known that the Euclidean algorithm is one of the oldest algorithms which appear in the computation of the greatest common divisor (GCD). Although the original Euclidean algorithm was presented to compute the positive greatest common divisor of two given positive integers, later it was generalized to polynomials in one variable over a field, and further to polynomials in any number of variables over any unique factorization domain in which the greatest common divisor can be computed.

S.M. Perera (✉)
Embry-Riddle Aeronautical University, Daytona Beach, USA
e-mail: pereras2@erau.edu

V. Olshevsky
University of Connecticut, Storrs, USA
e-mail: olshevsky@math.uconn.edu

I.S. Kotsireas and E. Martínez-Moro (eds.), *Applications of Computer Algebra*, Springer Proceedings in Mathematics & Statistics 198,
DOI 10.1007/978-3-319-56932-1_24

1.1 GCD Computing Algorithms

The Euclidean algorithm for computing polynomial GCD evolved with the early work of Brown (see, e.g., [14, 15]) and thereafter several other authors studied: the degree of the greatest common divisor of two polynomials in connection to a companion matrix [2], stable algorithms to compute polynomial ε-GCD using the displacement structure of Sylvester and Bezout matrices [10], generalization of the Euclidean algorithm (determining the greatest common left divisor) to polynomial matrices [1], estimation of the degree of ε-GCDs at a low computational cost [41], approximate factorization of multivariate polynomials with complex coefficients containing numerical noise [35], generalized Euclidean algorithm of the Routh–Hurwitz type [19], a numeric parameter for determining two prime polynomials under small perturbations with the help of an inversion formula for Sylvester matrices [5, 6], algorithms to approximate GCD for polynomials with coefficients of floating-point numbers [39], and so on. Our intention is not to consider the Euclidean algorithm via the above methods but to see it via the Hessenberg displacement structure of a Bezoutian over a system of polynomials $\{Q\} = \{Q_k(x)\}_{k=0}^{n}$ satisfying recurrence relations having $\deg Q_k(x) = k$ and to derive a fast algorithm based on quasiseparable polynomials.

1.2 Connection to Bezoutian

Given a pair of polynomials $a(x)$ and $b(x)$ with $\deg a(x) = n$ and $\deg b(x) \le n$, the classical Bezoutian of $a(x)$ and $b(x)$ is the bilinear form given by

$$\frac{a(x)b(y) - a(y)b(x)}{x - y} = \sum_{i=0}^{n-1}\sum_{j=0}^{n-1} s_{ij}x^i y^j$$

and the Bezoutian matrix of $a(x)$ and $b(x)$ is defined by the $n \times n$ symmetric matrix $Bez(a, b) = \left[s_{ij}\right]_{i,j=0}^{n-1}$.

In the eighteenth century, the Bezoutian was invented in order to build a bridge between polynomial and linear algebra. As it was remarked in [23, 44], the Bezoutian concept in principle already evolves from Euler's work in elimination theory for polynomials. Hermite was the first who studied Bezoutians in more detail to solve root localization problems for polynomials (Routh–Hurwitz), which are important in particular for the investigation of the stability of linear systems. Note that in the early stages of work related to Bezoutians, the language of quadratic forms was more common than matrix language. After the first observation of inversion of Bezoutians as Toeplitz or Hankel by L.T. Lander in 1974, there were significant results published to show that the inverse of Toeplitz is T-Bezoutian and Hankel is H-Bezoutian (see, e.g., [23, 26, 27, 30, 43]). These works show us great examples on the importance of matrix representations for the inverses of Hankel, Toeplitz, and more general types of structured matrices in the construction of fast algorithms for solving structured systems of equations and interpolation problems.

As this paper connects the generalized Bezoutian with confederate matrices via a Hessenberg displacement structure, we should remark heavily on Barnett's result [3] on showing the important relationship between a Bezoutian matrix and a matrix polynomial associated with the companion matrix. Apart from this, several others studied connections of Bezoutians to GCD including: computing the greatest common right divisor using Sylvester and generalized Bezoutian resultant matrices [13], matrix representations for generalized Bezoutians via generalized companion matrices [43], Bezoutians of Chebyshev polynomials of first and second kind [20], generalized Barnett factorization formula for polynomial Bezoutian matrices and reduction of Bezoutian via polynomial Vandermonde matrix [45], computation of polynomial GCD and coefficients of the polynomials generated in the Euclidean scheme via Bezoutian [11], computation of the GCD of two polynomials using Bernstein–Bezoutian matrix [12], and so on.

1.3 Connection to Displacement Structure

This paper describes a fast Euclid-type algorithm for quasiseparable polynomials via a fast Schur-type algorithm for a Bezoutian matrix preserving a Hessenberg displacement structure. The structured matrices like Toeplitz, Hankel, Toeplitz plus Hankel, Vandermonde, Cauchy, etc. belong to a more general family of matrices with low rank displacement structure and that can be used to design fast algorithms.

Definition 1 A linear displacement operator $\Theta_{\Omega,M,F,N}(.) : C^{n\times n} \to C^{n\times n}$ is a function which transforms each matrix $R \in C^{n\times n}$ to the matrix given by the displacement equation

$$\Theta_{\Omega,M,F,N}(R) = \Omega\, RM - FRN = GB \tag{1}$$

where $\Omega, M, F, N \in C^{n\times n}$ are given matrices and $G \in C^{n\times\alpha}$, $B \in C^{\alpha\times n}$. The pair $\{G, B\}$ on last right in (1) is called a minimal generator of R and

$$rank\ \left\{\Theta_{\Omega,M,F,N}(R)\right\} = \alpha. \tag{2}$$

Example 1 Toeplitz matrix $T = [t_{i-j}]_{1\le i,j\le n}$ satisfies the displacement equation

$$T - Z \cdot T \cdot Z^T = \begin{bmatrix} t_0 & t_{-1} & \cdots & t_{-n+1} \\ t_1 & 0 & \cdots & 0 \\ \vdots & \vdots & \ddots & \vdots \\ t_{n-1} & 0 & \cdots & 0 \end{bmatrix}$$

$$= \begin{bmatrix} \frac{t_0}{2} & 1 \\ t_1 & 0 \\ \vdots & \vdots \\ t_{n-1} & 0 \end{bmatrix} \begin{bmatrix} 1 & 0 & \cdots & 0 \\ \frac{t_0}{2} & t_{-1} & \cdots & t_{-n+1} \end{bmatrix}$$

where Z is a lower shift matrix. Thus, rank $\{\Theta_{I,I,Z,Z^T}(T)\} = 2$.

Example 2 Hankel matrix $H = [h_{i+j-2}]_{1 \le i,j \le n}$ satisfies the displacement equation

$$Z \cdot H - H \cdot Z^T = \begin{bmatrix} 0 & -h_0 & -h_1 & \cdots & -h_{n-2} \\ h_0 & 0 & 0 & \cdots & 0 \\ \vdots & \vdots & \vdots & \ddots & \vdots \\ h_{n-2} & 0 & 0 & \cdots & 0 \end{bmatrix}$$

$$= \begin{bmatrix} 1 & 0 \\ 0 & h_0 \\ \vdots & \vdots \\ 0 & h_{n-2} \end{bmatrix} \begin{bmatrix} 0 & -h_0 & \cdots & -h_{n-2} \\ 1 & 0 & \cdots & 0 \end{bmatrix}$$

where Z is a lower shift matrix. Thus, rank $\{\Theta_{Z,I,I,Z^T}(H)\} = 2$.

A fast algorithm for the structured matrices which preserve displacement structure first appeared in Morf's Thesis [38]. Thus the crucial shift-low-rank updating property was recognized by the author as the proper generalization of the Toeplitz and Hankel structured matrices. The algorithm was called Fast Cholesky decomposition. In June 1971, computer programs were successfully completed by the author. In the same Thesis he announced a divide-and-conquer algorithm but it was not shown how to design a super fast algorithm. Such an algorithm was obtained in Brent-Gustafson-Yun in 1979. Moreover the paper of Kailath et al. [31] proved crucial results demonstrating that the Schur complement inherits the displacement rank. This idea was the opening of a new chapter, as it filled in the missing link of proving a super fast complexity in Morf's Thesis. Delosme in [17] obtained formulas for generator updates for the Toeplitz case and claimed that those coincided with the classical Schur algorithm. Furthermore one can find (e.g., in [18, 32–34]) algorithms that connect structured matrices with displacement equations to derive fast algorithms. Many polynomial computations can be reduced to structured matrix computations. In this way, the matrix interpretation of many classical polynomial algorithms for determining GCD can be expressed.

The displacement equations of the structured matrices are used to design fast Schur-type algorithms having complexity $\mathcal{O}(n^2)$. The existence of fast Schur-type algorithms for Toeplitz and Toeplitz-like matrices having low displacement rank was shown in [31, 38]. It was shown in [25] that the low displacement rank Vandermonde-like and Cauchy-like matrices can be used to derive fast Schur-type algorithms. We should also recall the fast $\mathcal{O}(n^2)$ algorithm for Cauchy-like displacement structured matrices via Gaussian elimination with partial pivoting and fast algorithms for Toeplitz-like, Toeplitz-plus-Hankel-like, Vandermonde-like matrices via transferring those matrices to Cauchy-like matrices in [21]. Moreover in [40] a fast Schur-type algorithm with stability criteria was presented for the factorization of Hankel-like matrices. Finally, the crucial result of the Schur-type algorithm was presented by Heinig and Olshevsky [24] for the matrices with Hessenberg displacement structure.

While the displacement structure is considered we should recall some results on Schur-type algorithms in connection to the Bezoutian. At this point, we should mention the results on: computing Schur type and hybrid (Schular type and Levinson type) algorithms to solve the system of equations involving Toeplitz-plus-Hankel matrices [29], computing a Schur-type algorithm for LDU-decomposing the strongly regular Toeplitz-plus-Hankel matrix [46], solving a system of equations by split algorithms for skewsymmetric Toeplitz matrices, centrosymmetric Toeplitz-plus-Hankel matrices, and general Toeplitz-plus-Hankel matrices [28], and more importantly, Olshevsky's claim on Schur-type algorithms in connection to Euclid-type algorithms via the Bezoutian in the 10th ILAS Conference in 2002 and 16th International Symposium on Mathematical Theory of Networks and Systems in 2004.

1.4 Main Results

In [24], a Schur-type algorithm was presented to compute a recursive triangular factorization $R = LDU$ for a strongly nonsingular $n \times n$ matrix R satisfying the displacement equation:

$$RY - VR = GH^T$$

with upper and lower Hessenberg matrices Y and V, respectively, and $n \times \alpha$ matrices G and H where α is small compared to n. The Schur–Hessenberg algorithm in [24] will have complexity $\mathcal{O}(n^3)$ in general for dense and unstructured Hessenberg matrices Y and V. However, one can explore the structures of Y and V to derive a $\mathcal{O}(n^2)$ algorithm. Thus in this paper, we explore the structures of Y and V to derive a fast hybrid of Schur-type and Euclid-type algorithms in connection with Bezoutian and confederate matrices over the system of quasiseparable polynomials.

We observe a displacement equation of a Bezoutian matrix associated with the reverse polynomials in connection to a companion matrix over the system of monomial basis (this displacement equation (14) is a variant of the Lancaster–Tismenetsky equation in [36]). We then use this to derive and generalize a displacement equation for a generalized Bezoutian matrix with confederate matrix respect to the system of polynomials $\{Q\} = \{Q_k(x)\}_{k=0}^{n}$ satisfying recurrence relations having deg $Q_k(x) = k$. Then, the displacement equation for a generalized Bezoutian with confederate matrix, and characteristics of the Schur complement of the generalized Bezoutian and generator updates for confederate and generalized Bezoutian matrices are used to derive the Schur–Euclid–Hessenberg algorithm. Finally, to derive a fast $\mathcal{O}(n^2)$ complexity Schur–Euclid–Hessenberg algorithm we take quasiseparable polynomials as the main tool.

Definition 2 A matrix $A = \left[a_{ij}\right]$ is called $(H, 1)$-quasiseparable (i.e., Hessenberg-1-quasiseparable) if **(i)** it is strongly upper Hessenberg ($a_{i+1,i} \neq 0$ for $i = 1, 2, \cdots, n-1$ and $a_{i,j} = 0$ for $i > j+1$), and **(ii)** max (rank A_{12}) = 1 where the maximum is taken over all symmetric partitions of the form

$$A = \left[\begin{array}{c|c} \star & A_{12} \\ \hline \star & \star \end{array}\right]$$

• Let $A = [a_{ij}]$ be a $(H, 1)$-quasiseparable matrix. For $\alpha_i = \frac{1}{a_{i+1,i}}$, then the system of polynomials related to A via

$$r_k(x) = \alpha_1 \alpha_2 \cdots \alpha_k det(xI - A)_{(k \times k)}$$

is called a system of $(H, 1)$-quasiseparable polynomials. In the classification paper [9], the characterization of orthogonal polynomials (orthogonal with respect to a weighted inner product (definite or indefinite) on the real line) and Szegö polynomials (orthogonal on the unit circle with respect to a weighted inner product) via tridiagonal and unitary Hessenberg matrices, respectively, are observed to belong to a wider class of $(H, 1)$-quasiseparable polynomials and matrices, respectively. Hence once a fast $\mathcal{O}(n^2)$ Schur–Euclid-type algorithm for quasiseparable polynomials is established, we analyze the complexity of Schur–Euclid-type algorithms for orthogonal and Szegö polynomials.

1.5 Structure of the Paper

The structure of the paper is as follows. In the next Sect. 2, we state polynomial division in a matrix form, arithmetic complexity, and see the connection to the Euclidean algorithm in matrix forms. In Sect. 3, we express displacement equations and characterizations of the Bezoutian matrix in connection to the Schur complement and reverse polynomials. Then in Sect. 4, we generalize the displacement equation in the former section via generalized Bezoutians and confederate matrices over the system of polynomials $\{Q\} = \{Q_k(x)\}_{k=0}^{n}$ satisfying recurrence relations having deg $Q_k(x) = k$. At the end of the section, we present a hybrid of Euclid-type algorithm and Schur-type algorithm using the Hessenberg displacement structure of the Bezoutian and call it the Schur–Euclid–Hessenberg algorithm. Finally in Sect. 5, we establish a fast $\mathscr{O}(n^2)$ Schur–Euclid–Hessenberg algorithm for quasiseparable polynomials while addressing the complexity of the algorithm for its subclasses: orthogonal and Szegö polynomials.

2 One Way to Express Polynomial Division in Matrix Form

In this section, we state polynomial division in a matrix form. In the meantime, we discuss the arithmetic cost of computing the polynomial division and the matrix form of the Euclidean algorithm in connection to polynomials. Similar to this approach one can see the results in [42] to compute polynomial division efficiently. We first give the Euclidean algorithm for computing the GCD of polynomials.

Let $a(x)$ and $b(x)$ be given with $\deg a(x) \geq \deg b(x)$; then the Euclidean algorithm applies to $a(x)$ and $b(x)$ and generates a sequence of polynomials $r^{(i)}(x), q^{(i-1)}(x)$, such that

$$\begin{array}{c} r^{(0)}(x) = a(x), \qquad r^{(1)}(x) = b(x) \\ r^{(i-2)}(x) = q^{(i-1)}(x) r^{(i-1)}(x) + r^{(i)}(x), \qquad i = 2, 3, \cdots, t+1 \end{array} \tag{3}$$

where $r^{(i)}(x)$ is the remainder of the division of $r^{(i-2)}(x)$ by $r^{(i-1)}(x)$. The algorithm stops when a remainder $r^{(t+1)}(x) = 0$ is found; then $r^{(t)}(x)$ is the desired GCD of $a(x)$ and $b(x)$. Note that since the deg $r^{(0)}(x) >$ deg $r^{(1)}(x) > \cdots >$ deg $r^{(t+1)}(x)$, the algorithm must terminate in a finite number of steps. If $r^{(t)}(x)$ is a constant then $r^{(0)}(x)$ and $r^{(1)}(x)$ are said to be relatively prime.

The following results show polynomial division (one step of a variant of Euclidean algorithm) in matrix form. Here we have considered $-c(x)$ as the remainder of the polynomial division of $a(x)$ by $b(x)$ just to be compatible with future discussions.

Lemma 1 *Let* $a(x) = a_n x^n + a_{n-1} x^{n-1} + \cdots + a_0$ *and* $b(x) = b_{n-k} x^{n-k} + b_{n-k-1} x^{n-k-1} + \cdots + b_0$ *where* $k \geq 1$. *Then the polynomial division of* $a(x)$ *by* $b(x)$ *can be seen via:*

$$-\begin{bmatrix} 0 \\ c_{n-k-1} \\ c_{n-k-2} \\ \vdots \\ c_0 \end{bmatrix} + q_0 \begin{bmatrix} b_{n-k} \\ b_{n-k-1} \\ b_{n-k-2} \\ \vdots \\ b_0 \end{bmatrix} = \begin{bmatrix} a_{n-k} \\ a_{n-k-1} \\ a_{n-k-2} \\ \vdots \\ a_0 \end{bmatrix} - \widehat{B}_k B_k^{-1} \begin{bmatrix} a_n \\ a_{n-1} \\ \vdots \\ a_{n-k+1} \end{bmatrix} \tag{4}$$

where $-c(x) = -c_{n-k-1} x^{n-k-1} - c_{n-k-2} x^{n-k-2} - \cdots - c_0$ *is the remainder,* q_0 *is the constant term of the quotient of polynomial division,*

$$B_k := toeplitz\left([b_{n-k} : b_{n-2k+1}], [b_{n-k}, \ zeros(1, k-1)]\right),$$

and

$$
\widehat{B_k} = \begin{bmatrix}
b_{n-2k} & b_{n-2k+1} & b_{n-2k+2} & & \cdots & b_{n-k-1} \\
b_{n-2k-1} & b_{n-2k} & b_{n-2k+1} & & \cdots & \vdots \\
b_{n-2k-2} & b_{n-2k-1} & b_{n-2k} & & & \\
\vdots & \vdots & \vdots & & & \vdots \\
b_0 & b_1 & \vdots & & & \\
0 & b_0 & b_1 & & & \\
0 & 0 & \ddots & \ddots & & \vdots \\
 & & \ddots & \ddots & & \\
\vdots & & & 0 \; b_0 & & b_1 \\
 & & & 0 & & b_0 \\
0 & & \cdots & & & 0
\end{bmatrix}.
$$

Proof If $q(x) = q_k x^k + q_{k-1}x^{k-1} + \cdots + q_0$ and $-c(x)$ are the quotient and remainder of the polynomial division of $a(x)$ by $b(x)$ then we can say

$$
\begin{aligned}
a_n x^n + a_{n-1}x^{n-1} + \cdots + a_0 = {} & (q_k x^k + q_{k-1}x^{k-1} + \cdots + q_0) \\
& \cdot (b_{n-k}x^{n-k} + b_{n-k-1}x^{n-k-1} + \cdots + b_0) \\
& -(c_{n-k-1}x^{n-k-1} + c_{n-k-2}x^{n-k-2} + \cdots + c_0)
\end{aligned} \tag{5}
$$

By equating the coefficients of x^n to x^{n-k+1} in (5);

$$
\begin{bmatrix} a_n \\ a_{n-1} \\ \vdots \\ a_{n-k+2} \\ a_{n-k+1} \end{bmatrix}
=
\begin{bmatrix}
b_{n-k} & 0 & \cdots & \cdots & 0 \\
b_{n-k-1} & b_{n-k} & \ddots & & \vdots \\
b_{n-k-2} & b_{n-k-1} & b_{n-k} & \ddots & \vdots \\
\vdots & & \ddots & \ddots & 0 \\
b_{n-2k+1} & b_{n-2k+2} & \cdots & b_{n-k-1} & b_{n-k}
\end{bmatrix}
\cdot
\begin{bmatrix} q_k \\ q_{k-1} \\ \vdots \\ q_2 \\ q_1 \end{bmatrix}
\tag{6}
$$

Note that the first matrix in the RHS of (6) is a lower Toeplitz matrix (say B_k where $B_k := \mathit{toeplitz}\,([b_{n-k} : b_{n-2k+1}], [b_{n-k},\ \mathit{zeros}(1, k-1)])$ so q_k's (except q_0) can be recovered from:

$$
\begin{bmatrix} q_k \\ q_{k-1} \\ \vdots \\ q_2 \\ q_1 \end{bmatrix}
= [B_k]^{-1} \cdot
\begin{bmatrix} a_n \\ a_{n-1} \\ \vdots \\ a_{n-k+2} \\ a_{n-k-1} \end{bmatrix}
\tag{7}
$$

Equating coefficients of x^{n-k} to the constant term in (5);

$$-\begin{bmatrix}0\\c_{n-k-1}\\c_{n-k-2}\\\vdots\\c_0\end{bmatrix} = \begin{bmatrix}a_{n-k}\\a_{n-k-1}\\\vdots\\a_0\end{bmatrix} - \begin{bmatrix} b_{n-2k} & b_{n-2k+1} & b_{n-2k+2} & & \cdots & b_{n-k}\\ b_{n-2k-1} & b_{n-2k} & b_{n-2k+1} & & \cdots & b_{n-k-1}\\ \vdots & \vdots & \vdots & & & \vdots\\ b_1 & \vdots & \vdots & & & \vdots\\ b_0 & b_1 & \vdots & & & \\ 0 & b_0 & b_1 & & & \vdots\\ 0 & 0 & b_0 & b_1 & & \vdots\\ \vdots & \ddots & \ddots & \ddots & \ddots & \\ \vdots & & \ddots & \ddots & \ddots & b_1\\ 0 & \cdots & \cdots & 0 & 0 & b_0 \end{bmatrix} \cdot \begin{bmatrix}q_k\\q_{k-1}\\q_{k-2}\\\vdots\\q_0\end{bmatrix}$$

By rearranging the above system we get

$$-\begin{bmatrix}0\\c_{n-k-1}\\c_{n-k-2}\\\vdots\\c_0\end{bmatrix} + q_0\begin{bmatrix}b_{n-k}\\b_{n-k-1}\\\vdots\\b_0\end{bmatrix}$$

$$= \begin{bmatrix}a_{n-k}\\a_{n-k-1}\\\vdots\\a_0\end{bmatrix} - \begin{bmatrix} b_{n-2k} & b_{n-2k+1} & b_{n-2k+2} & & \cdots & b_{n-k-1}\\ b_{n-2k-1} & b_{n-2k} & b_{n-2k+1} & & \cdots & b_{n-k-2}\\ \vdots & \vdots & \vdots & & & \vdots\\ b_1 & \vdots & \vdots & & & \vdots\\ b_0 & b_1 & \vdots & & & \\ 0 & b_0 & b_1 & & & \vdots\\ 0 & 0 & b_0 & b_1 & & \vdots\\ \vdots & \ddots & \ddots & \ddots & \ddots & \\ \vdots & & \ddots & \ddots & \ddots & b_1\\ 0 & \cdots & \cdots & 0 & 0 & b_0\\ 0 & \cdots & \cdots & 0 & 0 & 0 \end{bmatrix} \cdot \begin{bmatrix}q_k\\q_{k-1}\\q_{k-2}\\\vdots\\q_1\end{bmatrix}.$$

However we can now use (7) to rewrite the right side of the above equation and which yields the result (4).

Corollary 1 *Let $a(x)$ and $b(x)$ be two polynomials such that $\deg a(x) = \deg b(x) + 1$ with $\deg a(x) = n$. Then the polynomial division of $a(x)$ by $b(x)$ can be seen via:*

$$
-\begin{bmatrix} 0 \\ c_{n-2} \\ \vdots \\ c_1 \\ c_0 \end{bmatrix} + q_0 \begin{bmatrix} b_{n-1} \\ b_{n-2} \\ \vdots \\ b_0 \end{bmatrix} = \begin{bmatrix} a_{n-1} \\ a_{n-2} \\ \vdots \\ a_0 \end{bmatrix} - q_1 Z^T \begin{bmatrix} b_{n-1} \\ b_{n-2} \\ \vdots \\ b_1 \\ b_0 \end{bmatrix}
$$

where $q_1 = a_n\, b_{n-1}^{-1}$, *and Z is the lower shift matrix.*

The following gives the Toeplitz matrix-based calculation of the quotient of the polynomial division and its arithmetic cost.

Corollary 2 *If* $a(x)$ *and* $b(x)$ *are two polynomials such that deg* $a(x) = n$ *and deg* $b(x) = n - k$ *where* $k \geq 1$, *then the arithmetic cost of computing the quotient of the polynomial division is* $\mathcal{O}(n \log n)$ *operations.*

Proof Let $q(x)$ be the quotient of the polynomial division of $a(x)$ by $b(x)$ stated via (5). Then the coefficients of quotient, q_k, can be recovered via

$$
\begin{bmatrix} q_k \\ q_{k-1} \\ \vdots \\ q_1 \\ q_0 \end{bmatrix} = \begin{bmatrix} b_{n-k} & 0 & \cdots & \cdots & 0 \\ b_{n-k-1} & b_{n-k} & \ddots & & \vdots \\ b_{n-k-2} & b_{n-k-1} & b_{n-k} & \ddots & \vdots \\ \vdots & & \ddots & \ddots & 0 \\ b_{n-2k} & b_{n-2k+1} & \cdots & b_{n-k-1} & b_{n-k} \end{bmatrix}^{-1} \cdot \begin{bmatrix} a_n \\ a_{n-1} \\ \vdots \\ a_{n-k+1} \\ a_{n-k} \end{bmatrix}
$$

$$
= B_{k+1}^{-1} \cdot \begin{bmatrix} a_n \\ a_{n-1} \\ \vdots \\ a_{n-k+1} \\ a_{n-k} \end{bmatrix} \tag{8}
$$

Note that $B_{k+1} = b_{n-k}I + b_{n-k-1}Z + \cdots + b_{n-2k}Z^k = \bar{b}(Z)$ where Z is the lower shift matrix and $Z^{k+1} = 0$. Thus $B_{k+1}^{-1} = \bar{b}(Z)^{-1} mod\ Z^{k+1}$. Note that B_{k+1}^{-1} is also a lower triangular Toeplitz matrix which is defined by its first column. Now by following [42], one can apply a divide-and-conquer technique for the block form of the B_{k+1}^{-1} to calculate the first column of B_{k+1}^{-1}. This yields the cost of computing B_{k+1}^{-1} and also a Toeplitz matrix times a vector is of order $\mathcal{O}(n \log n)$.

The following shows the cost of computing sequences of remainder polynomials via a variant of the Euclidean algorithm which corresponds to polynomial division in matrix form.

Corollary 3 *If the sequence of remainders of the polynomial division is computed via Lemma* 1, *then the cost of computing one division is* $\mathcal{O}(n \log^2 n)$ *and* $\mathcal{O}(n\, t\, \log^2 n)$ *for generating the full sequence where n is the degree of the divisor and t is the number of steps.*

Proof As stated in Corollary 2, the cost of computing the quotient of the polynomial division is $\mathcal{O}(n \log n)$. Thus computing $B_k^{-1}\bar{a}$ where $\bar{a} = \begin{bmatrix} a_n \ a_{n-1} \ \cdots \ a_{n-k+1} \end{bmatrix}^T$ costs $\mathcal{O}(n \log n)$ operations. Now the multiplication of $B_k^{-1}\bar{a}$ by a tall sparse matrix $\widehat{B}_k$ together with vector subtraction yields $\mathcal{O}(n \log^2 n)$ operations for one division or one step in calculating the remainder. Thus to generate t steps or for a full sequence it costs $\mathcal{O}(nt \log^2 n)$ operations.

The next section shows the displacement equations of Bezoutian and Schur complement in connection to reverse polynomials.

3 Displacement Structures and Characterizations of Bezoutian

This section describes two types of displacement equations of the Bezoutian while introducing characterization of the Bezoutian via Gaussian elimination and Schur complement. These displacement equations of Bezoutians are elaborated in connection to a lower shift matrix and a companion matrix but associated with the reverse polynomials.

Definition 3 Let $a(x) = a_n x^n + a_{n-1}x^{n-1} + \cdots + a_0$. Then, the reverse polynomial of a(x) is defined as $a^\sharp(x) = x^n a(x^{-1}) = a_0 x^n + a_1 x^{n-1} + \ldots + a_{n-1}x + a_n$.

We define the Bezoutian associated with the reverse polynomials as follows.

Definition 4 Let $P = \{1, x, x^2, ..., x^n\}$ be a monomial basis and let $a(x)$ and $b(x)$ be polynomials of degree not greater than n. Then a matrix $S^\sharp = [s_{ij}]$ is the Bezoutian associated with the reverse polynomials $a^\sharp(x)$ and $b^\sharp(x)$, say $S^\sharp = Bez_P(a^\sharp, b^\sharp)$, if

$$S(a^\sharp, b^\sharp) = \frac{a^\sharp(x) \cdot b^\sharp(y) - b^\sharp(x) \cdot a^\sharp(y)}{x - y} = \sum_{i,j=0}^{n-1} s_{ij} x^i y^j$$

$$= \begin{bmatrix} 1 \ x \ x^2 \ \cdots \ x^{n-1} \end{bmatrix} S^\sharp \begin{bmatrix} 1 \\ y \\ y^2 \\ \vdots \\ y^{n-1} \end{bmatrix}. \quad (9)$$

3.1 Displacement Structures of Bezoutian

Here we obtain two types of displacement equations of the Bezoutian associated with the reverse polynomials. Once we establish the displacement structures, we state connections of the displacement equations to the Gohberg, Kailath, and Olshevsky algorithm (GKO algorithm) [21] and the Heinig and Olshevsky algorithm (HO algorithm) [24].

Below we give the GKO algorithm for matrix R_1 satisfying the Sylvester displacement equation.

Lemma 2 *Let matrix* $R_1 = \begin{bmatrix} r_1 & R_{12} \\ R_{21} & R_{22} \end{bmatrix}$ *satisfy:*

$$\Delta_{F_1,A_1}(R_1) = \begin{bmatrix} f_1 & 0 \\ * & F_2 \end{bmatrix} \cdot R_1 - R_1 \cdot \begin{bmatrix} a_1 & * \\ 0 & A_2 \end{bmatrix} = G^{(1)}B^{(1)} = \begin{bmatrix} g_1 \\ G_1 \end{bmatrix} \begin{bmatrix} b_1 & B_1 \end{bmatrix}$$

If r_1 *(i.e., (1,1) entry of* R_1*)* $\neq 0$*, then the Schur complement* $R_2 = R_{22}^{(1)} - R_{21}\frac{1}{r_1}R_{12}$ *satisfies the Sylvester type displacement equation*

$$F_2 \cdot R_2 - R_2 \cdot A_2 = G^{(2)}B^{(2)}$$

with

$$G_2^{(2)} = G_1 - R_{21}\frac{1}{r_1}g_1, \quad B_2^{(2)} = B_1 - b_1\frac{1}{r_1}R_{12}$$

where g_1 *and* b_1 *are the first row of* $G^{(1)}$ *and the first column of* $B^{(1)}$ *respectively.*

The Lemma 2 shows that if R_1 satisfies a Sylvester type displacement equation then so does its Schur complement. Thus the displacement equation of the Schur complement of the matrix will also yield the factorization. One can recover the first row and column of R_1, and R_2, by using generator updates. Proceeding recursively one finally obtains the LU factorization of R_1. Moreover authors in [21] note that one can obtain a fast Gaussian elimination algorithm with partial pivoting for Cauchy-like, Vandermonde-like, and Chebyshev-like displacement structures.

Lemma 3 *Let the matrix* R_1 *satisfy:*

$$\Delta_{F_1,A_1}(R_1) = F_1 \cdot R_1 - R_1 \cdot A_1 = G^{(1)}B^{(1)}$$

and let the matrices be partitioned as

$$R_1 = \begin{bmatrix} R_{11} & R_{12} \\ R_{21} & R_{22} \end{bmatrix}, \quad F_1 = \begin{bmatrix} F_{11} & F_{12} \\ F_{21} & F_{22} \end{bmatrix}, \quad A_1 = \begin{bmatrix} A_{11} & A_{12} \\ A_{21} & A_{22} \end{bmatrix},$$

$$G^{(1)} = \begin{bmatrix} G_1 \\ G_2 \end{bmatrix}, \quad B^{(1)} = \begin{bmatrix} B_1 & B_2 \end{bmatrix}.$$

If R_{11} *is nonsingular, then the Schur complement of* R_1*, i.e.,* $R_2 = R_{22} - R_{21}R_{11}^{-1}R_{12}$ *satisfies*

$$F_2 \cdot R_2 - R_2 \cdot A_2 = G^{(2)}B^{(2)}$$

with

$$G^{(2)} = G_2 - R_{21}R_{11}^{-1}G_1, \;\; B^{(2)} = B_2 - B_1R_{11}^{-1}R_{12}$$
$$A_2 = A_{22} - A_{21}R_{11}^{-1}R_{12}, \;\; F_2 = F_{22} - R_{21}R_{11}^{-1}F_{12}.$$

From the Lemma 3, one can observe that a Schur-type algorithm can be designed for nontriangular matrices $\{F_1, A_1\}$. Authors in [24] specialize this crucial result by deriving a Gaussian elimination algorithm for matrices with Hessenberg displacement structure. Although the algorithm has complexity $\mathcal{O}(n^3)$, in general one can explore the structures of F_1 and A_1 to derive fast algorithms.

We first observe a Sylvester type displacement equation for the Bezoutian associated with reverse polynomials in connection to the lower shift matrix and see it as a GKO-type displacement equation, but in our case it is for the Bezoutian.

Lemma 4 *Let $S^\sharp = Bez_P(a^\sharp, b^\sharp)$ be the Bezoutian associated with the reverse polynomials where $a(x) = a_nx^n + a_{n-1}x^{n-1} + \cdots + a_0$ and $b(x) = b_nx^n + b_{n-1}x^{n-1} + \cdots + b_0$ then $S^\sharp$ satisfies the displacement equation:*

$$ZS^\sharp - S^\sharp Z^T = GJG^T \tag{10}$$

where $G = \begin{bmatrix} a_n & b_n \\ a_{n-1} & b_{n-1} \\ \vdots & \vdots \\ a_1 & b_1 \end{bmatrix}$, $J = \begin{bmatrix} 0 & 1 \\ -1 & 0 \end{bmatrix}$ *and Z is the lower shift matrix.*

Proof By following the Definition 4 of the Bezoutian associated with the reverse polynomials, we get:

$$(x - y)S(a^\sharp, b^\sharp) = a^\sharp(x) \cdot b^\sharp(y) - b^\sharp(x) \cdot a^\sharp(y). \tag{11}$$

Observing the RHS:

$$\begin{aligned} a^\sharp(x)b^\sharp(y) - b^\sharp(x)a^\sharp(y) &= \begin{bmatrix} 1 & x & \cdots & x^n \end{bmatrix} \begin{bmatrix} a_n & b_n \\ a_{n-1} & b_{n-1} \\ \vdots & \vdots \\ a_0 & b_0 \end{bmatrix} \begin{bmatrix} b_n & b_{n-1} & \cdots & b_0 \\ -a_n & -a_{n-1} & \cdots & -a_0 \end{bmatrix} \begin{bmatrix} 1 \\ y \\ \vdots \\ y^n \end{bmatrix} \\ &= \begin{bmatrix} 1 & x & x^2 & \cdots & x^n \end{bmatrix} \widetilde{G}J\widetilde{G}^T \begin{bmatrix} 1 \\ y \\ y^2 \\ \vdots \\ y^n \end{bmatrix} \end{aligned} \tag{12}$$

where $\widetilde{G} = \begin{bmatrix} a_n & b_n \\ a_{n-1} & b_{n-1} \\ \vdots & \vdots \\ a_0 & b_0 \end{bmatrix}$ and $J = \begin{bmatrix} 0 & 1 \\ -1 & 0 \end{bmatrix}$. Let's pad $S^\sharp$ with zeros such that $\begin{bmatrix} S^\sharp & 0 \\ 0 & 0 \end{bmatrix} = \widetilde{S}$. We can always use $\widetilde{S}$ instead of $S^\sharp$ because

$$\begin{bmatrix} 1 \; x \; x^2 \; \cdots \; x^{n-1} \end{bmatrix} S^\sharp \begin{bmatrix} 1 \\ y \\ y^2 \\ \vdots \\ y^{n-1} \end{bmatrix} = \begin{bmatrix} 1 \; x \; x^2 \; \cdots \; x^n \end{bmatrix} \widetilde{S} \begin{bmatrix} 1 \\ y \\ y^2 \\ \vdots \\ y^n \end{bmatrix}$$

By following (9) together with $\widetilde{S}$ we get:

$$(x-y)S(a^\sharp, b^\sharp) = \begin{bmatrix} x \; x^2 \; \cdots \; x^{n+1} \end{bmatrix} \widetilde{S} \begin{bmatrix} 1 \\ y \\ \vdots \\ y^n \end{bmatrix} - \begin{bmatrix} 1 \; x \; \cdots \; x^n \end{bmatrix} \widetilde{S} \begin{bmatrix} y \\ y^2 \\ \vdots \\ y^{n+1} \end{bmatrix}$$

$$= \begin{bmatrix} 1 \; x \; x^2 \; \cdots \; x^n \end{bmatrix} (Z\widetilde{S} - \widetilde{S}Z^T) \begin{bmatrix} 1 \\ y \\ y^2 \\ \vdots \\ y^n \end{bmatrix}$$

Following the immediate result with (11) and (12)

$$Z\widetilde{S} - \widetilde{S}Z^T = \widetilde{G}J\widetilde{G}^T. \tag{13}$$

In the relation (13) one can peel off the last row of the generator $\widetilde{G}$, and peel off the last row and column of $\widetilde{S}$ resulting in $S^\sharp$, hence the result.

The following result is an immediate consequence of the Bezoutian satisfying the displacement equation (10) in connection to the displacement rank.

Corollary 4 *If the Bezoutian* $S^\sharp$ *satisfies the displacement equation* (10), *then* *rank* $\left\{\Theta_{Z,I,I,Z^T}(S^\sharp)\right\} = 2$.

By following the GKO algorithm [21], one can claim that the displacement equation (10) is of GKO-type but for the Bezoutian having low displacement rank.

Next, we see the second displacement equation of the Bezoutian associated with the reverse polynomials satisfying Hessenberg displacement structure.

Lemma 5 *A matrix $S^\sharp$ is a Bezoutian for reverse polynomials $a^\sharp(x)$ and $b^\sharp(x)$ if and only if it satisfies the equation*

$$C_a^T S^\sharp - S^\sharp C_a = 0. \tag{14}$$

for a matrix C_a of the form

$$C_a = \begin{bmatrix} -\frac{a_{n-1}}{a_n} & -\frac{a_{n-2}}{a_n} & -\frac{a_{n-3}}{a_n} & \cdots & -\frac{a_0}{a_n} \\ 1 & 0 & 0 & \cdots & 0 \\ 0 & 1 & 0 & & \vdots \\ \vdots & \ddots & \ddots & \ddots & 0 \\ 0 & \cdots & 0 & 1 & 0 \end{bmatrix} \tag{15}$$

where $a(x) = a_n x^n + a_{n-1}x^{n-1} + \cdots + a_0$ and $b(x) = b_n x^n + b_{n-1}x^{n-1} + \cdots + b_0$.

Proof Let $S^\sharp$ be the Bezoutian associated with the reverse polynomials $a^\sharp(x)$ and $b^\sharp(x)$ and C_a have the above structure (entries in the first row are extracted from the coefficients of $a(x)$) then it can easily be seen by matrix multiplication that

$$C_a^T S^\sharp - S^\sharp C_a = 0. \tag{16}$$

Notice that this is a variant of the Lancaster–Tismenetsky equation in [36].

Now let $C_a^T S^\sharp - S^\sharp C_a = 0$ where $S^\sharp = [s_{ij}]$ and C_a is the matrix of the given form so one can recover the columns of $S^\sharp$ as follows.

the second column of $S^\sharp$ by:

$$C_a^T s_{i,1} + \frac{a_{n-1}}{a_n} s_{i,1} - s_{i,2} = 0 \Rightarrow s_{i,2} = C_a^T s_{i,1} + \frac{a_{n-1}}{a_n} s_{i,1}$$

$$s_{i,2} = s_{1,1}\,\mathbf{a} + \left(Z^T + \frac{a_{n-1}}{a_n}\right) s_{i,1}$$

the third column of $S^\sharp$ by:

$$C_a^T s_{i,2} + \frac{a_{n-2}}{a_n} s_{i,1} - s_{i,3} = 0 \Rightarrow s_{i,3} = C_a^T s_{i,2} + \frac{a_{n-2}}{a_n} s_{i,1}$$

$$\begin{aligned} s_{i,3} &= s_{1,2}\,\mathbf{a} + Z^T s_{i,2} + \frac{a_{n-2}}{a_n} s_{i,1} \\ &= \left(Z^T s_{1,1} + s_{1,2} I_n\right)\mathbf{a} + \left(\left(Z^T\right)^2 + \frac{a_{n-1}}{a_n} Z^T + \frac{a_{n-2}}{a_n}\right) s_{i,1} \end{aligned}$$

proceeding recursively the kth column of $S^\sharp$ by:

$$s_{i,k} = \left(\left(Z^T\right)^{k-2} s_{1,1} + \left(Z^T\right)^{k-3} s_{1,2} + \cdots + s_{1,k-1} I_n \right) \mathbf{a}$$
$$+ \left(\left(Z^T\right)^{k-1} + \frac{a_{n-1}}{a_n} \left(Z^T\right)^{k-2} + \frac{a_{n-2}}{a_n} \left(Z^T\right)^{k-3} + \cdots + \frac{a_{n-k+1}}{a_n} \right) s_{i,1}$$

where $\mathbf{a} = \left[-\frac{a_{n-1}}{a_n} \; -\frac{a_{n-2}}{a_n} \; \cdots \; -\frac{a_0}{a_n} \right]^T$ and I_n is the identity matrix. Hence once all columns are recovered it should be clear that since $S^\sharp$ satisfies (16) there is no other matrix satisfying (16) and having the same first column s_{i1}. Thus $S^\sharp = Bez_P(a^\sharp, b^\sharp)$.

The displacement equation (16) is of HO-type but for the Bezoutian associated with reverse polynomials satisfying Hessenberg displacement structure in connection to the companion matrix C_a.

3.2 Characterization of Bezoutian

In this section, we perform Gaussian elimination on a Bezoutian satisfying displacement structure (10) and then elaborate on the relationship between a Bezoutian with its Schur complement.

Before we see the Schur complement of a Bezoutian as a Bezoutian, let us provide a supportive result to see how the displacement equation (13) helps us to address the rank of a Bezoutian as it is padded with zeros.

Lemma 6 *A matrix* $S^\sharp = Bez_P(a^\sharp, b^\sharp) \in R^{n,n}$ *is a Bezoutian if and only if* $\widetilde{S} = \begin{bmatrix} S^\sharp & 0 \\ 0 & 0 \end{bmatrix}$ *has displacement rank* 2.

Proof Lemma 4 suggests that if $S^\sharp$ is a Bezoutian then $\widetilde{S}$ has displacement rank 2. Conversely, if $\widetilde{S}$ has displacement rank 2, then $Z\widetilde{S} - \widetilde{S}Z^T = \widetilde{G}J\widetilde{G}^T$ for some $\widetilde{G} \in R^{n+1,2}$. Now, assume we have two polynomials and we wish to compute the Bezoutian associated with them. Lemma 4 suggests that we can do this by writing the polynomials as the columns of the generator and recovering $S^\sharp$ from the displacement equation $Z\widetilde{S} - \widetilde{S}Z^T = \widetilde{G}J\widetilde{G}^T$. Let $a^\sharp(x)$ and $b^\sharp(x)$ be the polynomials defined by the first and second columns of $\widetilde{G}$, respectively. Then, the Bezoutian generated by these two polynomials will be exactly $S^\sharp$.

Lemma 7 *If we perform Gaussian elimination on a Bezoutian, then the result will still be a Bezoutian.*

Proof First, it is easy to see that Gaussian elimination on the matrix $S^\sharp$ corresponds to Gaussian elimination on its generator. Second, if we perform Gaussian elimination on $S^\sharp$ and then pad the resultant matrix with a row and a column of zeros, the result will be the same as the result of padding $S^\sharp$ first to obtain $\widetilde{S}$ and performing the same steps of Gaussian elimination on $\widetilde{S}$. This is because the corresponding steps of Gaussian elimination will not alter a column or a row of zeros. Therefore, if we have an arbitrary Bezoutian $S^\sharp$, we know that $\widetilde{S}$ has displacement rank 2. Let us say Gaussian elimination is performed on $S^\sharp$ to obtain a different matrix $S^{(1)}$. From the discussion above, we can infer that the same steps of Gaussian elimination performed on $\widetilde{S}$ will result in $\begin{bmatrix} S^{(1)} & 0 \\ 0 & 0 \end{bmatrix}$. Let $\widetilde{G}$ be the generator of $\widetilde{S}$ and $G^{(1)}$ be the result after applying steps of Gaussian elimination to $\widetilde{G}$. It is clear that $G^{(1)}$ is the generator of $\begin{bmatrix} S^{(1)} & 0 \\ 0 & 0 \end{bmatrix}$, which in turn implies $S^{(1)}$ is a Bezoutian, hence the result.

The above result immediately implies the following statement.

Corollary 5 *The Schur complement of a Bezoutian is a Bezoutian.*

Proof This follows because Schur complementation is equivalent to Gaussian elimination.

4 Schur–Euclid-Type Algorithm via Bezoutian Having Hessenberg Structure

This section describes Hessenberg displacement structure of the Bezoutian associated with reverse polynomials expanded in a monomial $P = \{1, x, \cdots, x^n\}$ basis and then generalizes it to the basis $Q = \{Q_0, Q_1, \cdots, Q_n\}$ where deg $Q_k(x) = k$. The main idea is to explore the transformation of Hessenberg displacement structure of a Bezoutian from a monomial basis P and to the generalized basis Q. Once it is established we see the connection of the Schur complement of a Bezoutian to the Schur–Euclid–Hessenberg algorithm via generator updates of the generalized Bezoutian and confederate matrix.

4.1 Hessenberg Displacement Structure of Bezoutian Over Monomial Basis

Here we use the displacement equation of a Bezoutian (14) associated with reverse polynomials over a monomial basis to address the connection among generator updates of the companion matrix, the Schur complement of the Bezoutian, and polynomial division.

Definition 5 A matrix R_1 is said to have Hessenberg displacement structure if it satisfies

$$F_1 R_1 - R_1 A_1 = G^{(1)} B^{(1)} \tag{17}$$

where A_1 is an upper Hessenberg matrix and F_1 is a lower Hessenberg Matrix.

In Lemma 5, we have seen the connection of the Bezoutian $S^\sharp = Bez_P(a^\sharp, b^\sharp)$ to the displacement equation $C_a^T S^\sharp - S^\sharp C_a = 0$ and vice-versa. Since C_a is an upper Hessenberg matrix, by following the Definition 5, we can say that the Bezoutian has the Hessenberg displacement structure associated with reverse polynomial over a monomial basis.

In the following, we use the Hessenberg displacement structure of a Bezoutian over a monomial basis to see a connection to generator updates of the companion matrix to polynomial division.

Lemma 8 *Let C_a (15) be the companion matrix of the polynomial $a(x)$ satisfying $C_a^T S^\sharp - S^\sharp C_a = 0$ where $S^\sharp = Bez_P(a^\sharp, b^\sharp)$ and deg $a(x) >$ deg $b(x)$. Then the generator update of C_a is the companion matrix of the polynomial $b(x)$.*

Proof Let deg $a(x) = n$ and deg $b(x) = n - k$. Since the Bezoutian satisfies the Hessenberg displacement structure $C_a^T S^\sharp - S^\sharp C_a = 0$ one can apply Lemma 3 to update the generators. Thus the updated companion matrix results in:

$$C_{new} = \begin{bmatrix} 0 & 0 & 0 & \cdots & 0 \\ 1 & 0 & 0 & \cdots & 0 \\ 0 & 1 & 0 & \cdots & 0 \\ \vdots & \ddots & \ddots & \ddots & \vdots \\ 0 & \cdots & 0 & 1 & 0 \end{bmatrix} - \begin{bmatrix} 0 & \cdots & 0 & 1 \\ 0 & \cdots & 0 & 0 \\ \vdots & & \vdots & \vdots \\ 0 & \cdots & 0 & 0 \end{bmatrix} \cdot R_{11}^{-1} \cdot R_{12} \tag{18}$$

However, from another displacement equation of the Bezoutian, i.e., following the formula (13) one can state:

$$ZR - RZ^T = G^{(1)} \bar{J} G^{(1)^T} \tag{19}$$

where $R = \begin{bmatrix} Bez(b^\sharp, a^\sharp) & 0 \\ 0 & 0 \end{bmatrix} = -\begin{bmatrix} S^\sharp & 0 \\ 0 & 0 \end{bmatrix}$, $\bar{J} = \begin{bmatrix} 0 & -1 \\ 1 & 0 \end{bmatrix}$, Z is the lower shift matrix,

and $G^{(1)} = \begin{bmatrix} a_n & 0 \\ \vdots & \vdots \\ a_{n-k+1} & 0 \\ a_{n-k} & b_{n-k} \\ \vdots & \vdots \\ a_0 & b_0 \end{bmatrix}$.

With the help of the above equation one can obtain the following equations by considering the first k columns of R:

$$\left[Zr_1 |\ Zr_2 - r_1 |\ Zr_3 - r_2 |\ \dots\ | Zr_k - r_{k-1} \right] = \left[a_n\mathbf{b} |\ a_{n-1}\mathbf{b} |\ a_{n-2}\mathbf{b} |\ \dots\ | a_{n-k+1}\mathbf{b} \right] \tag{20}$$

From these it is possible to obtain the following relations:

$$\begin{aligned}
Zr_1 &= a_n\mathbf{b} \\
Zr_2 &= r_1 + a_{n-1}\mathbf{b} = Z^T a_n\mathbf{b} + a_{n-1}\mathbf{b} \\
Zr_3 &= r_2 + a_{n-2}\mathbf{b} = (Z^T)^2 a_n\mathbf{b} + Z^T a_{n-1}\mathbf{b} + a_{n-2}\mathbf{b} \\
&\vdots \\
Zr_k &= r_{k-1} + a_{n-k+1}\mathbf{b} = (Z^T)^{k-1} a_n\mathbf{b} + (Z^T)^{k-2} a_{n-1}\mathbf{b} + \ldots + Z^T a_{n-k+2}\mathbf{b} + a_{n-k+1}\mathbf{b}
\end{aligned}$$

Let R_{11} be the $k \times k$ upper block of R. Then, from the above equations it is possible to express R_{11} as:

$$\begin{bmatrix}
0 & & \cdots & 0 & a_n b_{n-k} \\
0 & & \cdot & a_n b_{n-k} & a_n b_{n-k-1} + a_{n-1} b_{n-k} \\
\vdots & \cdot & & & \vdots \\
0 & a_n b_{n-k} & & & \\
a_n b_{n-k} & a_n b_{n-k-1} + a_{n-1} b_{n-k} & \cdots & & a_n b_{n-(2k-1)} + a_{n-1} b_{n-(2k)} + \ldots + a_{n-k+1} b_{n-k}.
\end{bmatrix}$$

Let $\widetilde{I}$ be the antidiagonal matrix. Multiplying the above (R_{11}) by $\widetilde{I}$ from the right:

$$\begin{bmatrix}
a_n b_{n-k} & 0 & \cdots & 0 & 0 \\
a_n b_{n-k-1} + a_{n-1} b_{n-k} & a_n b_{n-k} \ 0 & & \vdots & 0 \\
\vdots & & \cdot & 0 & \vdots \\
 & & & a_n b_{n-k} & 0 \\
a_n b_{n-(2k-1)} + a_{n-1} b_{n-2k} + \ldots + a_{n-k+1} b_{n-k} & & \cdots & a_n b_{n-k-1} + a_{n-1} b_{n-k} & a_n b_{n-k}
\end{bmatrix}$$

However this can easily be factored as:

$$R_{11} \cdot \widetilde{I} = \begin{bmatrix}
b_{n-k} & & & \\
b_{n-k-1} & b_{n-k} & & \\
\vdots & b_{n-k-1} & \ddots & \\
 & & \ddots & \ddots \\
b_{n-(2k-1)} & \cdots & b_{n-k-1} & b_{n-k}
\end{bmatrix} \cdot \begin{bmatrix}
a_n & & & \\
a_{n-1} & a_n & & \\
\vdots & a_{n-1} & \ddots & \\
 & & \ddots & \ddots \\
a_{n-k+1} & \cdots & a_{n-1} & a_n
\end{bmatrix} \tag{21}$$

So, $R_{11} \cdot \widetilde{I} = B_k A_k$ where we denote the Toeplitz matrix composed of the coefficients of $a(x)$ on the lower diagonals in the above Eq. (21) as A_k and similarly for B_k. Therefore, $R_{11}^{-1} = \widetilde{I} A_k^{-1} B_k^{-1}$.

Let R_{21} be the $(n - k + 1) \times k$ block of R right below R_{11}, i.e.,

$$R_{21} = \begin{bmatrix} a_nb_{n-k-1} & a_nb_{n-k-2}+a_{n-1}b_{n-k-1} & \dots & a_nb_{n-2k}+a_{n-1}b_{n-2k+1}+\dots+a_{n-k+1}b_{n-k-1} \\ a_nb_{n-k-2} & a_nb_{n-k-3}+a_{n-1}b_{n-k-2} & \dots & \vdots \\ \vdots & \vdots & & \\ a_nb_2 & a_nb_1+a_{n-1}b_2 & \dots & a_{n-k+3}b_0+a_{n-k+2}b_1+a_{n-k+1}b_2 \\ a_nb_1 & a_nb_0+a_{n-1}b_1 & \dots & a_{n-k+2}b_0+a_{n-k+1}b_1 \\ a_nb_0 & a_{n-1}b_0 & \dots & a_{n-k+1}b_0 \\ 0 & 0 & \dots & 0 \end{bmatrix}$$

Let us compute $R_{21} \cdot \widetilde{I}$:

$$R_{21}\cdot\widetilde{I} = \begin{bmatrix} a_nb_{n-2k}+a_{n-1}b_{n-2k+1}+\dots+a_{n-k+1}b_{n-k-1} & \dots & a_nb_{n-k-2}+a_{n-1}b_{n-k-1} & a_nb_{n-k-1} \\ \vdots & \dots & a_nb_{n-k-3}+a_{n-1}b_{n-k-2} & a_nb_{n-k-2} \\ & & \vdots & \vdots \\ a_{n-k+3}b_0+a_{n-k+2}b_1+a_{n-k+1}b_2 & \dots & a_nb_1+a_{n-1}b_2 & a_nb_2 \\ a_{n-k+2}b_0+a_{n-k+1}b_1 & \dots & a_nb_0+a_{n-1}b_1 & a_nb_1 \\ a_{n-k+1}b_0 & \dots & a_{n-1}b_0 & a_nb_0 \\ 0 & \dots & 0 & 0 \end{bmatrix}$$

It is possible to factor the above as follows:

$$R_{21}\cdot\widetilde{I} = \begin{bmatrix} b_{n-2k} & b_{n-2k+1} & b_{n-2k+2} & \dots & b_{n-k-1} \\ b_{n-2k-1} & b_{n-2k} & b_{n-2k+1} & \dots & b_{n-k-2} \\ \vdots & & & & \\ b_0 & & & & \vdots \\ 0 & \ddots & & & \\ & \ddots & & & \vdots \\ \vdots & & \ddots & & \\ & & & \ddots & b_0 \\ 0 & & \dots & & 0 \end{bmatrix} \cdot \begin{bmatrix} a_n & & & \\ a_{n-1} & a_n & & \\ \vdots & a_{n-1} & \ddots & \\ & & \ddots & \ddots \\ a_{n-k+1} & & \dots & a_{n-1} & a_n \end{bmatrix}$$

From this it follows that $R_{21} = \widehat{B}_k \cdot A_k \cdot \widetilde{I}$ where $\widehat{B}_k$ is the first matrix in the RHS of the above equation. Therefore

$$R_{21}R_{11}^{-1} = (\widehat{B}_kA_k\widetilde{I})\cdot(\widetilde{I}A_k^{-1}B_k^{-1}) = \widehat{B}_k \cdot B_k^{-1}.$$

Moreover one can say that $(R_{11}^{-1}R_{12})^T = R_{21}R_{11}^{-1}$. Thus

$$(R_{11}^{-1}R_{12})^T = \begin{bmatrix} b_{n-2k} & b_{n-2k+1} & \cdots & b_{n-k-1} \\ b_{n-2k-1} & b_{n-2k} & \cdots & b_{n-k-2} \\ b_{n-2k-2} & b_{n-2k-1} & \cdots & b_{n-k-3} \\ \vdots & & & \vdots \\ b_0 & & & \\ 0 & \ddots & & \\ & \ddots & & \\ \vdots & & \ddots & \\ & & \ddots & b_0 \\ 0 & & \cdots & 0 \end{bmatrix} \cdot \begin{bmatrix} b_{n-k} & & & & \\ b_{n-k-1} & b_{n-k} & & & \\ b_{n-k-2} & b_{n-k-1} & b_{n-k} & & \\ \vdots & & & \ddots & \\ b_{n-(2k-1)} & b_{n-(2k-2)} & \cdots & b_{n-k-1} & b_{n-k} \end{bmatrix}^{-1} \tag{22}$$

Since the generator update of C_a in Eq. (18) uses only the last row of $R_{11}^{-1}R_{12}$, one has to consider only the last column of its transpose (after peeling off the last entry) which is $\begin{bmatrix} \frac{b_{n-k-1}}{b_{n-k}} \\ \frac{b_{n-k-2}}{b_{n-k}} \\ \frac{b_{n-k-3}}{b_{n-k}} \\ \vdots \\ \frac{b_0}{b_{n-k}} \end{bmatrix}$. Thus the updated matrix C_{new} in (18) is given by:

$$\begin{bmatrix} -\frac{b_{n-k-1}}{b_{n-k}} & -\frac{b_{n-k-2}}{b_{n-k}} & \cdots & -\frac{b_1}{b_{n-k}} & -\frac{b_0}{b_{n-k}} \\ 1 & 0 & \cdots & 0 & 0 \\ 0 & 1 & \cdots & 0 & 0 \\ \vdots & \ddots & \ddots & & \vdots \\ 0 & \cdots & 0 & 1 & 0 \end{bmatrix}$$

which is the companion matrix for $b(x)$ and hence the result.

Corollary 6 *The first column of the Bezoutian $S^\sharp = Bez_P(a^\sharp, b^\sharp)$ where deg $a(x) >$ deg $b(x)$ contains scalar multiples of the coefficients of $b(x)$.*

Proof This result is trivial as the first column of R, i.e., $r_1 = Z^T a_n \mathbf{b}$ where $\mathbf{b} = \begin{bmatrix} b_{n-k} & b_{n-k-1} & \cdots & b_0 \end{bmatrix}^T$ and $R = -\begin{bmatrix} S^\sharp & 0 \\ 0 & 0 \end{bmatrix}$.

The next result shows the updated first column of the Schur complement of a Bezoutian in the kth step.

Corollary 7 *The first column of the Schur complement of $S^\sharp = Bez_P(a^\sharp, b^\sharp)$ where deg $a(x) >$ deg $b(x)$, contains scalar multiples of the coefficients of the polynomial $-c(x)$ which is the remainder of the polynomial division of $a(x)$ by $b(x)$.*

Proof Let us partition the generator $G^{(1)}$ in (19) as $G^{(1)} = \begin{bmatrix} G_1 \\ G_2 \end{bmatrix}$ where $G_1 = \begin{bmatrix} a_n & 0 \\ a_{n-1} & 0 \\ \vdots & \vdots \\ a_{n-k+1} & 0 \end{bmatrix}$ and $G_2 = \begin{bmatrix} a_{n-k} & b_{n-k} \\ a_{n-k-1} & b_{n-k-1} \\ \vdots & \vdots \\ a_0 & b_0 \end{bmatrix}$. Since $S^\sharp$ satisfies the displacement equation (19) with lower shift matrix Z, one can apply the block form of Lemma 2 to update the generator $G^{(1)}$ via:

$$\begin{bmatrix} d_{n-k} & b_{n-k} \\ d_{n-k-1} & b_{n-k-1} \\ \vdots & \vdots \\ d_0 & b_0 \end{bmatrix} = \begin{bmatrix} a_{n-k} & b_{n-k} \\ a_{n-k-1} & b_{n-k-1} \\ \vdots & \vdots \\ a_0 & b_0 \end{bmatrix} - R_{21}R_{11}^{-1} \begin{bmatrix} a_n & 0 \\ a_{n-1} & 0 \\ \vdots & \vdots \\ a_{n-k+1} & 0 \end{bmatrix}$$

which results in the formula:

$$\begin{bmatrix} d_{n-k} \\ d_{n-k-1} \\ \vdots \\ d_0 \end{bmatrix} = \begin{bmatrix} a_{n-k} \\ a_{n-k-1} \\ \vdots \\ a_0 \end{bmatrix} - R_{21}R_{11}^{-1} \begin{bmatrix} a_n \\ a_{n-1} \\ \vdots \\ a_{n-k+1} \end{bmatrix}$$

The matrices R_{11} and R_{21} in the above system are expressed explicitly in Lemma 8 and are the same as the matrices B_k and $\widehat{B_k}$ respectively defined in Lemma 1. Thus combining both Lemmas results in:

$$\begin{bmatrix} d_{n-k} \\ d_{n-k-1} \\ d_{n-k-2} \\ \vdots \\ d_0 \end{bmatrix} = - \begin{bmatrix} 0 \\ c_{n-k-1} \\ c_{n-k-2} \\ \vdots \\ c_0 \end{bmatrix} + q_0 \begin{bmatrix} b_{n-k} \\ b_{n-k-1} \\ b_{n-k-2} \\ \vdots \\ b_0 \end{bmatrix}$$

where from Lemma 1, $-c(x) = -\sum_{i=k+1}^{n} c_{n-i}x^{n-i}$ is the remainder and q_0 is the constant term in the quotient of the polynomial division of $a(x)$ by $b(x)$. Thus the generator update of $G^{(1)}$ via the Bezoutian $Bez_P(a^\sharp, b^\sharp)$ corresponds to the polynomial division of $a(x)$ by $b(x)$. Moreover following Lemma 2, the Schur complement of the Bezoutian, which is the new Bezoutian $Bez_P(b^\sharp, c^\sharp)$, also satisfies the displacement equation (19) with the above generator updates. Thus, as in Lemma 8, one can recover the scalar multiple of the coefficients of $c(x)$ from the first column of $Bez_P(b^\sharp, c^\sharp)$.

Theorem 1 *Let the Bezoutian* $S^\sharp = Bez_P(a^\sharp, b^\sharp)$ *where* $\deg a(x) > \deg b(x)$ *and* C_a *is the companion matrix defined via* (15). *Then the generator update of the Bezoutian*

$S^\sharp$ over the displacement equation $C_a^T S^\sharp - S^\sharp C_a = 0$ coincides with the polynomial division of $a(x)$ by $b(x)$.

Proof Lemma 5 shows that the Bezoutian $S^\sharp$ satisfies $C_a^T S^\sharp - S^\sharp C_a = 0$. Since C_a has upper Hessenberg structure one can apply Lemma 3 to update $S^\sharp$, C_a and C_a^T. Here there is no need to update $G^{(1)}$ and $B^{(1)}$ since they are both 0. Moreover, updates for C_a and C_a^T via Lemma 3 will preserve the upper and lower Hessenberg structures. As we know the Bezoutian $S^\sharp$ is completely determined by polynomials $a^\sharp(x)$ and $b^\sharp(x)$. By Lemma 8 the polynomial $b(x)$ can be recovered from the generator update of the companion matrix C_a, i.e., after generator updates the companion matrix of the polynomial $a(x)$ becomes the companion matrix of the polynomial $b(x)$. Now by Corollary 7, one can recover the scalar multiple of the coefficients of $-c(x)$ which is the remainder of the polynomial division of $a(x)$ by $b(x)$ via the first column of the Schur complement of the Bezoutian $Bez_P(b^\sharp, c^\sharp)$. Hence the result.

The next section generalizes the result in this section having polynomials expanded over the basis $\{Q_k(x)\}_{k=0}^n$ where $\deg Q_k(x) = k$.

4.2 *Hessenberg Displacement Structure of Bezoutian Over Generalized Basis*

In this section, we first generalize Hessenberg displacement structure of a Bezoutian over monomial basis $P = \{x^k\}_{k=0}^n$ to an arbitrary basis $Q = \{Q_k(x)\}_{k=0}^n$ where $\deg Q_k(x) = k$. As a result of this, we will have a new displacement equation with the generalized Bezoutian and confederate matrix. Next we elaborate the Schur complement of the generalized Bezoutian over $\{Q\}$ and use this to analyze the generator updates of a generalized Bezoutian with the polynomial division over $\{Q\}$. Finally, we state the Schur–Euclid–Hessenberg algorithm.

Definition 6 Let $\{Q\} = \{Q_0(x), Q_1(x), \cdots, Q_n(x)\}$ be a system of polynomials satisfying $\deg Q_k(x) = k$ and, $a(x)$ and $b(x)$ be polynomials of degree not greater than n. Then a matrix $S_Q = [\hat{s}_{ij}]$ is the generalized Bezoutian associated with the reverse polynomials $a^\sharp(x)$ and $b^\sharp(x)$ over $\{Q\}$ say $S_Q = Bez_Q(a^\sharp, b^\sharp)$ if

$$\frac{a^\sharp(x) \cdot b^\sharp(y) - b^\sharp(x) \cdot a^\sharp(y)}{x - y} = \sum_{i,j=0}^{n-1} \hat{s}_{ij} Q_i(x) Q_j(y)$$

$$= \begin{bmatrix} Q_0(x) & Q_1(x) & \cdots & Q_{n-1}(x) \end{bmatrix} S_Q \begin{bmatrix} Q_0(y) \\ Q_1(y) \\ \vdots \\ Q_{n-1}(y) \end{bmatrix} \qquad (23)$$

The next result shows a relationship between a Bezoutian for polynomials over the monomial basis $\{P\}$ and the generalized basis $\{Q\} = \{Q_0(x), Q_1(x), \cdots, Q_n(x)\}$ having deg $Q_k(x) = k$.

Lemma 9 *Let B_{PQ} be uni upper triangular basis transformation matrix corresponding to passing basis $\{P\} = \{x^k\}_{k=0}^n$ to basis $\{Q\} = \{Q_k(x)\}_{k=0}^n$ where* deg $Q_k(x) = k$ *via:*

$$\left[Q_0(x)\ Q_1(x)\ \cdots\ Q_{n-1}(x)\right] B_{PQ} = \left[1\ x\ \cdots\ x^{n-1}\right].$$

Then

$$S_Q = B_{PQ}\, S^{\sharp}\, B_{PQ}^T \tag{24}$$

where $S^{\sharp} = Bez_P(a^{\sharp}, b^{\sharp})$ and $S_Q = Bez_Q(a^{\sharp}, b^{\sharp})$.

Proof Recall from the Definition 4 of the Bezoutian associated with the reverse polynomials over basis $\{P\}$

$$\frac{a^{\sharp}(x)\cdot b^{\sharp}(y) - b^{\sharp}(x)\cdot a^{\sharp}(y)}{x-y} = \left[1\ x\ x^2\ \cdots\ x^{n-1}\right] S^{\sharp} \begin{bmatrix} 1 \\ y \\ y^2 \\ \vdots \\ y^{n-1} \end{bmatrix}$$

We can revise the RHS of the above system:

$$\begin{aligned} \frac{a^{\sharp}(x)\cdot b^{\sharp}(y) - b^{\sharp}(x)\cdot a^{\sharp}(y)}{x-y} &= \left[\,1\ x\ \cdots\ x^{n-1}\,\right] B_{PQ}^{-1} B_{PQ}\, S^{\sharp}\, B_{PQ}^T B_{PQ}^{-T} \begin{bmatrix} 1 \\ y \\ \vdots \\ y^{n-1} \end{bmatrix} \\ &= \left[\,Q_0(x)\ Q_1(x)\ \cdots\ Q_{n-1}(x)\,\right] B_{PQ}\, S^{\sharp}\, B_{PQ}^T \begin{bmatrix} Q_0(y) \\ Q_1(y) \\ \vdots \\ Q_{n-1}(y) \end{bmatrix}. \end{aligned}$$

Following Definition 6 gives the result.

To generalize the displacement equation for a Bezoutian we have to explore the Hessenberg structured confederate matrix. Thus we will give the definition of a confederate matrix introduced in [37] next.

Definition 7 Let polynomials $\{Q\} = \{Q_0(x), Q_1(x), Q_2(x), ..., Q_n(x)\}$ with deg $Q_k(x) = k$ be specified by the recurrence relation

$$Q_k(x) = \alpha_k x Q_{k-1}(x) - r_{k-1,k} Q_{k-1}(x) - r_{k-2,k} Q_{k-2}(x) - \ldots - r_{0,k} Q_0(x),\ \alpha_k \neq 0$$

for $k > 0$ and $Q_0(x)$ is a constant. Define for the polynomial

$$a(x) = a_0Q_0(x) + a_1Q_1(x) + \ldots + a_nQ_n(x)$$

its confederate matrix (with respect to the polynomial system Q) by

$$C_Q(a) = \begin{bmatrix} \frac{r_{0,1}}{\alpha_1} & \frac{r_{0,2}}{\alpha_2} & \frac{r_{0,3}}{\alpha_3} & \cdots & \frac{r_{0,n}}{\alpha_n} - \frac{a_0}{\alpha_n a_n} \\ \frac{1}{\alpha_1} & \frac{r_{1,2}}{\alpha_2} & \frac{r_{1,3}}{\alpha_3} & \cdots & \frac{r_{1,n}}{\alpha_n} - \frac{a_1}{\alpha_n a_n} \\ 0 & \frac{1}{\alpha_2} & \frac{r_{2,3}}{\alpha_3} & \cdots & \frac{r_{2,n}}{\alpha_n} - \frac{a_2}{\alpha_n a_n} \\ \vdots & \vdots & \ddots & & \vdots \\ 0 & 0 & \cdots & \frac{1}{\alpha_{n-1}} & \frac{r_{n-1,n}}{\alpha_n} - \frac{a_{n-1}}{\alpha_n a_n} \end{bmatrix}.$$

In the special case where $a(x) = Q_n(x)$, we have $a_0 = a_1 = \cdots = a_{n-1} = 0$. We refer to [37] for many useful properties of the confederate matrix and only recall here that

$$Q_k(x) = \alpha_0\alpha_1 \cdots \alpha_k \cdot det(xI - [C_Q(a)]_{k\times k}),$$

and

$$a(x) = \alpha_0\alpha_1 \cdots \alpha_n a_n \cdot det(xI - [C_Q(a)]),$$

where $[C_Q(a)]_{k\times k}$ denotes the $k \times k$ leading submatrix of $C_Q(a)$.

As we have seen the generalized Bezoutian associated with reverse polynomials and confederate matrix capturing recurrence relations over $\{Q\}$, we will next generalize the displacement equation $C_a^T S^\sharp - S^\sharp C_a = 0$ passing $Bez_P(a^\sharp, b^\sharp)$ to the generalized Bezoutian $S_Q = Bez_Q(a^\sharp, b^\sharp)$ and companion matrix C_a to the confederate matrix C_Q.

Theorem 2 *Let* $\{Q\} = \{Q_0(x), Q_1(x), Q_2(x), ..., Q_n(x)\}$ *where* $\deg Q_k(x) = k$ *be the system of polynomials satisfying recurrence relations*

$$\begin{aligned} &Q_0(x) = 1 \\ &Q_k(x) = xQ_{k-1}(x) - r_{k-1,k}Q_{k-1}(x) - r_{k-2,k}Q_{k-2}(x) - \ldots - r_{0,k}Q_0(x) \end{aligned} \quad (25)$$

$a(x) = a_0Q_0(x) + a_1Q_1(x) + \ldots + a_nQ_n(x)$ *and similarly for* $b(x)$. *Then a matrix* $S_Q = Bez_Q(a^\sharp, b^\sharp)$ *is a Bezoutian associated with reverse polynomials* $a^\sharp(x)$ *and* $b^\sharp(x)$ *if and only if* S_Q *satisfies the equation*

$$C_Q^T S_Q - S_Q C_Q = 0 \quad (26)$$

for some confederate matrix

$$C_Q = \tilde{I} C_Q^T \tilde{I} \quad (27)$$

where

$$C_{Q'} = \begin{bmatrix} r_{0,1} & r_{0,2} & r_{0,3} & \cdots & r_{0,n} - \frac{a_0}{a_n} \\ 1 & r_{1,2} & r_{1,3} & \cdots & r_{1,n} - \frac{a_1}{a_n} \\ 0 & 1 & r_{2,3} & \cdots & r_{2,n} - \frac{a_2}{a_n} \\ \vdots & \vdots & \ddots & & \vdots \\ 0 & 0 & \cdots & 1 & r_{n-1,n} - \frac{a_{n-1}}{a_n}. \end{bmatrix} \tag{28}$$

Proof In the monomial basis $\{P\}$ it has been proven that

$$C_a^T S^\sharp - S^\sharp C_a = 0$$

where $S^\sharp = Bez_P(a^\sharp, b^\sharp)$ and $C_a = \begin{bmatrix} -\frac{a_{n-1}}{a_n} & -\frac{a_{n-2}}{a_n} & -\frac{a_{n-3}}{a_n} & \cdots & -\frac{a_0}{a_n} \\ 1 & 0 & 0 & \cdots & 0 \\ 0 & 1 & 0 & \cdots & 0 \\ \vdots & \vdots & \ddots & & \vdots \\ 0 & 0 & \cdots & 1 & 0 \end{bmatrix}$. Thus it is possible to rewrite the above system as

$$S_1 C_1^T - C_1 S_1 = 0 \tag{29}$$

where $S_1 = \tilde{I} S^\sharp \tilde{I} = Bez_P(a, b)$ and $C_1 = \tilde{I} C_a^T \tilde{I}$. Now with the help of the structure of the uni upper triangular basis transformation matrix B_{PQ} together with the result [32] one can revise the above system as

$$S_{Q'}\, C_{Q'}^T - C_{Q'}\, S_{Q'} = 0$$

where $S_{Q'} = B_{PQ}\, S_1\, B_{PQ}^T = Bez_Q(a, b)$ and $C_{Q'} = B_{PQ}\, C_1\, B_{PQ}^{-1}$ is the confederate matrix given by (28). By rearranging the above system we get

$$(\tilde{I} S_{Q'} \tilde{I})(\tilde{I} C_{Q'}^T \tilde{I}) - (\tilde{I} C_{Q'} \tilde{I})(\tilde{I} S_{Q'} \tilde{I}) = 0$$

yields the result:

$$C_Q^T\, S_Q - S_Q\, C_Q = 0.$$

Conversely if $C_Q^T\, S_Q - S_Q\, C_Q = 0$ and $S_Q = [\hat{s}_{ij}]$ then the second column of S_Q is given by:

$$C_Q^T\, \hat{s}_{i,1} - \left(r_{n-1,\,n} - \frac{a_{n-1}}{a_n}\right) \hat{s}_{i,1} - \hat{s}_{i,2} = 0 \Rightarrow \hat{s}_{i,2} = C_Q^T\, \hat{s}_{i,1} - \left(r_{n-1,\,n} - \frac{a_{n-1}}{a_n}\right) \hat{s}_{i,1}$$

the third column of S_Q is given by:

$$C_Q^T\, \hat{s}_{i,2} - \left(r_{n-2,\,n} - \frac{a_{n-2}}{a_n}\right) \hat{s}_{i,1} - r_{n-2,\,n-1}\, \hat{s}_{i,2} - \hat{s}_{i,3} = 0$$

$$\hat{s}_{i,3} = C_Q^T \hat{s}_{i,2} - \left(r_{n-2,\,n} - \frac{a_{n-2}}{a_n} \right) \hat{s}_{i,1} - r_{n-2,\,n-1}\, \hat{s}_{i,2}$$

proceeding recursively the kth column of S_Q is given by:

$$\hat{s}_{i,k} = C_Q^T \hat{s}_{i,k-1} - \left(r_{n-k+1,\,n} - \frac{a_{n-k+1}}{a_n} \right) \hat{s}_{i,1} - r_{n-k+1,\,n-1}\, \hat{s}_{i,2} - \cdots - r_{n-k+1,\,n-k+2}\, \hat{s}_{i,k-1}.$$

Thus one can recover all columns of S_Q and it should be clear that since S_Q satisfies (26) there is no other matrix which satisfies (26) and has the same first column $\hat{s}_{i,1}$. Thus $S_Q = Bez_Q(a^\sharp, b^\sharp)$. □

It has been shown in Lemma 7 and Corollary 5 that the Schur complement of a Bezoutian is Bezoutian. Thus in the following result, we will generalize the result obtained in Sect. 3. We show that the Schur complement of a Bezoutian S_Q with respect to the generalized basis $\{Q\}$ is congruent to the Schur complement of the Bezoutian $S^\sharp$ with respect to the monomial basis $\{P\}$.

Theorem 3 *The Schur complement of the generalized Bezoutian S_Q over the basis $\{Q\} = \{Q_k(x)\}_{k=0}^n$ where deg $Q_k(x) = k$ is congruent to the Schur complement of the Bezoutian $S^\sharp$ over monomials $\{P\} = \{x^k\}_{k=0}^n$.*

Proof Let us partition the Bezoutian matrix: $S^\sharp = \begin{bmatrix} S_{11} & s_{12} \\ s_{12}^T & s_{22} \end{bmatrix}$. From Lemma 9, we know that $S_Q = B_{PQ}\, S^\sharp\, B_{PQ}^T$ so the basis transformation matrix which is an upper triangular matrix having 1's along the diagonal can be partitioned as $B_{PQ} = \begin{bmatrix} U_{11} & u_{12} \\ 0 & 1 \end{bmatrix}$. Now analyze the block products of $S_Q = B_{PQ}\, S^\sharp\, B_{PQ}^T$ to find its Schur complement.

$$\begin{aligned} S_Q &= \begin{bmatrix} U_{11} & u_{12} \\ 0 & 1 \end{bmatrix} \begin{bmatrix} S_{11} & s_{12} \\ s_{12}^T & s_{22} \end{bmatrix} \begin{bmatrix} U_{11}^T & 0 \\ u_{12}^T & 1 \end{bmatrix} \\ &= \begin{bmatrix} U_{11} & u_{12} \\ 0 & 1 \end{bmatrix} \begin{bmatrix} I & \frac{1}{s_{22}} s_{12} \\ 0 & 1 \end{bmatrix} \begin{bmatrix} S_{11} - \frac{1}{s_{22}} s_{12} s_{12}^T & 0 \\ 0 & s_{22} \end{bmatrix} \begin{bmatrix} I & 0 \\ \frac{1}{s_{22}} s_{12}^T & 1 \end{bmatrix} \begin{bmatrix} U_{11}^T & 0 \\ u_{12}^T & 1 \end{bmatrix} \\ &= \begin{bmatrix} * & s_{12} u_{11} + s_{22} u_{12} \\ s_{12}^T u_{11}^T + s_{22} u_{12}^T & s_{22} \end{bmatrix} \end{aligned}$$

where $* := U_{11}\left(S_{11} - \frac{1}{s_{22}} s_{12} s_{12}^T\right) U_{11}^T + (s_{12} u_{11} + s_{22} u_{12}) \left(\frac{1}{s_{22}} s_{12}^T u_{11}^T + u_{12}^T\right)$. Thus Schur complement of S_Q say S_{Qs} is given by:

$$\begin{aligned} S_{Qs} &= \left(U_{11} \left(S_{11} - \frac{1}{s_{22}} s_{12} s_{12}^T \right) U_{11}^T + (s_{12} u_{11} + s_{22} u_{12}) \left(\frac{1}{s_{22}} s_{12}^T u_{11}^T + u_{12}^T \right) \right) \\ &\quad - \frac{1}{s_{22}} (s_{12} u_{11} + s_{22} u_{12}) (s_{12}^T u_{11}^T + s_{22} u_{12}^T) \\ &= U_{11} \left(S_{11} - \frac{1}{s_{22}} s_{12} s_{12}^T \right) U_{11}^T \end{aligned}$$

Hence the result.

The next result shows the connection of generator updates of the generalized Bezoutian to polynomial division over basis $\{Q\}$.

Theorem 4 *Let* $\{Q\} = \{Q_0, Q_1, Q_2, ..., Q_n\}$ *where* $deg\, Q_k(x) = k$ *be the system of polynomials satisfying recurrence relations* (25) *and* $S_Q = Bez_Q(a^\sharp, b^\sharp) = [\hat{s}_{ij}]$. *If* $a(x) = a_0Q_0(x) + a_1Q_1(x) + \ldots + a_nQ_n(x)$ *and* $b(x) = b_0Q_0(x) + b_1Q_1(x) + \ldots + b_{n-k}Q_{n-k}(x)$ *then the coefficients of the remainder* $-c(x)$ *of the polynomial division* $a(x)$ *by* $b(x)$ *can be recovered from:*

$$-\begin{bmatrix} c_{n-k-1} \\ c_{n-k-2} \\ \vdots \\ c_0 \end{bmatrix} = \left[\left[C_Q^T - \left(r_{n-1,\,n} - \frac{a_{n-1}}{a_n}\right) I_n\right] \hat{s}_{i,1}\right]' - \frac{\hat{s}_{12}}{\hat{s}_{11}} \left[\hat{s}_{i,1}\right]' \tag{30}$$

where

$$C_Q = \begin{bmatrix} r_{n-1,\,n} - \frac{a_{n-1}}{a_n} & r_{n-2,\,n} - \frac{a_{n-2}}{a_n} & r_{n-3,\,n} - \frac{a_{n-3}}{a_n} & \cdots & r_{0,\,n} - \frac{a_0}{a_n} \\ 1 & r_{n-2,\,n-1} & r_{n-3,\,n-1} & \cdots & r_{0,\,n-1} \\ 0 & 1 & r_{n-3,\,n-2} & \cdots & r_{0,\,n-2} \\ \vdots & \ddots & \ddots & & \\ 0 & \cdots & 0 & 1 & r_{0,\,1} \end{bmatrix} \tag{31}$$

and prime means peeling off the first k components.

Proof We have shown in Theorem 2 that $C_Q^TS_Q - S_QC_Q = 0$ and it is the same as $S_QC_Q - C_Q^TS_Q = 0$. One can clearly see that C_Q is an upper Hessenberg matrix, so with that said, S_Q has Hessenberg displacement structure. Hence we can apply Lemma 3 to update C_Q. As deg $b(x)$ is $n-k$ let us partition matrices: $C_Q = \begin{bmatrix} C_q(k,k) & C_q(k,n-k) \\ C_q(n-k,k) & C_q(n-k,n-k) \end{bmatrix}$ and $S_Q = \begin{bmatrix} S_q(k,k) & S_q(k,n-k) \\ S_q(n-k,k) & S_q(n-k,n-k) \end{bmatrix}$ where $S_q(n-k,k) = \left[S_q(k,n-k)\right]^T$ and apply the block form of Lemma 3

$$\begin{aligned} C_{new} &= C_q(n-k,n-k) - C_q(n-k,k)\left[S_q(k,k)\right]^{-1} S_q(k,n-k) \\ &= \begin{bmatrix} r_{n-k-1,\,n-k} & r_{n-k-2,\,n-k} & r_{n-k-3,\,n-k} & \cdots & r_{0,\,n-k} \\ 1 & r_{n-k-2,\,n-k-1} & r_{n-k-3,\,n-k-1} & \cdots & r_{0,\,n-k-1} \\ 0 & 1 & r_{n-k-3,\,n-k-2} & \cdots & r_{0,\,n-k-2} \\ \vdots & \ddots & \ddots & & \vdots \\ 0 & \cdots & 0 & 1 & r_{0,\,1} \end{bmatrix} \\ &\quad - \begin{bmatrix} 0 & \cdots & 0 & 1 \\ 0 & \cdots & 0 & 0 \\ \vdots & & \vdots & \vdots \\ 0 & \cdots & 0 & 0 \end{bmatrix} \left[S_q(k,k)\right]^{-1} S_q(k,n-k). \end{aligned}$$

Thus when updating C_Q only the first row of $C_q(n-k, n-k)$ changes with respect to the last row of the product $\left[S_q(k,k)\right]^{-1} S_q(k, n-k)$. Let us restate the (1,1) block of C_{new} and (1,1), (1,2) blocks of S_Q in terms of a basis transformation matrix which can be partitioned as $B_{PQ} = \begin{bmatrix} B_{pq}(k,k) & B_{pq}(k, n-k) \\ B_{pq}(n-k, k) & \widehat{B}_{PQ} \end{bmatrix}$. Thus the above system can be seen as:

$$C_{new} = \tilde{I} \begin{bmatrix} r_{0,1} & r_{0,2} & r_{0,3} & \cdots & r_{0,n-k} \\ 1 & r_{1,2} & r_{1,3} & \cdots & r_{1,n-k} \\ 0 & 1 & r_{2,3} & \cdots & r_{2,n-k} \\ \vdots & \ddots & \ddots & & \vdots \\ 0 & \cdots & 0 & 1 & r_{n-k-1,n-k} \end{bmatrix}^T \tilde{I}$$

$$- \begin{bmatrix} 0 \cdots 0 & 1 \\ 0 \cdots 0 & 0 \\ \vdots & \vdots \\ 0 \cdots 0 & 0 \end{bmatrix} \left[S_q(k,k)\right]^{-1} \left[B_{pq}(k,k)\right]^T \left[B_{pq}(k,k)\right]^{-T} S_q(k, n-k)$$

$$= \widehat{B}_{PQ} \left[\begin{bmatrix} 0 & 0 & 0 & \cdots & 0 \\ 1 & 0 & 0 & \cdots & 0 \\ 0 & 1 & 0 & \cdots & 0 \\ \vdots & \ddots & \ddots & \ddots & \vdots \\ 0 & \cdots & 0 & 1 & 0 \end{bmatrix} - \begin{bmatrix} 0 \cdots 0 & 1 \\ 0 \cdots 0 & 0 \\ \vdots & \vdots \\ 0 \cdots 0 & 0 \end{bmatrix} \cdot R_{11}^{-1} \cdot R_{12} \right] \left[\widehat{B}_{PQ}\right]^{-1}.$$

Hence going back to the block form of the Bezoutian (22) we get

$$C_{new} =$$
$$\begin{bmatrix} r_{n-k-1,n-k} - \frac{b_{n-k-1}}{b_{n-k}} & r_{n-k-2,n-k} - \frac{b_{n-k-2}}{b_{n-k}} & r_{n-k-3,n-k} - \frac{b_{n-k-3}}{b_{n-k}} & \cdots & r_{0,n-k} - \frac{b_0}{b_{n-k}} \\ 1 & r_{n-k-2,n-k-1} & r_{n-k-3,n-k-1} & \cdots & r_{0,n-k-1} \\ 0 & 1 & r_{n-k-3,n-k-2} & \cdots & r_{0,n-k-2} \\ \vdots & & \ddots & \ddots & \\ 0 & \cdots & 0 & 1 & r_{0,1}. \end{bmatrix}$$

Thus when a generalized Bezoutian S_Q satisfies the displacement equation $C_Q^T S_Q - S_Q C_Q = 0$ with an upper Hessenberg matrix C_Q, the generator update of the confederate matrix C_Q corresponding to polynomial $a(x)$ over $\{Q\}$ results in a confederate matrix C_{new} (say C_{Qb}) and that corresponds to the polynomial $b(x) = b_0 Q_0(x) + b_1 Q_1(x) + \ldots + b_{n-k} Q_{n-k}(x)$.

Now following Lemma 3, one can see a new system with generator updates as

$$S_{Qs} C_{Qb} - C_{Qb}^T S_{Qs} = 0$$

where C_{Qb} is the confederate matrix of $b(x)$ over basis $\{Q\}$ and S_{Qs} is the Schur complement of S_Q. We have shown in Theorem 3 that the Schur complement of S_Q, which is S_{Qs}, is congruent to the Schur complement of $S^\sharp$, which is $S^{(1)}$, i.e.,

$$S_{Qs} = U_{11} S^{(1)} U_{11}^T \tag{32}$$

where U_{11} is the (1,1) block of the uni upper triangular basis transformation matrix B_{PQ}. Moreover from Corollary 7, we have shown that the coefficients of the remainder over monomials can be retrieved from the first column of the Schur complement of $S^\sharp$ which is $S^{(1)}$. This together with the system (32) tells us that coefficients of the remainder over basis $\{Q\}$ can be retrieved from the first column of the S_{Qs}, the Schur complement of S_Q.

Before computing the coefficients of the remainder over $\{Q\}$ let us observe the second column of S_Q. This can be seen directly following Theorem 2 so the second column of S_Q is given by:

$$\hat{s}_{i,2} = \left[C_Q^T - \left(r_{n-1,\,n} - \frac{a_{n-1}}{a_n} \right) \right] \hat{s}_{i,1}.$$

The coefficients of the remainder $-c(x)$ over $\{Q\}$ can be retrieved from the first column of the Schur complement $S_{Qs} = [\tilde{s}_{ij}]$ so those can be retrieved from:

$$s_{i,1} = \left[\hat{s}_{i,2} \right]' - \hat{s}_{12} \left[\frac{1}{\hat{s}_{11}} \hat{s}_{i,1} \right]'$$

where prime means peeling off the first k components. Hence the result.

Remark 1 The above Theorem further shows that the generator update of a Bezoutian S_Q satisfying displacement equation $C_Q^T S_Q - S_Q C_Q = 0$ coincides with polynomial division over basis $\{Q\}$.

As we have the generalized result for the Bezoutian satisfying Hessenberg displacement structure $C_Q^T S_Q - S_Q C_Q = 0$ let us state the Schur–Euclid–Hessenberg algorithm to recover the coefficients of the remainder $c(x)$ of the polynomial division of $a(x)$ by $b(x)$ over basis $\{Q\} = \{Q_k\}_{k=0}^n$ where $\deg Q_k(x) = k$. In the meantime we will be providing triangular factorization of the Bezoutian ($S_Q = [\hat{s}_{ij}] = LDU$) over basis $\{Q\}$. The k-th row u_k of U and the k-th column l_k of L are given by $u_k = \frac{1}{\hat{s}_{11}^k} \left[0 \; \hat{s}_{1,\cdot}^{(k)} \right]$ and $l_k = u_k^T$ where "0" stands for a zero vector of appropriate length. The diagonal factor is given by $D = diag \, [\hat{s}_{11}^{(k)}]_{k=1}^n$.

The Schur–Euclid–Hessenberg Algorithm

Input: Coefficients of $a(x)$ and $b(x)$, say $a_0, a_1, \cdots, a_n$ and $b_0, b_1, \cdots, b_{n-1}$, and if the degree of $b(x)$ is $n-k < n-1$ then list its coefficients as $b_0, b_1, \cdots, b_{n-k}, 0$. Here "0" means a zero vector of appropriate length up to $n-1$.

Initialization: $\hat{s}_{1,\cdot}^{(1)} = \hat{s}_{1,\cdot}, \hat{s}_{\cdot,1}^{(1)} = \hat{s}_{1,\cdot}^{T}, \ C_Q^{(1)} = C_Q$ in (27)

Recursion: For $k = 1, \cdots, n-1$ compute

1. The k-th entry D, the k-th row of U, and k-th column of L by
   $d^{(k)} = \hat{s}_{11}^{(k)}, \ \ u^{(k)} = \frac{1}{d^{(k)}} \hat{s}_{1,\cdot}^{(k)}, \ \ l^{(k)} = [u^{(k)}]^T$
2. The second row and column of $S_Q^{(k)}$ by
   If $k = 1$
   $\hat{s}_{2,\cdot}^{(k)} = \hat{s}_{1,\cdot}^{(k)} \cdot \left[C_Q^{(1)} - \left(r_{n-1,n} - \frac{a_{n-1}}{a_n} \right) I \right], \ \ \hat{s}_{\cdot,2}^{(k)} = [\hat{s}_{2,\cdot}^{(k)}]^T$
   else
   $\hat{s}_{2,\cdot}^{(k)} = \hat{s}_{1,\cdot}^{(k)} \cdot \left[C_Q^{(k)} - r_{n-k,n-k+1} I \right], \ \ \hat{s}_{\cdot,2}^{(k)} = [\hat{s}_{2,\cdot}^{(k)}]^T$
3. The first row and column of $S_Q^{(k+1)}$ which is the Schur complement of $S_Q^{(k)}$ by
   $\hat{s}_{1,\cdot}^{(k+1)} = \hat{s}_{2,\cdot}^{(k)'} - \hat{s}_{21}^{(k)} \left(\frac{1}{\hat{s}_{11}^{(k)}} \hat{s}_{1,\cdot}^{(k)} \right)', \ \ \hat{s}_{\cdot,1}^{(k+1)} = [\hat{s}_{1,\cdot}^{(k+1)}]^T$
   Here the prime means the first component is peeled off. (If deg $b(x) = n - k$ peel off the first k components for the first step).
4. Coefficients of the remainder $c(x)$ by
   $c_{\cdot,1}^{(k+1)} = \frac{1}{\hat{s}_{11}^{(k+1)}} \hat{s}_{\cdot,1}^{(k+1)}$
5. New Confederate matrix generated by
   $C_Q^{(k+1)} = C_Q^{(k)''} - (e_1)' \left(\frac{1}{\hat{s}_{11}^{(k)}} \hat{s}_{1,\cdot}^{(k)'} \right)$

Here double prime means peel off the top row and the first column of the matrix. (If deg $b(x) = n - k$ peel off the first k rows and columns of the matrix for the first step).
Output: Coefficients of the remainder and triangular factorization of the generalized Bezoutian

Proposition 1 *The arithmetic cost of computing the Schur–Euclid–Hessenberg algorithm for a generalized Bezoutian is $\mathcal{O}(M(n)n)$ where $M(n)$ is the cost of multiplying the confederate matrix C_Q by vectors ($k = 1, 2, \cdots, n$).*

Due to the upper Hessenberg structure of the confederate matrix C_Q in the above scenario $M(n) = n^2$. Thus to derive a fast Schur–Euclid–Hessenberg algorithm one has to analyze the confederate matrices based on quasiseparable polynomials or their subclasses as orthogonal and Szegö polynomials as defined in the next section.

5 A Fast Schur–Euclid Algorithm for Quasiseparable Polynomials

We have seen the Schur–Euclid–Hessenberg algorithm for generalized Bezoutian associated with the polynomials expanded over the basis $\{Q_k(x)\}_{k=0}^{n}$ where deg $Q_k(x) = k$ and its arithmetic complexity. The cost of Schur–Euclid–Hessenberg algorithm is determined by the cost of the multiplication of the confederate matrix by vectors. Thus the arithmetic complexity of the algorithm can be reduced having sparse, banded, or structured confederate matrices. Hence in this section, we discuss the complexity of Schur–Euclid–Hessenberg algorithm for quasiseparable polynomials and therefore its sub classes [7]: orthogonal and Szegö polynomials while elaborating a fast Schur–Euclid–Hessenberg algorithm for quasiseparable polynomials.

5.1 Schur–Euclid–Hessenberg Algorithm for Quasiseparable Polynomials

In this section, we analyze the matrix decomposition for the confederate matrix over quasiseparable polynomials and use the decomposition to derive a fast Schur–Euclid–Hessenberg algorithm for a Bezoutian associated with the quasiseparable polynomials. The ultimate idea is to explore the confederate matrix over quasiseparable polynomials to reduce the cost of computing $M(n)$ in Proposition 1.

Let us start with the generator definition of $(H, 1)$-quasiseparable matrix which is equivalent to the rank Definition 2. We use quasiseparable generators to define a system of quasiseparable polynomials.

Definition 8 A matrix A is called $(H, 1)$-quasiseparable if **(i)** it is strongly upper Hessenberg, and **(ii)** it can be represented in the form

$$A = \begin{bmatrix} d_1 & & & & \\ q_1 & d_2 & & g_i b_{ij}^{\times} h_j & \\ 0 & q_2 & \ddots & & \\ \vdots & \ddots & \ddots & \ddots & \\ 0 & \cdots & 0 & q_{n-1} & d_n \end{bmatrix}$$

where $b_{ij}^{\times} = b_{i+1} b_{i+2} \cdots b_{j-1}$ for $j > i+1$ and $b_{ij}^{\times} = 1$ for $j = i+1$. The scalar elements $\{q_k, d_k, g_k, b_k, h_k\}$ are called the generators of the matrix A.

The results of [7, 37] allow one to observe a bijection between the set of strongly upper Hessenberg matrices $\mathscr{H}$ (say $A = [a_{ij}] \in \mathscr{H}$) and the set of polynomials system $\mathscr{P}$ (say $R = \{r_k(x)\} \in \mathscr{P}$ with deg $r_k(x) = k$) via

$$f : \mathscr{H} \to \mathscr{P}, \text{ where } r_k(x) = \frac{1}{a_{2,1} a_{3,2} \cdots a_{k,k-1}} det(xI - A)_{k \times k}. \tag{33}$$

The following lemma is given in [7, 8] and is a consequence of Definition 8 and [37].

Lemma 10 *Let A be an $(H, 1)$-quasiseparable matrix specified by its generators as in Definition 8. Then a system of polynomials $\{r_k(x)\}$ satisfies the recurrence relations*

$$r_k(x) = \frac{1}{q_k}\left[(x-d_k)r_{k-1}(x) - \sum_{j=0}^{k-2} g_{j+1}b^{\times}_{j+1,k}h_k\, r_j(x)\right] \tag{34}$$

if and only if $\{r_k(x)\}$ *is related to* A *via* (33).

With the help of the system of quasiseparable polynomials $\{r_k(x)\}_{k=0}^n$ satisfying k-term recurrence relations (34), we will define a confederate matrix for the polynomial

$$a(x) = a_0 r_0(x) + a_1 r_1(x) + \cdots + a_n r_n(x) \tag{35}$$

by

$$C_R(a) = \begin{bmatrix} d_1 & g_1h_2 & g_1b_2h_3 & \cdots & \cdots & g_1b^{\times}_{1,n}h_n - \frac{a_0}{a_n} \\ q_1 & d_2 & g_2h_3 & \cdots & \cdots & g_2b^{\times}_{2,n}h_n - \frac{a_1}{a_n} \\ 0 & q_2 & d_3 & \cdots & \cdots & g_3b^{\times}_{3,n}h_n - \frac{a_1}{a_n} \\ \vdots & \ddots & \ddots & \ddots & \ddots & \vdots \\ \vdots & \ddots & \ddots & q_{n-2} & d_{n-1} & g_{n-1}h_n - \frac{a_{n-2}}{a_n} \\ 0 & \cdots & \cdots & 0 & q_{n-1} & d_n - \frac{a_{n-1}}{a_n}. \end{bmatrix} \tag{36}$$

The matrix (36) is an upper Hessenberg matrix so following Theorem 4 the Bezoutian associated with quasiseparable polynomials satisfies $C_R^T S_R - S_R C_R = 0$, where R is the system of quasiseparable polynomials satisfying recurrence relations (34), $C_R = \tilde{I}[C_R(a)]^T\tilde{I}$, and $a(x)$ is defined via (35). Though one can state a Schur–Euclid–Hessenberg algorithm for a Bezoutian associated with quasiseparable polynomials using the Schur–Euclid–Hessenberg algorithm in Sect. 4.2 it is not cheap because the structure of the confederate matrix C_R is not sparse. Thus to reduce the cost of the Schur–Euclid–Hessenberg algorithm for a Bezoutian associated with quasiseparable polynomials one has to explore the structure of the confederate matrix C_R via matrix decomposition.

Theorem 5 *Let* C_R *be the matrix specified by generators* $\{q_k, d_k, g_k, b_k, h_k\}$ *and coefficients* a_k *via*

$$C_R = \begin{bmatrix} d_n - \frac{a_{n-1}}{a_n} & g_{n-1}h_n - \frac{a_{n-2}}{a_n} & g_{n-2}b_{n-1}h_n - \frac{a_{n-3}}{a_n} & \cdots\cdots & g_1b^{\times}_{1,n}h_n - \frac{a_0}{a_n} \\ q_{n-1} & d_{n-1} & g_{n-2}h_{n-1} & \cdots\cdots & g_1b^{\times}_{1,n-1}h_{n-1} \\ 0 & q_{n-2} & d_{n-2} & \cdots\cdots & g_1b^{\times}_{1,n-2}h_{n-2} \\ \vdots & \ddots & \ddots & \ddots\;\ddots & \vdots \\ \vdots & \ddots & \ddots & q_2\;\; d_2 & g_1h_2 \\ 0 & \cdots & \cdots & 0\;\; q_1 & d_1 \end{bmatrix} \tag{37}$$

then the following decomposition holds:

$$C_R = \left[\tilde{\theta}_n\left(\cdots\left(\tilde{\theta}_3\left(\tilde{\theta}_2\tilde{\theta}_1 + \tilde{\Delta}_2\right) + \tilde{\Delta}_3\right)\cdots\right) + \tilde{\Delta}_n\right] + \frac{1}{a_n}\cdot \tilde{A}_1\tilde{A}_2\cdots\tilde{A}_n \tag{38}$$

where

$$\tilde{\theta}_1 = \begin{bmatrix} I_{n-2} & & \\ & d_2 & g_1 \\ & q_1 & d_1 \end{bmatrix}, \quad \tilde{\theta}_k = \begin{bmatrix} I_{n-k-1} & & & \\ & d_{k+1} & b_k & \\ & q_k & h_k & \\ & & & I_{k-1} \end{bmatrix}, \quad \tilde{\theta}_n = \begin{bmatrix} h_n & \\ & I_{n-1} \end{bmatrix}$$

$$\tilde{\Delta}_k = \begin{bmatrix} 0_{n-k-1} & & & \\ & 0 & g_k - d_k b_k & \\ & 0 & d_k - d_k h_k & \\ & & & 0_{k-1} \end{bmatrix}, \quad \tilde{\Delta}_n = \begin{bmatrix} d_n - d_n h_n & \\ & 0_{n-1} \end{bmatrix} \tag{39}$$

and

$$\tilde{A}_k = \begin{bmatrix} I_{k-1} & & & \\ & -a_{n-k} & 1 & \\ & 0 & 0 & \\ & & & I_{n-k-1} \end{bmatrix}, \quad \tilde{A}_n = \begin{bmatrix} I_{n-1} & \\ & -a_0. \end{bmatrix} \tag{40}$$

Proof Let us split the matrix C_R into $C_R = \tilde{H} + \frac{1}{a_n} C$ where

$$\tilde{H} = \begin{bmatrix} d_n & g_{n-1}h_n & g_{n-2}b_{n-1}h_n & \cdots\cdots & g_1 b^{\times}_{1,n} h_n \\ q_{n-1} & d_{n-1} & g_{n-2}h_{n-1} & \cdots\cdots & g_1 b^{\times}_{1,n-1} h_{n-1} \\ 0 & q_{n-2} & d_{n-2} & \cdots\cdots & g_1 b^{\times}_{1,n-2} h_{n-2} \\ \vdots & \ddots & \ddots & \ddots \; \ddots & \vdots \\ \vdots & \ddots & \ddots & q_2 \;\; d_2 & g_1 h_2 \\ 0 & \cdots & \cdots & 0 \;\; q_1 & d_1 \end{bmatrix}$$

and $C = \begin{bmatrix} -a_{n-1} & -a_{n-2} & \cdots & -a_1 & -a_0 \\ 0 & 0 & \cdots & 0 & 0 \\ 0 & 0 & \cdots & 0 & 0 \\ \cdot & \cdot & \cdot & \cdot & \cdot \\ 0 & 0 & \cdots & 0 & 0 \end{bmatrix}$. One can easily show by matrix multiplication that the latter matrix admits the factorization $C = \tilde{A}_1 \tilde{A}_2 \cdots \tilde{A}_n$ with the given $\tilde{A}_k$ for $k = 1, 2, \cdots, n$. Thus we only have to prove the decomposition for $\tilde{H}$ in terms of $\tilde{\theta}_k$'s and $\tilde{\Delta}_k$'s.

Showing the decomposition for $\tilde{H} = \tilde{\theta}_n \left(\cdots \left(\tilde{\theta}_3 \left(\tilde{\theta}_2 \tilde{\theta}_1 + \tilde{\Delta}_2 \right) + \tilde{\Delta}_3 \right) \cdots \right) + \tilde{\Delta}_n$ is equivalent to showing that the matrix $\tilde{H}$ satisfies the iteration:

$$\tilde{H}_0 = I_n, \; \tilde{H}_k = \tilde{\theta}_k \tilde{H}_{k-1} + \tilde{\Delta}_k, \; k = 1, 2, \cdots, n, \; \tilde{H} = \tilde{H}_n. \tag{41}$$

Let us show by induction that for every $k = 3, 4, \cdots, n$:

$$\tilde{H}_{k-1}(n-k+1 : n, n-k+1 : n) = \tilde{H}(n-k+1 : n, n-k+1 : n) \tag{42}$$

The basis of induction (k=3) is trivial

$$\tilde{H}(n-2:n, n-2:n) = \begin{bmatrix} d_2 & g_1h_2 \\ q_1 & d_1 \end{bmatrix} = \tilde{H}_2(n-2:n, n-2:n).$$

Assume that (42) holds for all indices up to k. Consider the matrix $\tilde{H}_k(n-k+2 : n, n-k+2:n)$:

$$\begin{bmatrix} d_{k+1} & b_k & \\ q_k & h_k & \\ & & I_{k-1} \end{bmatrix} \left[\begin{array}{c|c} 1 & 0 \quad \cdots \quad 0 \\ \hline 0 & \\ \vdots & \tilde{H}_{k-1}(n-k+1:n, n-k+1:n) \\ 0 & \end{array}\right] + \begin{bmatrix} 0 & g_k - d_k b_k & \\ 0 & d_k - d_k h_k & \\ & & 0_{k-1\cdot} \end{bmatrix} \tag{43}$$

The first row of the matrix $\tilde{H}_{k-1}(n-k+1 : n, n-k+1 : n)$ equals the first row of the matrix $\tilde{H}(n-k+1 : n, n-k+1 : n)$. Therefore, performing the matrix product in (43) we get:

$$\left[\begin{array}{c|c|c} d_{k+1} & d_k b_k + (g_k - d_k b_k) & \tilde{H}_{k-1}(n-k-2, n-k:n) b_k \\ \hline q_k & d_k h_k + (d_k - d_k h_k) & \tilde{H}_{k-1}(n-k-1, n-k:n) h_k \\ \hline 0 & & \\ \vdots & \tilde{H}_{k-1}(n-k:n, n-k-1) & \tilde{H}_{k-1}(n-k:n, n-k:n) \\ 0 & & \end{array}\right]$$

which is equal to $\tilde{H}(n-k+2 : n, n-k+2 : n)$. By induction we get $\tilde{H}_{n-1}(1 : n, 1:n) = \tilde{H}(1:n, 1:n)$. Substituting this into recursion (41) we get

$$\tilde{H}_n = \begin{bmatrix} h_n & \\ & I_{n-1} \end{bmatrix} \tilde{H}_{n-1} + \begin{bmatrix} d_n - d_n h_n & \\ & 0_{n-1} \end{bmatrix} = \tilde{H}.$$

Now we have the decomposition of the confederate matrix C_R (38) over quasiseparable polynomials and Hessenberg-1-quasiseparable displacement structure of the Bezoutian $C_R^T S_R - S_R C_R = 0$. Thus the following Schur–Euclid–Hessenberg algorithm for a Bezoutian associated with quasiseparable polynomials can be used to recover coefficients of the remainder $c(x)$ of the polynomial division $a(x)$ by $b(x)$ over the basis $\{R\} = \{r_k\}_{k=0}^n$ where deg $r_k(x) = k$ and $\{R\}$ satisfies the recurrence relations (34), and also to obtain the triangular factorization of Bezoutian over the system of quasiseparable polynomials $\{R\}$.

The Schur–Euclid–Hessenberg Algorithm for Bezoutian Over Quasiseparabale polynomials

Input: Generators $\{q_k, d_k, g_k, b_k, h_k\}$. Coefficients of $a(x)$ and $b(x)$, say $a_0, a_1, \cdots, a_n$ and $b_0, b_1, \cdots, b_{n-1}$, and if the degree of $b(x)$ is $n-k < n-1$ then list its coefficients as $b_0, b_1, \cdots, b_{n-k}, 0$ here "0" means a zero vector of appropriate length up to $n-1$.

Initialization: Set $\tilde{\theta}_k$ and $\tilde{\Delta}_k$ in terms of generators $\{q_k, d_k, g_k, b_k, h_k\}$, and $\tilde{A}_k$ in terms of a_k and $C_R^{(1)} = C_R$ via (38). Set $\hat{s}_{.,1}^{(1)} = \hat{s}_{.,1}$.

Recursion: For $k = 1, \cdots, n-1$ compute

1. The k-th entry D, the k-th row of U, and k-th column of L by
   $d^{(k)} = \hat{s}_{11}^{(k)}, \quad u^{(k)} = \frac{1}{d^{(k)}}\hat{s}_{1,\cdot}^{(k)}, \quad l^{(k)} = [u^{(k)}]^T$
2. The second row and column of $S_R^{(k)}$ by
   If $k = 1$
   $\hat{s}_{2,\cdot}^{(k)} = \frac{1}{q_{n-1}}\hat{s}_{1,\cdot}^{(k)}\left[C_R^{(1)} - \left(d_n - \frac{a_{n-1}}{a_n}\right)I\right], \quad \hat{s}_{\cdot,2}^{(k)} = [\hat{s}_{2,\cdot}^{(k)}]^T$
   else
   $\hat{s}_{2,\cdot}^{(k)} = \frac{1}{q_{n-k}}\hat{s}_{1,\cdot}^{(k)}\left[C_R^{(k)} - d_{n-k+1}I\right], \quad \hat{s}_{\cdot,2}^{(k)} = [\hat{s}_{2,\cdot}^{(k)}]^T$
3. The first row and column of $S_R^{(k+1)}$ which is the Schur complement of $S_R^{(k)}$ by
   $\hat{s}_{1,\cdot}^{(k+1)} = \hat{s}_{2,\cdot}^{(k)\prime} - \hat{s}_{21}^{(k)}\left(\frac{1}{\hat{s}_{11}^{(k)}}\hat{s}_{1,\cdot}^{(k)}\right)', \quad \hat{s}_{\cdot,1}^{(k+1)} = [\hat{s}_{1,\cdot}^{(k+1)}]^T$
   Here the prime means the first component is peeled off. (If deg $b(x) = n - k$, then peel off the first k components for the first step).
4. Coefficients of the remainder $c(x)$ by
   $c_{\cdot,1}^{(k+1)} = \frac{1}{\hat{s}_{11}^{(k+1)}}\hat{s}_{\cdot,1}^{(k+1)}$
5. Confederate matrix after peeling off row(s) and column(s) by
   $\tilde{C}_R^{(k)} = \tilde{\theta}_{n-k}''\left(\cdots\left(\tilde{\theta}_3''\left(\tilde{\theta}_2''\tilde{\theta}_1'' + \tilde{\Delta}_2''\right) + \tilde{\Delta}_3''\right)\cdots\right) + \tilde{\Delta}_{n-k}''$
   Here the double prime means peel off the top row and the first column of matrices. (If deg $b(x) = n - k$, then peel off the first k rows and columns of the matrix for the first step).
6. New confederate matrix generated by
   $C_R^{(k+1)} = \tilde{C}_R^{(k)} - q_{n-k}(e_1)'\left(\frac{1}{\hat{s}_{11}^{(k)}}\hat{s}_{1,\cdot}^{(k)\prime}\right)$

Output: Coefficients of the remainder and triangular factorization of the Bezoutian associated with quasiseparable polynomials.

As we have seen in Proposition 1, the cost of the Schur–Euclid–Hessenberg algorithm is dominated by $M(n)$ which is the cost of multiplication of a confederate matrix by vectors and this occurs in step 2 of the algorithm. Also note that for the multiplication of $C_R^{(k)}$ by vectors, i.e., for $k = 1$ we have to multiply n factors of $\tilde{\theta}_k$, $n - 1$ factors of $\tilde{\Delta}_k$, and n factors of $\tilde{A}_k$ together with the scaling factor $\frac{1}{a_n}$ (note that we do not have to scale for the monic polynomial case) by a vector and for $k = 2, 3, \cdots, n-1$ we have to multiply at most $n - k$ factors of $\tilde{\theta}$ and $n - k - 1$ factors of $\tilde{\Delta}$ by a vector. Thus the most expensive step in the recursion is when $k = 1$. Now for $k = 1$ in step 2 of the Schur–Euclid–Hessenberg algorithm in the quasiseparable case, we have at most 4 multiplications and 2 additions corresponding to multiplication of $\tilde{\theta}_k$, at most 2 multiplications corresponding to multiplication of $\tilde{\Delta}_k$, and at most 1 multiplication and 1 addition corresponding to multiplication of $\tilde{A}_k$ by the first row of the Bezoutian. Thus the arithmetic cost of computing $C_R^{(1)}$ by a vector is at most $11n - 2$ operations as opposed to Hessenberg structured matrix C_R (37) by a vector, which is $n^2 - 2$ operations. Hence $\forall n > 11$, one can design a fast Schur–Euclid–Hessenberg algorithm for quasiseparable polynomials.

Proposition 2 *The arithmetic cost of computing the Schur–Euclid–Hessenberg algorithm for the Bezoutian associated with the quasiseparable polynomials satisfying recurrence relations* (34) *is* $\mathcal{O}(n^2)$.

5.2 Cost of Schur–Euclid–Hessenberg Algorithm for Orthogonal Polynomials

In this section, we observe the cost of computing the Schur–Euclid–Hessenberg Algorithm for a Bezoutian associated with the orthogonal polynomials. The main idea here is to explore the confederate matrix with respect to the orthogonal polynomial system to reduce the cost of computing $M(n)$ in Proposition 1.

It is well known [16] that systems of polynomials $R = \{r_k(x)\}_{k=0}^n$ orthogonal with respect to an inner product of the form

$$< p(x), q(x) >= \int_a^b p(x)q(x)w^2(x)dx$$

satisfy a three-term recurrence relation of the form

$$r_k(x) = \frac{1}{q_k}(x - d_k)r_{k-1}(x) - \frac{g_{k-1}}{q_k} \cdot r_{k-2}(x), \quad q_k \neq 0. \tag{44}$$

Define for the polynomial

$$a(x) = a_0 r_0(x) + a_1 r_1(x) + \cdots + a_{n-1} r_{n-1}(x) + a_n r_n(x) \tag{45}$$

its confederate matrix, given by

$$C(a) = \begin{bmatrix} d_1 & g_1 & 0 & \cdots & 0 & -\frac{a_0}{a_n} \\ q_1 & d_2 & g_2 & \ddots & \vdots & -\frac{a_1}{a_n} \\ 0 & q_2 & d_3 & \ddots & 0 & -\frac{a_2}{a_n} \\ \vdots & \ddots & \ddots & \ddots & \ddots & \vdots \\ 0 & \cdots & 0 & q_{n-2} & d_{n-1} & g_{n-1} - \frac{a_{n-2}}{a_n} \\ 0 & 0 & \cdots & 0 & q_{n-1} & d_n - \frac{a_{n-1}}{a_n} \end{bmatrix} \tag{46}$$

which has been called a comrade matrix in [4].

The matrix (46) is an upper Hessenberg matrix so by Theorem 4 the Bezoutian associated with orthogonal polynomials satisfies $C_R^T S_R - S_R C_R = 0$, where R is the system of orthogonal polynomials satisfying recurrence relations (44), $C_R = \tilde{I}C(a)^T\tilde{I}$, and $a(x)$ is defined via (45). Moreover orthogonal polynomials are a sub class of quasiseparable polynomials [7] so to express Schur–Euclid–Hessenberg algorithm for a Bezoutian associated with the orthogonal polynomials one has to revise the Schur–Euclid–Hessenberg Algorithm in the quasiseparable case in Sect. 5.1. To do so one has to consider the Bezoutian associated with orthogonal polynomials and initialize the algorithm with the comrade matrix $C_R = \tilde{I}C(a)^T\tilde{I}$ having generators $\{q_k, d_k, g_k\}$ with the coefficients a_k. Due to the sparse structure of the comrade matrix, we could ignore step 5 of the Schur–Euclid–Hessenberg Algorithm in the quasiseparable case and revise the first matrix in the RHS of step 6 to be the comrade matrix $C_R = \tilde{I}C(a)^T\tilde{I}$.

The matrix C_R is almost tridiagonal. Thus the cost of multiplication of C_R by vectors is only $M(n) = \mathcal{O}(n)$ operations. Hence one can design a fast Schur–Euclid–Hessenberg algorithm for a Bezoutian associated with orthogonal polynomials having complexity $\mathcal{O}(n^2)$ operations.

Proposition 3 *The arithmetic cost of computing the Schur–Euclid–Hessenberg algorithm for the Bezoutian associated with the orthogonal polynomials satisfying* (44) *is* $\mathcal{O}(n^2)$.

5.3 Cost of Schur–Euclid–Hessenberg Algorithm for Szegö Polynomials

In this section, we observe the cost of computing the Schur–Euclid–Hessenberg Algorithm for a Bezoutian associated with the Szegö polynomials. The main idea here is to explore the confederate matrix with respect to the Szegö polynomial system to reduce the cost of computing $M(n)$ in Proposition 1.

Szegö polynomials $S = \{\phi_k^\sharp(x)\}_{k=0}^n$ or polynomials orthonormal on the unit circle with respect to an inner product of the form

$$< p(x), q(x) >= \frac{1}{2\pi}\int_{-\pi}^{\pi} p(e^{i\theta})[q(e^{i\theta})]^* w^2(\theta)d\theta,$$

for any such inner product, it is known [22] that there exist a set of reflection coefficients $\{\rho_k\}$ satisfying

$$\rho_0 = -1, \quad |\rho_k| < 1, \quad k = 1, 2, \cdots, n-1, \quad |\rho_n| \leq 1,$$

and complementary parameters $\{\mu_k\}$ defined by the reflection coefficients via

$$\mu_k = \begin{cases} \sqrt{1-|\rho_k|^2}, & |\rho_k| < 1 \\ 1, & |\rho_k| = 1 \end{cases}$$

such that the corresponding Szegö polynomials satisfying the two-term recurrence relations

$$\begin{bmatrix} \phi_k(x) \\ \phi_k^\sharp(x) \end{bmatrix} = \frac{1}{\mu_0}\begin{bmatrix} 1 \\ 1 \end{bmatrix}, \quad \begin{bmatrix} \phi_k(x) \\ \phi_k^\sharp(x) \end{bmatrix} = \frac{1}{\mu_k}\begin{bmatrix} 1 & -\rho_k^\star \\ -\rho_k & 1 \end{bmatrix}\begin{bmatrix} \phi_{k-1}(x) \\ x\,\phi_{k-1}^\sharp(x) \end{bmatrix} \tag{47}$$

where $\{\phi_k(x)\}$ is a system of auxiliary polynomials. Define for the polynomial

$$a(x) = a_0\phi_0^\sharp(x) + a_1\phi_1^\sharp(x) + \cdots + a_{n-1}\phi_{n-1}^\sharp(x) + a_n\phi_n^\sharp(x) \tag{48}$$

its confederate matrix is given by

$$C_S(a) = \begin{bmatrix} -\rho_0^*\rho_1 & -\rho_0^*\mu_1\rho_2 & -\rho_0^*\mu_1\mu_2\rho_3 & \cdots & -\rho_0^*\mu_1\mu_2\cdots\mu_{n-1}\rho_n - \frac{a_0}{a_n} \\ \mu_1 & -\rho_1^*\rho_2 & -\rho_1^*\mu_2\rho_3 & \cdots & -\rho_1^*\mu_2\mu_3\cdots\mu_{n-1}\rho_n - \frac{a_1}{a_n} \\ 0 & \mu_1 & -\rho_2^*\rho_3 & \cdots & -\rho_2^*\mu_3\mu_4\cdots\mu_{n-1}\rho_n - \frac{a_1}{a_n} \\ \ddots & \ddots & \ddots & \ddots & \vdots \\ 0 & \cdots & 0 & \mu_{n-1} & -\rho_{n-1}^*\rho_n - \frac{a_{n-1}}{a_n}. \end{bmatrix} \tag{49}$$

The matrix (49) is an upper Hessenberg matrix so following Theorem 4 the Bezoutian associated with Szegö polynomials satisfies $C_S^T S_S - S_S C_S = 0$ where S is the system of Szegö polynomials satisfying recurrence relations (47), $C_S = \tilde{I}[C_S(a)]^T\tilde{I}$, and $a(x)$ is defined via (48). The matrix C_S is not sparse like the orthogonal polynomial case so to reduce the cost of multiplication of C_S by vectors one has to use the nested factorization of C_S.

Lemma 11 *Let C_S be the matrix specified by generators $\{\rho_k^*, \rho_k, \mu_k\}$ and coefficients a_k via*

$$C_S = \tilde{I}[C_S(a)]^T\tilde{I} \tag{50}$$

then the following decomposition holds:

$$C_S = \tilde{\Gamma}_0\tilde{\Gamma}_1\tilde{\Gamma}_2\cdots\tilde{\Gamma}_n + \frac{1}{a_n}\cdot\tilde{A}_1\tilde{A}_2\cdots\tilde{A}_n \tag{51}$$

where

$$\tilde{\Gamma}_0 = \begin{bmatrix} -\rho_n & \\ & I_{n-1} \end{bmatrix}, \quad \tilde{\Gamma}_k = \begin{bmatrix} I_{k-1} & & & \\ & \rho_{n-k}^* & \mu_{n-k} & \\ & \mu_{n-k} & -\rho_{n-k} & \\ & & & I_{n-k-1} \end{bmatrix}, \quad \tilde{\Gamma}_n = \begin{bmatrix} I_{n-1} & \\ & \rho_0^* \end{bmatrix} \tag{52}$$

and

$$\tilde{A}_k = \begin{bmatrix} I_{k-1} & & & \\ & -a_{n-k} & 1 & \\ & 0 & 0 & \\ & & & I_{n-k-1} \end{bmatrix}, \quad \tilde{A}_n = \begin{bmatrix} I_{n-1} & \\ & -a_0. \end{bmatrix}$$

Proof The matrix C_S can be split into $C_S = \tilde{U} + \frac{1}{a_n}C$ where

$$\tilde{U} = \begin{bmatrix} -\rho_{n-1}^{*}\rho_n & -\rho_{n-2}^{*}\mu_{n-1}\rho_n & -\rho_{n-3}^{\star}\mu_{n-2}\mu_{n-1}\rho_n & \cdots & -\rho_0^{*}\mu_1\mu_2\cdots\mu_{n-1}\rho_n \\ \mu_{n-1} & -\rho_{n-2}^{\star}\rho_{n-1} & -\rho_{n-3}^{\star}\mu_{n-2}\rho_{n-1} & \cdots & -\rho_0^{*}\mu_1\mu_2\cdots\mu_{n-2}\rho_{n-1} \\ 0 & \mu_{n-2} & -\rho_{n-3}^{\star}\rho_{n-2} & \cdots & -\rho_0^{*}\mu_1\mu_2\cdots\mu_{n-3}\rho_{n-2} \\ \ddots & \ddots & \ddots & \ddots & \vdots \\ 0 & \cdots & 0 & \mu_1 & -\rho_0^{\star}\rho_1 \end{bmatrix}$$

and $C = \begin{bmatrix} -a_{n-1} & -a_{n-2} & \cdots & -a_1 & -a_0 \\ 0 & 0 & \cdots & 0 & 0 \\ 0 & 0 & \cdots & 0 & 0 \\ \cdot & \cdot & \cdot & \cdot & \cdot \\ 0 & 0 & \cdots & 0 & 0 \end{bmatrix}$. Let U be the unitary Hessenberg matrix [9] corresponding to Szegö polynomials $\{\phi_k^{\sharp}(x)\}$ satisfying 2-term recurrence relations (47), then we can see that

$$\tilde{U} = \tilde{I}U^T\tilde{I}$$

It is well known that unitary Hessenberg matrix U can be written as the product $U = \Gamma_0\Gamma_1\Gamma_2\cdots\Gamma_n$ where

$$\Gamma_0 = \begin{bmatrix} \rho_0^{*} & \\ & I_{n-1} \end{bmatrix}, \quad \Gamma_k = \begin{bmatrix} I_{k-1} & & & \\ & -\rho_k & \mu_k & \\ & \mu_k & \rho_k^{*} & \\ & & & I_{n-k-1} \end{bmatrix}, \quad \Gamma_n = \begin{bmatrix} I_{n-1} & \\ & -\rho_n \end{bmatrix}.$$

Thus $\tilde{U}$ has the factorization

$$\tilde{\Gamma}_0\tilde{\Gamma}_1\tilde{\Gamma}_2\cdots\tilde{\Gamma}_n$$

where $\tilde{\Gamma}_k = \tilde{I}\Gamma_k^T\tilde{I}$ for $k = 0, 1, \cdots, n$.

One can see clearly by the matrix multiplication that the matrix C admits the factorization

$$C = \tilde{A}_1\tilde{A}_2\cdots\tilde{A}_n$$

where $\tilde{A}_k = \begin{bmatrix} I_{k-1} & & & \\ & -a_{n-k} & 1 & \\ & 0 & 0 & \\ & & & I_{n-k-1} \end{bmatrix}$, $\quad \tilde{A}_n = \begin{bmatrix} I_{n-1} & \\ & -a_0 \end{bmatrix}$, hence the result.

Szegö polynomials are a sub class of quasiseparable polynomials [7] so to express the Schur–Euclid–Hessenberg algorithm for the Bezoutian associated with the Szegö polynomials one has to revise the Schur–Euclid–Hessenberg Algorithm in the quasiseparable case, in Sect. 5.1. To do so one has to consider the Bezoutian associated with Szegö polynomials and initialize the algorithm with the confederate matrix (51) having generators $\{\rho_k^*, \rho_k, \mu_k\}$ with the coefficients a_k. But to reduce the complexity $M(n)$ one has to revise step 5 of the Schur–Euclid–Hessenberg Algorithm for the quasiseparable case with the decomposition $\tilde{C}_S^{(k)} = \tilde{\Gamma}_k'' \tilde{\Gamma}_{k+1}'' \cdots \tilde{\Gamma}_n''$ where $\tilde{\Gamma}_k$'s are given in (52) and the double prime means peel off the top k column(s) and row(s) if $\deg b(x) = n - k$ where $\deg a(x) = n$ while revising the first matrix in the RHS of step 6 to be the confederate matrix $\tilde{C}_S^{(k)}$ in step 5.

Due to the decomposition of C_S, in step 2 of the Schur–Euclid–Hessenberg algorithm for Szegö polynomials, we have at most 4 multiplications and 2 additions corresponding to multiplication of $\tilde{\Gamma}_k$ by the first row of the Bezoutian, and at most 1 multiplication and 1 addition corresponding to multiplication of $\tilde{A}_k$ by the first row of the Bezoutian. Thus the total cost of multiplication of $n+1$ factors of $\tilde{\Gamma}_k$ which is $\tilde{U}$ by the vector is $6(n+1)$ operations, and n factors of $\tilde{A}_k$ which is C by the vector costs $2n$ operations. Together with the multiplication of C by quantity $\frac{1}{a_n}$ gives the overall cost of multiplication of C_S by vectors as $M(n) = \mathcal{O}(n)$ complexity. Hence one can design a fast Schur–Euclid–Hessenberg algorithm for Szegö polynomials.

Proposition 4 *The arithmetic cost of computing the Schur–Euclid–Hessenberg algorithm for a Bezoutian associated with the Szegö polynomials satisfying recurrence relations* (47) *is* $\mathcal{O}(n^2)$.

6 Conclusion

In this paper, we have derived a Schur–Euclid–Hessenberg algorithm to compute the triangular factorization of a generalized Bezoutian. In this case, it is associated with the system of polynomials $\{Q\} = \{Q_k(x)\}_{k=0}^n$, where $\deg Q_k(x) = k$ and satisfies k-term recurrence relations while recovering the coefficients of the remainder of the polynomial division over basis $\{Q\}$. Once the generalization results were established, we explore a fast Schur–Euclid–Hessenberg algorithm for quasiseparable polynomials. This algorithm generalizes the result for fast Schur–Euclid–Hessenberg algorithm for orthogonal polynomials and Szegö polynomials. To derive the fast algorithm we exploit the decomposition of confederate matrices over quasiseparable and Szegö polynomials and use the sparse comrade matrix for orthogonal polynomials. The presented Schur–Euclid–Hessenberg algorithm enables us to compute a fast triangular factorization of the Bezoutian associated with quasiseparable, Szegö, and orthogonal polynomials, and to recover coefficients of the remainder in polynomial division over quasiseparable, Szegö, and orthogonal basis with complexity $\mathcal{O}(n^2)$ operations.

References

1. Allen, B.M., Rosenthal, J.: A matrix Euclidean algorithm induced by state space realization. Linear Algebra Appl. **288**, 105–121 (1999)
2. Barnett, S.: Greatest common divisor of two polynomials. Linear Algebra Appl. **3**(1), 7–9 (1970)
3. Barnett, S.: A note on the Bezoutian matrix. SIAM J. Appl. Math. **22**(1), 84–86 (1972)
4. Barnett, S.: A companion matrix analogue for orthogonal polynomials. Linear Algebra Appl. **12**(3), 197–202 (1975)
5. Beckermann, B., Labahn, G.: When are two numerical polynomials relatively prime? J. Symb. Comput. **26**(6), 677–689 (1998)
6. Beckermann, B., Labahn, G.: A fast and numerically stable Euclidean-like algorithm for detecting relatively prime numerical polynomials. J. Symb. Comput. **26**(6), 691–714 (1998)
7. Bella, T., Eidelman, Y., Gohberg, I., Olshevsky V.: Classifications of three-term and two-term recurrence relations via subclasses of quasiseparable matrices. SIAM J. Matrix Anal.
8. Bella, T., Eidelman, Y., Gohberg, I., Olshevsky, V., Tyrtyshnikov, E.: Fast inversion of polynomial-Vandermonde matrices for polynomial systems related to order one quasiseparable matrices. In: Kaashoek, M.A., Rodman, L., Woerdeman, H.J. (eds.) Advances in Structured Operator Theory and Related Areas, Operator Theory: Advances and Applications, vol. 237, pp. 79–106. Springer, Basel (2013)
9. Bella, T., Olshevsky, V., Zhlobich, P.: Classifications of recurrence relations via subclasses of (H, k)-quasiseparable matrices. In: Van Dooren, P., Bhattacharyya, S.P., Chan, R.H., Olshevsky, V., Routray, A. (eds.) Numerical Linear Algebra in Signals, Systems and Control. Lecture Notes in Electrical Engineering, vol. 80, pp. 23–54. Springer, Netherlands (2011)
10. Bini, D.A., Boito, P.: Structured matrix-based methods for polynomial ε-GCD: analysis and comparisons. In: D'Andrea, C., Mourrain, B. (eds.) Proceedings of the 2007 International Symposium on Symbolic and Algebraic Computation, Waterloo, Canada, July 29–August 01 2007 (ISAAC '07), pp. 9–16. ACM, New York (2007)
11. Bini, D.A., Gemignani, L.: Fast parallel computation of the polynomial remainder sequence via Bezout and Hankel matrices. SIAM J. Comput. **24**(1), 63–77 (1995)
12. Bini, D.A., Gemignani, L.: Bernstein–Bezoutian matrices. Theor. Comput. Sci. **315**(2–3), 319–333 (2004)
13. Bitmead, R.R., Kung, S.Y., Anderson, B.D., Kailath, T.: Greatest common divisor via generalized Sylvester and Bezout matrices. IEEE Trans. Autom. Control **23**(6), 1043–1047 (1978)
14. Brown, W.S.: On Euclid's algorithm and the computation of polynomial greatest common divisors. J. Assoc. Comput. Mach. **18**(4), 478–504 (1971)
15. Brown, W.S., Traub, J.F.: On Euclid's algorithm and the theory of subresultants. J. Assoc. Comput. Mach. **18**(4), 505–514 (1971)
16. Calvetti, D., Reichel, L.: Fast inversion of vandermondelike matrices involving orthogonal polynomials. BIT Numer. Math. **33**(3), 473–484 (1993)
17. Delosme, J.M., Morf, M.: Mixed and minimal representations for Toeplitz and related systems, In: Proceedings 14th Asilomar Conference on Circuits, Systems, and Computers, Monterey, California (1980)
18. Dym, H.: Structured matrices, reproducing kernels and interpolation. In: Olshevsky, V. (ed.) Structured Matrices in Mathematics, Computer Science and Engineering I: Contemporary Mathematics, vol. 280, p. 329. American Mathematical Society, Providence (2001)
19. Genin, Y.V.: Euclid algorithm, orthogonal polynomials, and generalized Routh–Hurwitz algorithm. Linear Algebra Appl. **246**, 131–158 (1996)
20. Gohberg, I., Olshevsky, V.: Fast inversion of Chebyshev–Vandermonde matrices. Numer. Math. **67**(1), 71–92 (1994)
21. Gohberg, T., Kailath, T., Olshevsky, V.: Fast Gaussian elimination with partial pivoting for matrices with displacement structure. Math. Comput. **64**(212), 1557–1576 (1995)
22. Grenander, U., Szegö, G.: Toeplitz Forms and Applications. University of California Press, Berkeley (1958)

23. Heinig, G.: Matrix representations of Bezoutians. Linear Algebra Appl. **223–224**, 337–354 (1995)
24. Heinig, G., Olshevsky, V.: The Schur algorithm for matrices with Hessenberg displacement structure. In: Olshevsky, V. (ed.) Structured Matrices in Mathematics, Computer Science, and Engineering II: Contemporary Mathematics, vol. 281, pp. 3–16. American Mathematical Society, Providence (2001)
25. Heinig, G., Rost, K.: Algebraic Methods for Toeplitz-Like Matrices and Operators. Operator Theory: Advances and Applications, vol. 13. Birkhäuser, Basel (1984)
26. Heinig, G., Rost, K.: On the inverses of Toeplitz-plus-Hankel matrices. Linear Algebra Appl. **106**, 39–52 (1988)
27. Heinig, G., Rost, K.: Matrix representations of Toeplitz-plus-Hankel matrix inverses. Linear Algebra Appl. **113**, 65–78 (1989)
28. Heinig, G., Rost, K.: Split algorithm and ZW-factorization for Toeplitz and Toeplitz-plus-Hankel matrices. In: Proceedings of the Fifteenth International Symposium on Mathematical Theory of Networks and Systems, Notre Dame, August 12–16 (2002)
29. Heinig, G., Rost, K.: New fast algorithms for Toeplitz-plus-Hankel matrices. SIAM J. Matrix Anal. Appl. **25**(3), 842857 (2004)
30. Heinig, G., Rost, K.: Fast algorithms for Toeplitz and Hankel matrices. Linear Algebra Appl. **435**(1), 159 (2011)
31. Kailath, T., Kung, S.Y., Morf, M.: Displacement ranks of matrices and linear equations. J. Math. Anal. Appl. **68**(2), 395407 (1979)
32. Kailath, T., Olshevsky, V.: Displacement-structure approach to polynomial Vandermonde and related matrices. Linear Algebra Appl. **261**, 49–90 (1997)
33. Kailath, T.: Displacement structure and array algorithms. In: Kailath, T., Sayed, A.H. (eds.) Fast Reliable Algorithms for Matrices with Structure. SIAM, Philadelphia (1999)
34. Kailath, T., Sayed, A.H.: Displacement structure: theory and applications. SIAM Rev. **37**(3), 297–386 (1995)
35. Kaltofen, E., May, J., Yang, Z., Zhi, L.: Approximate factorization of multivariate polynomials using singular value decomposition. J. Symb. Comput. **43**(5), 359–376 (2008)
36. Lancaster, P., Tismenetsky, M.: The Theory of Matrices with Applications, 2nd edn. Academic Press INC., Orlando (1985)
37. Maroulas, J., Barnett, S.: Polynomials with respect to a general basis. I. Theory. J. Math. Anal. Appl. **72**, 177–194 (1979)
38. Morf, M.: Fast Algorithms for Multivariable Systems. Ph. D. Thesis, Stanford University (1974)
39. Noda, M.T., Sasaki, T.: Approximate GCD and its application to ill-conditioned algebraic equations. J. Comput. Appl. Math. **38**, 335–351 (1991)
40. Olshevsky, V., Stewart, M.: Stable factorization of Hankel and Hankel-like matrices. Numer. Linear Algebra **8**(6–7), 401–434 (2001)
41. Pan, V.Y.: Computation of approximate polynomial GCDs and an extension. Inf. Comput. **167**(2), 71–85 (2001)
42. Pan, V.Y., Tsigaridas, E.: Nearly optimal computations with structured matrices. In: Watt, S.M., Verschelde, J., Zhi, L. (eds.) Proceedings of the International Conference on Symbolic Numeric Computation, China, July 2014, pp. 21–30. ACM Digital Library, New York (2014)
43. Rost, K.: Generalized companion matrices and matrix representations for generalized Bezoutians. Linear Algebra Appl. **193**(1), 151–172 (1993)
44. Wimmer, H.K.: On the history of the Bezoutian and the resultant matrix. Linear Algebra Appl. **128**, 27–34 (1990)
45. Yang, Z.H.: Polynomial Bezoutian matrix with respect to a general basis. Linear Algebra Appl. **331**, 165–179 (2001)
46. Zarowski, C.J.: A Schur Algorithm for Strongly Regular Toeplitz-plus-Hankel Matrices. In: Proceedings of the 33rd Midwest Symposium on Circuits and Systems, Alta, Aug, 1990. IEEE Xplore Digital Library, vol. 1, pp. 556–559. IEEE, New York (1990)

The Use of CAS Piranha for the Construction of Motion Equations of the Planetary System Problem

A.S. Perminov and E.D. Kuznetsov

Abstract In this paper, we consider the using of the computer algebra system Piranha as applied to the study of the planetary problem. Piranha is an echeloned Poisson series processor, which is written in C++ language. It is new, specified, high-efficient program for analytical transformations of polynomials, Fourier and Poisson series. We used Piranha for the expansion of the Hamiltonian of four-planetary problem into the Poisson series and the construction of motion equations by the Hori–Deprit method. Both of these algorithms are briefly presented in this work. Different properties of the series representation of the Hamiltonian and motion equations are discussed.

Keywords Planetary problem · Motion equations · Second system of Poincare elements · Echeloned series · Poisson series processor · Hori-Deprit method

1 Introduction

The investigation of the orbital evolution of planetary systems is one of the fundamental problems of celestial mechanics. For the simple case of two-body problem (the Sun and one planet), the planetary orbit can be described as an ellipse, which is usually named a Keplerian orbit. Planetary orbits are perturbed if we have more than one planet around the Sun. The perturbed motion is described using of time-dependent osculating orbital elements. Here the osculating orbit of the planet is the Keplerian orbit that it would have around the Sun if perturbations were not present. As is known from celestial mechanics, the planetary problem of three or more bodies (two or more planets around the Sun) does not have the exact analytical solution. Methods of the perturbation theory are used to find an approximate solution of the problem of the planetary motion.

A.S. Perminov (✉) · E.D. Kuznetsov
Ural Federal University, Yekaterinburg, Russia
e-mail: perminov12@yandex.ru

E.D. Kuznetsov
e-mail: Eduard.Kuznetsov@urfu.ru

I.S. Kotsireas and E. Martínez-Moro (eds.), *Applications of Computer Algebra*, Springer Proceedings in Mathematics & Statistics 198,
DOI 10.1007/978-3-319-56932-1_25

Using of the perturbation theory, we can start with a simplified form of the original problem. In our case, the gravitational interaction between planets is excluded. Planetary orbits are Keplerian. Perturbation methods set the iterative process in which the previous solution is improved on each step. So, the perturbation theory leads to the solution in the form of a power series in a small parameter of the problem. The small parameter quantifies the deviation from the exactly solved problem and it is proportional to the ratio of the sum of planet masses and the mass of the Sun. As showed Poincare in [1] series of perturbation methods are asymptotic expansions and not convergent in common case.

Our main objective is the construction of semi-analytical motion theory of the second order of planetary masses. In this paper, we consider the construction of equations of the motion theory. Further these equations can be numerically integrated for the investigation of the orbital evolution of various planetary systems. The first stage of our work is the expansion of the planetary system Hamiltonian into the Poisson series. The second stage is the construction of motion equations in time-averaged elements by the Hori–Deprit method. We consider the problem for the case of planetary systems with four planets. It is sufficient to study the orbital evolution of giant-planets of the Solar system and the most of extrasolar systems also.

For our purposes, the Hamiltonian of four-planetary system is written in Jacobi coordinates [2]. It is the hierarchical coordinate system which is more preferable for the study of the planetary motion. The position of each following body in these coordinates is determined relative to the center of the mass of the previously including bodies set.

We used the second system of Poincare elements for the construction of the Hamiltonian expansion. It allows simplify the angular part of the series. In this case only one angular element—the mean longitude—is defined [3]. Elements of the second Poincare system are defined through classical Keplerian orbital elements by the following way

$$L_i = M_i\sqrt{\kappa_i^2 a_i}, \quad \lambda_i = \Omega_i + \omega_i + l_i,$$

$$\xi_{1i} = \sqrt{2L_i(1-\sqrt{1-e_i^2})}\cos(\Omega_i+\omega_i), \; \xi_{2i} = \sqrt{2L_i\sqrt{1-e_i^2}(1-\cos I_i)}\cos\Omega_i,$$

$$\eta_{1i} = -\sqrt{2L_i(1-\sqrt{1-e_i^2})}\sin(\Omega_i+\omega_i), \; \eta_{2i} = -\sqrt{2L_i\sqrt{1-e_i^2}(1-\cos I_i)}\sin\Omega_i,$$

where M_i is normalized mass, κ_i^2 is normalized gravitational parameter, a_i is the semi-major axis of the orbital ellipse, e_i is the eccentricity of this ellipse, I_i is the inclination of the orbital plane relative to the reference plane of the coordinate system, quantities Ω_i, ω_i, l_i are longitude of the ascending node of the orbit, argument of the pericenter of the orbit and mean anomaly of the planet on the orbit respectively. The longitude of the ascending node and inclination define position of the orbital plane. The argument of the pericenter defines position the ellipse in the orbital plane. The mean anomaly defines position of the planet on the orbital ellipse. Index i is a sequence number of the planet. Six classical Keplerian elements or canonical Poincare elements completely define the position and the velocity of the planet relative to the origin of the coordinate system.

It should be noted that Poincare elements ξ_1 and η_1 are called eccentric; it means that these elements are proportional to the eccentricity of the orbit. Elements ξ_2 and η_2 are called oblique because they are proportional to the inclination of the orbit. Due to Poincare elements are canonical, three pairs of these are canonical conjugated as the momentum and its the corresponding coordinate, namely L and λ, ξ_1 and η_1, ξ_2 and η_2.

We need to obtain motion equations in time-averaged elements. The use of these elements allows to eliminate short-periodic perturbations in the planetary motion and to construct the motion theory for a long-time period. In this case, only secular and long-periodic perturbations are taken into account. Short-periodic perturbations are excluded. Therefore we can increase the step of integration with respect to time. We used the Hori–Deprit method [4, 5] for the construction of these equations. This method is based on the Poisson brackets formalism and it is characterized by efficiency and very ease for the computer implementation. The first the Hamiltonian of the problem is averaged and the second the generating function of the transformation between osculating and averaging elements is constructed. After that, right hands of motion equations are calculated in averaged elements.

All analytical transformations performed by means of the computer algebra system Piranha [6]. Let us explore its main features and possibilities.

2 Overview of Piranha

The computer algebra system Piranha is new, specified, high-efficient program for analytical manipulations with different series. It is an echeloned Poisson series processor, which written in C++ programming language. This program was written by Francesco Biscani from Max Planck Institute for Astronomy (Heidelberg, Germany). Piranha is freeware, object-oriented, and cross-platform software. For the convenience Piranha has Python user-interface which is the set of some Python libraries and called Pyranha. Standard Python environment can be used for Pyranha access through the terminal or executable scripts.

Piranha can works with different series types, such as

1. Multivariable polynomials.
2. Poisson series.
3. Echeloned Poisson series (Poisson series with denominators).

It is possible to use various types of series coefficients and powers of variables. Real types with different precision and rational type are available for series coefficients. Powers of series variables can be chosen as integer or rational types. Using of rational coefficients in series eliminates rounding errors and provides arbitrary precision of resulting series. In this work, we used echeloned Poisson series with rational coefficients and powers.

Let us consider the basic functionality of Piranha.

1. The summation and the multiplication of series by using of standard operators '+,' '−,' and'∗' in Python-terminal or scripts.
2. The automatically calculation of binomial expansions up to terms of the chosen order. For example, if at the input of Python-terminal we have some expression like $(1 + x)^q$, at the output we will obtain its representation as the series truncated up the chosen order. Here q is an arbitrary rational number, x is a symbol variable or another series.
3. The truncation of the series up to terms with given order of powers of the series variables. The function `truncate_degree(arg, max_degree, names)` is used for the degree-based truncation of the series `arg`. All items which degree is greater than `max_degree` will be eliminated. The argument `names` is a list of names of variables which are chosen for the truncation.
4. The substitution into the series of some numerical values or another series. The function `subs(arg, name, x)` is used for the substitution of the quantity `x` in the variable `name` of the series `arg`.
5. The estimation of the series by different numerical values using of the function `evaluate(arg, eval_dict)`. All symbolic variable of the series `arg` will be replaced by corresponding numerical values. The argument `eval_dict` is the evaluation dictionary consisting of pairs (the name of the variable—corresponding numerical value).
6. It is possible to save/load of the resulting series in/from text file or zipped text file by functions `save(arg, path_to_file)/load(path_to_file)`. Using of this functions, the series `arg` will be saved or loaded.
7. The functions `integrate(arg, name)` and `partial(arg, name)` calculate the integral and the partial derivative of the series `arg` with respect to the variable `name`.
8. The Poisson bracket of arguments `f` and `g` can be calculated using of the function `pbracket(f, g, p_list, q_list)`, where `p_list` and `q_list` are lists of names of momentum and coordinates correspondingly.
9. Such functions as `cos(x)` and `sin(x)` are implemented in Piranha also and they can be used for the construction of the angular part of Poisson series. The argument `x` can be any numerical or series type.

We implemented the simple Poincare processor for the construction of the classical celestial mechanics expansions, which are needed for our transformations. It can be used for the calculation of following the base series.

- x_k, y_k, z_k are rectangular coordinates of k-th planet.
- r_k, $1/r_k$ are the distance to the k-th planet and its inverse value.
- Next, using of the previously mentioned series it is possible to take the expansion of $\cos H_{ij}$, where H_{ij} is the angle between radius vectors r_i and r_j.
- The ratio r_i/r_j can be obtained from expansions of r_i and $1/r_j$.
- The inverse absolute value of radius vectors difference, which is denoted below as $1/\Delta_{ij}$, can be expanded into the series as follows.

$$1/\Delta_{ij} = |\mathbf{r}_i - \mathbf{r}_j|^{-1} = \frac{1}{r_j}\sum_{n=0}^{\infty}\left(\frac{r_i}{r_j}\right)^n P_n(\cos H_{ij}), \tag{1}$$

where $1 \le j < i \le 4$, P_n is Legendre polynomial of n-th degree, H_{ij} is the angle between two radius vectors. The series in Legendre polynomials absolutely converges when $|r_i/r_j| < 1$. Then we expressed each Legendre polynomials through cosine of the angle.

All expansions are expressed through the second system of Poincare elements. Algorithms for the construction of these expansions are available in our work [7] and implemented as Python scripts. It is important to note that in [7] we used the old version of Piranha system and Legendre polynomials are saved as symbol variables. The current version of Piranha has more performance and it more efficiently uses operating memory.

In the process, Piranha showed a high speed of calculations. Calculations were performed on Quad-core PC with 3400 MHz Core i5 processor and 32 Gb available memory. Unix-like OS Ubuntu 14 and Python 2.7 is used. Table 1 presents a time of the series calculation, a number of its items and a series truncation error for base series. The series truncation error is determined as the relative difference between the series expansion and the exact expression. Parameter n in the first column is the limit of degrees of eccentric and oblique Poincare elements. Results in the last column are correspond to the series for $1/\Delta_{ij}$ with maximum degree of cosines is equal to 25.

The accuracy estimation of the base series is determined for giant-planets of the Solar System. Indexed quantities were calculated for all planetary pairs. The value

Table 1 Calculation time, the number of terms and the series truncation error for the base series

n	Feature	x/a, y/a	z/a	r/a	a/r	r_i/r_j	$\cos\theta_{ij}$	$1/\Delta_{ij}$
5	Length	96	116	46	41	400	2438	79688
	Error	10^{-7}	10^{-7}	10^{-8}	10^{-8}	10^{-8}	10^{-7}	10^{-6}–10^{-11}
	Time	0.1^s	0.1^s	0.1^s	0.1^s	0.1^s	0.8^s	2^s
6	Length	154	516	66	61	847	6342	168984
	Error	10^{-9}	10^{-9}	10^{-9}	10^{-9}	10^{-9}	10^{-9}	10^{-8}–10^{-10}
	Time	0.1^s	0.1^s	0.1^s	0.1^s	0.1^s	1.4^s	4^s
8	Length	333	616	132	127	3004	32035	595450
	Error	10^{-10}	10^{-10}	10^{-11}	10^{-11}	10^{-11}	10^{-11}	10^{-10}–10^{-13}
	Time	0.2^s	0.2^s	0.2^s	0.2^s	0.2^s	4^s	14^s
9	Length	460	966	178	173	5158	64691	1021422
	Error	10^{-10}	10^{-10}	10^{-11}	10^{-11}	10^{-11}	10^{-11}	10^{-11}–10^{-14}
	Time	0.3^s	0.3^s	0.2^s	0.2^s	0.2^s	7^s	110^s

of $1/\Delta_{ij}$ for the planetary pair 'Uranus–Neptune' has the lowest accuracy due to the ratio r_i/r_j has the largest value for this pair. The highest accuracy gives the planetary pair 'Jupiter–Neptune.'

3 The Expansion of the Hamiltonian

The algorithm of the Hamiltonian expansion is discussed more detail in [7]. Let us briefly review here the basic formulas of the algorithm. The Hamiltonian can be expressed as the sum of the undisturbed Hamiltonian and the disturbing part

$$h = -\sum_{i=1}^{4} \frac{M_i \kappa_i^2}{2a_i} + \mu \times Gm_0 \Big\{ \sum_{i=2}^{4} \frac{m_i(2\mathbf{r}_i \mathbf{R}_i + \mu R_i^2)}{r_i \tilde{R}_i (r_i + \tilde{R}_i)} - \sum_{i=1}^{4} \sum_{j=1}^{i-1} \frac{m_i m_j}{|\rho_i - \rho_j|} \Big\}. \quad (2)$$

Here

$$\mathbf{R}_i = \sum_{k=1}^{i} \frac{m_k}{\bar{m}_k} \mathbf{r}_k, \quad \tilde{R}_i = \sqrt{r_i^2 + 2\mu \mathbf{r}_i \mathbf{R}_i + \mu^2 R_i^2}, \quad (3)$$

and

$$|\rho_i - \rho_j| = \mathbf{r}_i - \mathbf{r}_j + \mu \sum_{k=j}^{i-1} \frac{m_k}{\bar{m}_k} \mathbf{r}_k, \quad (4)$$

where numbers i and j satisfy a condition $1 \le j < i \le 4$; ρ_k is the barycentric radius vector of k-th planet, $\mathbf{r}_k$ is Jacobi radius vector of the same planet; μm_k is the mass of the planet in items of the Sun mass m_0, $\bar{m}_k = 1 + \mu m_1 + \cdots + \mu m_k$, $M_i = m_i \bar{m}_{i-1}/\bar{m}_i$, $\kappa_i^2 = Gm_0 \bar{m}_i / \bar{m}_{i-1} \mu$, G is the gravitational parameter and μ is the small parameter. If we take into account the Solar system then the value of μ can take equal to 0.001.

The first sum in (2) is the undisturbed part of the Hamiltonian, which describes the Keplerian motion of planets around the Sun. The expression in figure brackets is the disturbing function. Double sum in (2) is the main part of the disturbing function. The main part describes the interaction between planets.

The Hamiltonian of the planetary problem can be expressed as

$$h = h_0 + \mu h_1 = h_0 + \sum_{k,n} A_{kn} x^k \cos(n\lambda), \quad (5)$$

where h_0 is the undisturbed Hamiltonian, μh_1 is the disturbing function, A_{kn} is numerical coefficients, x^k is the product of Poincare elements with corresponding degrees, cosine is represent the angular part of the series, $n\lambda$ is the linear combination of mean longitudes of planets.

Common form of the expansion of the main part up to the second degree of the small parameter is shown here

$$\frac{1}{|\rho_i - \rho_j|} = \frac{1}{\Delta_{ij}} - \mu \frac{A_{ij}}{\Delta_{ij}^3} + \mu^2 \Big(\frac{3}{2} \frac{A_{ij}^2}{\Delta_{ij}^5} - \frac{1}{2} \frac{B_{ij}}{\Delta_{ij}^3} \Big) + \ldots, \tag{6}$$

and here for items of the second part of the disturbing function

$$\frac{2\mathbf{r}_i \mathbf{R}_i + \mu \mathbf{R}_i^2}{r_i \tilde{R}_i (r_i + \tilde{R}_i)} = \frac{C_i}{r_i^3} + \mu \Big(-\frac{3}{2} \frac{C_i^2}{r_i^5} + \frac{1}{2} \frac{D_i}{r_i^3} \Big) + \mu^2 \Big(\frac{5}{2} \frac{C_i^3}{r_i^7} - \frac{3}{2} \frac{C_i D_i}{r_i^5} \Big) + \ldots, \tag{7}$$

where

$$A_{ij} = (\mathbf{r}_i - \mathbf{r}_j) \sum_{k=j}^{i-1} \frac{m_k}{\bar{m}_k} \mathbf{r}_k, \; B_{ij} = \Big(\sum_{k=j}^{i-1} \frac{m_k}{\bar{m}_k} \mathbf{r}_k \Big)^2, \; C_i = \mathbf{r}_i \sum_{k=1}^{i-1} \frac{m_k}{\bar{m}_k} \mathbf{r}_k, \; D_i = B_{i1}. \tag{8}$$

Scalar products are expressed here through cosines of angles. Such quantities as the small parameter μ and masses ratio $m_k/\bar{m}_k$ are used as symbol variables also.

We have two expansions of the Hamiltonian into the Poisson series up to the second and the third degrees of the small parameter. The properties of these series are presented in Tables 2 and 3. For items of the disturbing function and the Hamiltonian are given the following properties: maximum degrees of eccentric and oblique Poincare elements (n_1, n_2, n_3), maximum degrees of cosines of angles (p_1, p_2, p_3), and the number of terms (N_1, N_2, N_3) for various degrees of the small parameter. The total number of terms in each groups is showed. For the length of the Hamiltonian expansion four terms from the undisturbed Hamiltonian are taken into account also.

Table 2 Properties of the Hamiltonian expansion up to the second degree of the small parameter

Indexes of items	Terms with μ^1		Terms with μ^2		Length of series
	n_1, p_1	N_1	n_2, p_2	N_2	
i, j	The main part of the disturbing function				
2, 1	5, 25	79688	3, 10	20347	100035
3, 2	5, 25	79688	3, 10	20347	100035
4, 3	5, 25	79688	3, 10	20347	100035
3, 1	5, 20	52787	2, 10	30112	82899
4, 2	5, 20	52787	2, 10	30112	82899
4, 1	5, 15	31178	2, 5	14782	45960
i	The second part of the disturbing function				
2, 3, 4	5, –	2646	3, –	2436	5082
	The Hamiltonian				
	5, 25	375816	3, 10	141129	516949

Table 3 Properties of the Hamiltonian expansion up to the third degree of a small parameter

indexes of items	Terms with μ^1		Terms with μ^2		Terms with μ^3		Length of series
	n_1, p_1	N_1	n_2, p_2	N_2	n_3, p_3	N_3	
i, j	The main part of the disturbing function						
2, 1	8, 32	947908	4, 20	197168	3, 10	55279	1200355
3, 2	8, 32	947908	4, 20	197168	3, 10	55279	1200355
4, 3	8, 32	947908	4, 20	197168	3, 10	55279	1200355
3, 1	6, 20	111949	3, 10	121488	2, 5	51320	284757
4, 2	6, 20	111949	3, 10	121488	2, 5	51320	284757
4, 1	6, 10	32474	3, 5	61046	1, 5	22141	115661
i	The second part of the disturbing function						
2, 3, 4	8, –	19542	4, –	8076	3, –	7140	34758
	The Hamiltonian						
	8, 40	3119638	5, 15	903602	3, 15	297758	4321002

Table 4 The estimation accuracy of the Hamiltonian expansion for the Solar system

Indexes	The expansion up to μ^2	Error	The expansion up to μ^3	Error
i, j	The main part of the disturbing function			
2, 1	$-7.00587 \cdot 10^{-2}$	$3 \cdot 10^{-6}$	$-7.00588747 \cdot 10^{-2}$	$9 \cdot 10^{-9}$
3, 2	$-6.6832878 \cdot 10^{-4}$	$6 \cdot 10^{-9}$	$-6.68328783 \cdot 10^{-4}$	$1 \cdot 10^{-12}$
4, 3	$-1.876166 \cdot 10^{-4}$	$2 \cdot 10^{-6}$	$-1.87616949 \cdot 10^{-4}$	$3 \cdot 10^{-8}$
3, 1	$-2.31483506 \cdot 10^{-3}$	$5 \cdot 10^{-9}$	$-2.31483505 \cdot 10^{-3}$	$3 \cdot 10^{-10}$
4, 2	$-4.92903269 \cdot 10^{-4}$	$1 \cdot 10^{-10}$	$-4.92903269 \cdot 10^{-4}$	$1 \cdot 10^{-13}$
4, 1	$-1.65422172 \cdot 10^{-3}$	$2 \cdot 10^{-10}$	$-1.65422172 \cdot 10^{-3}$	$7 \cdot 10^{-10}$
i	The second part of the disturbing function			
2, 3, 4	$1.7504193 \cdot 10^{-5}$	$1 \cdot 10^{-7}$	$1.75041953 \cdot 10^{-5}$	$1 \cdot 10^{-11}$
	The Hamiltonian in general			
	$-4.56917912 \cdot 10^{-5}$	$1 \cdot 10^{-9}$	$-4.56917913 \cdot 10^{-2}$	$4 \cdot 10^{-12}$

The approximation accuracy for both Hamiltonian was calculated for the Solar system. Poincare elements for the Solar system are taken on 01/01/2000 and correspond to the mean ecliptic of the Solar system. The estimation accuracy of the series approximation is presented in Table 4 for the whole Hamiltonian. Columns 'Error' consist of absolute values of relative differences between the series expansion and the exact expression. The estimation was carried out for Poincare elements corresponding to the elements of giant-planets of the Solar system.

4 The Hori–Deprit Method

The averaged Hamiltonian was constructed by the Hori–Deprit method. Let us denote slow variables of the problem as $x = (L, \xi_1, \eta_1, \xi_2, \eta_2)$ and fast variables as λ. Rate of change for slow variables is much less than rate of change for fast variables. The rates of slow variables are proportionally the small parameter while the rates of fast variables are in proportion to the mean motions. Averaged variables are denoted as X and Λ. After averaging transformation with respect to the mean longitudes λ, the Hamiltonian is written as the series of the small parameter

$$H(X) = H_0 + \sum_{m=1}^{\infty} \mu^m H_m(X), \tag{9}$$

where quantities H_m are obtained from the main equation of the Hori–Deprit method

$$H_m(X) = h_m + \sum \frac{1}{r!}\{T_{j_r}, \{\cdots, \{T_{j_1}, h_{j_0}\}\}\}. \tag{10}$$

The summation is over the domain $0 \leq j_0 \leq m-1$; $j_1, j_2, \cdots, j_r \geq 1$; $\sum_{s=0}^{k} j_s = m$; $1 \leq r \leq m$. The figure brackets is Poisson brackets with respect to the Poincare elements. h_m are items of not averaged Hamiltonian h, and the generating function of the transformation to averaging elements is defined as

$$T(X, \Lambda) = \sum_{m=1}^{\infty} \mu^m T_m(X, \Lambda). \tag{11}$$

On each step of the method the Eq. (10) can be written in the next form

$$H_m(X) = \{T_m, h_0\} + \Phi_m, \tag{12}$$

or, as it is shown in [4]

$$\Phi_m = H_m + \sum_{k=1}^{4} \nu_k \frac{\partial T_m}{\partial \Lambda_m}, \tag{13}$$

where Φ_m is defined on the previous step of the method. In the general case Φ_m is the echeloned Poisson series

$$\Phi_m(X, \Lambda) = \sum B_{pn} X^p \cos n\Lambda, \tag{14}$$

where B_{pn} are the coefficients of the echeloned Poisson series (it includes the denominator as the linear combination of frequencies ν_k of fast variables).

If $H_m(X) = \sum B_{pn} X^p, n \in \{n_1 = \cdots = n_4 = 0\}$ then the solution of the Eq. (13) can be written as $T_m(X, \Lambda) = \sum \frac{B_{pn}}{n\nu} X^p \sin n\Lambda$, $n \notin \{n_1^2 + \cdots + n_4^2 \neq 0\}$.

Averaged motion equations can be obtained using Poisson brackets

$$\frac{\mathrm{d}X}{\mathrm{d}t} = \{H, X\}, \quad \frac{\mathrm{d}\Lambda}{\mathrm{d}t} = \{H, \Lambda\}. \tag{15}$$

The transformation from osculating to averaged elements gives by functions for the change of variables u_m, v_m

$$X = x + \sum_{m=1}^{\infty} (-1)^m \mu^m u_m(x, \lambda), \quad u_m = \sum \frac{1}{r!}\{T_{j_r}, \{\cdots, \{T_{j_1}, X\}\}\} \tag{16}$$

$$\Lambda = \lambda + \sum_{m=1}^{\infty} (-1)^m \mu^m v_m(x, \lambda), \quad v_m = \sum \frac{1}{r!}\{T_{j_r}, \{\cdots, \{T_{j_1}, \Lambda\}\}\} \tag{17}$$

where the summation over the domain $j_1, j_2, \cdots, j_r \geq 1; \sum_{s=0}^{k} j_s = m; 1 \leq r \leq m$.

We have constructed two sets of motion equations for the motion theory based on both expansions of the Hamiltonian in osculating elements.

The averaged Hamiltonian of the problem and the generating function of the transformation are constructed up to terms with the second (the first) degree of the small parameter for the first (the second) expansion of the Hamiltonian in osculating elements. Motion equations and functions for the change of variables are given on the second (the first) approximation of the Hori–Deprit method.

Table 5 The number of terms of the averaged Hamiltonian and the generating function

	H_0	H_1	H_2	T_1	T_2
The first expansion	4	6 393	379 859	2 774 983	2 926 631 639
The second expansion	4	207 258	6 607 811		

Table 6 The number of terms of motion equations and functions for the change of variables for the motion theory of the first order (based on the first expansion of the Hamiltonian in osculating elements)

Motion equations				Functions for the change			
El.	Length	El.	Length	El.	Length	El.	Length
L_1	0	ξ_{11}, η_{11}	906	L_1	1155789	ξ_{11}, η_{11}	490889
L_2	0	ξ_{12}, η_{12}	1063	L_2	1515779	ξ_{12}, η_{12}	639527
L_3	0	ξ_{13}, η_{13}	1060	L_3	1516684	ξ_{13}, η_{13}	639647
L_4	0	ξ_{14}, η_{14}	897	L_4	1158504	ξ_{14}, η_{14}	491129
λ_1	2936	ξ_{21}, η_{21}	911	λ_1	2357878	ξ_{21}, η_{21}	444825
λ_2	3452	ξ_{22}, η_{22}	1071	λ_2	3085823	ξ_{22}, η_{22}	582066
λ_3	3453	ξ_{23}, η_{23}	1071	λ_3	3087298	ξ_{23}, η_{23}	582061
λ_4	2939	ξ_{24}, η_{24}	911	λ_4	2362303	ξ_{24}, η_{24}	444810

Table 7 The number of terms of motion equations and functions for the change of variables for the motion theory of the first order (based on the second expansion of the Hamiltonian in osculating elements)

Motion equations				Functions for the change			
El.	Length	El.	Length	El.	Length	El.	Length
L_1	0	ξ_{11}, η_{11}	126588	L_1	6629954	ξ_{11}, η_{11}	6254429
L_2	0	ξ_{12}, η_{12}	140799	L_2	7801278	ξ_{12}, η_{12}	7171783
L_3	0	ξ_{13}, η_{13}	140835	L_3	7801342	ξ_{13}, η_{13}	7171812
L_4	0	ξ_{14}, η_{14}	126696	L_4	6630146	ξ_{14}, η_{14}	6254516
λ_1	147362	ξ_{21}, η_{21}	105814	λ_1	6824308	ξ_{21}, η_{21}	4896428
λ_2	175749	ξ_{22}, η_{22}	105814	λ_2	8131632	ξ_{22}, η_{22}	4896428
λ_3	175786	ξ_{23}, η_{23}	105814	λ_3	8133236	ξ_{23}, η_{23}	4896428
λ_4	147473	ξ_{24}, η_{24}	105814	λ_4	6829120	ξ_{24}, η_{24}	4896428

Table 8 The number of terms of motion equations and functions for the change of variables for the motion theory of the second order (based on the first expansion of the Hamiltonian in osculating elements)

Motion equations				Functions for the change			
El.	Length	El.	Length	El.	Length	El.	Length
L_1	0	ξ_{11}, η_{11}	33269	L_1	$108 \cdot 10^6$	ξ_{11}, η_{11}	$116 \cdot 10^6$
L_2	0	ξ_{12}, η_{12}	42130	L_2	$144 \cdot 10^6$	ξ_{12}, η_{12}	$145 \cdot 10^6$
L_3	0	ξ_{13}, η_{13}	42386	L_3	$144 \cdot 10^6$	ξ_{13}, η_{13}	$145 \cdot 10^6$
L_4	0	ξ_{14}, η_{14}	33627	L_4	$108 \cdot 10^6$	ξ_{14}, η_{14}	$116 \cdot 10^6$
λ_1	301702	ξ_{21}, η_{21}	21538	λ_1	$138 \cdot 10^6$	ξ_{21}, η_{21}	$53 \cdot 10^6$
λ_2	360535	ξ_{22}, η_{22}	27549	λ_2	$176 \cdot 10^6$	ξ_{22}, η_{22}	$71 \cdot 10^6$
λ_3	346285	ξ_{23}, η_{23}	27841	λ_3	$176 \cdot 10^6$	ξ_{23}, η_{23}	$71 \cdot 10^6$
λ_4	254644	ξ_{24}, η_{24}	22216	λ_4	$138 \cdot 10^6$	ξ_{24}, η_{24}	$53 \cdot 10^6$

The number of terms of the averaged Hamiltonian and the generating function is shown in Table 5. The number of terms of motion equations and functions for the change of variables for the first order motion theory is given in Tables 6 and 7 for all Poincare elements. The number of terms of motion equations and functions for the change of variables for the second order motion theory is given in Table 8 for all Poincare elements.

5 Conclusion

The expansion of the Hamiltonian of the four-planetary problem into the Poisson series is constructed for two cases—up to the second and the third degree of the small parameter. The error estimation of series truncation is 10^{-9} for the Hamiltonian expansion in the first case. The second expansion was constructed with error of series truncation is 10^{-12}.

The averaged Hamiltonian and the generating function of the transformation are constructed up to terms with the first degree of the small parameter. Motion equations and functions for the change of variables are obtained on the first step of the Hori–Deprit method. The number of terms of resulting series are given.

In the process of our calculations Piranha shows the ability to work with very large series.

Acknowledgements The authors are grateful to Francesco Biscani for the implementation of echeloned Poisson series in Piranha and his help with technical questions. This work is supported by Act 211 Government of the Russian Federation, agreement 02.A03.21.0006.

References

1. Poincare, H.: Les Methodes nouvelles de la Mecanique Celeste. Grauthier-Viltars, Paris (1892)
2. Murray, C.D., Dermott, S.F.: Solar System Dynamics. Cambridge University Press, Cambridge (1999)
3. Charlier, C.L.: Die Mechanik des Himmels. Walter de Gruyter & Co., Berlin (1927)
4. Kholshevnikov, K.V.: Asimptoticheskie metody nebesnoi mekhaniki. Asymptotic Methods in Celestial Mechanics. Leningr. Gos. Univ., Leningrad (1985)
5. Kuznetsov, E.D., Kholshevnikov, K.V.: Dynamical evolution of weakly disturbed two-planetary system on cosmogonic time scales: the Sun–Jupiter–Saturn system. Solar Syst. Res. **40**(3), 239–250 (2006)
6. Biscani, F.: The Piranha computer algebra system. https://github.com/bluescarni/piranha (2015)
7. Perminov, A.S., Kuznetsov, E.D.: Expansion of the Hamiltonian of the planetary problem into the Poisson series in elements of the second poincare system. Solar Syst. Res. **49**(6), 430–441 (2015)

Code-Based Cryptosystems Using Generalized Concatenated Codes

Sven Puchinger, Sven Müelich, Karim Ishak and Martin Bossert

Abstract The security of public-key cryptosystems is mostly based on number-theoretic problems like factorization and the discrete logarithm. There exists an algorithm which solves these problems in polynomial time using a quantum computer. Hence, these cryptosystems will be broken as soon as quantum computers emerge. Code-based cryptography is an alternative which resists quantum computers since its security is based on an NP-complete problem, namely decoding of random linear codes. The McEliece cryptosystem is the most prominent scheme to realize code-based cryptography. Many code classes were proposed for the McEliece cryptosystem, but most of them are broken by now. Sendrier suggested to use ordinary concatenated codes, however, he also presented an attack on such codes. This work investigates generalized concatenated codes to be used in the McEliece cryptosystem. We examine the application of Sendrier's attack on generalized concatenated codes and present alternative methods for both partly finding the code structure and recovering the plaintext from a cryptogram. Further, we discuss modifications of the cryptosystem making it resistant against these attacks.

Keywords Post-Quantum cryptography · Code-based cryptosystems · McEliece cryptosystem · Generalized concatenated codes

---

S. Puchinger (✉) · S. Müelich · K. Ishak · M. Bossert
Institute of Communications Engineering, Ulm University, Ulm, Germany
e-mail: sven.puchinger@uni-ulm.de

S. Müelich
e-mail: sven.mueelich@uni-ulm.de

K. Ishak
e-mail: karim.ishak@uni-ulm.de

M. Bossert
e-mail: martin.bossert@uni-ulm.de

I.S. Kotsireas and E. Martínez-Moro (eds.), *Applications of Computer Algebra*, Springer Proceedings in Mathematics & Statistics 198,
DOI 10.1007/978-3-319-56932-1_26

1 Introduction

Public-key cryptography was introduced in 1976 by [10]. The advantage in comparison to classical cryptosystems is that sender and receiver do not have to share a common secret key, since two different keys are used for encryption and decryption. The receiver (Bob) publishes a public key, which is used by the sender (Alice) to encrypt messages she wants to send to Bob. When Bob receives an encrypted message, he uses his private key for decryption. Nowadays, the security of public-key cryptosystems is usually based on number theoretic problems, like factorization of large numbers (RSA [24]) or the discrete logarithm (Elgamal [11]). For solving these two problems there are no efficient algorithms known so far. However, Shor's algorithm solves these problems in polynomial time on quantum computers [27]. As soon as quantum computers will exist in the future, the aforementioned cryptosystems are broken and will become useless. Hence, there is a need for so-called post-quantum cryptography, i.e., new methods which resist the quantum computer. One candidate for this purpose is code-based cryptography.

The first code-based cryptosystem was proposed by McEliece only 2 years after the emerge of public-key cryptography. The security of this system is based on the NP-complete problem of decoding random linear codes [3]. Using the McEliece cryptosystem, encryption and decryption can be performed very efficiently. The main problem is the large size of the public key. For this reason, code-based cryptography was forgotten for a long time and now becomes interesting again due to quantum computer resistance. Initially, McEliece suggested to use binary Goppa codes in his cryptosystem. Later, other code classes were suggested. However, in most cases it was also shown that there are attacks which break them. This work investigates generalized concatenated codes for use in the McEliece cryptosystem.

This paper is structured as follows. In Sect. 2, we summarize the McEliece cryptosystem and recall a general attack on the system. In Sect. 3, we present coding theory fundamentals, such as ordinary and generalized concatenated codes. Furthermore, we discuss the use of generalized concatenated codes in the McEliece cryptosystem in Sect. 4 and describe some generalized concatenated codes that are not ordinary concatenated codes. Section 5 is about Sendrier's attack, which recovers the structure of an ordinary concatenated code used in the McEliece cryptosystem. We examine under which conditions the attack can be modified in order to work also with generalized concatenated codes. In Sect. 6, we give alternatives for parts of Sendrier's attack in order to apply it on GC codes. Also, an attack which recovers the plaintext from a cryptogram instead of finding the structure of the underlying code is explained. Section 7 presents methods which can be used in order to prevent the attacks explained before. Finally, Sect. 8 concludes the paper.

2 McEliece Cryptosystem

The McEliece cryptosystem, introduced in [19], is the first public-key cryptosystem based on coding theory. For generating private and public key, Bob first selects an error-correcting code of length n and dimension k which can correct up to t errors. He then computes a $(k \times n)$ generator matrix $\mathbf{G}$ for this code. Furthermore, he randomly produces two matrices, $\mathbf{S}$, which is a $(k \times k)$ invertible matrix and $\mathbf{P}$, which is a $(n \times n)$ permutation matrix. These matrices are used in order to obfuscate $\mathbf{G}$ and hence to hide the structure of the code. Therefore he calculates $\tilde{\mathbf{G}} = \mathbf{S} \cdot \mathbf{P}$ and publishes the pair $(\tilde{\mathbf{G}}, t)$ as public key. The code as well as the matrices $\mathbf{G}$, $\mathbf{S}$ and $\mathbf{P}$ he keeps secret as private key. In order to send a message to Bob, Alice makes use of Bob's public key to encrypt her message. She breaks her message into k-bit blocks and multiplies each of these blocks to the obfuscated generator matrix $\tilde{\mathbf{G}}$. To each of the blocks, she then adds a random vector $\mathbf{e}$ of length n and weight $\leq t$, which can be interpreted as error. Hence, the calculation $\mathbf{r} = \mathbf{m} \cdot \tilde{\mathbf{G}} + \mathbf{e}$ can directly be compared to the mapping of information blocks to codewords in a typical channel coding scenario. In order to decrypt the cipher $\mathbf{r}$, Bob needs the matrices $\mathbf{P}$ and $\mathbf{S}$ and a decoding algorithm for the used code. He calculates

$$\begin{aligned}
\hat{\mathbf{r}} = \mathbf{r} \cdot \mathbf{P}^{-1} &= (\mathbf{m} \cdot \tilde{\mathbf{G}} + \mathbf{e}) \cdot \mathbf{P}^{-1} \\
&= (\mathbf{mS} \cdot \mathbf{G} \cdot \mathbf{P} + \mathbf{e}) \cdot \mathbf{P}^{-1} \\
&= \mathbf{m} \cdot \mathbf{S} \cdot \mathbf{G} \cdot \mathbf{P} \cdot \mathbf{P}^{-1} + \mathbf{e} \cdot \mathbf{P}^{-1} \\
&= \mathbf{m} \cdot \mathbf{S} \cdot \mathbf{G} + \mathbf{e} \cdot \mathbf{P}^{-1}.
\end{aligned}$$

In analogy to channel coding, $\hat{\mathbf{r}}$ has the form of a received word consisting of the information word $\mathbf{m} \cdot \mathbf{S}$ and the error $\mathbf{e} \cdot \mathbf{P}^{-1}$. Bob uses the decoding algorithm on $\hat{\mathbf{r}}$ to obtain $\hat{\mathbf{m}} = \mathbf{m} \cdot \mathbf{S}$. Finally he can multiply $\hat{\mathbf{m}}$ with $\mathbf{S}^{-1}$ to retrieve $\mathbf{m}$.

In order to be used in the McEliece cryptosystem, a code class needs to fulfill two requirements. An efficient decoding algorithm has to exist for the used code class, and the code has to be indistinguishable from a random code. In the original proposal, binary Goppa codes were used. For suitable parameters they are unbroken until today, since they fulfill both requirements. Other code classes were suggested (cmp. Table 1).

Encryption and decryption in the McEliece scheme is competitive with number-theoretic methods like RSA in terms of complexity and easiness of implementation. Since the security is based on the NP-complete problem of decoding random linear codes, the McEliece cryptosystem is a candidate for post-quantum cryptography. The system's main drawback is a large key size because generator matrices are used as keys. For this reason the cryptosystem was not applicable for a long time.

Another code-based cryptosystem similar to McEliece is the Niederreiter cryptosystem introduced in [22]. In contrast to the McEliece cryptosystem, which uses a codeword with an added error as ciphertext, the Niederreiter cryptosystem represents the ciphertext as a syndrome and the error vector is the message. Instead of a

Table 1 Proposed code classes for the McEliece cryptosystem and suggested attacks

Code class	Proposal	Attacks
GRS codes	1986: Niederreiter [22]	1992: Sidelnikov and Shestakov [29]
Ordinary concatenated codes	1995: Sendrier [25]	1998: Sendrier [26]
Reed–Muller codes	1994: Sidelnikov [28]	2007: Minder and Shokrollahi [21]
		2013: Chizhov and Borodin [6]
Algebraic geometry codes	1996: Janwa and Moreno [14]	2008: Faure and Minder [14]
		2014: Couvreur et al. [8]
Subcodes of GRS codes	2005: Berger and Loidreau [2]	2010: Wieschebrink [31]

generator matrix, Niederreiter uses a parity check matrix as public key, and hence is also called a dual version of McEliece. It was shown in [17] that the cryptosystems of McEliece and Niederreiter are equivalent when set up for corresponding choices of parameters. This means, that an attack on McEliece cryptosystem also breaks the Niederreiter cryptosystem and vice versa.

There are two kinds of possible attacks to the McEliece cryptosystem. In a *structural attack*, the adversary tries to retrieve the code structure and hence to recover $\mathbf{S}', \mathbf{G}', \mathbf{P}'$, or an efficient decoder of the code generated by $\mathbf{S}' \cdot \mathbf{G}' \cdot \mathbf{P}'$. Structural attacks were for example successfully applied to (subcodes of) generalized Reed–Solomon codes, Reed–Muller codes, Algebraic geometry codes, and ordinary concatenated codes. A *nonstructural attack* tries to recover the message from the cryptogram r and the public-key $(\tilde{\mathbf{G}}, t)$. This is equivalent to the problem of decoding random linear codes.

Information Set Decoding Attack

In the following, we give an example for a message attack called *information set decoding*, as described in [1, 16, 19], of which we present an efficient modification for concatenated codes in Sect. 6.2.

Given a code with parameters (n, k, d) and generator matrix $\tilde{\mathbf{G}}$. In order to recover $\mathbf{m}$ in $\mathbf{r} = \mathbf{m} \cdot \tilde{\mathbf{G}} + \mathbf{e}$ we randomly choose δ coordinates of $\mathbf{r}$ and $\tilde{\mathbf{G}}$. With $\mathbf{r}_\delta$ we denote the vector we get by only taking the δ chosen coordinates from the vector $\mathbf{r}$. Similarly, $\tilde{\mathbf{G}}_\delta$ denotes the matrix obtained from $\tilde{\mathbf{G}}$ by extracting the δ chosen columns. Restricting our vectors to the δ chosen coordinates we obtain $\mathbf{r}_\delta = \mathbf{m} \cdot \tilde{\mathbf{G}}_\delta + \mathbf{e}_\delta$. If we are lucky and choose δ error-free coordinates, $\mathbf{e}_\delta$ is the zero vector. Thus, the system of linear equations $\mathbf{r}_\delta = \mathbf{m} \cdot \tilde{\mathbf{G}}_\delta$, with known $\tilde{\mathbf{G}}_\delta$, $\mathbf{r}_\delta$ and unknown $\mathbf{m}$, has a solution, which is unique as long as $\tilde{\mathbf{G}}_\delta$ has rank k.

Obviously, we must choose $\delta \geq k$. For *MDS* codes, we know that any set of k columns of $\tilde{\mathbf{G}}$ is linearly independent [20]. Other codes do not have this property, however, to our knowledge, this has been an open problem for many years. For most practically good codes, a linearly independent set of columns is obtained with high probability already for values of δ slightly larger than k.

Algorithm 1: Information Set Decoding Attack

Input: $\tilde{\mathbf{G}}$ and $\mathbf{r} = \mathbf{m} \cdot \tilde{\mathbf{G}} + \mathbf{e}$ with $\mathrm{wt_H}(\mathbf{e}) = t$
Output: $\mathbf{m}$

1 **do**
2 | Choose δ many coordinates at random // $O(1)$
3 | Solve $\mathbf{r}_\delta = \hat{\mathbf{m}} \cdot \tilde{\mathbf{G}}_\delta$ for $\hat{\mathbf{m}}$ // $O(k^3)$
4 **while** $\nexists \hat{\mathbf{m}}$ *or* $\mathrm{d_H}(\hat{\mathbf{m}} \cdot \tilde{\mathbf{G}}, \mathbf{r}) \geq \frac{d}{2}$
5 **return** $\hat{\mathbf{m}}$

Theorem 1 *If $t < \frac{d}{2}$, the algorithm is correct. Its expected complexity is*

$$\frac{\binom{n}{\delta}}{\binom{n-t}{\delta}} \cdot O(\delta^3).$$

Proof From coding theory we know that if $t < \frac{d}{2}$, there is a unique $\hat{\mathbf{m}}$ such that the Hamming distance of $\hat{\mathbf{m}} \cdot \tilde{\mathbf{G}}$ and $\mathbf{r}$ is $\mathrm{d_H}(\hat{\mathbf{m}} \cdot \tilde{\mathbf{G}}, \mathbf{r}) \geq \frac{d}{2}$. This $\hat{\mathbf{m}}$ is also the unique solution of $\mathbf{r} - \mathbf{e} = \hat{\mathbf{m}} \cdot \tilde{\mathbf{G}}$. If δ is chosen large enough, a random submatrix $\tilde{\mathbf{G}}_\delta$ of rank k with error-free positions is found in a step with a nonzero probability, and thus, by the lemma of Borel–Cantelli, the algorithm terminates in finite time with probability 1.

Concerning the complexity, we assume that δ is chosen sufficiently large such that the probability that a submatrix $\tilde{\mathbf{G}}_\delta$ has rank $<k$ can be neglected. Thus, the number of loops required to terminate the algorithm is geometrically distributed with parameter

$$p = \frac{\binom{n-t}{\delta}}{\binom{n}{\delta}},$$

which is exactly the probability of choosing δ out of $n - t$ correct positions in Line 2 of Algorithm 1. Thus, the expected number of loops required is $\frac{1}{p}$ and together with the complexity of Line 3, which is $O(\delta)$, we obtain the expected complexity

$$\frac{\binom{n}{\delta}}{\binom{n-t}{\delta}} \cdot O(\delta^3).$$

■

We can thus conclude that in case of practical codes, where $\delta \approx k$, we obtain an upper bound on the work factor[1] of

[1]Estimation of the complexity up to a constant factor which is not depending on the parameters of the system.

$$\frac{\binom{n}{k}}{\binom{n-t}{k}} \cdot O(k^3),$$

where $t = \lfloor \frac{d-1}{2} \rfloor$. According to [13] the parameters have to be chosen such that a work factor of 2^{128} (for mid-term security) or 2^{256} (for long-term security) is obtained. There are several speed-ups [1, 7, 23], and generalizations [16] of the information set decoding attack.

3 Coding Theory Fundamentals

In this section, we present notations and known results which we do not assume all readers to know.

3.1 Basics

The *Hamming weight* $\mathrm{wt_H}(\mathbf{c})$ of $\mathbf{c} \in \mathbb{F}_q^n$ is defined as the number of nonzero positions of $\mathbf{c}$. The *Hamming distance* of $\mathbf{c}_1, \mathbf{c}_2 \in \mathbb{F}_q^n$ is $\mathrm{d_H}(\mathbf{c}_1, \mathbf{c}_2) := \mathrm{wt_H}(\mathbf{c}_1 - \mathbf{c}_2)$. An $\mathbb{F}_q$-linear code $\mathscr{C}(q^m; n, k, d)$ of length n, dimension k and minimum distance $d = \min_{\mathbf{c}_1 \neq \mathbf{c}_2}\{\mathrm{d_H}(\mathbf{c}_1, \mathbf{c}_2)\}$ over $\mathbb{F}_{q^m}$ is a k-dimensional $\mathbb{F}_q$-subspace of $\mathbb{F}_{q^m}^n$. Often, the field is clear from the context, so we write $\mathscr{C}(n, k, d)$.

We will need the following lemma in Sect. 5.1. More precisely, we require the slightly weaker statement that for any linear code $\mathscr{C}$ with dual distance $d^\perp$, for any $r < d^\perp$ positions, we can find a codeword in $\mathscr{C}$ in which we can choose the r positions arbitrarily.

Lemma 1 [20] *Any set of $r \leq d^\perp - 1$ columns of $[\mathscr{C}]$ contains each r-tuple exactly $\frac{2^k}{2^r}$ times, and $d^\perp$ is the largest number with this property, where $[\mathscr{C}]$ is a $2^k \times n$ array of codewords of the code $\mathscr{C}(n, k, d)$.*

Definition 1 (*Support and Connection of Vectors* [26, Definition 9–11]). The following notation will be used in Sect. 5.1.

- The *support* of a vector $\mathbf{c}$ is given as $\mathrm{supp}(\mathbf{c}) = \{i : c_i \neq 0\}$.
- The *support of a set* is the union of the supports of its elements.
- A codeword $\mathbf{c} \in \mathscr{C}$ is called *minimal support codeword* if there is no other codeword $\mathbf{c}' \in \mathscr{C}$ with $\mathrm{supp}(\mathbf{c}') \subseteq \mathrm{supp}(\mathbf{c})$.
- The set of all minimal support codewords in $\mathscr{C}$ is called $\mathscr{P}(\mathscr{C})$
- Two vectors $\mathbf{c}, \mathbf{c}'$ are called *connected* if their supports intersect.
- Positions i, j are connected in a set $S \in \mathscr{C}$ if there is a sequence words $\mathbf{c}_1, \ldots, \mathbf{c}_r \in S$ such that
  - $i \in \mathrm{supp}(\mathbf{c}_1)$ and $j \in \mathrm{supp}(\mathbf{c}_r)$

- $\mathbf{c}_k$ and $\mathbf{c}_{k+1}$ are connected $\forall\, k = 1, \ldots, r-1$.

- A set $S \in \mathscr{C}$ *connects a set of positions* I if any two elements in I are connected in S.

3.2 Concatenated Codes

We distinguish between ordinary concatenated codes (OC codes or OCC) introduced by [12]), and generalized concatenated codes (GC codes or GCC) introduced by [5]. Concatenated codes can be used to construct long codes by only using short codes. The main advantage of such a construction is a comparatively short decoding time, since we only have to decode short codes.

3.2.1 Ordinary Concatenated Codes

We describe OC codes as in [25, 26]. The following codes and mappings uniquely determine an OC code.

- Linear *inner code* $\mathscr{B}(q; n_B, k_B, d_B)$.
- Linear *outer code* $\mathscr{A}(q^{k_B}; n_A, k_A, d_A)$.
- $\mathbb{F}_q$-linear map $\theta : \mathbb{F}_{q^{k_B}} \to \mathscr{A}$.

We define the mapping

$$\Theta : \mathbb{F}_{q^{k_B}}^{n_A} \to \mathscr{B}^{n_A}$$
$$\begin{bmatrix} a_1 \\ \vdots \\ a_{n_A} \end{bmatrix} \mapsto \begin{bmatrix} \theta(a_1) \\ \vdots \\ \theta(a_{n_A}) \end{bmatrix}.$$

Definition 2 (*Ordinary Concatenated Code*) Let n_A, Θ, $\mathscr{A}$ and $\mathscr{B}$ be as above. Then, the corresponding *ordinary concatenated* (OC) code, or OCC, is given as

$$\mathscr{C}_{\mathrm{OC}} = \Theta(\mathscr{A}) \subseteq \mathscr{B}^{n_A}$$

Due to its construction, an OCC is $\mathbb{F}_q$-linear since $\mathscr{A}$ is $\mathbb{F}_{q^{k_B}}$-linear, implying $\mathbb{F}_q$-linearity, and θ is $\mathbb{F}_q$-linear. The code has $(q^{k_B})^{k_A} = q^{k_B \cdot k_A}$ codewords, each of it consisting of n_A many codewords from $\mathscr{B}$, resulting in a codelength of $n_A \cdot n_B$ elements of $\mathbb{F}_q$. Thus, the code has parameters

$$\mathscr{C}_{\mathrm{OC}}(q; n_{\mathrm{OC}} = n_A \cdot n_B, k_{\mathrm{OC}} = k_B \cdot k_A, d_{\mathrm{OC}}),$$

where d_{OC} is the minimum distance, whose value we do not consider here.

3.2.2 Generalized Concatenated Codes

GC codes are a generalization of OC codes, introduced by [5]. Here, we give a definition which is similar to the above-mentioned definition of OC codes by [25], which was not given in this form before. A comprehensive overview of GC codes can be found in [4] and we explain in Appendix A why our definition matches [4]. Similar to OC codes, we require the following parameters, codes and mappings.

- $k_1, k_2, \ldots, k_\ell \in \mathbb{N}$ with $k_B = \sum_{i=1}^{\ell} k_i$.
- $\mathbb{F}_{q^{k_i}}$-linear outer codes $\mathscr{A}^{(i)}(q^{k_i}; n_A, k_A^{(i)}, d_A^{(i)})$ for $i = 1, \ldots, \ell$
- $\mathbb{F}_q$-linear inner code $\mathscr{B}(q; n_B, k_B, d_B)$.
- $\mathbb{F}_q$-linear map $\theta : \bigoplus_{i=1}^{\ell} \mathbb{F}_{q^{k_i}} \to \mathscr{B}$.

Again, we define a mapping

$$\Theta : \bigoplus_{i=1}^{\ell} (\mathbb{F}_{q^{k_i}})^{n_A} \to \mathscr{B}^{n_A}$$

$$\left(\begin{bmatrix} a_{1,1} \\ a_{1,2} \\ \vdots \\ a_{1,n_A} \end{bmatrix}, \begin{bmatrix} a_{2,1} \\ a_{2,2} \\ \vdots \\ a_{2,n_A} \end{bmatrix}, \ldots, \begin{bmatrix} a_{\ell,1} \\ a_{\ell,2} \\ \vdots \\ a_{\ell,n_A} \end{bmatrix} \right) \mapsto \begin{bmatrix} \theta(a_{1,1}, \ldots, a_{\ell,1}) \\ \theta(a_{1,2}, \ldots, a_{\ell,2}) \\ \vdots \\ \theta(a_{1,n_A}, \ldots, a_{\ell,n_A}) \end{bmatrix}.$$

Definition 3 (*Generalized Concatenated Code*) Let n_A, Θ, $\mathscr{A}^{(i)}$ and $\mathscr{B}$ be as above. Then, the corresponding *generalized concatenated* (GC) code, or GCC, is given by

$$\mathscr{C}_{\mathrm{GC}} = \Theta\left(\bigoplus_{i=1}^{\ell} \mathscr{A}^{(i)}\right) \subseteq \mathscr{B}^{n_A}$$

with parameters $\mathscr{C}_{\mathrm{GC}}(q; n_{\mathrm{GC}} = n_A \cdot n_B, k_{\mathrm{GC}} = \sum_{i=1}^{\ell} k_A^{(i)}, d_{\mathrm{GC}})$, see Appendix A.

In our definition, $\mathscr{C}_{\mathrm{GC}} \subseteq \mathscr{B}^{n_A}$, which is often written as an $n_A \times n_B$ matrix over $\mathbb{F}_q$. We can also write it as an $n_A \cdot n_B$ vector over $\mathbb{F}_q$ and the information words from the set $\bigoplus_{i=1}^{\ell} \mathbb{F}_{q^{k_i}}$ as vectors of dimension $k_{\mathrm{GC}} = \sum_{i=1}^{\ell} k_A^{(i)}$ over $\mathbb{F}_q$, which we need in order to define a generator matrix $\mathbf{G}$ of the code. The advantage of GC compared to OC codes is that we allow several outer codes with different dimensions and are hence able to obtain better codes, cf. Appendix A. Bounds on the minimum distance d_{GC} of GC codes can also be found in Appendix A.

4 The McEliece Cryptosystem Using GCC

The motivation to use concatenated codes in the McEliece cryptosystem has the big advantage of very low decoding complexity, which is retained when going from OC to GC codes. OC codes have the drawback of possibly larger key sizes at the same security level compared to codes without concatenated structure. This disadvantage is not present in the GCC case since its construction admits larger overall dimension at the same minimum distance, or a better decoding performance at the same dimension compared to OC codes [4].

4.1 Assumption that θ Is Linear

In the McEliece cryptosystem, only the use of linear codes is reasonable because the existence of a generator matrix is required. In the original definition of GC codes, it was not assumed that the mapping θ is $\mathbb{F}_q$-linear. However, $\theta : \bigoplus_{i=1}^{\ell} \mathbb{F}_{q^{k_i}} \to \mathscr{B}$ must be bijective and thus, its image is a linear subspace of $\mathbb{F}_q^{n_B}$.

One can now ask the question whether a GC code with a nonlinear θ can be a linear code. And if yes, is there an alternative GCC construction using a linear θ' and possibly different other outer codes, yielding the same code. Both questions are open problems. If the first one is true and the second one is not, these codes might resist the attacks presented in this paper.

However, since most good GC code constructions having low decoding complexity, which motivated the use of GC codes here, use an $\mathbb{F}_q$-linear θ (cf. Appendix A), we make the assumption that θ is linear.

4.2 Some GC Codes that Are No OC Codes

In general, it is well known that GC codes are a generalization of OC codes [4]. Obviously, any OCC is a GCC, given by only one outer code. On the other hand, it is mentioned in [9] that any GCC can be viewed as an OCC. However, in general, one must admit nonlinear inner and outer codes in the definition of OCC to make this statement become true. Since there is a known structural attack on OCC [25, 26], we would like to know which GCC cannot be constructed as OCC. We are not able to give a complete classification of the set of GC codes not containing OC codes. However, we are able to prove the following statement for the special case of

$$k_1 = k_2 = \cdots = k_\ell,$$

which is a very important sub-class of GC codes as described in Appendix A.

Theorem 2 *If* $k_1 = k_2 = \cdots = k_\ell$, *a GCC is an OCC* $\Leftrightarrow \mathscr{A}^{(i)} = \mathscr{A}^{(j)}\ \forall i, j$.

Proof This proof can be found in Appendix B due to its technicality. ■

Theorem 2 provides an exact statement which GC codes can be constructed as OC codes in this case. In particular, it shows that only for very few choices of the outer codes $\mathscr{A}^{(i)}$, we obtain an OC code. For instance, it follows directly that the dimensions of the outer codes must all be the same, which would correspond to rather suboptimal GC codes, see Appendix A.

Hence, we see that the set of linear GC codes has a much larger cardinality than the set of linear OC codes. Note, that Theorem 2 can be used to practically estimate the number of linear GC codes which Alice can choose from, namely by counting the possible choices of outer codes $\mathscr{A}^{(i)}$.

It is an open problem to prove a similar statement as in Theorem 2 for the general case where the k_i's are not all the same.

5 Applying Parts of Sendrier's Attack

Sendrier's attack [25, 26] was proposed to find the structure of a concatenated code from a given obfuscated generator matrix. In this section, we deal with the question how we have to modify the attack to work also with GC codes. We generally divide Sendrier's attack into three steps, where the first two try to revert the permutations done to the generator matrix up to a certain level, and the third step attempts to find possible generator matrices of the inner and outer codes.

It turns out that the first two steps and the first half of the third step can be directly applied to GC codes without modification, partly under different conditions (see Corollary 1).

5.1 First Step

The first step aims to find the inner blocks of a GC code, which are the positions corresponding to the same set $\mathscr{B}$ in $\mathbf{c} \in \mathscr{C}_{\mathrm{GC}} \subseteq \mathscr{B}^{n_A}$. Recall Definition 1. As in [26], we can formally define an inner block as

Definition 4 The i-th *inner block* of a GCC is the support $\mathrm{supp}(\{\Theta(a \cdot \mathbf{e}_i) : a \in \mathbb{F}_{q^{k_B}}\})$, where $\mathbf{e}_i$ is the i-th unit vector.

Let now $\mathscr{C}_{\mathrm{GC}}$ be a given GCC. Similar to the statement of [26, Proposition 16], we can prove the following well-known result.

Theorem 3 *The support of every* $\mathbf{c} \in \mathscr{P}(\mathscr{C}_{\mathrm{GC}}^{\perp})$ *with* $\mathrm{wt}_{\mathrm{H}}(\mathbf{c}) < \min(d_A^{(1)\perp}, \ldots, d_A^{(\ell)\perp}, 2 \cdot d_B{}^{\perp})$ *is contained in a single inner block of* $\mathscr{C}_{\mathrm{GC}}$.

Proof Let $\mathbf{c} \in \mathscr{P}(\mathscr{C}_{\mathrm{GC}}^{\perp})$ with $\mathrm{wt_H}(\mathbf{c}) < \min(d_A^{(1)\perp}, \ldots, d_A^{(\ell)\perp}, 2 \cdot d_B{}^{\perp})$. Then the support of $\mathbf{c}$ is contained in $r \leq \mathrm{wt_H}(\mathbf{c})$ inner blocks. We want to show that $r < 2$.

Since $\mathrm{wt_H}(\mathbf{c}) < d_A^{(i)\perp}$, the positions of $\mathscr{A}^{(i)}$ corresponding to these inner blocks can be chosen arbitrarily from $\mathbb{F}_{q^{k_i}}^r$ due to Lemma 1. Thus, the positions of $\bigoplus_{i=1}^{\ell} \mathscr{A}^{(i)}$ corresponding to these inner blocks can be chosen arbitrarily from $\bigoplus_{i=1}^{\ell} \mathbb{F}_{q^{k_i}}^r$.

Due to $\theta(\bigoplus_{i=1}^{\ell} \mathbb{F}_{q^{k_i}}) = \mathscr{B}$, $\mathscr{C}_{\mathrm{GC}}$ contains codewords that have any element of $\mathscr{B}^r$ in the r inner blocks. Any element of $\mathscr{C}_{\mathrm{GC}}^{\perp}$ with support contained in these r inner blocks must have codewords of $\mathscr{B}^{\perp}$ in all its inner blocks because for any of the r inner blocks j, one can construct a codeword that has an arbitrary element of $\mathscr{B}$ in inner block j and the zero codewords in the other $r-1$ blocks. Hence, $r < 2$ due to $\mathrm{wt_H}(\mathbf{c}) < 2 \cdot d_B{}^{\perp}$ and the pigeonhole principle. ■

The following statement is taken from [26].

Lemma 2 [26, Proposition 15] *Let $\mathscr{C}$ be a linear code. $\mathscr{P}(\mathscr{C})$ connects* supp($\mathscr{C}$) *iff $\mathscr{C}$ is not the direct sum of two disjoint support codes.*

Using Theorem 3 and Lemma 2, we can prove the following corollary.

Corollary 1 *If $\mathscr{B}$ is not the direct sum of two disjoint support codewords, the set*

$$\Xi = \left\{\mathbf{c} \in \mathscr{P}(\mathscr{C}_{\mathrm{GC}}^{\perp}) : \mathrm{wt_H}(\mathbf{c}) < \min(d_A^{(1)\perp}, \ldots, d_A^{(\ell)\perp}, 2 \cdot d_B{}^{\perp})\right\}$$

connects the inner blocks of $\mathscr{C}_{\mathrm{GC}}$.

Proof Due to Theorem 3, every minimal support vector $\mathbf{c} \in \mathscr{P}(\mathscr{C}_{\mathrm{GC}}^{\perp})$ with weight $\mathrm{wt_H}(\mathbf{c}) < \min(d_A^{(1)\perp}, \ldots, d_A^{(\ell)\perp}, 2 \cdot d_B{}^{\perp})$ is contained in a single inner block. This imples, that $\mathbf{c}$, restricted to this inner block, is in a minimal support vector of $\mathscr{P}(\mathscr{B}^{\perp})$. Since $\mathscr{B}$ is not the direct sum of two disjoint support codes, by Lemma 2, $\mathscr{P}(\mathscr{B})$ connects supp($\mathscr{B}$), which corresponds to the entire inner block. Hence, if we find enough $\mathbf{c} \in \Xi$, we obtain the supports of all inner blocks. ■

Corollary 1 gives us the tools for finding the supports of the inner blocks of $\mathscr{C}_{\mathrm{GC}}$. We simply exploit [26, Propositions 13 and 14] to find as many minimal support codewords as necessary to identify the inner blocks, as described in [26]. However, the sufficient condition that this method works is a bit more strict compared to the OC case since we can only use minimal support words $\mathbf{c}$ of weight $\mathrm{wt_H}(\mathbf{c}) < \min(d_A^{(1)\perp}, \ldots, d_A^{(\ell)\perp}, 2 \cdot d_B{}^{\perp})$.

If the method works, we obtain the supports of the inner blocks from which we can construct a permutation matrix $\mathbf{P}_{\text{Step 1}}$ which re-orders the columns of $\tilde{\mathbf{G}}$ such that columns corresponding to the same inner bock are grouped together, meaning that they form a $k_{\mathrm{GC}} \times n_B$ submatrix of $\tilde{\mathbf{G}} \cdot \mathbf{P}_{\text{Step 1}}$.

5.2 Second Step

Codewords of $\mathscr{C}_{\mathrm{GC}}$ are elements of $\mathscr{B}^{n_A}$, exactly as codewords of OC codes. In Sect. 5.1, we saw how to identify the inner blocks of the code, which are the positions that correspond to the same $\mathscr{B}$ in $\mathscr{B}^{n_A}$. However, in order to identify the structure of the code, we also need to know the permutations between the inner blocks. That means, we want to re-order the positions such that we obtain a codeword in $[\sigma(\mathscr{B})]^{n_A}$, where $\sigma(\mathscr{B}) = \{\sigma(\mathbf{c}) := (c_{\sigma(1)}, c_{\sigma(2)}, \ldots, c_{\sigma(n)} : \mathbf{c} \in \mathscr{B}\}$.

This part of Sendrier's attack only depends on properties of the inner code $\mathscr{B}$. To be exact, Sendrier uses the i-th *signature* of a code $\mathscr{B}$ [26], which is the weight distribution of $\mathscr{B}$ punctured at position i, to identify the permutations between two codes $\mathscr{B}$ and $\mathscr{B}'$. Thus, it is directly applicable to GC codes.

Using this method, it is possible to extract the relative permutations of the different inner blocks and to re-order them to be in the same order as one specific block, which is permuted from the original code $\mathscr{B}$ by some permutation σ. It is mentioned in [25] that this part of the attack only works if the automorphism group of $\mathscr{B}$ is reduced to the identity element, which, if this condition is not fulfilled would yield a bad overall code.

Figure 1 illustrates how the first two steps of Sendrier's attack recover the structure of the permutation matrix $\mathbf{P}$ used in the obfuscated generator matrix $\tilde{\mathbf{G}} = \mathbf{S} \cdot \mathbf{G} \cdot \mathbf{P}$. Here, $\mathbf{P}_{\text{Step 1}}$ and $\mathbf{P}_{\text{Step 2}}$ denote the matrices that we obtain by Steps 1 and 2, respectively, and which we can multiply to $\tilde{\mathbf{G}}$ from the right to structure the inner blocks. The $\mathbf{P}_{i,j}$'s are $n_B \times n_B$ submatrices of a permutation matrix and the $\mathbf{P}_i$'s are $n_B \times n_B$ permutation matrices. $\mathbf{P}_1$ is the permutation matrix that transforms $\mathscr{B}$ into $\sigma(\mathscr{B})$.

Note, that after applying Step 2, the effective permutation matrix $\tilde{\mathbf{G}} \cdot \mathbf{P}_{\text{Step 1}} \cdot \mathbf{P}_{\text{Step 2}}$ is still in a form in which inner blocks are permuted among each other and within the inner blocks, positions are permuted. However, the first kind of permutation simply corresponds to a permutation of the outer codes (all the same) and the latter is a permutation of the inner code.

Thus, we recovered the permutations such that $\tilde{\mathbf{G}} \cdot \mathbf{P}_{\text{Step 1}} \cdot \mathbf{P}_{\text{Step 2}}$ is a generator matrix of a GC code with outer codes equivalent to the original outer codes (equivalent by the same permutation τ) and with the inner code (or equivalently, the image of θ) being permuted by σ.

$$
\tilde{\mathbf{G}} = \mathbf{S}\cdot\mathbf{G}\cdot
\begin{bmatrix}
\mathbf{P}_{1,1} & \mathbf{P}_{1,2} & \mathbf{P}_{1,3} & \cdots & \mathbf{P}_{1,n_A} \\
\mathbf{P}_{2,1} & \mathbf{P}_{2,2} & \mathbf{P}_{2,3} & \cdots & \mathbf{P}_{2,n_A} \\
\mathbf{P}_{3,1} & \mathbf{P}_{3,2} & \mathbf{P}_{3,3} & \cdots & \mathbf{P}_{3,n_A} \\
\vdots & \vdots & \vdots & \ddots & \vdots \\
\mathbf{P}_{n_A,1} & \mathbf{P}_{n_A,2} & \mathbf{P}_{n_A,3} & \cdots & \mathbf{P}_{n_A,n_A}
\end{bmatrix}
\xrightarrow{\text{Step 1}}
\tilde{\mathbf{G}}\cdot\mathbf{P}_{\text{Step 1}} = \mathbf{S}\cdot\mathbf{G}\cdot
\begin{bmatrix}
\mathbf{0} & \mathbf{P}_2 & \mathbf{0} & \cdots & \mathbf{0} \\
\mathbf{0} & \mathbf{0} & \mathbf{0} & \cdots & \mathbf{P}_{n_A} \\
\mathbf{P}_1 & \mathbf{0} & \mathbf{0} & \cdots & \mathbf{0} \\
\vdots & \vdots & \vdots & \ddots & \vdots \\
\mathbf{0} & \mathbf{0} & \mathbf{P}_3 & \cdots & \mathbf{0}
\end{bmatrix}
\xrightarrow{\text{Step 2}}
\tilde{\mathbf{G}}\cdot\mathbf{P}_{\text{Step 1}}\cdot\mathbf{P}_{\text{Step 2}} = \mathbf{S}\cdot\mathbf{G}\cdot
\begin{bmatrix}
\mathbf{0} & \mathbf{P}_1 & \mathbf{0} & \cdots & \mathbf{0} \\
\mathbf{0} & \mathbf{0} & \mathbf{0} & \cdots & \mathbf{P}_1 \\
\mathbf{P}_1 & \mathbf{0} & \mathbf{0} & \cdots & \mathbf{0} \\
\vdots & \vdots & \vdots & \ddots & \vdots \\
\mathbf{0} & \mathbf{0} & \mathbf{P}_1 & \cdots & \mathbf{0}
\end{bmatrix}
$$

Fig. 1 Illustration of permutation recovery in Steps 1 and 2 of Sendrier's attack

5.3 Third Step

The subsequent steps are applied after obtaining the permutation matrices $\mathbf{P}_{\text{Step 1}}$ and $\mathbf{P}_{\text{Step 2}}$ of the first two steps of Sendrier's attack. This step, we subdivide into two Substeps 3.1 and 3.2.

Step 3.1

By transforming the matrix $\tilde{\mathbf{G}} \cdot \mathbf{P}_{\text{Step 1}} \cdot \mathbf{P}_{\text{Step 2}}$ into reduced row echelon form, the first $k_B \times n_B$ submatrix is a generator matrix of a permuted version $\sigma(\mathscr{B})$ of the code $\mathscr{B}$. This part of Sendrier's attack is directly applicable to GC codes since the first $k_B \times n_B$ submatrix of the reduced row echelon form of $\tilde{\mathbf{G}} \cdot \mathbf{P}_{\text{Step 1}} \cdot \mathbf{P}_{\text{Step 2}}$ is a basis of the row space $\mathscr{V}$ of $\tilde{\mathbf{G}} \cdot \mathbf{P}_{\text{Step 1}} \cdot \mathbf{P}_{\text{Step 2}}$, restricted to the first n_B columns. This subspace equals the span of all codewords restricted to an inner block j, permuted by a permutation σ, and thus the image of

$$\sigma\left(\theta\left(\left\{\left[a_{j,1} \dots a_{j,\ell}\right] : a_{j,i} \text{ is } j\text{-th position of } \mathbf{a}_i \in \mathscr{A}^{(i)}\right\}\right)\right).$$

Due to Theorem 1, it holds that

$$\left\{\left[a_{j,1} \dots a_{j,\ell}\right] : a_{j,i} \text{ is } j\text{-th position of } \mathbf{a}_i \in \mathscr{A}^{(i)}\right\} = \bigoplus_{i=1}^{\ell} \mathbb{F}_{q^{k_i}}$$

and we obtain

$$\mathscr{V} = \sigma\left(\theta\left(\bigoplus_{i=1}^{\ell} \mathbb{F}_{q^{k_i}}\right)\right) = \sigma\left(\mathscr{B}\right).$$

Step 3.2

The remaining part of the third step of Sendrier's attack on OC codes is responsible for obtaining the structure of the outer code up to a permutation and a *Frobenius field automorphism* applied componentwise [26]. This method works because the generator matrix of obtained by row-reducing $\tilde{\mathbf{G}} \cdot \mathbf{P}_{\text{Step 1}} \cdot \mathbf{P}_{\text{Step 2}}$ is highly structured in the OC case. However, it remains an open problem whether similar arguments can be used to find ways of utilizing the structure of this matrix in the GC case.

6 Alternatives to Parts of Sendrier's Attack

In Sect. 5, we saw that parts of Sendrier's structural attack on OC codes can be applied to GCC directly. However, it remains an open problem to recover a complete structure of the code. Also, the steps of the attacks are not always guaranteed to work since the sufficient conditions of the steps are not always fulfilled. Thus, we are interested in replacing as many parts of Sendrier's attack as possible by an alternative.

6.1 Sendrier's Second and First Part of Third Step

Assume that Step 1 of Sendrier's attack was successful and we obtained the permutation matrix $\mathbf{P}_{\text{Step1}}$ such that the code generated by the matrix $\tilde{\mathbf{G}} \cdot \mathbf{P}_{\text{Step1}}$ is a subset of

$$\bigoplus_{i=1}^{n_A} \sigma_i(\mathscr{B}),$$

so we know which positions correspond to the same inner block, although the blocks are in a different order than in the original code and also within the blocks, positions are permuted arbitrarily (by a permutation σ_i). By computing $\mathbf{r} \cdot \mathbf{P}_{\text{Step1}}$, we also know which positions of the cipher $\mathbf{r}$ correspond to the same inner blocks.

We can find generator matrices $\mathbf{G}_{\sigma_i(\mathscr{B})}$ of all codes given by the positions of the i-th inner block by the following method, which is similar to Sendrier's Step 3.1 (Sect. 5.3) but more general: Restrict $\tilde{\mathbf{G}}$ to the columns corresponding to the i-th inner block and just extract a linearly independent subset vectors in the row span, e.g., by Gaussian elimination in $O({n_B}^3)$ time. This method works because the positions of the outer codes corresponding to the j-th inner block attain all values of $\bigoplus_{i=1}^{\ell} \mathbb{F}_{q^{k_i}}$, cf. Lemma 1 with $k_A^{(i)} < n_A$ for all i, and thus, the span of the rows of $\tilde{\mathbf{G}}$ restricted to the j-th inner block are equal to $\theta(\bigoplus_{i=1}^{\ell} \mathbb{F}_{q^{k_i}}) = \mathscr{B}$, with positions permuted by the permutation σ_i.

The advantage of this approach is that the second step of Sendrier's attack is not required. Also, we can replace $\sigma_i(\mathscr{B})$ by a code $\mathscr{B}_i$ and the method will still work. This fact is important in Sect. 7.

6.2 Nonstructural Attack

Let $\mathbf{r} = \mathbf{m}\tilde{\mathbf{G}} + \mathbf{e}$. In this section, we present an attack that does not find a structure of the code, but is able to recover the message $\mathbf{m}$ from the received word $\mathbf{r}$ if not too many errors $\text{wt}_{\text{H}}(\mathbf{e})$ occurred. Such attacks are called nonstructural.[2] The method works for both OC and GC codes.

We need to know the positions of the inner blocks, which we can obtain by Step 1 of Sendrier's attack. Thus, we also know which positions of $\mathbf{r}$ correspond to the same inner blocks (by compuing $\mathbf{r} \cdot \mathbf{P}_{\text{Step 1}}$). The number of errors is not changed by this operation since $\text{wt}_{\text{H}}(\mathbf{e} \cdot \mathbf{P}_{\text{Step 1}}) = \text{wt}_{\text{H}}(\mathbf{e})$. Also, we require that either Step 3.1 of Sendrier's attack or our alternative presented in Sect. 6.1 worked. This means that we know generator matrices $\mathbf{G}_{\sigma_i(\mathscr{B})}$ of the codes in the inner blocks i.

[2]With "nonstructural", we do not mean "generic". The method assumes that there is a specific structure, but it does not try to recover it. Therefore, it is not applicable to a McEliece cryptosystem using an arbitrary code class.

We can divide our nonstructural attack into two parts:

6.2.1 Part 1

The goal of this part is to decode the positions of $\mathbf{r}$ that correspond to the i-th inner block of $\mathscr{C}_{\mathrm{GC}}$ in the code $\sigma_i(\mathscr{B})$. In general, there are several possible methods to decode in $\sigma_i(\mathscr{B})$, e.g.,

- If we know a structural attack on the McEliece cryptosystem using the inner code $\mathscr{B}$, we can use this attack in combination with the known generator matrix $\mathbf{G}_{\sigma_i(\mathscr{B})}$ of $\sigma_i(\mathscr{B})$ to obtain an efficient decoder of the code $\sigma_i(\mathscr{B})$ for any i. Since $\mathscr{B}$ has much smaller length than the entire code $\mathscr{C}_{\mathrm{GC}}$ (in most cases, $n_B \in \Theta(\sqrt{n_{\mathrm{GC}}})$), such structural attacks have much smaller work factors than direct attacks on a code of length n_{GC}.
- We can apply the information set decoder described in Sect. 2 on the generator matrix $\mathbf{G}_{\sigma_i(\mathscr{B})}$ and the corresponding part of $\mathbf{r}$. Due to Theorem 1, the attack finds the correct part of the codeword $\mathbf{c} = \mathbf{r} - \mathbf{e}$ if the number of errors in this block does not exceed half the minimum distance of the inner code. Otherwise, the decoding result is wrong (another codeword is found) or decoding fails (for instance, after some finite time without result, one aborts the algorithm).

In both cases, we can[3] obtain decoders that find a codeword $\tilde{\mathbf{c}}_i$ from a received word $\mathbf{r}_i = \mathbf{c}_i + \mathbf{e}_i$ if and only if $\mathrm{wt}_{\mathrm{H}}(\mathbf{r}_i - \tilde{\mathbf{c}}_i) < \frac{d_B}{2}$, where $\mathbf{c}_i \in \sigma_i(\mathscr{B})$ is the part of $\mathbf{c} \cdot \mathbf{P}_{\mathrm{Step}\ 1}$ corresponding to the i-th inner block. This type of decoder is called *bounded minimum distance* decoder [4].

If $\mathbf{c}_i = \tilde{\mathbf{c}}_i$, we say that decoding is *correct*. If $\mathbf{c}_i \neq \tilde{\mathbf{c}}_i$ decoding is *wrong* and if the decoder does not have a result, decoding *failed*. Suppose that n_{c} inner blocks were correctly and n_{w} were wrongly decoded, and in n_{f} inner blocks, decoding failed.

6.2.2 Part 2

The second part of our nonstructural attack can be seen as a speed-up of the information set decoding attack (cf. Sect. 2) utilizing the results of Part 1. As in information set decoding, we are looking for δ error-free positions of $\mathbf{r}$, where $\delta \geq k_{\mathrm{GC}}$. We make use of the fact that inner blocks which were decoded correctly in Part 1 do not contain errors. Thus, instead of finding δ single error-free positions in $\mathbf{r}$, we simply try to find $\tau \geq \frac{\delta}{n_B}$ inner blocks that were correctly decoded in Part 1 and thus obtain $\tau \cdot n_B \geq \delta$ error-free positions. Also, we can ignore blocks in which decoding failed. These tricks reduce the overall complexity of the attack significantly. The method is illustrated in Fig. 2. Here, the received word, which is an element of $(\mathbb{F}_q^{n_B})^{n_A}$, is seen

[3]If the obtained algorithms that can correct more than half the minimum distance of errors, we can simply declare a decoding failure if the distance of codeword to received word is greater than half the minimum distance.

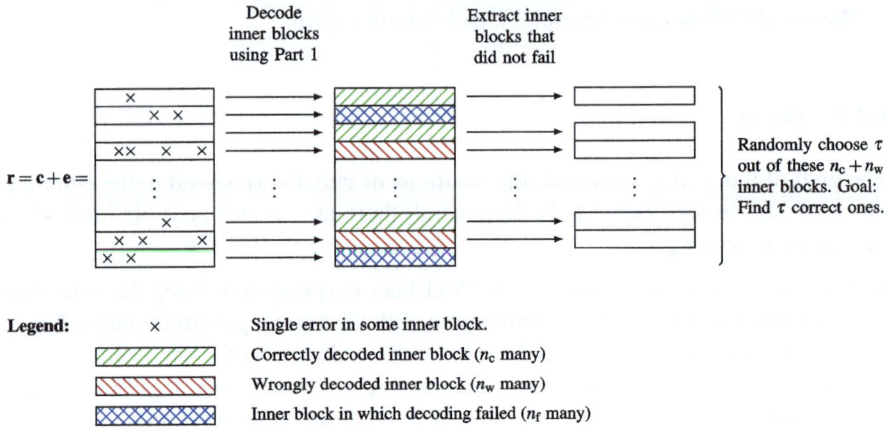

Fig. 2 Illustration of nonstructural attack

as an $n_A \times n_B$ matrix over $\mathbb{F}_q$, where each inner block corresponds to a row of the matrix.

We denote by $\mathbf{r}_\tau$, $\mathbf{e}_\tau$ and $\tilde{\mathbf{G}}_\tau$ the parts of $\mathbf{r}$, $\mathbf{e}$ and $\tilde{\mathbf{G}}$ restricted to the columns corresponding to the τ chosen inner blocks. If we find τ of the n_c correctly decoded blocks, the system

$$\mathbf{r}_\tau = \hat{\mathbf{m}} \cdot \tilde{\mathbf{G}}_\tau + \underbrace{\mathbf{e}_\tau}_{=\mathbf{0}} = \hat{\mathbf{m}} \cdot \tilde{\mathbf{G}}_\tau$$

has a solution $\hat{\mathbf{m}}$. If τ is chosen large enough, $\tilde{\mathbf{G}}_\tau$ has full rank k_{OC}, the solution $\hat{\mathbf{m}}$ is unique and fulfills $\mathrm{d_H}(\hat{\mathbf{m}} \cdot \tilde{\mathbf{G}}, \mathbf{r}) < \frac{d_{GC}}{2}$. For most practical codes, we conjecture that it is not necessary to choose τ much larger than $\frac{\delta}{n_B} \approx \frac{k_B}{n_B}$.

The entire nonstructural attack is summarized in Algorithm 2.

Algorithm 2: NonStructural Attack

Input: $\mathbf{r} = m \cdot \tilde{\mathbf{G}} + \mathbf{e}$ with $\mathrm{wt_H}(\mathbf{e}) = t$, $\mathbf{P}_{\text{Step 1}}$ and $\mathbf{G}_{\sigma_i(\mathscr{B})}$ for all $i = 1, \ldots, n_A$.
Output: $\mathbf{m}$

```
Decode inner blocks of r as described in Part 1, using P_Step 1 and G_σi(B).
do
    Choose τ out of n_A − n_f inner blocks, in which decoding did not fail.
    Solve r_τ = m̂ · G̃_τ for m̂.
while ∄m̂ or d_H(m̂ · G̃, r) ≥ d_GC/2
return m̂
```

Theorem 4 *If* $t < \frac{d_{GC}}{2}$, *Algorithm 2 is correct with high probability.*

Proof Line 1 corresponds to Part 1 of the nonstructural attack. Its correctness follows from the arguments in Sect. 6.2.1. Due to $t < \frac{d_{GC}}{2}$, $\tau \leq n_A - n_f$ with high probability. Thus, Line 3 finds τ correct blocks with nonzero probability in any loop and hence, it must find them in finite time with probability 1. When τ correct blocks are found and τ is chosen large enough, the system $\mathbf{r}_\tau = \hat{\mathbf{m}} \cdot \tilde{\mathbf{G}}_\tau$ has a unique solution, again with high probability. Since the number of errors is less than half the minimum distance, it holds that $\mathbf{m} = \hat{\mathbf{m}}$ and $\mathbf{c} = \hat{\mathbf{m}} \cdot \tilde{\mathbf{G}}$. Thus, $d_H(\hat{\mathbf{m}} \cdot \tilde{\mathbf{G}}, \mathbf{r}) = wt_H(\mathbf{e}) = t < \frac{d_{GC}}{2}$ and the algorithm terminates. ■

6.2.3 Complexity of the Nonstructural Attack

In general, the work factor of the nonstructural attack is the sum of the work factors of the two parts:

$$W = W_1 + W_2$$

Assume that in the first part, the decoding was done using the information set decoding attack. Thus, we have to apply n_A many small attacks, each of work factor

$$\frac{{k_B}^3 \cdot \binom{n_B}{k_B}}{\binom{n-t_B}{k_B}},$$

where $t_B = \lfloor \frac{d_B - 1}{2} \rfloor$ is half the minimum distance of $\mathcal{B}$. Thus,

$$W_1 = n_A \cdot \frac{{k_B}^3 \cdot \binom{n_B}{k_B}}{\binom{n-t_B}{k_B}}$$

In the second part, the probability of choosing a subset of τ correctly decoded inner blocks is

$$p = \frac{\binom{n_c}{\tau}}{\binom{n_c+n_w}{\tau}}.$$

Solving the system of linear equations can be done in ${k_{GC}}^3$ operations, yielding an expected work factor of

$$W_2 = \frac{{k_{GC}}^3}{p} = {k_{GC}}^3 \cdot \frac{\binom{n_c+n_w}{\tau}}{\binom{n_c}{\tau}}.$$

In Appendix C, it is shown that for the parameters proposed for an OC code construction by [25], we obtain a work factor of

$$W \approx 2^{29.7},$$

which is considered to be insecure [13]. We conclude that we have found a nonstructural attack whose work factor is significantly reduced compared to a naive structural attack on $\mathscr{C}_{\mathrm{GC}}$ directly. Thus, parameters of a GCC or OCC construction must be chosen much larger than nonconcatenated codes in order to compensate the security level. This increases the size of the public key considerably and probably implies that GC codes are not practically relevant to the McEliece cryptosystem, which already struggles with the disadvantage of large key sizes.

7 Methods of Preventing Parts of the Attacks

In the previous sections, we saw that Sendrier's attack for OC codes is partially applicable to GC codes. Also, we were able to give a nonstructural attack which is efficient for practical GC codes. In this section, we present methods for preventing parts of these attacks.

7.1 *Preventing the Second Step of Sendrier's Attack*

Sendrier's second step tries to synchronize the permutations of the inner blocks. As already mentioned in Sect. 5.2, this method only works if the permutation group of the code $\mathscr{B}$ is reduced to the identity element. Thus, one possibility would be to choose $\mathscr{B}$ with a nontrivial permutation group. However, it is already mentioned in [26] that such codes yield bad OC codes, implying that also GC codes would not be good.

Another possibility would be to change the definition of OC or GC codes such that we use different codes in each inner block. This corresponds to having several mappings

$$\theta_i : \bigoplus_{i=1}^{\ell} \to \mathscr{B}_i$$

with $i = 1, \ldots, n_A$ and $\mathscr{B}_i(q; n_B, k_{Bi}, d_{Bi})$ pairwise distinct, such that

$$\Theta : \bigoplus_{i=1}^{\ell} (\mathbb{F}_{q^{k_i}})^{n_A} \to \bigoplus_{i=1}^{n_A} \mathscr{B}_i$$

$$\left(\begin{bmatrix} a_{1,1} \\ a_{1,2} \\ \vdots \\ a_{1,n_A} \end{bmatrix}, \begin{bmatrix} a_{2,1} \\ a_{2,2} \\ \vdots \\ a_{2,n_A} \end{bmatrix}, \dots, \begin{bmatrix} a_{\ell,1} \\ a_{\ell,2} \\ \vdots \\ a_{\ell,n_A} \end{bmatrix} \right) \mapsto \begin{bmatrix} \theta_1(a_{1,1}, \dots, a_{\ell,1}) \\ \theta_2(a_{1,2}, \dots, a_{\ell,2}) \\ \vdots \\ \theta_{n_A}(a_{1,n_A}, \dots, a_{\ell,n_A}) \end{bmatrix}$$

in the definition of OC or GC codes. This construction is similar to the one used to define Justesen codes [15], which are certain OC codes with different inner codes. If the codes $\mathscr{B}_i$ have pairwise different j-th signatures (cf. Sect. 5.2) for all $j = 1, \dots, n_B$, Step 2 of Sendrier's attack does not work for either modified OC or modified GC codes. However, it can easily be seen that the alternative method described in Sect. 6.1 still works for different inner codes and thus, also the nonstructural attack can be applied in this case.

7.2 Preventing the First Step of Sendrier's Attack

Any attack described in this paper relies on the success of the first step of Sendrier's attack. Therefore, it is an important question whether we can find a large sub-class of GC codes which are resistant against this part of the attack.

The necessary condition for this method to work is that the inner code $\mathscr{B}$ is not the union of two disjoint support codewords. Sendrier [26] already mentioned that codes violating this condition are rather bad code. Also, if the inner code was exactly the union of r disjoint support codes which cannot be further splitted, the attack would give us $r \cdot n_A$ connected disjoint subsets of the code positions. We thus need to try the subsequent parts of the attack for all combinations of r subsets grouped to an inner block each. For small r and n_A, the number of possibilities might still be small enough to not increase the overall work factor much.

As proven in Sect. 5.1, a sufficient condition that Sendrier's attack works is that the set

$$\Xi := \left\{ \mathbf{c} \in \mathscr{P}(\mathscr{C}_{\mathrm{GC}}^{\perp}) : \mathrm{wt_H}(\mathbf{c}) < \min({d_A^{(1)}}^{\perp}, \dots, {d_A^{(\ell)}}^{\perp}, 2 \cdot {d_B}^{\perp}) \right\}$$

is not empty. Every $\mathbf{c} \in \Xi$ is in $\mathscr{C}_{\mathrm{GC}}^{\perp}$ and thus, $\mathrm{wt_H}(\mathbf{c}) \geq {d_{\mathrm{GC}}}^{\perp} \geq {d_B}^{\perp}$. Therefore, it follows that

$$\Xi \neq \emptyset \quad \Rightarrow \quad {d_B}^{\perp} < \min({d_A^{(1)}}^{\perp}, \dots, {d_A^{(\ell)}}^{\perp}, 2 \cdot {d_B}^{\perp}).$$

Hence, if any of the outer codes $\mathscr{A}^{(i)}$ has dual distance ${d_A^{(i)}}^{\perp} \leq {d_B}^{\perp}$, $\Xi = \emptyset$ and Sendrier's first step is not guaranteed to work. It needs to be mentioned that $\Xi \neq \emptyset$

is a sufficient condition and someone might find a modification of the first step that can handle the case $\Xi = \emptyset$. This problem needs further investigation.

In the OC case, if ${d_A}^{\perp}$ is decreased, the dimension k_A is also decreased and thus, the OC code might become bad. The advantage of GC codes is that only one of the outer codes needs to have this property and we can still obtain a good GC code satisfying $\Xi = \emptyset$. This fact makes us believe that there is the possibility of a large sub-class of practically relevant GC codes that resist the first step of Sendrier's attack.

8 Conclusion

In this work, we studied the suitability of generalized concatenated codes in the McEliece cryptosystem, motivated by the advantage of faster decoding than codes without concatenated structure. First, we gave a partial classification of GC codes that cannot be described as OC codes, for which a complete structural attack is known [26]. We analyzed Sendrier's structural attack on OC codes for applicability in the GC case. Step 1 of this attack can be directly applied, however with a stricter sufficient condition. Steps 2 and 3.1 were proven to work in exactly the same cases as for OC codes. However, it remains an open problem whether Step 3.2 of Sendrier's attack can be modified to work with GC codes.

We further gave an alternative method of obtaining the result of Step 3.1, only requiring the output of Step 1 of Sendrier's attack. In contrast to Step 2, this method works for all outer codes and can be performed in polynomial time. We were able to improve the complexity of the information set decoding attack significantly, using the result of Step 1. This gives us a nonstructural attack which we showed to be efficient for code parameters similar to the original McEliece Goppa codes construction. Hence, we can conclude that if Sendrier's first step works, this nonstructural method forces code parameters to be chosen so large that key sizes become impractical compared to other code constructions. Figure 3 summarizes the attacks discussed in the paper.

We proposed several methods which have the potential to prevent parts of Sendrier's attack, especially Step 1, and which only work in the GC case. This fact shows that GC codes, in contrast to OC codes, are still candidates for the use in the McEliece cryptosystem. It needs to be studied whether the methods of preventing Sendrier's first step cannot be circumvented by any efficient method. Other open problems are finding a necessary condition for Sendrier's first step to work. Also, Step 3.2 requires further studies in order to give a complete structural attack on GC codes.

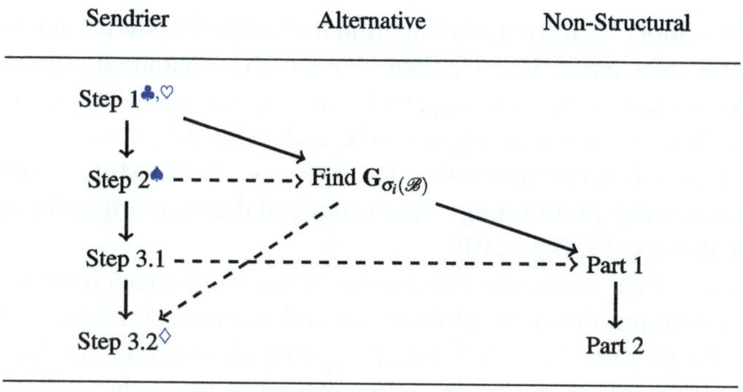

A ⟶ B: B needs result of A.

A - - → B: B can alternatively use the result of A.

♣ Sufficient condition: $d_B < \min(d_A^{(1)\perp}, \ldots, d_A^{(\ell)\perp}, 2 \cdot d_B{}^{\perp})$

♡ Necessary condition: $\mathscr{B}$ is not a direct sum of two disjoint support codes

♠ Works only if the permutation group of $\mathscr{B}$ is reduced to the identity element.

♢ So far: Only works for OCC. Open problem if also applicable to GCC.

Fig. 3 Summary of attacks

Appendix

A GCC Construction and Decoding

Generalized concatenated (GC) codes were introduced by [5]. This appendix presents construction and decoding of GC codes according to [4, Chap. 9]. Code concatenation is used in order to obtain long codes with low decoding complexity. The advantage of a GC code in comparison to an OC code with same length and dimension is, that the GC code can correct more errors, see [4]. A GC code consists of one inner and several outer codes of different dimensions. If we only use one outer code, we obtain an OC code.

The idea of generalized code concatenation is to partition the inner code into several levels of subcodes. We generate a partition tree as follows. The inner code becomes the root of the tree. We partition the inner code into subcodes which form the second level of the tree. We again partition each of the subcodes and continue until we end up at a level in which each subcode consists of only one codeword. These subcodes become the leaves of the tree. Let $\mathscr{B}_i^{(j)}(q; n_B, k_{B_i}{}^{(j)}, d_{B_i}{}^{(j)})$ denote the inner codes at level j. The partitioning should be done such that the minimum distance of the subcodes increases strictly monotonically from level to level in the partition tree. Each codeword can be uniquely identified by enumerating the branches of the partition tree and following this enumeration from the root to the corresponding leaf. The numeration from level j to level $j+1$ is protected by an outer code $\mathscr{A}^{(j)}(q^{k_j}; n_A, k_A^{(j)}, d_A^{(j)})$. This encoding scheme matches the definition of GC codes in Sect. 3.2.2 by simply taking θ as the function that maps the enumeration of a codeword from the root to a leaf to the codeword of $\mathscr{B}$ which is contained in this leaf. Note, that for many linear codes $\mathscr{B}$, there is a partitioning which corresponds to an $\mathbb{F}_q$-linear mapping θ [4]. Also, most practically good GC codes fulfill $k_1 = k_2 = \cdots = k_\ell = 1$ due to the existence of many linear subcodes of $\mathscr{B}$ (e.g., Reed–Muller codes), which helps constructing many partitionings. An example of the encoding and transmission process is visualized in [4, Fig. 9.10].

To obtain a good GC code, the dimensions of the outer codes have to be different. Also, the minimum distances of the outer codes should decrease from level to level. Keeping the product $d_A^{(j)} \cdot d_{B_i}^{(j)}$ for all i, j roughly constant also leads to good properties. The latter follows from a decoding procedure that reduces the problem of decoding GC codes to a sequence of ℓ decoders of OC codes with minimum distances $d_A^{(j)} \cdot d_{B_{i_j}}^{(j)}$ for some sequence of i_j's for all $j = 1, \ldots, \ell$. We refer to the example presented in [4, Fig. 9.11]. The length of the constructed GC code is $n_{\mathrm{GC}} = n_A \cdot n_B$, the dimension is $k = \sum_{i=1}^{\ell} k_A^{(i)}$, and the minimum distance is lower bounded by $d_{\mathrm{GC}} \geq \min_{i,j} \left(d_A^{(j)} \cdot d_{B_i}^{(j)}\right)$.

B Proof of Theorem 2

In this appendix, we prove Theorem 2. We first recall some useful and well-known facts about vector and matrix representations of extension fields.

Every finite field $\mathbb{F}_{q^m}$ is an $\mathbb{F}_q$-vector space of dimension m. Thus, there is a basis $B = \{\beta_1, \ldots, \beta_m\} \subseteq \mathbb{F}_{q^m}$ in which every element $a \in \mathbb{F}_{q^m}$ has a unique representation $a = \sum_{i=1}^{m} a_i \beta_i$ with $a_i \in \mathbb{F}_q$. Define the vector space isomorphism

$$\mathrm{ext}_B : \mathbb{F}_{q^m} \to \mathbb{F}_q{}^m, a \mapsto \mathbf{a} = [a_1, \ldots, a_m].$$

We call $\mathrm{ext}_B(a)$ the *vector representation* of a with respect to the basis B. It is well known that the set $\{\mathrm{ext}_B(\cdot) : B \text{basis of } \mathbb{F}_{q^m} \text{over} \mathbb{F}_q\}$ is equal to all vector space isomorphisms ($= \mathbb{F}_q$-linear maps) $\mathbb{F}_{q^m} \to \mathbb{F}_q^m$. This implies that for any $b \in \mathbb{F}_{q^m}$ and $\mathbf{b} \in \mathbb{F}_q^m$, there is a basis B such that $\mathrm{ext}_B(b) = \mathbf{b}$.

Lemma 3 *Some facts about vector and matrix representation of finite extensions of finite fields* $\mathbb{F}_{q^m}/\mathbb{F}_q$*:*

(i) Every finite field $\mathbb{F}_{q^m}$ *is isomorphic to a subfield* $\mathscr{M}_{q^m}$ *of the matrix ring* $\mathbb{F}_q^{m\times m}$*. We write* $\mathrm{mr}(a) \in \mathscr{M}_{q^m}$ *to denote the matrix representation of an element* $a \in \mathbb{F}_{q^m}$.

(ii) Every column or row of a matrix representation of $\mathbb{F}_{q^m}$ *can be used to uniquely represent elements of* $\mathbb{F}_{q^m}$*. We denote the vector representation of an element* $a \in \mathbb{F}_{q^m}$*, given by this column or row, by* $\mathrm{vr}(a) \in \mathbb{F}_q^m$. $\mathrm{vr}(\cdot) = \mathrm{ext}_B(\cdot)$ *for some basis* B.

(iii) If a specific column or row as in (ii) is chosen, the set of representative vectors of all elements in $\mathbb{F}_{q^m}$ *is equal to* $\mathbb{F}_q^m$.

(iv) *If a specific row as in (ii) is chosen, the multiplication of two elements* $a, b \in \mathbb{F}_{q^m}$ *corresponds to* $\mathrm{vr}(a \cdot b) = \mathrm{vr}(a) \cdot \mathrm{mr}(b)$.
(v) *If a specific column as in (ii) is chosen, the multiplication of two elements* $a, b \in \mathbb{F}_{q^m}$ *corresponds to* $\mathrm{vr}(a \cdot b) = \mathrm{mr}(a) \cdot \mathrm{vr}(b)$.
(vi) *For a specific row or column as in (ii) and an arbitrary basis B of* $\mathbb{F}_{q^m}$ *over* $\mathbb{F}_q$, $\mathscr{M}_{q^m}$ *can be chosen such that the vector representation of* $a \in \mathbb{F}_{q^m}$ *is* $\mathrm{ext}_B(a)$.

Proof (i) This statement is well known and can be found in [18] or [30].
(ii) Since the operations multiplication, addition and inversion in $\mathbb{F}_{q^m}$ correspond to the same operations of matrices in $\mathscr{M}_{q^m}$, all matrices except for the zero matrix in $\mathscr{M}_{q^m}$ are invertible. Now choose an arbitrary row (column) index i. We show that the rows (columns) of matrices in $\mathscr{M}_{q^m}$ of this index are distinct. Choose two matrices $\mathbf{M}_1, \mathbf{M}_2 \in \mathscr{M}_{q^m}$. Assume that their i-th rows (columns) are the same. Then the i-th row (column) of $\mathbf{M}_1 - \mathbf{M}_2 \in \mathscr{M}_{q^m}$ is the zero vector. Thus, $\mathbf{M}_1 - \mathbf{M}_2$ is not invertible and must be the zero matrix, implying that $\mathbf{M}_1 = \mathbf{M}_2$. Let $\phi(a)$ be the operation of extracting a specific row (column) from $a \in \mathscr{M}_{q^m}$. Since

$$\begin{aligned}\phi(\mathrm{mr}(\alpha \cdot a + \beta \cdot b)) &= \phi(\alpha \cdot \mathrm{mr}(a) + \beta \cdot \mathrm{mr}(b)) \\ &= \alpha \cdot \phi(\mathrm{mr}(a)) + \beta \cdot \phi(\mathrm{mr}(b))\end{aligned}$$

for all $\alpha, \beta \in \mathbb{F}_q$ and $a, b \in \mathbb{F}_{q^m}$, $\phi(\mathrm{mr}(\cdot))$ is a vector space isomorphism and is therefore equal to $\mathrm{ext}_B(\cdot)$ for some basis B.
(iii) This follows from a simple counting argument. Due to (ii), a specific row (column) represents all elements from $\mathscr{M}_{q^m}$, thus also from $\mathbb{F}_{q^m}$, uniquely. Hence, $|\{\mathrm{mr}(a) : a \in \mathbb{F}_{q^m}\}| = |\mathbb{F}_{q^m}| = q^m = |\mathbb{F}_q^m|$. Since $\{\mathrm{mr}(a) : a \in \mathbb{F}_{q^m}\} \subseteq \mathbb{F}_q^m$, $\{\mathrm{mr}(a) : a \in \mathbb{F}_{q^m}\} = \mathbb{F}_q^m$.
(iv) It is clear by $\mathrm{mr}(a \cdot b) = \mathrm{mr}(a) \cdot \mathrm{mr}(b)$ and by looking at the operations necessary to calculate the i-th column ($= \mathrm{vr}(a \cdot b)$) of the result on the right-hand side.
(v) Analog statement as in (iv).
(vi) The statement is clear since we can simply change the basis of the matrix representation, by setting $\mathrm{mr}_{\mathrm{new}}(a) = B \cdot \mathrm{mr}(a) \cdot B^{-1}$ for all $a \in \mathbb{F}_{q^m}$.

■

Using these definitions and statements, we are able prove Theorem 2. We also recall its statement.

Theorem 2 *If* $k_1 = k_2 = \cdots = k_\ell$, *a GCC is an OCC* $\Leftrightarrow \mathscr{A}^{(i)} = \mathscr{A}^{(j)}\ \forall i, j$.

Proof "$\Rightarrow$": Let $\mathscr{C}_{\mathrm{GC}} = \Theta(\bigoplus_{i=1}^{\ell} \mathscr{A}^{(i)})$, with θ $\mathbb{F}_q$-linear, be a GC code. Assume that $\mathscr{C}_{\mathrm{GC}}$ is an OCC. Then, there are an $\mathbb{F}_q$-linear θ' and an $\mathbb{F}_{q^{k_B}}$-linear code $\mathscr{A}$ such that $\Theta'(\mathscr{A}) = \Theta\left(\bigoplus_{i=1}^{\ell} \mathscr{A}^{(i)}\right)$ and thus,

$$\mathscr{A} = \Theta'^{-1}\left(\Theta\left(\bigoplus_{i=1}^{\ell} \mathscr{A}^{(i)}\right)\right).$$

Hence, $\Theta'^{-1}\left(\Theta\left(\bigoplus_{i=1}^{\ell} \mathscr{A}^{(i)}\right)\right)$ must be an $\mathbb{F}_{q^{k_B}}$-linear code. The mapping $\Theta'^{-1}(\Theta(\cdot)) : \bigoplus_{i=1}^{\ell} \mathbb{F}_{q^{k_i}}^{n_A} \to \mathbb{F}_{q^{k_B}}^{n_A}$ is componentwise $\mathbb{F}_q$-linear, i.e., there is an $\mathbb{F}_q$-linear mapping $\tilde{\theta} : \bigoplus_{i=1}^{\ell} \mathbb{F}_{q^{k_i}} \to \mathbb{F}_{q^{k_B}}$ such that

$$\Theta'^{-1}\left(\Theta\left(\mathbf{a}_1, \mathbf{a}_2, \dots, \mathbf{a}_\ell\right)\right) := \Theta'^{-1}\left(\Theta\left(\begin{bmatrix} a_{1,1} \\ a_{1,2} \\ \vdots \\ a_{1,n_A} \end{bmatrix}, \begin{bmatrix} a_{2,1} \\ a_{2,2} \\ \vdots \\ a_{2,n_A} \end{bmatrix}, \dots, \begin{bmatrix} a_{\ell,1} \\ a_{\ell,2} \\ \vdots \\ a_{\ell,n_A} \end{bmatrix}\right)\right)$$

$$= \begin{bmatrix} \tilde{\theta}(a_{1,1}, \dots, a_{\ell,1}) \\ \tilde{\theta}(a_{1,2}, \dots, a_{\ell,2}) \\ \vdots \\ \tilde{\theta}(a_{\ell,n_A}, \dots, a_{\ell,n_A}) \end{bmatrix} =: \begin{bmatrix} \tilde{a}_1 \\ \tilde{a}_2 \\ \vdots \\ \tilde{a}_{n_A} \end{bmatrix} \in \mathbb{F}_{q^{k_B}}^{n_A}$$

for all $\mathbf{a}_i \in \mathbb{F}_{q^{k_i}}$. Due to $k_B = \sum_{i=1}^{\ell} k_i = \sum_{i=1}^{\ell} k_1 = \ell \cdot k_1$, $k_i = k_1 | k_B$ and $\mathbb{F}_{q^{k_B}}$ can be seen as an extension field of $\mathbb{F}_{q^{k_1}}$ with extension degree $[\mathbb{F}_{q^{k_B}} : \mathbb{F}_{q^{k_1}}] = \ell$.

$$\begin{bmatrix} \mathbf{a}_1 \\ \mathbf{a}_2 \\ \vdots \\ \mathbf{a}_\ell \end{bmatrix} = \begin{bmatrix} a_{1,1} & a_{1,2} & \dots & a_{1,n_A} \\ a_{2,1} & a_{2,2} & \dots & a_{2,n_A} \\ \vdots & \vdots & \ddots & \vdots \\ a_{\ell,1} & a_{\ell,2} & \dots & a_{\ell,n_A} \end{bmatrix} = \mathrm{ext}_B\left(\left[\tilde{a}_1\ \tilde{a}_2\ \dots\ \tilde{a}_{n_A}\right]\right).$$

Choosing the corresponding matrix representation of $\alpha \in \mathbb{F}_{q^{k_B}}$ over $\mathbb{F}_{q^{k_1}}$ as in Lemma 3, we can write

$$\mathrm{ext}_B\left(\alpha \cdot \left[\tilde{a}_1\ \tilde{a}_2\ \dots\ \tilde{a}_{n_A}\right]\right) = \mathrm{mr}(\alpha) \cdot \mathrm{ext}_B\left(\left[\tilde{a}_1\ \tilde{a}_2\ \dots\ \tilde{a}_{n_A}\right]\right) = \mathrm{mr}(\alpha) \cdot \begin{bmatrix} \mathbf{a}_1 \\ \mathbf{a}_2 \\ \vdots \\ \mathbf{a}_\ell \end{bmatrix}$$

Since $\mathbf{a}_i \in \mathscr{A}^{(i)}$, $\left[\tilde{a}_1\ \tilde{a}_2\ \dots\ \tilde{a}_{n_A}\right] \in \mathscr{A}$ also $\alpha \cdot \left[\tilde{a}_1\ \tilde{a}_2\ \dots\ \tilde{a}_{n_A}\right]$ must be in $\mathscr{A}$ for all $\alpha \in \mathbb{F}_{q^{k_B}}$ and thus, $\mathrm{ext}_B\left(\alpha \cdot \left[\tilde{a}_1\ \tilde{a}_2\ \dots\ \tilde{a}_{n_A}\right]\right) \in \bigoplus_{i=1}^{\ell} \mathscr{A}^{(i)}$. Due to Lemma 3, we can choose α such that the i-th row of $\mathrm{mr}(\alpha)$ is can be an arbitrary $\left[\alpha_1\ \alpha_2\ \dots\ \alpha_\ell\right] \in \mathbb{F}_{q^{k_i}}^{\ell}$ and thus, the i-th row of $\mathrm{ext}_B\left(\alpha \cdot \left[\tilde{a}_1\ \tilde{a}_2\ \dots\ \tilde{a}_{n_A}\right]\right)$ is $\sum_{j=1}^{\ell} \alpha_i \mathbf{a}_i$. This implies that $\mathscr{A}^{(j)} \subseteq \mathscr{A}^{(i)}$ for all $j, i = 1, \dots, \ell$ and thus all outer codes $\mathscr{A}^{(i)}$ are the same.

"$\Leftarrow$": If $\mathscr{A}^{(j)} = \mathscr{A}^{(i)}$ for all i, j, we can choose any basis B of $\mathbb{F}_{q^{k_B}}$ over $\mathbb{F}_{q^{k_i}}$. Since multiplying elements of $\mathrm{ext}_B^{-1}(\bigoplus_{i=1}^{\ell} \mathscr{A}^{(i)})$ by $\mathbb{F}_{q^{k_B}}$ scalars corresponds to a left multiplication by a matrix in $\mathscr{M}_{q^{k_B}}$, and any $\mathbb{F}_{q^{k_i}}$-linear combination of elements of different $\mathscr{A}^{(i)}$'s again is contained in any $\mathscr{A}^{(i)}$, the set $\mathscr{A} := \text{ext}_B^{-1}(\bigoplus_{i=1}^{\ell} \mathscr{A}^{(i)})$ is an $\mathbb{F}_{q^{k_B}}$-linear code. The $\mathbb{F}_q$-linear map is given by $\theta \circ \text{ext}_B$. ■

C Work Factor of Nonstructural Attack on Code Ex. in [25]

This appendix computes the work factor of our nonstructural attack presented in Sect. 6.2 when applied to the OC code example which was proposed by Sendrier in [25] with parameters $(2048, 308, \geq 425)$.

The inner code is a random code $\mathscr{B}(16, 7, 5)$ over $\mathbb{F}_2$ and the outer code is a GRS code $\mathscr{A}(128, 44, 85)$ over $\mathbb{F}_{2^7}$. A simulation was performed using Matlab on 1500 random codes ($\mathscr{B}(16, 7, 5)$) by adding errors with a probability of $\frac{212}{2048}$ to each codeword of $\mathscr{B}$ and then decoding it. 1,000,000 codewords for each code were used. The estimations for the probabilities of correct decoding, wrong decoding and failure in decoding are $p_\mathsf{c} = 0.7741$, $p_\mathsf{w} = 0.0441$ and $p_\mathsf{f} = 0.1818$, respectively. The corresponding standard deviation values are 0.00042, 0.0043 and 0.0043. The expected number of correctly and wrongly decoded, and failed inner blocks are then given by

$$\begin{aligned} n_\mathsf{c} &= n_A \cdot p_\mathsf{c} = 128 \cdot 0.7741 \approx 99, \\ n_\mathsf{w} &= n_A \cdot p_\mathsf{w} = 128 \cdot 0.0441 \approx 6, \\ n_\mathsf{f} &= n_A \cdot p_\mathsf{f} = 128 \cdot 0.1818 \approx 23. \end{aligned}$$

By choosing $m = k_A = 44$ inner blocks, we obtain the work factor

$$W_2 = \frac{308^3}{p} \approx \frac{308^3}{0.0345} \approx 8.4686 \cdot 10^8 \approx 2^{29.7}.$$

With

$$W_1 = 128 \cdot \frac{7^3 \cdot \binom{16}{7}}{\binom{16-\lfloor\frac{5-1}{2}\rfloor}{7}} \approx 1.4635 \cdot 10^5,$$

$W_1 \ll W_2$, and the overall work factor is then equal to

$$W \approx 2^{29.7}.$$

This work factor is considered to be insecure [13].

References

1. Becker, A., Joux, A., May, A., Meurer, A.: Decoding random binary linear codes in $2^{n/20}$: how 1+ 1= 0 improves information set decoding. In: Advances in Cryptology—EUROCRYPT 2012, pp. 520–536. Springer, Berlin (2012)
2. Berger, T.P., Loidreau, P.: How to mask the structure of codes for a cryptographic use. Des. Codes Cryptogr. **35**(1), 63–79 (2005)
3. Berlekamp, E.R., McEliece, R.J., Van Tilborg, H.C.A.: On the inherent intractability of certain coding problems. IEEE Trans. Inf. Theory **24**(3), 384–386 (1978)
4. Bossert, M.: Channel Coding for Telecommunications. Wiley, New York (1999)
5. Blokh, È.L., Zyablov, V.V.: Coding of generalized concatenated codes. Problemy Peredachi Informatsii **10**(3), 45–50 (1974)
6. Chizhov, I.V., Borodin, M.A.: The failure of McEliece PKC based on Reed–Muller codes. IACR Cryptol. ePrint Arch. **2013**, 287 (2013)
7. Coffey, J.T., Goodman, R.M.: The complexity of information set decoding. IEEE Trans. Inf. Theory **36**(5), 1031–1037 (1990)
8. Couvreur, A., Márquez-Corbella, I., Pellikaan, R.: A polynomial time attack against algebraic geometry code based public key cryptosystems (2014). arXiv:1401.6025
9. Chabanne, H., Sendrier, N.: On the concatenated structures of a [49, 18, 12] binary abelian code. Discret. Math. **112**(1), 245–248 (1993)
10. Diffie, W., Hellman, M.E.: New directions in cryptography. IEEE Trans. Inf. Theory **22**(6), 644–654 (1976)
11. ElGamal, T.: A public key cryptosystem and a signature scheme based on discrete logarithms. IEEE Trans. Inf. Theory **31**(4), 469–472 (1985)
12. Forney, D.G.: Concatenated codes. vol. 11 MIT press, Cambridge (1966)
13. Heyse, S.: Post quantum cryptography: implementing alternative public key schemes on embedded devices. PhD thesis, dissertation for the degree of doktor-ingenieur: 10.2013/Stefan Heyse.–Bochum, 2013.–235 p.–Bibliogr (2013)
14. Janwa, H., Moreno, O.: McEliece public key cryptosystems using algebraic-geometric codes. Des. Codes Cryptogr. **8**(3), 293–307 (1996)
15. Justesen, J.: Class of constructive asymptotically good algebraic codes. IEEE Trans. Inf. Theory **18**(5), 652–656 (1972)
16. Lee, P.J., Brickell, E.F.: An observation on the security of McEliece's public-key cryptosystem. Workshop on the Theory and Application of Cryptographic Techniques. Springer, Berlin, Heidelberg. (1988)
17. Li, Y.X., Deng, R.H., Wang, X.M.: On the equivalence of McEliece's and Niederreiter's public-key cryptosystems. IEEE Trans. Inf. Theory **40**(1), 271–273 (1994)
18. Lidl, R., Niederreiter, H.: Finite Fields, vol. 20. Cambridge University Press, Cambridge (1997)
19. McEliece, R.J.: A public-key cryptosystem based on algebraic coding theory. DSN Prog. Rep. **42**(44), 114–116 (1978)
20. MacWilliams, F.J., Sloane, N.J.A.: The Theory of Error Correcting Codes. Elsevier, Amsterdam (1977)
21. Minder, L., Shokrollahi, A.: Cryptanalysis of the Sidelnikov cryptosystem. In: Annual International Conference on the Theory and Applications of Cryptographic Techniques, pp. 347–360. Springer, Berlin (2007)
22. Niederreiter, H.: Knapsack-type cryptosystems and algebraic coding theory. Probl. Control Inf. Theory (Problemy Upravleniya I Teorii Informatsii) **15**(2), 159–166 (1986)
23. Peters, C.: Information-set decoding for linear codes over $\mathbf{F}_q$. In: Post-Quantum Cryptography, pp. 81–94. Springer, Berlin (2010)
24. Rivest, R.L., Shamir, A., Adleman, L.: A method for obtaining digital signatures and public-key cryptosystems. Commun. ACM **21**(2), 120–126 (1978)
25. Sendrier, N.: On the structure of randomly permuted concatenated code. [Research Report] RR-2460, INRIA (1995). <inria-00074216>

26. Sendrier, N.: On the concatenated structure of a linear code. Appl. Algebra Eng. Commun. Comput. **9**(3), 221–242 (1998)
27. Shor, P.W.: Algorithms for quantum computation: discrete Logarithms and factoring. In: 1994 Proceedings of the 35th Annual Symposium on Foundations of Computer Science, pp. 124–134. IEEE (1994)
28. Sidelnikov, V.M.: Public-key cryptosystem based on binary Reed–Muller codes. Discret. Math. Appl. **4**(3), 191–207 (1994)
29. Sidelnikov, V.M., Shestakov, S.O.: On the insecurity of cryptosystems based on generalized Reed–Solomon codes. Discret. Math. Appl. **2**(4), 439–444 (1992)
30. Wardlaw, W.P.: Matrix representation of finite fields. Math. Mag. **67**, 289–293 (1994)
31. Wieschebrink, C.: Cryptanalysis of the Niederreiter public key scheme based on GRS subcodes. In: International Workshop on Post-Quantum Cryptography, pp. 61–72. Springer, Berlin (2010)

Univariate Real Root Isolation over a Single Logarithmic Extension of Real Algebraic Numbers

Adam Strzeboński and Elias P. Tsigaridas

Abstract We present algorithmic, complexity, and implementation results for the problem of isolating the real roots of a univariate polynomial $B \in L[x]$, where $L = \mathbb{Q}[\lg(\alpha)]$ and α is a positive real algebraic number. The algorithm approximates the coefficients of B up to a sufficient accuracy and then solves the approximate polynomial. For this we derive worst-case (aggregate) separation bounds. We also estimate the expected number of real roots when we draw the coefficients from a specific distribution and illustrate our results experimentally. A generalization to bivariate polynomial systems is also presented. We implemented the algorithm in C as part of the core library of MATHEMATICA for the case $B \in \mathbb{Z}[\lg(q)][x]$ where q is positive rational number and we demonstrate its efficiency over various data sets.

Keywords Real root isolation · Logarithm · Algebraic number · Separation bound

1 Introduction

We consider the problem of isolating the real roots of a univariate polynomial the coefficients of which are polynomials in the logarithm of a positive real algebraic number. We consider two variants of the problem. In the first variant the argument of the logarithm is a positive real algebraic number. In the second the argument is a bivariate homogeneous polynomial evaluated at two real algebraic numbers. The reader can refer to the end of the introduction for a detailed presentation of the notation that we use. The first problem that we consider is the following:

Problem 1 Consider the square-free polynomial

A. Strzeboński
Wolfram Research Inc., 100 Trade Centre Drive, Champaign, IL 61820, USA
e-mail: adams@wolfram.com

E.P. Tsigaridas (✉)
POLSYS Project, INRIA Paris-Rocquencourt UPMC, Univ Paris 06, LIP6, Paris, France
e-mail: elias.tsigaridas@inria.fr

I.S. Kotsireas and E. Martínez-Moro (eds.), *Applications of Computer Algebra*, Springer Proceedings in Mathematics & Statistics 198,
DOI 10.1007/978-3-319-56932-1_27

$$B_\alpha = \sum_{i=0}^{d} b_i\, x^i, \qquad \text{where} \quad b_i = \sum_{j=0}^{\nu_i} b_{i,j}\, (\lg(\alpha))^j \ ,$$

$b_{i,j} \in \mathbb{Z}$, the bitsize of $b_{i,j}$ is bounded by τ, and α is a positive real root of a polynomial $A \in \mathbb{Z}[x]$ of degree m and maximum coefficient bitsize τ. Finally, let $\nu = \max_i \nu_i$. What is the Boolean complexity of isolating the real roots of B_α?

The problem of isolating the real roots of a univariate polynomial is a well-studied problem. However, most of the results focus on polynomials with rationals or algebraic numbers as coefficients. We are not aware of any complexity results that consider polynomials with transcendental numbers as coefficients. We present the first complexity bounds for the real solving problem for a family of polynomials with coefficients involving logarithms of algebraic numbers.

In addition, our implementation is the first complete one for solving exactly polynomial with such transcendental numbers as coefficients.

We tackle the problem by approximating the coefficients of B_α up to a sufficient precision. In this way we relate it to numerical univariate real solving algorithms [30, 34], see also [21, 32], and to algorithms based on the bitstream model, e.g., [15, 26, 33]. For a detailed treatment of numerical solvers we refer the reader to [25, Chapter 15]. Problem 1 is a generalization of the problem of solving polynomials with coefficients in an extension field, [22, 35, 36], see also [9, 23, 39, 40] and references therein. We also refer to the recent work of Bates and Sottile [4] on Khovanskii–Rolle continuation algorithm that exploits logarithms of polynomial expressions. For the close-related problem of computing the zeros of analytic functions using inclusion and exclusion predicates we refer the reader to [11, 20, 41, 42].

To obtain the various bounds we have to combine several algebraic techniques in a novel way and to provide new evaluation and perturbation bounds; the latter turn out to be useful in other applications as well. Our analysis is based on effective lower bounds of linear forms in two logarithms; a result due to Mignotte and Waldschmidt (Theorem 3). We combine this bound with univariate and multivariate separation and evaluation bounds of polynomials and polynomial systems. The idea is to approximate the coefficients of B_α up to a sufficient precision and then isolate the real roots of the approximate polynomial. The precision is such that the number of the real roots remains the same and from the isolated intervals of the approximate polynomial we can derive isolating intervals for the real roots of B_α.

First, we need to quantify "sufficient accuracy". We treat the logarithm as a parameter and the separation bound of B_α turns out to be a univariate polynomial in this parameter. We estimate a lower bound on this evaluation by proving that it depends only on the closest root and the separation bound of the polynomial (Lemma 2) and combining it with Theorem 3. This approach saves us a factor compared to the straightforward one of factoring the polynomial in linear factors and bounding the separation using Theorem 3 directly.

This approach turns out to be applicable for tackling a more general problem, Problem 2, where the argument of the logarithm is a homogeneous bivariate polynomial evaluated at two real algebraic numbers. It is a simplified version of Problem 1.

However, while the resolution of the latter depends on combinations of univariate separation bounds, Problem 2 depends on successive applications of aggregate multivariate separation bounds and applications of Theorem 3. For this and for making the presentation easier for the reader we present both approaches.

We also estimate the expected number of real roots of B_α in the case where all the polynomials b_i have the same degree ν and their coefficients, $b_{i,j}$, are Gaussian random variables with mean zero and variance $\binom{d}{i}$. In this case the expected number of real roots is $\sqrt{d}$. We implemented our algorithms in C as part of the core library of MATHEMATICA for the case $B \in \mathbb{Z}[\lg(q)][x]$ where q is positive rational number and we demonstrate its efficiency over various data sets. Our results support experimentally the $\sqrt{d}$ bound for the number of roots of random polynomials of this kind. Finally, we generalize our bounds to handle bivariate polynomial systems. We prove a perturbation bound for the roots of a bivariate polynomial system that is applicable to a broader context.

The rest of the paper is structured as follows. First we introduce our notation and in Sect. 2 we present the main tools that we will use throughout the paper. In Sect. 3 we present an algorithm for tackling Problem 1 as well as its complexity analysis, experimental results and the bound for the expected number of real roots. We present a more general version of Problem 1 in Sect. 5 and the extension to bivariate polynomial systems in Sect. 6.

Notation. In what follows $\mathcal{O}_B$, resp. $\mathcal{O}$, means bit, resp. arithmetic, complexity and the $\widetilde{\mathcal{O}}_B$, resp. $\widetilde{\mathcal{O}}$, notation means that we are ignoring logarithmic factors, see [38, Definition 25.8]. For a polynomial $A = \sum_{i=0}^{d} a_i\, x^i \in \mathbb{Z}[x]$, $\mathsf{deg}(A) = d$ denotes its degree and $\mathcal{L}\,(A) = \tau$ the maximum bitsize of its coefficients, including a bit for the sign. For $a \in \mathbb{Q}$, $\mathcal{L}\,(a) \geq 1$ is the maximum bitsize of the numerator and the denominator. We write $\Delta_\alpha(A)$ to denote the minimum distance between a root α of a polynomial A and any other root; we also use $\Delta(\alpha)$ where A is clear from the context; We also use Δ_i instead of $\Delta(\alpha_i)$, where α_i is a root of A and $1 \leq i \leq \mathsf{deg}(A)$. $\Delta(A) = \min_\alpha \Delta_\alpha(A)$ is the *separation bound*, that is the minimum distance between all the roots of A, and $\Sigma(A) = -\sum_{i=1}^{n} \lg \Delta_i(A)$. The Mahler measure of A is $\mathcal{M}\,(A) = a_d \prod_{|\alpha|\geq 1} |\alpha|$, where α runs through the complex roots of A. If $A \in \mathbb{Z}[x]$ and $\mathcal{L}\,(A) = \tau$, then $\mathcal{M}\,(A) \leq \|A\|_2 \leq \sqrt{d+1}\|A\|_\infty = 2^\tau\sqrt{d+1}$ [28, p. 152]. We denote by $\lg(\cdot)$, resp. $\ln(\cdot)$, the logarithm with base 2, resp. e. Let $L_\alpha = \lg(\alpha)$, where α is a positive algebraic number, and $L_H = \lg A(\gamma_1, \gamma_2)$, where $\gamma_{\{1,2\}}$ are real algebraic and A is a bivariate homogeneous polynomial and $A(\gamma_1, \gamma_2) > 0$.

2 Preliminaries

Real algebraic numbers are the real roots of univariate polynomials with integer coefficients; we denote their set by $\mathbb{R}_{\mathsf{alg}}$. We represent them using the *isolating interval representation.* If $\alpha \in \mathbb{R}_{\mathsf{alg}}$ then the representation consists of a square-free polynomial with integer coefficients, $A \in \mathbb{Z}[x]$, that has α as a real root, and an isolating interval with rational endpoints, $\mathcal{I} = [\mathsf{a}_1, \mathsf{a}_2]$, that contains α and no other root of the polynomial. We write $\alpha \cong (A, \mathcal{I})$. Such a representation could be also used to represent the real roots of polynomials with real numbers as coefficients, provided that there is an algorithm for isolating them.

The following proposition provides upper and aggregate bounds for the roots of a univariate polynomial. Various versions of the proposition could be found, e.g. [10, 13, 37]. The aggregate version of Eq. (2) comes from a simplified version of [19, Theorem 11].

Proposition 1 (DMM$_1$) *Let $f = \sum_{i=0}^{d} a_i x^i \in \mathbb{R}[x]$ be a univariate polynomial of degree d such that $a_d a_0 \neq 0$. The distinct roots of f are $\alpha_1, \ldots, \alpha_r$. For any root α_k it holds*

$$\frac{|a_0|}{2\,\|f\|_\infty} \leq |\alpha_k| \leq 2\frac{\|f\|_\infty}{|a_d|} . \tag{1}$$

Let K be any subset of $\{1, \ldots, r\}$ with cardinality $|K|$. Then

$$\prod_{k \in K} \Delta_k \geq 2^{-4d \lg d}\, \mathcal{M}(f)^{-2(r-1)}\, |\mathbf{sr}_r(f, f')| , \tag{2}$$

where $\mathbf{sr}_r(f, f')$ is the r-th subresultant coefficient of the subresultant sequence of f and its derivative f'.

The following lemma provides a lower bound on the evaluation of a polynomial that depends on the closest root and on the aggregate separation bound of the polynomial. For another proof with slightly different bounds, suggested by one of the reviewers, we refer the reader to the appendix.

Lemma 2 *Let $L \in \mathbb{C}$ and γ_1 the root of the square-free polynomial f that is closest to L. Then*

$$|f(L)| \geq |a_d|^7\, |L - \gamma_1|^6\, \mathcal{M}(f)^{-6}\, 2^{\lg \prod_i \Delta_i - 6} .$$

Proof There are at most six roots of f such that $|L - \gamma_i| \leq |\gamma_i - \gamma_{c_i}| = \Delta_i$, where γ_{c_i} is the root closest to γ_i. This is a consequence of the vertex degree of planar nearest neighbor graphs [18]. Wlog let them be the first six ones. Then

$$\begin{aligned} |f(L)| &= |a_d| \prod_{i=1}^{d} |L - \gamma_i| = |a_d| \prod_{i=1}^{6} |L - \gamma_i| \prod_{j=7}^{d} |L - \gamma_j| \\ &\geq |a_d||L - \gamma_1|^6 \frac{1}{\prod_{i=1}^{6} \Delta_i} \prod_{j=1}^{d} \Delta_j \\ &\geq |a_d|^7 |L - \gamma_1|^6 \mathcal{M}(f)^{-6}\, 2^{\lg \prod_i \Delta_i - 6} . \end{aligned}$$

For the last inequality we use $\Delta_i \leq 2\,\mathcal{M}(f)\,/|a_d|$, that in turn relies on $\Delta_i = |\gamma_i - \gamma_{c_i}| \leq |\gamma_i| + |\gamma_{c_i}| \leq 2\,\mathcal{M}(f)\,/|a_d|$. □

In the sequel we will use the previous lemma in conjunction with Theorem 3 and almost always L will be the logarithm of an algebraic number. It might be the case that L is a root of f and thus the evaluation $f(L)$ is zero. However, we omit this case as it can be detected rather easily and does not affect in any case the complexity of the algorithms that we consider.

We will also need the following theorem, due to Mignotte and Waldschmidt [29]. It provides an effective lower bound on a homogeneous linear form with two logarithms of (real) algebraic numbers with algebraic coefficients. This result generalizes a result by Gel'fond. A generalization that handles general linear forms is due to Baker, e.g. [3]. In what follows by the height of an algebraic number, α, we mean the height of the minimum polynomial of α.

Theorem 3 [29] *Let* $\Lambda = \beta \log(\alpha_1) - \log(\alpha_2)$, *where* log *is any determination of the logarithm, and* $\beta, \alpha_1, \alpha_2$ *are three nonzero algebraic numbers of degrees* D_0, D_1, D_2, *respectively. Let* A_i *be a bound on the height of* α_i *such that* $\exp(|\log(\alpha_i)|) \leq A_i$, *for* $i \in \{1, 2\}$. *B is an upper bound on the height of* β *and* e^{D_0}. *If D is the degree over* $\mathbb{Q}$ *of the field* $\mathbb{Q}(\beta, \alpha_1, \alpha_2)$, *and* $T = \ln(B) + \ln\ln(A_1) + \ln\ln(A_2) + \ln(D)$, *then if* $\Lambda \neq 0$, *then* $|\Lambda| > \exp(-5 \cdot 10^{10} \cdot D^4 \cdot \ln(A_1) \cdot \ln(A_2) \cdot T^2)$.

3 An Algorithm for B_α

In what follows we assume that L_α is indeed a transcendental number. This could be tested using Lindemann–Weierstrass theorem. The following lemma is based on arguments in [2].

Lemma 4 *Let* α *be a positive real root of a univariate polynomial* $A \in \mathbb{Z}[x]$ *that has degree m and maximum coefficient bitsize* τ. *Then* $2^{-2\tau-m-2} \leq |\lg(\alpha)| \leq \tau + 1$.

Proof The right inequality follows from Cauchy's bound, since $|\alpha| \leq 2^{\tau+1}$.

For the left inequality, first we need to bound $|\alpha - 1|$. Notice that $\alpha - 1$ is a root of $\bar{A}(x) = A(x + 1)$. The coefficients of $\bar{A}(x)$ are bounded by $2^{m+\tau}$. Using Cauchy's bound

$$|\alpha - 1| \geq 2^{-\tau-m-1} .$$

Using the inequality $|e^z - 1| \leq |z|e^{|z|}$, we get

$$|\alpha - 1| \leq |e^{\ln(\alpha)} - 1| \leq \frac{|\lg(\alpha)|}{\lg(e)} |\alpha| ,$$

and thus $|\lg(\alpha)| \geq 2^{-2\tau-m-2}$, which concludes the proof. □

Lemma 5 *Let* b_i *be as in Problem* 1. *If* $b_i(L_\alpha) \neq 0$, *then*

$$2^{-\widetilde{O}(m^4 \nu^4 \tau (\tau^2+\nu^2))} \leq |b_i(L_\alpha)| \leq 2^{\widetilde{O}(\nu+\tau)}.$$

Proof To bound b_i we proceed as follows:

$$\begin{aligned}|b_i(L_\alpha)| = |\sum_{j=0}^{\nu} b_{i,j} L_\alpha^j| \leq 2^\tau \sum_{j=0}^{\nu} |L_\alpha|^j \leq 2^\tau \sum_{j=0}^{\nu} (\tau+1)^j \\ \leq 2^{\tau+\nu+1} \tau^{\nu+1} \leq 2^{\tau+2\nu \lg(2\tau)} .\end{aligned} \tag{3}$$

To compute a lower bound for $|b_i(L_\alpha)|$ we assume that $\beta_{i,1}$ is the root of $b_i(y)$ closest to L_α and we apply Lemma 2, i.e.,

$$|b_i(L_\alpha)| > |b_{i,\nu}|^7 \, |L_\alpha - \beta_{i,1}|^6 \, \mathcal{M}(b_i)^{-6} \, 2^{\lg \prod_j \Delta_j(b_i) - 6} .$$

It holds $|b_{i,\nu}| \geq 1$; Theorem 3 implies

$$|L_\alpha - \beta_{i,1}| \geq \exp(c_1 \, m^4 \, \nu^4 \, \tau \, (\tau + \nu + \ln(m\tau\nu))^2) ,$$

where c_1 is constant that can be computed explicitly.

Landau's inequality gives $\mathcal{M}(b_i) \leq (\nu+1)\|b_i\|_\infty \leq 2^{\tau+\lg\nu+1}$. Finally, using Proposition 1 we have $\lg \prod_j \Delta_j(b_i) \geq -\mathcal{O}(\nu^2 + \nu\tau + \nu \lg \nu)$. Combining all the inequalities we get

$$|b_i(L_\alpha)| \geq \exp(c_2 \, m^4 \, \nu^4 \, \tau \, (\tau + \nu + \ln(m\tau\nu))^2) ,$$

$$\text{or} \quad |b_i(L_\alpha)| \geq \exp(-\widetilde{\mathcal{O}}(m^4 \, \nu^4 \, \tau \, (\tau^2 + \nu^2))) ,$$

where c_2 is constant that can be computed explicitly. □

The previous lemma allows us to bound $\| B_\alpha \|_2$. Using Eq. (3) from the proof of the previous lemma we get $\| B_\alpha \|_2^2 = \sum_{i=0}^{d} |b_i(L_\alpha)|^2 \leq (d+1)\, 2^{2\tau+2} \, \tau^{2\nu+2}$, which results to

$$\| B_\alpha \|_2 \leq d \, 2^{\tau+1} \, \tau^{\nu+1} . \tag{4}$$

Lemma 6 *Let B_α be as in Problem 1, then*

$$2^{-\widetilde{\mathcal{O}}(d^6 \nu^4 m^4 \tau (\nu^2+\tau^2))} \leq |\mathsf{disc}(B_\alpha)| \leq 2^{\widetilde{\mathcal{O}}(d\nu + d\tau + m^4 \, \nu^4 \, \tau \, (\tau^2+\nu^2))} .$$

Proof We consider B_α as a bivariate polynomial in $\mathbb{Z}[L_\alpha, x]$. To bound $|\mathsf{disc}(B_\alpha)|$ we consider the identity

$$\begin{aligned}|\mathsf{disc}(B_\alpha)| &= \left| \frac{1}{b_d(L_\alpha)} \mathtt{res}_x(B_\alpha(L_\alpha, x), \partial B_\alpha(L_\alpha, x)/\partial x) \right| \\ &= \left| \frac{1}{b_d(L_\alpha)} R_B(L_\alpha) \right| ,\end{aligned} \tag{5}$$

where the resultant, $R_B \in \mathbb{Z}[L_\alpha]$, can be computed as the determinant of the Sylvester matrix of $B_\alpha(L_\alpha, x)$ and $\partial B_\alpha(L_\alpha, x)/\partial x$, evaluated at L_α.

The Sylvester matrix is of size $(2d-1) \times (2d-1)$, the elements of which belong to $\mathbb{Z}[L_\alpha]$. The determinant consists of $(2d-1)!$ terms. Each term is a product of $d-1$ polynomials in L_α of degree at most ν and bitsize at most τ, times a product of d polynomials in L_α of degree at most $\nu - 1$ and bitsize at most $\tau + \lg d$. The first product results a polynomial of degree $(d-1)\nu$ and bitsize $(d-1)\tau + (d-1)\lg d$. The second product results polynomials of degree $d(\nu-1)$ and bitsize $d\tau + d\lg(d(\nu-1))$. Thus, any term in the determinant expansion is a polynomial in L_α of degree less than $2d\nu$ and bitsize at most $2d\tau + 6d\lg(d\nu)$. The determinant itself is a polynomial in L_α of degree at most $2d\nu$ and of bitsize $2d\tau + 10d\lg(d\nu)$.

We compute an upper bound of $|R_B(L_q)|$ as follows:

$$|R_B(L_q)| \leq 2^{2d\tau+10d\lg(d\nu)} \sum_{k=0}^{2d\nu} |L_\alpha|^k \leq 2^{2d\tau+10d\lg(d\nu)} \tau^{2d\nu+1} .$$

For the lower bound, we consider R_B as a univariate polynomial, say in z, and let r be its leading coefficient. By ρ_k we denote its roots. If apply Lemma 2, by assuming that ρ_1 is closest root to L_α, then

$$|R_B(L_\alpha)| > |r|^7 \, |L_\alpha - \rho_1|^6 \, \mathcal{M}(R_B)^{-6} \, 2^{\lg \prod_k \Delta_k(R_B) - 6} .$$

It holds $|r| \geq 1$, $\mathcal{M}(R_B) \leq 2^{\widetilde{O}(d\tau)}$, and $-\lg \prod_k \Delta_k(R_B) = O(d^2\nu\tau + d^2\nu \lg(d\nu))$. We also use Theorem 3

$$|L_\alpha - \rho_1| \geq \exp(-O(d^4\nu^4 m^4 \tau(d\nu + d\tau + \lg(d\nu m\tau))^2)) .$$

By combining all the inequalities we get

$$|R_B(L_\alpha)| \geq \exp(-O(d^4\nu^4 m^4 \tau(d\nu + d\tau + \lg(d\nu m\tau))^2)) .$$

Equation (5) with the previous inequality and Lemma 5 imply

$$2^{-\widetilde{O}(d^6\nu^4 m^4 \tau(\nu^2+\tau^2))} \leq |\mathsf{disc}(B_\alpha)| \leq 2^{\widetilde{O}(d\nu + d\tau + m^4\nu^4\tau(\tau^2+\nu^2))} ,$$

which concludes the proof. □

We combine Lemma 5, 6, and Eq. (4) with Proposition 1 to derive the following (separation) bounds for B_α.

Lemma 7 *Let B_α be as in Problem* 1 *and β_i be its roots. Let K be any subset of the roots of B_α, then*

$$2^{-\widetilde{O}(m^4\nu^4\tau(\tau^2+\nu^2))} \leq |\beta_i| \leq 2^{\widetilde{O}(m^4\nu^4\tau(\tau^2+\nu^2))} ,$$

$$\Sigma(B_\alpha) = -\lg \prod_{i \in K} \Delta(\beta_i) = \widetilde{\mathcal{O}}(d^6 \nu^4 m^4 \tau(\nu^2 + \tau^2)) \ .$$

3.1 Isolating the Real Roots of B_α

The main idea behind the algorithm for isolating the real roots of B_α is to approximate its coefficients up to a specified accuracy so that the resulting approximate polynomial, $\widetilde{B}_\alpha$, has real roots that are close to the real roots of B_α. We isolate the real roots of $\widetilde{B}_\alpha$ and the approximation is such that it guarantees that the resulting isolating intervals are also isolating intervals for the real roots of B_α. Several approaches are known in this context [26, 30, 33, 34], we follow [27, Theorem 3].

We divide by the leading coefficient to make the polynomial monic. As stated in Problem 1, the polynomials $b_i \in \mathbb{Z}[y]$ have coefficients of maximum bitsize bounded by τ and degree bounded by ν.

Let σ be such that $\left|\frac{b_i(L_\alpha)}{b_d(L_\alpha)}\right| \leq 2^\sigma$ and ρ such that $\rho = \max_j\{1, \max\{1, |\log|\beta_j||\}\}$, that is a logarithmic root bound for the roots of B_α.

If we approximate the coefficients of B_α up to accuracy $\mathcal{O}(d\rho + \Sigma(B_\alpha))$, then we can approximate the roots (of $\widetilde{B}_\alpha$) in $\widetilde{\mathcal{O}}_B(d^3 + d^2\sigma + d\Sigma(B_\alpha))$. In this way the number of real roots of $\widetilde{B}_\alpha$ is the same as the number of real roots of B_α. Moreover, from the isolating intervals of $\widetilde{B}_\alpha$ we can derive isolating intervals for the roots of B_α. We refer the reader to [27] for a comprehensive treatment.

We bound the various quantities. Lemma 7 indicates that

$$\Sigma(B_\alpha) = \widetilde{\mathcal{O}}(d^6 \nu^4 m^4 \tau(\nu^2 + \tau^2)) \ . \tag{6}$$

To bound σ we use Lemma 5 and so, for all i, $\left|\frac{b_i(L_\alpha)}{b_d(L_\alpha)}\right| \leq 2^{\widetilde{\mathcal{O}}(m^4 \nu^4 \tau (\tau^2 + \nu^2))}$. And thus

$$\sigma = \widetilde{\mathcal{O}}(m^4 \nu^4 \tau (\tau^2 + \nu^2)) \ . \tag{7}$$

The same bound holds for ρ. Hence we need to approximate the coefficients of B_α up to accuracy

$$\widetilde{\mathcal{O}}(d\rho + \Sigma(B_\alpha)) = \widetilde{\mathcal{O}}(d^6 \nu^4 m^4 \tau(\nu^2 + \tau^2)) \ .$$

We can isolate the real roots in $\widetilde{\mathcal{O}}(d^3 + d^7 \nu^4 m^4 \tau(\nu^2 + \tau^2))$.

It remains to estimate the cost of obtaining the approximation on the coefficients of B_α, that is successive approximations of $b_i(L_\alpha)/b_d(L_\alpha)$ up to accuracy of $\mathcal{O}(d\rho + \Sigma(B_\alpha))$ bits after the binary point. Since $|b_i(L_\alpha)/b_d(L_\alpha)| \leq 2^\sigma$, to approximate each fraction, for $0 \leq i \leq d-1$, to desired accuracy ℓ, it is sufficient to approximate $b_i(L_\alpha)$, for $0 \leq i \leq d$, up to precision $\mathcal{O}(\ell + \sigma)$.

The algorithm requires approximation of $b_i(L_\alpha)$, for $0 \leq i \leq d$, to precision $\mathcal{O}(d\rho + \Sigma(B_\alpha) + \sigma)$. Hence, it is sufficient to approximate $b_i(L_q)$ to accuracy $\widetilde{\mathcal{O}}(d^6\nu^4m^4\tau(\nu^2+\tau^2))$.

Approximation of L_α to accuracy of $t > 0$ bits yields an approximation of $b_{i,j}L_\alpha^j$ to accuracy of at least

$$t - \lg|b_{i,j}| - \lg(j) - (j-1)\lg|2L_\alpha| \geq t - \tau - \lg(\nu) - \nu(\lg(\tau)+1)$$

bits and an approximation of $b_i(L_\alpha)$ to accuracy of at least $t - \tau - 2\lg(\nu) - \nu(\lg(\tau)+1)$ bits. Therefore, we need an approximation of $\lg(\alpha)$ up to $t = \widetilde{\mathcal{O}}(d^6\nu^4m^4\tau(\nu^2+\tau^2))$ bits.

For this we need to approximate α up to this accuracy and then evaluate $\lg(\alpha)$. The cost of the first operation is $\widetilde{\mathcal{O}}_B(m^2\tau + m\,t)$ [31]. The cost of approximating the logarithm up to t bits is quasi-linear $\widetilde{\mathcal{O}}_B(t)$ [7], see also [8] and references therein.

After we have obtained the approximation of L_α, say $\tilde{L}$ we need construct the approximated coefficients of B_α by evaluating the polynomials b_i (of degree ν) at $\tilde{L}$; there are $d+1$ polynomials. Each evaluation costs $\widetilde{\mathcal{O}}_B(\nu t)$ [5] and so the overall cost is $\widetilde{\mathcal{O}}_B(d\nu t) = \widetilde{\mathcal{O}}_B(d^7\nu^5m^4\tau(\nu^2+\tau^2))$.

Theorem 8 *The Boolean complexity of isolating the real roots of B_α of Problem 1 is $\widetilde{\mathcal{O}}(d^7\nu^5m^4\tau(\nu^2+\tau^2))$.*

If we want to drop the assumption that the polynomial B_α is square-free, then we can apply a subresultant-based algorithm to compute its square-free part, or its square-free factorization [38]. To apply such algorithms we need to check if the leading coefficient of the polynomial in the subresultant sequence is zero or not. These coefficients are polynomials in $\mathbb{Z}[L_\alpha]$. Therefore, the basic operation needed is to compute the sign of a univariate polynomial evaluated at the logarithm of an algebraic number. To accomplish such an operation we need the bounds of Lemma 5. The exact complexity of the complete algorithm for square-free factorization of polynomials in $\mathbb{Z}[L_\alpha][x]$ is beyond the scope of this paper.

4 Experiments

We present experimental results for an implementation of the algorithm isolating roots of the polynomial in Problem 1 in the special case where the algebraic number α is a rational. The algorithm has been implemented in C as a part of the *Mathematica* system. We have implemented the modified version of Descartes' algorithm due to Sagraloff [33], that applies to polynomials with bitstream coefficients, see also [15, 26], and we adapted our bounds to it. The theoretical complexity of the algorithm is worse by factor than the complexity the algorithm that we presented in the previous section, but its implementation is much easier.

Table 1 Uniformly distributed coefficients, $\tau = 10$

d	ν						
	10	20	50	100	200	500	1000
10	0.006	0.011	0.027	0.060	0.122	0.358	0.857
20	0.015	0.025	0.058	0.110	0.235	0.678	1.53
50	0.042	0.068	0.142	0.272	0.581	1.61	3.56
100	0.116	0.164	0.339	0.640	1.19	3.14	7.65
200	0.496	0.516	0.900	1.65	2.76	6.41	16.7
500	3.43	4.53	5.30	6.52	10.4	21.5	54.6
1000	25.5	23.1	27.7	36.8	45.7	79.9	173

Table 2 Uniformly distributed coefficients, $\tau = 1000$

d	ν						
	10	20	50	100	200	500	1000
10	0.006	0.011	0.028	0.054	0.120	0.362	0.883
20	0.015	0.026	0.060	0.116	0.237	0.809	1.65
50	0.045	0.072	0.157	0.299	0.671	1.74	3.98
100	0.136	0.200	0.356	0.759	1.37	3.41	7.78
200	0.442	0.605	0.985	1.62	2.84	7.25	17.9
500	4.30	4.48	5.95	7.55	12.6	25.4	60.1
1000	20.5	30.4	30.4	34.8	44.7	81.4	183

The experiments have been run on a 64-bit Windows virtual machine with a 3 GHz Intel Core i7 processor and 6 GB of RAM. The timings are given in seconds. The MATHEMATICA code that we used to perform the experiments is publicly available.[1]

Example 9 (*Random polynomials with uniformly distributed coefficients*) For given values of d, ν, and τ each instance (polynomial) was generated by selecting integer coefficients $b_{i,j}$ randomly w.r.t. the uniform distribution in $\mathbb{Z} \cap [-2^{\tau-1}, 2^{\tau-1}]$ and a positive rational number $\alpha \neq 1$ with $\mathcal{L}(\alpha) \leq \tau$. Each timing is an average for 10 randomly generated problems. The results are in Tables 1 and 2.

Applying a least squares fit to the experimental data yields proportionality of the computation time to $d^{1.4}\nu^{0.8}$. There is very little dependence of the computation time on the value of τ (see also the next section).

Example 10 (*Random polynomials with Gaussian distribution of coefficients*) For given values of d and ν, each problem was generated by setting $\alpha = 3$ and selecting coefficients $b_{i,j}$ as nearest integers to real numbers selected randomly w.r.t. the Gaussian distribution with mean 0 and variance $\binom{d}{i}$. Each result is an average for 100

[1] http://members.wolfram.com/adams/LogRootIsolExamples.txt.

Table 3 Gaussian distribution of coefficients

d	ν						
	10	20	50	100	200	500	1000
10	0.004	0.005	0.013	0.028	0.072	0.290	0.992
	3.20	3.06	3.28	3.14	3.30	3.10	3.22
20	0.013	0.022	0.050	0.109	0.239	0.902	2.07
	4.40	4.18	4.56	4.48	4.66	4.28	4.14
50	0.080	0.118	0.191	0.406	0.794	2.34	5.33
	7.46	7.22	6.74	6.96	7.12	7.06	6.86
100	0.309	0.384	0.596	0.477	1.06	2.03	5.07
	9.92	10.12	9.98	10.12	9.90	10.44	10.02
200	1.75	2.19	2.49	4.10	6.56	9.42	18.8
	13.98	14.02	13.78	14.36	13.98	14.24	13.92
500	32.4	32.9	34.4	35.9	39.9	51.7	88.5
	22.92	22.50	22.46	22.10	21.92	22.72	22.80

randomly generated problems. For each value d and ν the upper section gives the computation time and the lower section gives the number of real roots. The results are in Table 3.

Applying a least squares fit to the experimental data yields proportionality of the computation time to $d^{1.7}\nu^{0.8}$. The average number of roots is, as expected, close to $\sqrt{d}$.

4.1 Random Polynomials

We were not able to construct polynomials that achieve the separation bounds of Lemma 7. It is not clear whether the effective lower bounds of Theorem 3 are tight. Our experimental results of the previous section suggest that this is not the case for random polynomials. In addition, this observation triggers the question of estimating the average behavior of the separation bounds. The first step is to estimate the expected number of real roots of B_q, when its coefficients are random variables.

Proposition 11 [14] *Let $v(t) = (f_o(t), \ldots, f_n(t))^\top$ be a vector of differentiable functions and $c_0, \ldots, c_n$ elements of a multivariate normal distribution with zero mean and covariance matrix C. The expected number of real zeros on an interval (or a measurable set) I of the equation $c_0 f_0(t) + \cdots + c_n f_n(t) = 0$, for $w(t) = C^{1/2}v(t)$, is*

$$\int_I \frac{1}{\pi}\|\mathbf{w}'(t)\|dt, \quad \mathbf{w} = w(t)/\|w(t)\|.$$

In logarithmic derivative notation it is

$$\frac{1}{\pi}\int_I \sqrt{\frac{\partial^2}{\partial x\,\partial y}\log\left(v(x)^\top C v(y)\right)|_{x=y=t}}dt.$$

We fix a logarithm L. For example $L = \lg(q)$ for a (fixed) positive rational number q, different from 0 and 1, or $L = \lg(\alpha)$, where α is a positive real algebraic number. Consider the polynomials $b_i = \sum_{j=0}^{\nu} b_{i,j} L^j$ where each of $b_{i,j}$ is a Gaussian random variable with mean zero and variance $\binom{d}{i}$. We denote this by $b_{i,j} \sim N(0, \binom{d}{i})$. Then

$$b_i \sim N(0, \binom{d}{i}\sum_{j=0}^{\nu} L^{2j}) = N(0, \binom{d}{i}\ell) .$$

In our case $v(x)^\top C v(y) = \ell(1 + x\,y)^d$, and the integral of Proposition 11 yields

$$\frac{1}{\pi}\int_{-\infty}^{\infty} \sqrt{\frac{\partial^2}{\partial x\,\partial y}\log \ell(1 + x\,y)^d|_{x=y=t}}\,dt = \sqrt{d}\ .$$

This leads to the following lemma:

Lemma 12 *Let B_α as in Problem* 1 *with a fixed α. Let all b_i have the same degree ν and $b_{i,j} \sim N(0, \binom{d}{i})$. Then the expected number of real roots of B_q is $\sqrt{d}$.*

Following, mutatis mutandis, the analysis of [16, Lemma 3.2] the previous lemma allows us to compute the distribution of the real roots and eventually to estimate the expected separation bound, which is $\mathrm{E}[-\lg \Delta(B_\alpha)] = \mathcal{O}(\lg d)$ (for the aforementioned distribution of the coefficients), *for the real roots*. This is far from the worst case proved in Lemma 7 but agrees with the running times of our implementation in Sect. 4. The bigger the (actual) separation bound, the less bits we need to isolate the real roots, and so the faster the algorithms perform. For estimating the expected separation bounds for the complex roots, we need to compute (expected) lower bounds on the discriminant. We are not aware of such bounds.

5 A Generalization

We present a generalization of Problem 1 where the argument of the logarithm is a homogeneous bivariate polynomial evaluated at two real algebraic numbers. As in the case of Problem 1 we rely on Theorem 3 for computing the various upper and lower bounds.

The precise problem definition is as follows:

Problem 2 Consider the square-free $B_H = \sum_{i=0}^{d} b_i\, x^i$, where $b_i = \sum_{j=0}^{\nu} b_{i,j}$ $(\lg(A(\gamma_1, \gamma_2)))^j$, $b_{i,j} \in \mathbb{Z}$, $\mathcal{L}\left(b_{i,j}\right) \leq \tau$, $A \in \mathbb{Z}[y_1, y_2]$ is a homogeneous polynomial of degree m and $\mathcal{L}(A) = \tau$ and γ_1, resp. γ_2, is a real root of a polynomial $C_1 \in \mathbb{Z}[y]$, resp. $C_2 \in \mathbb{Z}[y]$, of degree n and $\mathcal{L}\left(C_{\{1,2\}}\right) = \tau$. We assume $A(\gamma_1, \gamma) > 0$ and $A(\gamma_1, \gamma_2) \neq 1$. What is the Boolean complexity of isolating the real roots of B_H?

We should warn the reader that the constants in the various bounds in the sequel are not the best possible.

Lemma 13 *Let $A \in \mathbb{Z}[y_1, y_2]$ be a homogeneous polynomial of degree m and $\mathcal{L}(A) = \tau$ and γ_1, resp. γ_2, be the positive real root of a polynomial $C_1 \in \mathbb{Z}[y]$, resp. $C_2 \in \mathbb{Z}[y]$, that is of degree n and $\mathcal{L}(C) = \tau$. Then $2^{-3n^2\tau - 5n^2 \lg(mn) - 4m\tau} \leq |\lg A(\gamma_1, \gamma_2)| \leq 4m\tau$.*

Proof Assume for the moment that we know positive integers t and T such that $|A(\gamma_1, \gamma_2)| \leq 2^T$ and $|A(\gamma_1, \gamma_2) - 1| \geq 2^{-t}$. Then from the inequality $|e^z - 1| \leq |z| e^{|z|}$ we deduce

$$
\begin{aligned}
|A(\gamma_1, \gamma_2) - 1| &\leq |\ln A(\gamma_1, \gamma_2)| e^{|\ln A(\gamma_1, \gamma_2)|} \Rightarrow \\
|A(\gamma_1, \gamma_2) - 1| &\leq \frac{|\lg A(\gamma_1, \gamma_2)|}{\lg(e)} |A(\gamma_1, \gamma_2)| \Rightarrow \\
2^{-t-1} &\leq |A(\gamma_1, \gamma_2) - 1| \leq |\lg A(\gamma_1, \gamma_2)|\, 2^T \Rightarrow \\
2^{-t-T-1} &\leq |\lg A(\gamma_1, \gamma_2)| \ .
\end{aligned}
$$

It remains to specify t and T. For the real algebraic numbers γ_1 and γ_2 it holds

$$
2^{-\tau-1} \leq |\gamma_{\{1,2\}}| \leq 2^{\tau+1}.
$$

We bound T as follows:

$$
|A(\gamma_1, \gamma_2)| \leq |\sum_{i=0}^{m} a_i \gamma_1^i \gamma_2^{m-i}| \leq \sum_{i=0}^{m} 2^{\tau}\, 2^{m\,\tau} ,
$$

and so

$$
|\lg A(\gamma_1, \gamma_2)| \leq (m+1)\tau + \lg(m+1) = T \ .
$$

We choose $T = 4m\tau = \mathcal{O}(m\tau)$ to simplify the calculations.

To compute a bound for t we consider the polynomial $\bar{A}(y_1, y_2) = A(y_1, y_2) - 1$ and the following polynomial system:

$$
\begin{cases}
F_1 = z - [A(y_1, y_2) - 1] & = 0 \\
F_2 = C_1(y_1) & = 0 \\
F_3 = C_2(y_2) & = 0
\end{cases}
$$

We will use a similar system in the sequel so we present various quantities that are related to it. For further details on DMM we refer the reader to [17].

A lower bound on z provides us with a lower bound for t. To compute a bound for z we use the DMM bound from [17, Theorem 3].

Let $\mathcal{D}$ be the mixed volume of the system, MV_i the mixed volume of the system if we discard the i-th polynomial, and $\#(Q_i)$ the number of integer points of the Newton polytope of the i-th polynomial, for $1 \leq i \leq 3$, $\varrho = \prod_{i=1}^{3} (\#Q_i)^{\mathsf{MV}_i}$, and $\mathcal{C} = \prod_{i=1}^{3} \|F_i\|_\infty^{\mathsf{MV}_i}$.

The univariate polynomial that has the z-coordinates of the solution set of the system as roots, we call them ζ, has degree $\mathcal{D}$ and maximum coefficient bitsize $\varrho 2^{\mathcal{D}} \mathcal{C}$. It holds $|\zeta| \geq (\varrho 2^{\mathcal{D}} \mathcal{C})^{-1}$. In our case

$$\begin{aligned} \mathcal{D} = n^2, \mathsf{MV}_1 &= n^2, \mathsf{MV}_2 = \mathsf{MV}_3 = n, \\ (\#Q_1) = m+1, (\#Q_2) &= (\#Q_3) = n+1, \\ \varrho = (m+1)^{n^2}(n+1)^{2n}, \mathcal{C} &\leq 2^{\tau(n^2+2n)}. \end{aligned}$$

Notice that it is exactly the use of mixed volume that allows us to take $\mathcal{D} = n^2$ instead of mn^2 which is the Bézout bound.

The lower bound for ζ becomes

$$|\zeta| \geq 2^{-(n^2+n^2\lg(m+1)+2n\lg(n+1)+\tau(n^2+2n))} ,$$

and hence

$$t = n^2 + n^2\lg(m+1) + 2n\lg(n+1) + \tau(n^2+2n) .$$

We choose $t = 3n^2\tau + 5n^2\lg(mn) = \widetilde{\mathcal{O}}(n^2\tau)$. □

Lemma 14 *Let b_i be as in Problem 2. If $b_i(L_H) \neq 0$, then $2^{-\widetilde{\mathcal{O}}(n^{10}\nu^4\tau(\tau^2+\nu^2))} \leq |b_i(L_H)| \leq 2^{\widetilde{\mathcal{O}}(\nu+\tau)}$.*

Proof For all i it holds

$$|b_i(L_H)| = |\sum_{j=0}^{\nu} b_{i,j} L_H^j| \leq \sum_{j=0}^{\nu} 2^{\tau}(4m\tau)^j \leq (\nu+1)2^{\tau}(4m\tau)^{\nu} ,$$

and so

$$|b_i(L_H)| \leq 2^{\tau+8\nu\lg(m\tau)} .$$

We consider b_i as a univariate polynomial in y and so $b_i = \sum_{j=0}^{\nu} b_{i,j} y^i = b_{i,\nu}\prod_{j=1}^{\nu}(y-\beta_{i,j})$, where $\beta_{i,j}$ are its roots. Let $\beta_{i,1}$ be the root closest to L_H; we apply Lemma 2

$$|b_i(L_H)| > |b_{i,\nu}|^7 \, |L_H - \beta_{i,1}|^6 \, \mathcal{M}(b_i)^{-6} \, 2^{\lg \prod_j \Delta(b_i) - 6} .$$

It holds $|b_{i,\nu}| \geq 1$, $\mathcal{M}(b_i) \leq 2^{\tau+\lg\nu+1}$, and $-\lg\prod_j \Delta(b_i) = \mathcal{O}(\nu^2 + \nu\tau)$.

To bound $|L_H - \beta_{i,1}|$ we use Theorem 3. For this we need to identify the real algebraic number $A(\gamma_1, \gamma_2)$ represents. Consider the following polynomial system:

$$\begin{cases} F_1 = z - A(y_1, y_2) & = 0 \\ F_2 = C_1(y_1) & = 0 \\ F_3 = C_2(y_2) & = 0 \end{cases}$$

The system is almost identical to the one in the proof of Lemma 13 and so we get all the (worst case) bounds from that system. If we eliminate y_1 and y_2, then we get a univariate polynomial in z among the solutions of which is the real algebraic number $A(\gamma_1, \gamma_2)$. The polynomial has degree n^2 and maximum coefficient bitsize $n^2 + n^2\lg(m+1) + 2n\lg(n+1) + \tau(n^2+2n) = \widetilde{\mathcal{O}}(n^2\tau)$. Then, Theorem 3 implies that

$$|L_H - \beta_{i,j}| \geq \exp(-\mathcal{O}(n^{10}\,\nu^4\,\tau(\tau+\nu+\lg(n\nu\tau))^2)) \ .$$

By combining all the bounds we obtain the bound $|b_i(L_H)| > 2^{-\mathcal{O}(n^{10}\,\nu^4\,\tau(\tau+\nu+\lg(n\nu\tau))^2)}$, which concludes the proof. □

An upper bound for $\|B_H\|_2$ is $\|B_H\|_2^2 = \sum_{i=0}^{d} |b_i(L_H)|^2 \Rightarrow \|B_H\|_2 \leq 2^{\tau+8\nu\lg(m\tau)+\lg(d)}$.

Lemma 15 *Let B_H be as in Problem 2, then*

$$2^{-\widetilde{\mathcal{O}}(d^6 n^8 \nu^4 \tau(\nu^2+\tau^2))} \leq |\mathsf{disc}(B_H)| \leq 2^{\widetilde{\mathcal{O}}(d\nu+d\tau+n^{10}\nu^4\tau(\tau^2+\nu^2))}.$$

Proof As in the proof of Lemma 6 we consider B_H as a bivariate polynomial in $\mathbb{Z}[L_H, x]$, and

$$\begin{aligned} |\mathsf{disc}(B_H)| &= \left| \frac{1}{b_d(L_H)} \mathtt{res}_x(B_H(L_H, x), \partial B_H(L_H, x)/\partial x) \right| \\ &= \left| \frac{1}{b_d(L_H)} R_B(L_H) \right| \ . \end{aligned}$$

The resultant $R_B \in \mathbb{Z}[L_H]$ is a univariate polynomial of degree at most $2d\nu$ and maximum coefficient bitsize $2d\tau + 10d\lg(d\nu)$. Therefore

$$\begin{aligned} |R_B(L_H)| &\leq 2^{2d\tau+10d\lg(d\nu)} \sum_{k=0}^{2d\nu} |L_H|^k \\ &\leq 2^{2d\tau+10d\lg(d\nu)} (4m\tau)^{2d\nu+1} \ . \end{aligned}$$

For the lower bound, let r be the leading coefficient of R_B and ρ_k its roots. Let ρ_1 be the root closest to L_H. Then $|r| \geq 1$, $\mathcal{M}(R_B) \leq 2^{2d\tau+12d\lg(d\nu)}$, $-\lg\prod_k \Delta(R_B) = \mathcal{O}(d^2\nu^2 + d^2\nu\tau)$. The application of Theorem 3 gives us

$$|L_H - \rho_1| \geq \exp(-\widetilde{\mathcal{O}}(d^6 n^8 \nu^4 \tau(\nu^2 + \tau^2))) \ .$$

Using Lemma 2 we get

$$|R_B(L_H)| > |r|^7 \, |L_H - \rho_1|^6 \, \mathcal{M}\,(R_B)^{-6} \, 2^{\lg \prod_k \Delta_k(R_B) - 6} \ ,$$

and thus

$$|R_B(L_H)| \geq \exp(-\widetilde{\mathcal{O}}(d^6 n^8 \nu^4 \tau(\nu^2 + \tau^2))) \ .$$

Combining Eq. (5) with the previous inequality and Lemma 14 we get

$$2^{-\widetilde{\mathcal{O}}(d^6 n^8 \nu^4 \tau(\nu^2+\tau^2))} \leq |\mathsf{disc}(B_H)| \leq 2^{\widetilde{\mathcal{O}}(d\nu + d\tau + n^{10}\nu^4\tau(\tau^2+\nu^2))} \ ,$$

that concludes the proof. □

Lemma 16 *Let B_H be as in Problem 2 and let β_j be its roots. Then*

$$2^{-\widetilde{\mathcal{O}}(n^{10}\nu^4\tau(\tau^2+\nu^2))} \leq |\beta_j| \leq 2^{\widetilde{\mathcal{O}}(n^{10}\nu^4\tau(\tau^2+\nu^2))} \ ,$$

$$\Sigma(B_H) = -\lg \prod_{(i,j)\in\Omega} |\beta_i - \beta_j| \widetilde{\mathcal{O}}(d^6 n^8 \nu^4 \tau(\nu^2 + \tau^2)) \ .$$

When we have two or more logarithms and the polynomials are not homogeneous or if we have homogeneous polynomials and three or more logarithms then we are not able to compute separation bounds. In this case the separation bounds are closely connected to major open problems in number theory, like the *four exponentials conjecture*. For example, no effective lower bounds are known for the expression $|\lg(\alpha_1)\lg(\alpha_2) - \lg(\alpha_3)\lg(\alpha_4)|$, where $\alpha_{\{1,2,3,4\}}$ are (real) algebraic numbers.

5.1 Isolating the Real Roots of B_H

We proceed as in Sect. 3.1 and we use the same notation. We approximate the coefficients of B_H up to accuracy $\mathcal{O}(d\rho + \Sigma(B_H))$ and we isolate the real roots in $\widetilde{\mathcal{O}}_B(d^3 + d^2\sigma + d\Sigma(B_H))$. From Lemma 16 we get $\Sigma(B_H) = \widetilde{\mathcal{O}}(d^6 n^8 \nu^4 \tau(\nu^2 + \tau^2))$. Moreover, $\rho = \widetilde{\mathcal{O}}(n^{10}\nu^4\tau(\tau^2 + \nu^2))$ and $\sigma = \widetilde{\mathcal{O}}(n^{10}\nu^4\tau(\tau^2 + \nu^2))$.

We need to estimate the cost of approximating $b_i(L_H)/b_d(L_H)$ up to accuracy of $\mathcal{O}(d\rho + \Sigma(B_H))$ bits after the binary point. Working as in Sect. 3.1 we deduce that we should approximate $L_H = \lg A(\gamma_1, \gamma_2)$ up to precision 2^{-t}, where $t = \mathcal{O}(d\rho + \Sigma(B_H))$. The cost of this approximation is quasi-linear $\widetilde{\mathcal{O}}_B(t)$ [7].

In addition, we should also approximate $A(\gamma_1, \gamma_2)$ up to this accuracy. Assume that we have isolating intervals $[\gamma_1]$, resp. $[\gamma_2]$, for the real algebraic number γ_1, resp. γ_2. Let their widths be 2^{-s}, where s is a positive integer that we should determine. That is $\mathtt{wid}[\gamma_1] = \mathtt{wid}[\gamma_2] = 2^{-s}$.

Recall that $2^{-\tau} \leq |\gamma_{\{1,2\}}| \leq 2^{\tau}$ and that $A = \sum_{i=0}^{m} a_i y_1^i y_2^{m-i}$ is a homogeneous bivariate polynomial of degree m.

For an expression E, let $[E]$ be its evaluation using interval arithmetic. Using the properties of interval arithmetic [1] we get that $\mathtt{wid}[a_i \gamma_1^i \gamma_2^{m-i}] \leq m 2^{\tau(m-1)} 2^{-s}$, and $\mathtt{wid}[A(\gamma_1, \gamma_2)] \leq m^2 2^{m\tau} 2^{-s} \leq 2^{-t}$, which leads to $s = t + m\tau + 2\lg(m) = \widetilde{\mathcal{O}}(n^8 \nu^5 \tau^3 (n^2 + d^8))$.

We approximate γ_1 and γ_2 up to this accuracy in $\widetilde{\mathcal{O}}_B(n^2\tau + ns) = \widetilde{\mathcal{O}}_B(n^2\tau + nm\tau + nt)$ [31].

It remains to estimate the cost of computing the approximated coefficients of B_H. After we have computed a approximation of L_H, say $\tilde{L}_H$, we need perform the evaluation $b_i(\tilde{L}_H)$; there are $d+1$. Each costs $\widetilde{\mathcal{O}}_B(\nu s)$ and the overall cost is $\widetilde{\mathcal{O}}_B(d\nu s)$.

Combining all the bounds we have the following theorem.

Theorem 17 *The Boolean complexity of isolating the real roots of* B_H *of Problem 2 is* $\widetilde{\mathcal{O}}_B(n^9 \nu^4 d^2 \tau (\tau^2 + \nu^2)(n^2 + d^5) + m\tau(n + d\nu))$.

6 An Extension to Bivariate Polynomial Systems

In this section we consider bivariate polynomial systems. Let $L = L_q$ or $L = L_H$ (Sects. 3, 5, respectively). The problem statement is as follows:

Problem 3 Consider the zero-dimensional, polynomial system (S_L) $F_1(x, y) = F_2(x, y) = 0$, where $F_{1,2} \in (\mathbb{Z}[L])[x_1, x_2]$ and their total degree is bounded by d. Let $L = L_q = \lg(q)$, resp. $L = L_H = \lg A(\gamma_1, \gamma_2)$, be as in Problem 1, resp. Problem 2. The coefficients of F_1 and F_2 are polynomials in L of degree ν and maximum coefficient bitsize at most τ. What is the Boolean complexity of isolating the real roots of (S_L)?

The complexity of the algorithms for solving bivariate polynomial systems depends heavily on the separation bound of the system. We present the separation bounds and we sketch the analysis of isolation process. We use the DMM bound [17]. Consider the polynomial system

$$(S_0) \quad F_1(x_1, x_2) = F_2(x_1, x_2) = u - x_1 = 0,$$

where u is a parameter. If we eliminate x_1 and x_2 from (S_0), then we get a univariate polynomial in u, $R_1 \in (\mathbb{Z}[L])[u]$, which is called the u-resultant. The DMM bound bounds the separation of S_L using the separation bound of R_1. Asymptotically, the latter depends on a lower bound on the discriminant of its square-free part [17, Theorem 3]. Hence, it suffices to estimate this bound.

The coefficients of R_1 are of the form $\varrho_k\, c_1^d\, c_2^d\, u^k$, where $0 \leq k \leq d^2$, $c_{\{1,2\}}$ denotes a monomial in the coefficients of $F_{\{1,2\}}$ of total degree d, and ϱ_k is an integer that depends on the integer points of the Newton polytopes of the polynomials and in our case is bounded by $|\varrho_k| \leq (d^2+2)^{2d}$. The degree of R_1 wrt u is $\mathcal{O}(d^2)$.

Recall that the coefficients of $F_{\{1,2\}}$ are polynomials in L. Thus, the coefficients of R_1 are also polynomials in L of degree at most $2d\nu$ and maximum coefficient bitsize $\widetilde{\mathcal{O}}(d\tau)$. If we compute the square-free part of R_1, then its coefficients are polynomials of degree bounded by $2d\nu$ and of maximum coefficient bitsize bounded by $2d\tau + 10d\lg(d) = \widetilde{\mathcal{O}}(d\tau)$ [43]. If $L = L_\alpha$ then we apply Lemma 6 and the logarithm of the separation bound of the system is $\widetilde{\mathcal{O}}(d^7\nu^{10}m^4\tau(\nu^2 + d^2\tau^2))$. We can obtain a similar bound if $L = L_H = \lg A(\gamma_1, \gamma_2)$ and we apply Lemma 15. In both cases, it seems that the bounds are quite pessimistic. We can also obtain the bounds by modifying accordingly the DMM bound [17].

To compute R_1 (or R_2 if we choose to eliminate x_2) we treat L as a new variable. The projection on x_1, that is the computation of R_1 costs $\widetilde{\mathcal{O}}_B(d^5\nu\tau)$ [12, Prop. 8 and Lemma 9]. The cost is the same for projection on the x_2-axis. Using the previous bounds and the results of Sects. 3.1 and 5.1 we can isolate the roots of the two projections. For the L_α case this cost is $\widetilde{\mathcal{O}}_B(d^2\nu^4 + d^8\nu^{12}m^4\tau(\nu^2 + d^2\tau^2))$. It remains to match the x_1 and x_2 coordinates. For example, we can use one of the three strategies in [12]. The main operation needed is the computation of sign of a univariate polynomial like B_α evaluated at a real algebraic number. We postpone the detailed analysis for a future communication

Another way to solve the system is to approximate L up to an accuracy, substitute this value to the polynomials $F_{\{1,2\}}$, and then solve the system. We need a perturbation bound for the roots of a bivariate system, similar to the one(s) for univariate polynomials [34, Theorem 19.1].

Theorem 18 *Consider a zero-dimensional polynomial system $F = 0$, where $F = (F_1, F_2)$ and $F_{\{1,2\}}$ are bivariate polynomials of degree d. The roots of system are contained in a disc with center the origin and radius r. Let $\widetilde{F} = (\widetilde{F}_1, \widetilde{F}_2)$ be a λ approximation of F, that is $\|F_i - \widetilde{F}_i\|_\infty \leq 2^{-\lambda}$. Then the zeros of F, $\alpha_1, \ldots, \alpha_{d^2}$, and the zeros of $\widetilde{F}$, $\widetilde{\alpha}_1, \ldots, \widetilde{\alpha}_{d^2}$, could be numbered such that, for $j \in [n]$,*

$$|\alpha_j - \widetilde{\alpha}_j| \leq 2^{\eta+1},$$

where $\eta = -\lambda/d^2 + 2\tau/d^2 + 12\lg(2d))/d + 4\lg(d)/d^2 + \lg(r) + 2$.

Proof We consider the polynomial system (S_0) and its resultant, R, after eliminating x_1 and x_2. The coefficients of R are of the form $\varrho_k\, c_1^d\, c_2^d\, u^k$, where $0 \leq k \leq d^2$, $c_{\{1,2\}}$ denotes a monomial in the coefficients of $F_{\{1,2\}}$ of total degree d, and $|\varrho_k| \leq (d^2+2)^{2d}$. The degree of R wrt u is $\mathcal{O}(d^2)$.

If we replace the polynomials $F_{\{1,2\}}$ by it approximations $\widetilde{F}_{\{1,2\}}$ and compute the resultant of the perturbed system, $\widetilde{R}$, this is also a polynomial in u of degree $\mathcal{O}(d^2)$. Its terms are of the form $\varrho_k\, \widetilde{c}_1^d\, \widetilde{c}_2^d\, u^k$, where $0 \leq k \leq d^2$, $\widetilde{c}_{\{1,2\}}$ denotes a monomial in the coefficients of $\widetilde{F}_{\{1,2\}}$ of total degree d, and ϱ_k is as before.

The inequality $\|F_i - \widetilde{F}_i\|_\infty \leq 2^{-\lambda}$ implies $\|R - \widetilde{R}\|_\infty \leq 2^{-\lambda+2d\tau+12d\lg(2d)}$.

Let $\alpha_{j,1}$, for $j \in [d^2]$, be the roots of R, and respectively $\widetilde{\alpha}_{j,1}$ the roots of $\widetilde{R}$. Recall that the roots of R are the x_1 coordinates of the system. Using [34, Theorem 19.1] we have the following inequality: $|\alpha_{i,1} - \widetilde{\alpha}_{i,1}| \leq 2^\eta$ where $\eta = -\lambda/d^2 + 2\tau/d^2 + 12\lg(2d))/d + 4\lg(d)/d^2 + \lg(r) + 2$.

We obtain the same bound if we replace $u - x_1$ with $u - x_2$ in (S_0). Thus, for any root α_j of F and $\widetilde{\alpha}_i$ of $\widetilde{F}$ we have $|\alpha_i - \widetilde{\alpha}_i| \leq 2^{\eta+1}$. □

Using the previous theorem we can mimic the procedure of the univariate case. We estimate the separation bound of (S_0) as presented in the beginning of the section. Next, we approximate L to an accuracy of this order, and we obtain two approximate polynomials, and thus a perturbed system. We solve the approximate system using a numerical subdivision solver, e.g., [24], and from the isolating boxes of the perturbed system we can derive isolating boxes for the roots of (S_0) by applying Theorem 18.

A possible alternative way of solving the bivariate polynomial systems of Problem 3 could be based on the recent work [6] on solving bivariate polynomial system of polynomials having integer coefficients.

The following is an alternative version of Lemma 2.

Lemma 19 *Let $L \in \mathbb{C}$ and γ_1 the root of the square-free polynomial f that is closest to L. Then*

$$|f(L)| \geq |a_d|^2 \, |L - \gamma_1| \, 2^{-d} \, \mathcal{M}(f) \, 2^{\lg \prod_j \Delta_j} \, .$$

Proof As γ_1 is the root closest to L it holds $|L - \gamma_i| \geq |\gamma_1 - \gamma_i|/2$. Then

$$\begin{aligned}
|f(L)| &= |a_d| \prod_{j=1}^{d} |L - \gamma_j| = |a_d||L - \gamma_1| \prod_{j \neq 1} |L - \gamma_j| \\
&\geq |a_d||L - \gamma_1| \prod_{j \neq 1} |\gamma_1 - \gamma_j|/2 \\
&\geq |a_d| \, |L - \gamma_1| \, 2^{1-d} \prod_{j \neq 1} |\gamma_1 - \gamma_j| \\
&\geq |a_d| \, |L - \gamma_1| \, 2^{1-d} \prod_{j \neq 1} \Delta_j \\
&\geq |a_d| \, |L - \gamma_1| \, 2^{1-d} \frac{1}{\Delta_1} \prod_{j} \Delta_j \\
&\geq |a_d| \, |L - \gamma_1| \, 2^{1-d} \frac{|a_d|}{2\mathcal{M}(f)} 2^{\lg \prod_j \Delta_j} \, .
\end{aligned}$$

For the last inequality we use $\Delta_i \leq 2\,\mathcal{M}(f)\,/|a_d|$, that in turn relies on $\Delta_i = |\gamma_i - \gamma_{c_i}| \leq |\gamma_i| + |\gamma_{c_i}| \leq 2\,\mathcal{M}(f)\,/|a_d|$. □

Acknowledgements Both authors would like to thank an anonymous referee for her, or his, very detailed comments that greatly improved the presentation of the results. Elias Tsigaridas is partially supported by the French National Research Agency (ANR-09-BLAN-0371-01), GeoLMI (ANR 2011 BS03 011 06), HPAC (ANR ANR-11-BS02-013) and an FP7 Marie Curie Career Integration Grant.

References

1. Alefeld, G., Herzberger, J.: Introduction to Interval Computations. Academic Press, New York (1983)
2. Baker, A.: Linear forms in the logarithms of algebraic numbers (IV). Mathematika **15**(02), 204–216 (1968)
3. Baker, A.: The theory of linear forms in logarithms. In: Transcendence Theory: Advances and Applications, pp. 1–27 (1977)
4. Bates, D.J., Sottile, F.: Khovanskii-Rolle continuation for real solutions. Found. Comput. Math. **11**(5), 563–587 (2011)
5. Bodrato, M., Zanoni, A.: Long integers and polynomial evaluation with Estrin's scheme. In: Proceedings of the 11th International Symposium on Symbolic and Numeric Algorithms for Scientific Computing (SYNASC), pp. 39–46. IEEE (2011)
6. Bouzidi, Y., Lazard, S., Moroz, G., Pouget, M., Rouillier, F., Sagraloff, M.: Solving bivariate systems using rational univariate representations. J. Complex. **37**, 34–75 (2016)
7. Brent, R.: Fast multiple-precision evaluation of elementary functions. J. ACM **23**(2), 242–251 (1976)
8. Brent, R., Zimmermann, P.: Modern Computer Arithmetic. Cambridge University Press, Cambridge (2010)
9. Cheng, J.-S., Gao, X.-S., Yap, C.-K.: Complete numerical isolation of real roots in zero-dimensional triangular systems. J. Symb. Comput. **44**, 768–785 (2009)
10. Davenport, J.H.: Cylindrical algebraic decomposition. Technical Report 88–10, School of Mathematical Sciences, University of Bath, England (1988). http://www.bath.ac.uk/masjhd/
11. Dedieu, J., Yakoubsohn, J.: Computing the real roots of a polynomial by the exclusion algorithm. Numer. Algorithms **4**(1), 1–24 (1993)
12. Diochnos, D.I., Emiris, I.Z., Tsigaridas, E.P.: On the asymptotic and practical complexity of solving bivariate systems over the reals. J. Symb. Comput. **44**(7), 818–835 (2009)
13. Du, Z., Sharma, V., Yap, C.K.: Amortized bound for root isolation via Sturm sequences. In: Wang, D., Zhi, L. (eds.) International Workshop on Symbolic Numeric Computing (SNC), pp. 113–129. Birkhauser, Beihang University, Beijing (2005)
14. Edelman, A., Kostlan, E.: How may zeros of a random polynomial are real? Bull. AMS **32**(1), 1–37 (1995)
15. Eigenwillig, A., Kettner, L., Krandick, W., Mehlhorn, K., Schmitt, S., Wolpert, N.: A descartes algorithm for polynomials with bit-stream coefficients. In: Ganzha, V., Mayr, E., Vorozhtsov, E. (eds.) CASC. LNCS, vol. 3718, pp. 138–149. Springer (2005)
16. Emiris, I.Z., Galligo, A., Tsigaridas, E.P.: Random polynomials and expected complexity of bisection methods for real solving. In: Watt, S. (ed.) Proceedings of the 35th ACM International Symposium on Symbolic & Algebraic Computation (ISSAC), pages 235–242, Munich, Germany. ACM, New York (2010)
17. Emiris, I.Z., Mourrain, B., Tsigaridas, E.P.: The DMM bound: multivariate (aggregate) separation bounds. In: Proceedings of the 35th ACM International Symposium on Symbolic & Algebraic Computation (ISSAC), pp. 243–250, Munich, Germany. ACM (2010)
18. Eppstein, D., Paterson, M.S., Yao, F.F.: On nearest-neighbor graphs. Discret. Comput. Geom. **17**(3), 263–282 (1997)
19. Escorcielo, P., Perrucci, D.: On the davenport-mahler bound (2016). arXiv preprint arXiv:1606.06572
20. Giusti, M., Lecerf, G., Salvy, B., Yakoubsohn, J.-C.: On location and approximation of clusters of zeros of analytic functions. Found. Comput. Math. **5**(3), 257–311 (2005)
21. Johnson, J., Krandick, W.: Polynomial real root isolation using approximate arithmetic. In: Proceedings of the Intertnational Symposium on Symbolic and Algebraic Computation (ISSAC), pp. 225–232. ACM (1997)
22. Johnson, J.R.: Algorithms for polynomial real root isolation. PhD thesis, The Ohio State University (1991)

23. Lu, Z., He, B., Luo, Y., Pan, L.: An algorithm of real root isolation for polynomial system. In: Wang, D., Zhi, L. (eds.) Proceedings of the 1st ACM International Work. Symbolic Numeric Computation (SNC), pp. 94–107 (2005)
24. Mantzaflaris, A., Mourrain, B., Tsigaridas, E.P.: On continued fraction expansion of real roots of polynomial systems, complexity and condition numbers. Theor. Comput. Sci. **412**(22), 2312–2330 (2011)
25. McNamee, J.M., Pan, V.Y.: Numerical Methods for Roots of Polynomials (II), chapter 15. Elsevier (2013)
26. Mehlhorn, K., Sagraloff, M.: A deterministic algorithm for isolating real roots of a real polynomial. J. Symb. Comput. **46**(1), 70–90 (2011)
27. Mehlhorn, K., Sagraloff, M., Wang, P.: From approximate factorization to root isolation with application to cylindrical algebraic decomposition. J. Symb. Comput. **66**, 34–69 (2015)
28. Mignotte, M.: Mathematics for Computer Algebra. Springer, New York (1992)
29. Mignotte, M., Waldschmidt, M.: Linear forms in two logarithms and Schneider's method. Math. Ann. **231**(3), 241–267 (1978)
30. Pan, V.: Univariate polynomials: nearly optimal algorithms for numerical factorization and rootfinding. J. Symb. Comput. **33**(5), 701–733 (2002)
31. Pan, V.Y., Tsigaridas, E.P.: On the boolean complexity of real root refinement. In: Kauers, M. (ed.) Proc. Int'l Symp. on Symb. and Algebraic Comp. (ISSAC), pp. 299–306, Boston, USA. ACM (2013)
32. Rouillier, F., Zimmermann, Z.: Efficient isolation of polynomial's real roots. J. Comput. Appl. Math. **162**(1), 33–50 (2004)
33. Sagraloff, M.: On the complexity of real root isolation. abs/1011.0344v1 (2010)
34. Schönhage, A.: The fundamental theorem of algebra in terms of computational complexity. Manuscript. Univ. of Tübingen, Germany, 1982. http://www.iai.uni-bonn.de/~schoe/fdthmrep.ps.gz
35. Strzeboński, A., Tsigaridas, E.P.: Univariate real root isolation in an extension field. In: Leykin, A. (ed.) Proc. 36th ACM Int'l Symp. on Symbolic & Algebraic Comp. (ISSAC), pp. 321–328, San Jose, CA, USA. ACM (2011)
36. Strzeboński, A., Tsigaridas, E.P.: Univariate real root isolation in multiple extension fields. In: Proc. 37th ACM Int'l Symp. on Symbolic & Algebraic Comp. (ISSAC), pp. 343–350, Grenoble, France. ACM (2012)
37. Tsigaridas, E.P., Emiris, I.Z.: On the complexity of real root isolation using continued fractions. Theor. Comput. Sci. **392**, 158–173 (2008)
38. von zur Gathen, J., Gerhard, J.: Modern Computer Algebra, 3rd edn. Cambridge University Press, New York (2013)
39. Xia, B., Yang, L.: An algorithm for isolating the real solutions of semi-algebraic systems. J. Symb. Comput. **34**, 461–477 (2002)
40. Xia, B., Zhang, T.: Real solution isolation using interval arithmetic. Comput. Math. Appl. **52**, 853–860 (2006)
41. Yakoubsohn, J.: Approximating the zeros of analytic functions by the exclusion algorithm. Numer. Algorithms **6**(1), 63–88 (1994)
42. Yakoubsohn, J.-C.: Numerical analysis of a bisection-exclusion method to find zeros of univariate analytic functions. J. Complex. **21**(5), 652–690 (2005)
43. Yap, C.: Fundamental Problems of Algorithmic Algebra. Oxford University Press, New York (2000)

On the Complexity of Multivariate Polynomial Division

Joris van der Hoeven

Abstract In this paper, we present a new algorithm for reducing a multivariate polynomial with respect to an autoreduced tuple of other polynomials. In a suitable sparse complexity model, it is shown that the execution time is essentially the same (up to a logarithmic factor) as the time needed to verify that the result is correct.

Keywords Sparse reduction · Complexity · Division · Algorithm

A.M.S. Subject Classification: 68W30 · 12Y05 · 68W40 · 13P10

1 Introduction

Sparse interpolation [1, 2, 5, 13] provides an interesting paradigm for efficient computations with multivariate polynomials. In particular, under suitable hypothesis, multiplication of sparse polynomials can be carried out in quasi-linear time, in terms of the expected output size. More recently, other multiplication algorithms have also been investigated, which outperform naive and sparse interpolation under special circumstances [12, 14]. An interesting question is how to exploit such asymptotically faster multiplication algorithms for the purpose of polynomial elimination. In this paper, we will focus on the reduction of a multivariate polynomial with respect to an autoreduced set of other polynomials and show that fast multiplication algorithms can indeed be exploited in this context in an asymptotically quasi-optimal way.

Consider the polynomial ring $\mathbb{K}[x] = \mathbb{K}[x_1, \ldots, x_n]$ over an effective field $\mathbb{K}$ with an effective zero test. Given a polynomial

This work has been supported by the ANR-09-JCJC-0098-01 MAGIX project, the Digiteo 2009-36HD grant and Région Ile-de-France.

J. van der Hoeven (✉)
LIX, CNRS, École polytechnique, 91128 Palaiseau Cedex, France
e-mail: vdhoeven@lix.polytechnique.fr
URL: http://lix.polytechnique.frvdhoeven

I.S. Kotsireas and E. Martínez-Moro (eds.), *Applications of Computer Algebra*, Springer Proceedings in Mathematics & Statistics 198,
DOI 10.1007/978-3-319-56932-1_28

$$P = \sum_{i \in \mathbb{N}^n} P_i x^i = \sum_{i_1,\ldots,i_n \in \mathbb{N}} P_{i_1,\ldots,i_n} x_1^{i_1} \cdots x_n^{i_n},$$

we call $\operatorname{supp} P = \{i \in \mathbb{N}^n : P_i \neq 0\}$ the *support* of P. The naive multiplication of two sparse polynomials $P, Q \in \mathbb{K}[x]$ requires a priori $\mathcal{O}(|\operatorname{supp} P||\operatorname{supp} Q|)$ operations in $\mathbb{K}$. This upper bound is sharp if P and Q are very sparse, but pessimistic if P and Q are dense.

Assuming that $\mathbb{K}$ has characteristic zero, a better algorithm was proposed in [2] (see also [1, 5] for some background). The complexity of this algorithm can be expressed in the expected size $s = |\operatorname{supp} P + \operatorname{supp} Q|$ of the *output* (when no cancellations occur). It is shown that P and Q can be multiplied using only $\mathcal{O}(\mathsf{M}(s) \log s)$ operations in $\mathbb{K}$, where $\mathsf{M}(s) = \mathcal{O}(s \log s \log \log s)$ stands for the complexity of multiplying two univariate polynomials in $\mathbb{K}[z]$ of degrees $<s$. Unfortunately, the algorithm in [2] has two drawbacks:

1. The algorithm leads to a big growth for the sizes of the coefficients, thereby compromising its bit complexity (which is often worse than the bit complexity of naive multiplication).
2. It requires $\operatorname{supp} PQ \subseteq \operatorname{supp} P + \operatorname{supp} Q$ to be known beforehand. More precisely, whenever a bound $\operatorname{supp} PQ \subseteq \operatorname{supp} P + \operatorname{supp} Q \subseteq \mathcal{S}$ is known, then we really obtain a multiplication algorithm of complexity $\mathcal{O}(\mathsf{M}(|\mathcal{S}|) \log |\mathcal{S}|)$.

In practice, the second drawback is of less importance. Indeed, especially when the coefficients in $\mathbb{K}$ can become large, then the computation of $\operatorname{supp} P + \operatorname{supp} Q$ is often cheap with respect to the multiplication PQ itself, even if we compute $\operatorname{supp} P + \operatorname{supp} Q$ in a naive way.

Recently, several algorithms were proposed for removing the drawbacks of [2]. First of all, in [13] we proposed a practical algorithm with essentially the same advantages as the original algorithm from [2], but with a good bit complexity and a variant which also works in positive characteristic. However, it still requires a bound for $\operatorname{supp} PQ$ and it only works for special kinds of fields $\mathbb{K}$ (which nevertheless cover the most important cases such as $\mathbb{K} = \mathbb{Q}$ and finite fields). Even faster algorithms were proposed in [9, 14], but these algorithms only work for special supports. Yet another algorithm was proposed in [7, 12]. This algorithm has none of the drawbacks of [2], but its complexity is suboptimal (although better than the complexity of naive multiplication).

At any rate, these recent developments make it possible to rely on fast sparse polynomial multiplication as a building block, both in theory and in practice. This makes it natural to study other operations on multivariate polynomials with this building block at our disposal. One of the most important such operations is division.

The multivariate analogue of polynomial division is the reduction of a polynomial $A \in \mathbb{K}[x]$ with respect to an autoreduced tuple $B = (B_1, \ldots, B_b) \in \mathbb{K}[x]^b$ of other polynomials. This leads to a relation

$$A = Q_1 B_1 + \cdots + Q_b B_b + R, \tag{1}$$

such that none of the terms occurring in R can be further reduced with respect to B. In this paper, we are interested in the computation of R as well as $Q_1, \ldots, Q_b$. We will call this the problem of *extended reduction*, in analogy with the notion of an "extended g.c.d.".

Now in the univariate context, "relaxed power series" provides a convenient technique for the resolution of implicit equations [6–8, 10]. One major advantage of this technique is that it tends to respect most sparsity patterns which are present in the input data and in the equations. The main technical tool in this paper (see Sect. 3) is to generalize this technique to the setting of multivariate polynomials, whose terms are ordered according to a specific admissible ordering on the monomials. This will make it possible to rewrite (1) as a so-called recursive equation (see Sect. 4.2), which can be solved in a relaxed manner. Roughly speaking, the cost of the extended reduction then reduces to the cost of the relaxed multiplications $Q_1 B_1, \ldots, Q_b B_b$. Up to a logarithmic overhead, we will show (Theorem 4) that this cost is the same as the cost of checking the relation (1).

In order to simplify the exposition, we will adopt a simplified sparse complexity model throughout this paper. In particular, our complexity analysis will not take into account the computation of support bounds for products or results of the extended reduction. Bit complexity issues will also be left aside in this paper. We finally stress that our results are mainly of theoretical interest since none of the proposed algorithms have currently been implemented. Nevertheless, practical gains are not to be excluded, especially in the case of small n, high degrees and dense supports.

2 Notations

Let $\mathbb{K}$ be an effective field with an effective zero test and let $x_1, \ldots, x_n$ be indeterminates. We will denote

$$\begin{aligned} \mathbb{K}[x] &= \mathbb{K}[x_1, \ldots, x_n] \\ P_i &= P_{i_1,\ldots,i_n} \\ x^i &= x_1^{i_1} \cdots x_n^{i_n} \\ i \preccurlyeq j &\Leftrightarrow i_1 \leqslant j_1 \wedge \cdots \wedge i_n \leqslant j_n, \end{aligned}$$

for any $i, j \in \mathbb{N}^n$ and $P \in \mathbb{K}[x]$. In particular, $i \preccurlyeq j \Leftrightarrow x^i | x^j$. For any subset $E \subseteq \mathbb{N}^n$ we will denote by $\operatorname{Fin}(E) = \{j \in \mathbb{N}^n : \exists i \in E, i \preccurlyeq j\}$ the *final segment* generated by E for the partial ordering $\preccurlyeq$.

Let $\leqslant$ be a total ordering on $\mathbb{N}^n$ which is compatible with addition. Two particular such orderings are the lexicographical ordering $\leqslant^{\text{lex}}$ and the reverse lexicographical ordering $\leqslant^{\text{rlex}}$:

$$\begin{aligned} i <^{\text{lex}} j &\Leftrightarrow \exists k, i_1 = j_1 \wedge \cdots \wedge i_{k-1} = j_{k-1} \wedge i_k < j_k \\ i <^{\text{rlex}} j &\Leftrightarrow \exists k, i_k < j_k \wedge i_{k+1} = j_{k+1} \wedge \cdots \wedge i_n = j_n. \end{aligned}$$

In general, it can be shown [16] that there exist real vectors $\lambda_1, \ldots, \lambda_n \in \mathbb{R}^m$ with $m \leqslant n$, such that

$$i \leqslant j \Leftrightarrow (\lambda_1 \cdot i, \ldots, \lambda_m \cdot i) \leqslant^{\text{lex}} (\lambda_1 \cdot j, \ldots, \lambda_m \cdot j). \tag{2}$$

In what follows, we will assume that $\lambda_1, \ldots, \lambda_n \in \mathbb{N}^n$ and $\gcd((\lambda_i)_1, \ldots, (\lambda_i)_n) = 1$ for all i. We will also denote

$$\lambda \cdot i = (\lambda_1 \cdot i, \ldots, \lambda_n \cdot i).$$

For instance, the graded reverse lexicographical ordering $\leqslant^{\text{grlex}}$ is obtained by taking $\lambda_1 = (1, \ldots, 1)$, $\lambda_2 = (0, \ldots, 1)$, $\lambda_2 = (0, \ldots, 0, 1, 0)$, ..., $\lambda_n = (0, 1, 0, \ldots, 0)$.

Given $P \in \mathbb{K}[x]$, we define its *support* by

$$\operatorname{supp} P = \{i \in \mathbb{N}^n : P_i \neq 0\}.$$

If $P \neq 0$, then we also define its *leading exponent* l_P and *coefficient* c_P by

$$\begin{aligned} l_P &= \max_{\leqslant} \operatorname{supp} P \\ c_P &= P_{l_P}. \end{aligned}$$

Given a finite set E, we will denote its cardinality by $|E|$.

3 Relaxed Multiplication

3.1 Relaxed Power Series

Let us briefly recall the technique of relaxed power series computations, which is explained in more detail in [7]. In this computational model, a univariate power series $f \in \mathbb{K}[[z]]$ is regarded as a stream of coefficients $f_0, f_1, \ldots$. When performing an operation $g = \Phi(f_1, \ldots, f_k)$ on power series, it is required that the coefficient g_n of the result is output as soon as sufficiently many coefficients of the inputs are known, so that the computation of g_n does not depend on the further coefficients. For instance, in the case of a multiplication $h = fg$, we require that h_n is output as soon as $f_0, \ldots, f_n$ and $g_0, \ldots, g_n$ are known. In particular, we may use the naive formula $h_n = \sum_{i=0}^{n} f_i g_{n-i}$ for the computation of h_n.

The additional constraint on the time when coefficients should be output admits the important advantage that the inputs may depend on the output, provided that we add a small delay. For instance, the exponential $g = \exp f$ of a power series $f \in z\mathbb{K}[[z]]$ may be computed in a relaxed way using the formula

$$g = \int f' g.$$

Indeed, when using the naive formula for products, the coefficient g_n is given by

$$g_n = \frac{1}{n}(f_1 g_{n-1} + 2 f_2 g_{n-2} + \cdots + n f_n g_0),$$

and the right-hand side only depends on the previously computed coefficients $g_0, \ldots, g_{n-1}$. More generally, equations of the form $g = \Phi(g)$ which have this property are called *recursive* equations and we refer to [11] for a mechanism to transform fairly general implicit equations into recursive equations.

The main drawback of the relaxed approach is that we cannot directly use fast algorithms on polynomials for computations with power series. For instance, assuming that $\mathbb{K}$ has sufficiently many 2^p-th roots of unity and that field operations in $\mathbb{K}$ can be done in time $\mathcal{O}(1)$, two polynomials of degrees $<n$ can be multiplied in time $\mathsf{M}(n) = \mathcal{O}(n \log n)$, using FFT multiplication [3]. Given the truncations $f_{;n} = f_0 + \cdots + f_{n-1} z^{n-1}$ and $g_{;n} = g_0 + \cdots + g_{n-1} z^{n-1}$ at order n of power series $f, g \in \mathbb{K}[[z]]$, we may thus compute the truncated product $(fg)_{;n}$ in time $\mathsf{M}(n)$ as well. This is much faster than the naive $\mathcal{O}(n^2)$ relaxed multiplication algorithm for the computation of $(fg)_{;n}$. However, the formula for $(fg)_0$ when using FFT multiplication depends on all input coefficients $f_0, \ldots, f_{n-1}$ and $g_0, \ldots, g_{n-1}$, so the fast algorithm is not relaxed (we will say that FFT multiplication is a *zealous* algorithm). Fortunately, efficient relaxed multiplication algorithms do exist:

Theorem 1 [4, 6, 7] *Let $\mathsf{M}(n)$ be the time complexity for the multiplication of polynomials of degrees $< n$ in $\mathbb{K}[z]$. Then there exists a relaxed multiplication algorithm for series in $\mathbb{K}[[z]]$ at order n of time complexity $\mathsf{R}(n) = \mathcal{O}(\mathsf{M}(n) \log n)$.*

Remark 1 In fact, the algorithm from Theorem 1 generalizes to the case when the multiplication on $\mathbb{K}$ is replaced by an arbitrary bilinear "multiplication" $\mathbb{M}_1 \times \mathbb{M}_2 \to \mathbb{M}_3$, where $\mathbb{M}_1$, $\mathbb{M}_2$ and $\mathbb{M}_3$ are effective modules over an effective ring $\mathbb{A}$. If $\mathsf{M}(n)$ denotes the time complexity for multiplying two polynomials $P \in \mathbb{M}_1[z]$ and $Q \in \mathbb{M}_2[z]$ of degrees $<n$, then we again obtain a relaxed multiplication for series $f \in \mathbb{M}_1[[z]]$ and $g \in \mathbb{M}_2[[z]]$ at order n of time complexity $\mathcal{O}(\mathsf{M}(n) \log n)$.

Theorem 2 [10] *If $\mathbb{K}$ admits a primitive 2^p-th root of unity for all p, then there exists a relaxed multiplication algorithm of time complexity*

$$\mathsf{R}(n) = \mathcal{O}(n \log n e^{2\sqrt{\log 2 \log \log n}}).$$

In practice, the existence of a 2^{p+1}-th root of unity with $2^p \geqslant n$ suffices for multiplication up to order n.

3.2 Relaxed Multivariate Laurent Series

Let $\mathbb{A}$ be an effective ring. A power series $f \in \mathbb{A}[[z]]$ is said to be *computable* if there is an algorithm which takes $n \in \mathbb{N}$ on input and produces the coefficient f_n on output. We will denote by $\mathbb{A}[[z]]^{\text{com}}$ the set of such series. Then $\mathbb{A}[[z]]^{\text{com}}$ is an effective ring for relaxed addition, subtraction and multiplication.

A computable Laurent series is a formal product $f z^k$ with $f \in \mathbb{A}[[z]]^{\text{com}}$ and $k \in \mathbb{Z}$. The set $\mathbb{A}((z))^{\text{com}}$ of such series forms an effective ring for the addition, subtraction and multiplication defined by

$$\begin{aligned} f z^k + g z^l &= (f z^{k-\min(k,l)} + g z^{l-\min(k,l)}) z^{\min(k,l)} \\ f z^k - g z^l &= (f z^{k-\min(k,l)} - g z^{l-\min(k,l)}) z^{\min(k,l)} \\ (f z^k)(g z^l) &= (f g) z^{k+l}. \end{aligned}$$

If $\mathbb{A}$ is an effective field with an effective zero test, then we may also define an effective division on $\mathbb{A}((z))^{\text{com}}$, but this operation will not be needed in what follows.

Assume now that z is replaced by a finite number of variables $z = (z_1, \ldots, z_n)$. Then an element of

$$\mathbb{A}((z))^{\text{com}} := \mathbb{A}((z_n))^{\text{com}} \cdots ((z_1))^{\text{com}}$$

will also be called a "computable lexicographical Laurent series". Any nonzero $f \in \mathbb{A}((z))$ has a natural valuation $v_f = (v_1, \ldots, v_n) \in \mathbb{Z}^n$, by setting $v_1 = \text{val}_{z_1} f$, $v_2 = \text{val}_{z_2}([z_1^{v_1}] f)$, etc. The concept of recursive equations naturally generalizes to the multivariate context. For instance, for an infinitesimal Laurent series $\varepsilon \in \mathbb{A}((z))^{\text{com}}$ (that is, $\varepsilon = f z^k$, where $v_f >^{\text{lex}} -k$), the formula

$$g = 1 + \varepsilon g$$

allows us to compute $g = (1 - \varepsilon)^{-1}$ using a single relaxed multiplication in $\mathbb{A}((z))^{\text{com}}$.

Now take $\mathbb{A} = \mathbb{K}[x]$ and consider a polynomial $P \in \mathbb{A}$. Then we define the Laurent polynomial $\hat{P} \in \mathbb{K}[x z^{-\lambda}] \subseteq \mathbb{A}((z))^{\text{com}}$ by

$$\hat{P} = \sum_{i \in \mathbb{N}^n} P_i x^i z^{-\lambda \cdot i}.$$

Conversely, given $f \in \mathbb{K}[x z^{-\lambda}]$, we define $\check{f} \in \mathbb{K}[x]$ by substituting $z_1 = \cdots = z_n = 1$ in f. We will call the transformations $P \mapsto \hat{P}$ and $\hat{P} \mapsto P = \check{\hat{P}}$ *tagging* resp. *untagging*; they provide us with a relaxed mechanism to compute with multivariate polynomials in $\mathbb{K}[x]$, such that the admissible ordering $\leqslant$ on $\mathbb{N}^n$ is respected. For instance, we may compute the relaxed product of two polynomials $P, Q \in \mathbb{K}[x]$

by computing the relaxed product $\hat{P}\hat{Q}$ and substituting $z_1 = \cdots = z_n = 1$ in the result. We notice that tagging is an injective operation which preserves the size of the support.

3.3 Complexity Analysis

Assume now that we are given $P, Q \in \mathbb{K}[x]$ and a set $\mathcal{R} \subseteq \mathbb{N}^n$ such that $\operatorname{supp}(PQ) \subseteq \mathcal{R}$. We assume that $\mathsf{SM}(s)$ is a function such that the (zealous) product PQ can be computed in time $\mathsf{SM}(|\mathcal{R}|)$. We will also assume that $\mathsf{SM}(s)/s$ is an increasing function of s. In [2, 15], it is shown that we may take $\mathsf{SM}(s) = \mathcal{O}(\mathsf{M}(s) \log s)$.

Let us now study the complexity of sparse relaxed multiplication of P and Q. We will use the classical algorithm for fast univariate relaxed multiplication from [6, 7], of time complexity $\mathsf{R}(s) = \mathcal{O}(\mathsf{M}(s) \log s)$. We will also consider semi-relaxed multiplication as in [8], where one of the arguments $\hat{P}$ or $\hat{Q}$ is completely known in advance and only the other one is computed in a relaxed manner.

Given $X \subseteq \mathbb{N}^n$ and $i \in \{1, \ldots, n\}$, we will denote

$$\delta_i(X) = \max\{\lambda_i \cdot k : k \in X\} + 1$$
$$\delta(X) = \delta_1(X) \cdots \delta_n(X).$$

We now have the following:

Theorem 3 *With the above notations, the relaxed product of P and Q can be computed in time*

$$\mathcal{O}(\mathsf{SM}(|\mathcal{R}|) \log \delta(\mathcal{R})).$$

Proof In order to simplify our exposition, we will rather prove the theorem for a semi-relaxed product of $\hat{P}$ (relaxed) and $\hat{Q}$ (known in advance). As shown in [8], the general case reduces to this special case. We will prove by induction over n that the semi-relaxed product can be computed using at most $3\mathsf{SM}(|\mathcal{R}|) \log \delta(\mathcal{R})$ operations in $\mathbb{K}$ if $\mathcal{R}$ is sufficiently large. For $n = 0$, we have nothing to do, so assume that $n > 0$.

Let us first consider the semi-relaxed product of $\hat{P}$ and $\hat{Q}$ with respect to z_1. Setting $l = \lceil \log_2 \delta_1(\mathcal{R}) \rceil$, the computation of this product corresponds (see the right-hand side of Fig. 1) to the computation of $\leqslant 2$ zealous $2^{l-1} \times 2^{l-1}$ products (i.e. 2 products of polynomials of degrees $< 2^{l-1}$ in z_1), $\leqslant 4$ zealous $2^{l-2} \times 2^{l-2}$ products, and so on until $\leqslant 2^l$ zealous 1×1 products. We finally need to perform 2^l semi-relaxed 1×1 products of series in $z_2, \ldots, z_n$ only.

More precisely, assume that $\hat{P}$ and $\hat{Q}$ have valuations p resp. q in z_1 and let $\hat{P}_i$ stand for the coefficient of z_1^i in P. We also define

$$\hat{\mathcal{R}} = \{(a_1, \ldots, a_n, b_1, \ldots, b_n) \in \mathbb{N}^n \times \mathbb{Z}^n : (a_1, \ldots, a_n) \in \mathcal{R} \wedge (\forall i, b_i = -\lambda_i \cdot a)\}.$$

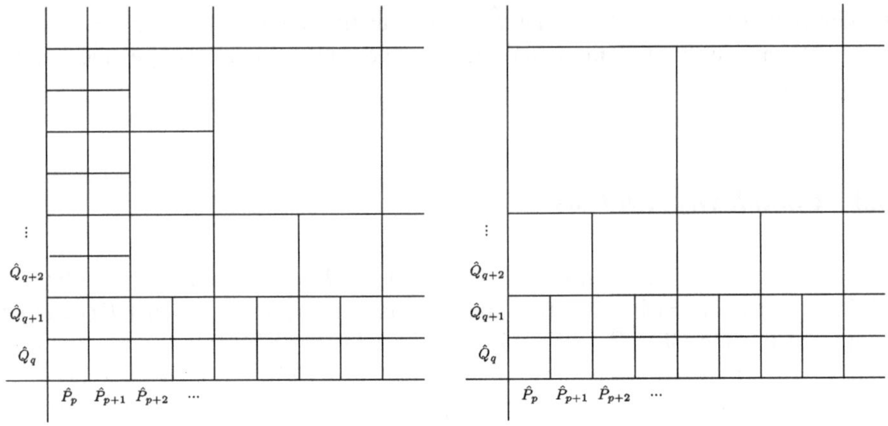

Fig. 1 Illustration of a fast relaxed product and a fast semi-relaxed product

Now consider a block size 2^k. For each i, we define

$$\begin{aligned}\hat{P}_{[i]} &= \hat{P}_{p+2^k i} z_1^{p+2^k i} + \cdots + \hat{P}_{p+2^k(i+1)-1} z_1^{p+2^k(i+1)-1}\\ \hat{Q}_{[i]} &= \hat{Q}_{q+2^k i} z_1^{q+2^k i} + \cdots + \hat{Q}_{q+2^k(i+1)-1} z_1^{q+2^k(i+1)-1}\\ \hat{\mathcal{R}}_{[i]} &= \{(a_1,\ldots,a_n,b_1,\ldots,b_n) \in \hat{\mathcal{R}} :\\ &\qquad 2^k i \leqslant a_1 - p - q \leqslant 2^k(i+1) - 1\},\end{aligned}$$

and notice that the $\hat{\mathcal{R}}_{[i]}$ are pairwise disjoint. In the semi-relaxed multiplication, we have to compute the zealous $2^k \times 2^k$ products $\hat{P}_{[i]}\hat{Q}_{[1]}$ for all $i \leqslant \lfloor (\delta_1(\mathcal{R}) + 1)/2^k \rfloor$. Since

$$\operatorname{supp} \hat{P}_{[i]}\hat{Q}_{[1]} \subseteq \hat{\mathcal{R}}_{[i+1]} \amalg \hat{\mathcal{R}}_{[i+2]},$$

we may compute all these products in time

$$\begin{aligned}&\mathsf{SM}(|\hat{\mathcal{R}}_{[1]} \amalg \hat{\mathcal{R}}_{[2]}|) + \cdots + \mathsf{SM}(|\hat{\mathcal{R}}_{[2^{l-k}]} \amalg \hat{\mathcal{R}}_{[2^{l-k}+1]}|)\\ &= (|\hat{\mathcal{R}}_{[1]} \amalg \hat{\mathcal{R}}_{[2]}|)\frac{\mathsf{SM}(|\hat{\mathcal{R}}_{[1]} \amalg \hat{\mathcal{R}}_{[2]}|)}{|\hat{\mathcal{R}}_{[1]} \amalg \hat{\mathcal{R}}_{[2]}|} + \cdots +\\ &\quad (|\hat{\mathcal{R}}_{[2^{l-k}]} \amalg \hat{\mathcal{R}}_{[2^{l-k}+1]}|)\frac{\mathsf{SM}(|\hat{\mathcal{R}}_{[2^{l-k}]} \amalg \hat{\mathcal{R}}_{[2^{l-k}+1]}|)}{|\hat{\mathcal{R}}_{[2^{l-k}]} \amalg \hat{\mathcal{R}}_{[2^{l-k}+1]}|}\\ &\leqslant (|\hat{\mathcal{R}}_{[1]} \amalg \hat{\mathcal{R}}_{[2]}| + \cdots + |\hat{\mathcal{R}}_{[2^{l-k}]} \amalg \hat{\mathcal{R}}_{[2^{l-k}+1]}|)\frac{\mathsf{SM}(|\hat{\mathcal{R}}|)}{|\hat{\mathcal{R}}|}\\ &\leqslant 2\mathsf{SM}(|\hat{\mathcal{R}}|) = 2\mathsf{SM}(|\mathcal{R}|).\end{aligned}$$

The total time spent in performing all zealous $2^k \times 2^k$ block multiplications with $2^k < 2^l$ is therefore bounded by $2\mathsf{SM}(|\mathcal{R}|) \log \delta_1(\mathcal{R})$.

Let us next consider the remaining 1×1 semi-relaxed products. If $n = 1$, then these are really scalar products, whence the remaining work can clearly be performed in time $\mathsf{SM}(|\mathcal{R}|) \log \delta_1(\mathcal{R})$ if $\mathcal{R}$ is sufficiently large. If $n > 1$, then for each i, we have

$$\operatorname{supp} \hat{P}_{[i]} \hat{Q}_{[0]} \subseteq \hat{\mathcal{R}}_{[i]}.$$

By the induction hypothesis, we may therefore perform this semi-relaxed product in time $3\mathsf{SM}(|\hat{\mathcal{R}}_{[i]}|)(\log \delta(\mathcal{R}) - \log \delta_1(\mathcal{R}))$. A similar argument as above now yields the bound $3\mathsf{SM}(|\mathcal{R}|)(\log \delta(\mathcal{R}) - \log \delta_1(\mathcal{R}))$ for performing all 1×1 semi-relaxed block products. The total execution time (which also takes into account the final additions) is therefore bounded by $3\mathsf{SM}(|\mathcal{R}|) \log \delta(\mathcal{R})$. This completes the induction.

Remark 2 In practice, the computation of zealous products of the form $\hat{P}_{[i]} \hat{Q}_{[j]}$ is best done in the untagged model, i.e. using the formula

$$\hat{P}_{[i]} \hat{Q}_{[j]} = \widehat{\widetilde{\hat{P}_{[i]}} \widetilde{\hat{Q}_{[j]}}}$$

Proceeding this way allows us to use any of our preferred algorithms for sparse polynomial multiplication. In particular, we may use [14] or [12].

4 Polynomial Reduction

4.1 Naive Extended Reduction

Consider a tuple $B = (B_1, \ldots, B_b) \in \mathbb{K}[x]^b$. We say that B is *autoreduced* if $B_i \neq 0$ for all i and $l_{B_i} \not\preccurlyeq l_{B_j}$ and $l_{B_j} \not\preccurlyeq l_{B_i}$ for all $i \neq j$. Given such a tuple B and an arbitrary polynomial $A \in \mathbb{K}[x]$, we say that A is *reduced* with respect to B if $l_{B_i} \not\preccurlyeq k$ for all i and $k \in \operatorname{supp} A$. An *extended reduction* of A with respect to B is a tuple $(Q_1, \ldots, Q_b, R)$ with

$$A = Q_1 B_1 + \cdots + Q_b B_b + R, \tag{3}$$

such that R is reduced with respect to B. The naive algorithm extended-reduce below computes an extended reduction of A.

Algorithm extended-reduce

INPUT: $A \in \mathbb{K}[x]$ and an autoreduced tuple $B \in \mathbb{K}[x]^b$

OUTPUT: an extended reduction of A with respect to B

Start with $Q := (0, \ldots, 0)$ and $R := A$

While R is not reduced with respect to B do

  Let i be minimal and such that $l_{B_i} \preccurlyeq k$ for some $k \in \operatorname{supp} R$

  Let $k \in \operatorname{supp} R$ be maximal with $l_{B_i} \preccurlyeq k$

Set $Q_i := Q_i + (R_k/c_{B_i})x^{k-l_{B_i}}$ and $R := R - (R_k/c_{B_i})x^{k-l_{B_i}} B_i$
Return $(Q_1, \ldots, Q_b, R)$

Remark 3 Although an extended reduction is usually not unique, the one computed by extended-reduce is uniquely determined by the fact that, in our main loop, we take i minimal with $l_{B_i} \preccurlyeq k$ for some $k \in \text{supp } R$. This particular extended reduction is also characterized by the fact that

$$\text{supp } Q_i + l_{B_i} \subseteq \text{Fin}(\{l_{B_i}\}) \setminus \text{Fin}(\{l_{B_1}, \ldots, l_{B_{i-1}}\})$$

for each i.

In order to compute $Q_1, \ldots, Q_b$ and R in a relaxed manner, upper bounds

$$\begin{aligned} \text{supp } Q_i &\subseteq \mathcal{Q}_i \\ \text{supp } Q_i B_i &\subseteq \mathcal{Q}_i + \text{supp } B_i \\ \text{supp } R &\subseteq \mathcal{R} \end{aligned}$$

need to be known beforehand. These upper bounds are easily computed as a function of $\mathcal{A} = \text{supp } A, \mathcal{B}_1 = \text{supp } B_1, \ldots, \mathcal{B}_b = \text{supp } B_b$ by the variant supp-extended-reduce of extended-reduce below. We recall from the end of the introduction that we do not take into account the cost of this computation in our complexity analysis. In reality, the execution time of supp-extended-reduce is similar to the one of extended-reduce, except that potentially expensive operations in $\mathbb{K}$ are replaced by boolean operations of unit cost. We also recall that support bounds can often be obtained by other means for specific problems.

Algorithm supp-extended-reduce
INPUT: subsets $\mathcal{A}$ and $\mathcal{B}_1, \ldots, \mathcal{B}_b$ of $\mathbb{N}^n$ as above
OUTPUT: subsets $\mathcal{Q}_1, \ldots, \mathcal{Q}_b$ and $\mathcal{R}$ of $\mathbb{N}^n$ as above

Start with $\mathcal{Q} := (\varnothing, \ldots, \varnothing)$ and $\mathcal{R} := \mathcal{A}$
While $\mathcal{R} \cap \text{Fin}(\{\max \mathcal{B}_1, \ldots, \max \mathcal{B}_b\}) \neq \varnothing$ do
Let i be minimal with $l_{\max \mathcal{B}_i} \preccurlyeq k$ for some $k \in \mathcal{R}$
Let $k \in \mathcal{R}$ be maximal with $l_{\max \mathcal{B}_i} \preccurlyeq k$
Set $\mathcal{Q}_i := \mathcal{Q}_i \cup \{k - \max \mathcal{B}_i\}$ and

$\mathcal{R} := \mathcal{R} \cup (\mathcal{B}_i + (k - \max \mathcal{B}_i)) \setminus \{k\}$
Return $(\mathcal{Q}_1, \ldots, \mathcal{Q}_b, \mathcal{R})$

4.2 Relaxed Extended Reduction

Using the relaxed multiplication from Sect. 3, we are now in a position to replace the algorithm extended-reduce by a new algorithm, which directly computes

$Q_1, \ldots, Q_b, R$ using the Eq. (3). In order to do this, we still have to put it in a recursive form which is suitable for relaxed resolution.

Denoting by e_i the i-th canonical basis vector of $\mathbb{K}[x]^{b+1}$, we first define an operator $\Phi : x_1^{\mathbb{N}} \cdots x_n^{\mathbb{N}} \to \mathbb{K}[x]^{b+1}$ by

$$\Phi(x^k) = \begin{cases} c_{B_i}^{-1} x^{k-l_{B_i}} e_i & \text{if } k \in \operatorname{Fin}(\{l_{B_i}, \ldots, l_{B_b}\}) \text{ and} \\ & i \text{ is minimal with } l_{B_i} \preccurlyeq k \\ e_{b+1} x^k & \text{otherwise} \end{cases}$$

By linearity, this operator extends to $\mathbb{K}[x]$

$$\Phi(P) = \sum_{i \in \operatorname{supp} P} P_i \Phi(x^i).$$

In particular, $\Phi(c_A x^{l_A})$ yields the "leading term" of the extended reduction $(Q_1, \ldots, Q_b, R)$. We also denote by $\hat{\Phi}$ the corresponding operator from $\mathbb{K}[xz^{-\lambda}]$ to $\mathbb{K}[xz^{-\lambda}]^{b+1}$ which sends $\hat{P}$ to $\widehat{\Phi(P)}$.

Now let $B_i^* = B_i - c_{B_i} x^{l_{B_i}}$ for each i. Then

$$(Q_i B_i)_k = (Q_i B_i^*)_k + (Q_i)_{k-l_{B_i}} c_{B_i}$$

for each $i \in \{1, \ldots, b\}$ and $k \in \mathbb{N}^n$. The equation

$$(Q_1 B_1 + \cdots + Q_b B_b + R)_k = A_k$$

can thus be rewritten as

$$\begin{aligned} &(Q_1)_{k-l_{B_1}} c_{B_1} + \cdots + (Q_i)_{k-l_{B_b}} c_{B_b} \\ &= (A - Q_1 B_1^* - \cdots - Q_b B_b^*)_k \end{aligned}$$

Using the operator Φ this equation can be rewritten in a more compact form as

$$(Q_1, \ldots, Q_b, R) = \Phi(A - Q_1 B_1^* - \cdots - Q_b B_b^*).$$

The tagged counterpart

$$(\hat{Q}_1, \ldots, \hat{Q}_b, \hat{R}) = \hat{\Phi}(\hat{A} - \hat{Q}_1 \hat{B}_1^* - \cdots - \hat{Q}_b \hat{B}_b^*)$$

is recursive, whence the extended reduction can be computed using b multivariate relaxed multiplications $\hat{Q}_1 \hat{B}_1^*, \ldots, \hat{Q}_b \hat{B}_b^*$. With $\mathcal{A}$, $\mathcal{B}_i$, $\mathcal{Q}_i$ and $\mathcal{R}$ as in the previous section, Theorem 3 therefore implies:

Theorem 4 *We may compute the extended reduction of A with respect to B in time*

$$\mathcal{O}\left(\mathsf{SM}(|\mathcal{B}_1 + \mathcal{Q}_1|) \log \delta(\mathcal{B}_1 + \mathcal{Q}_1) + \cdots + \mathsf{SM}(|\mathcal{B}_b + \mathcal{Q}_b|) \log \delta(\mathcal{B}_b + \mathcal{Q}_b) + |\mathcal{R}|\right).$$

Remark 4 Following Remark 1, we also notice that A, the Q_i and R may be replaced by vectors of polynomials in $\mathbb{K}[x]^m$ (regarded as polynomials with coefficients in $\mathbb{K}^m$), in the case that several polynomials need to be reduced simultaneously.

References

1. Ben-Or, M., Tiwari, P.: A deterministic algorithm for sparse multivariate polynomial interpolation. In: STOC '88: Proceedings of the Twentieth Annual ACM Symposium on Theory of Computing, pp. 301–309, New York, NY, USA, 1988. ACM Press
2. Canny, J., Kaltofen, E., Lakshman, Y.: Solving systems of non-linear polynomial equations faster. In: Proceedings of ISSAC '89, pp. 121–128, Portland, Oregon, A.C.M., New York, 1989. ACM Press
3. Cooley, J.W., Tukey, J.W.: An algorithm for the machine calculation of complex Fourier series. Math. Comput. **19**, 297–301 (1965)
4. Fischer, M.J., Stockmeyer, L.J.: Fast on-line integer multiplication. In: Proceedings of 5th ACM Symposium on Theory of Computing, vol. 9, pp. 67–72 (1974)
5. Grigoriev, D.Y., Karpinski, M.: The matching problem for bipartite graphs with polynomially bounded permanents is in NC. In: Proceedings of the 28th IEEE Symposium on the Foundations of Computer Science, pp. 166–172 (1987)
6. van der Hoeven, J.: Lazy multiplication of formal power series. In: Küchlin, W.W. (ed.) Proceedings of ISSAC '97, pp. 17–20. Maui, Hawaii, July 1997
7. van der Hoeven, J.: Relax, but don't be too lazy. JSC **34**, 479–542 (2002)
8. van der Hoeven, J.: Relaxed multiplication using the middle product. In: Bronstein, M. (ed.) Proceedings of ISSAC '03, pp. 143–147, Philadelphia, USA, August 2003
9. van der Hoeven, J.: The truncated Fourier transform and applications. In: Gutierrez, J. (ed.) Proceedings of ISSAC 2004, pp. 290–296, Univ. of Cantabria, Santander, Spain, July 4–7, 2004
10. van der Hoeven, J.: New algorithms for relaxed multiplication. JSC **42**(8), 792–802 (2007)
11. van der Hoeven, J.: From implicit to recursive equations. Technical report, HAL. http://hal.archives-ouvertes.fr/hal-00583125 (2011)
12. van der Hoeven, J., Lecerf, G.: On the complexity of blockwise polynomial multiplication. In: Proceedings of ISSAC '12, pp. 211–218, Grenoble, France, July 2012
13. van der Hoeven, J., Lecerf, G.: On the bit-complexity of sparse polynomial multiplication. JSC **50**, 227–254 (2013)
14. van der Hoeven, J., Schost, É.: Multi-point evaluation in higher dimensions. AAECC **24**(1), 37–52 (2013)
15. Kaltofen, E., Lakshman, Y.N.: Improved sparse multivariate polynomial interpolation algorithms. In: ISSAC '88: Proceedings of the International Symposium on Symbolic and Algebraic Computation, pp. 467–474. Springer (1988)
16. Robbiano, L.: Term orderings on the polynominal ring. Eur. Conf. Comput. Algebra **2**, 513–517 (1985)

Preserving Syntactic Correctness While Editing Mathematical Formulas

Joris van der Hoeven, Grégoire Lecerf and Denis Raux

Abstract GNU $\TeX_{\text{MACS}}$ is a free software for editing scientific documents with mathematical formulas, which can also be used as an interface for many computer algebra systems. We present the design of a new experimental mathematical editing mode which preserves the syntactic correctness of formulas during the editing process (i.e. all formulas can be parsed using a suitable, sufficiently rich grammar). The main constraint is to remain as closely as possible to the existing presentation-oriented formula editor, which has the advantage of being very user friendly.

Keywords Mathematical editing · Syntactic correctness · Packrat parsing · $\TeX_{\text{MACS}}$

A.M.S. Subject Classification 68U15 · 68U35 · 68N99

1 Introduction

Most mathematical formulas in current scientific papers only carry very poor semantics. For instance, consider the two formulas $f(x + y)$ and $a(b + c)$. People typically enter these formulas using the LaTeX pseudo-code `$f(x+y)$` and `$a(b+c)$`. Doing so, we do not transmit the important information that we probably meant to apply f to $x + y$ in the first formula and to multiply a with $b + c$ in the second one. The problem to automatically recover such information is very hard

J. van der Hoeven (✉) · G. Lecerf · D. Raux
Laboratoire d'informatique, UMR 7161 CNRS, Campus de l'École polytechnique,
1, rue Honoré d'Estienne d'Orves, Bâtiment Alan Turing, CS35003 91120 Palaiseau,
France
e-mail: vdhoeven@lix.polytechnique.fr

G. Lecerf
e-mail: lecerf@lix.polytechnique.fr

D. Raux
e-mail: raux@lix.polytechnique.fr

I.S. Kotsireas and E. Martínez-Moro (eds.), *Applications of Computer Algebra*,
Springer Proceedings in Mathematics & Statistics 198,
DOI 10.1007/978-3-319-56932-1_29

in general. For this reason, it would be desirable to have mathematical authoring tools in which it is easy to write formulas which systematically carry this type of information.

One important application where semantics matters is computer algebra. Popular computer algebra systems such as MATHEMATICA and MAPLE contain formula editors in which it is only possible to input formulas which can at least be understood from a syntactical point of view by the system. However, these systems were not really designed for writing scientific papers: they only offer a suboptimal typesetting quality, no advanced document preparation features, and no support for more informal authoring styles which are typical for scientific papers.

The GNU TEXMACS editor was designed to be a fully fledged wysiwyg alternative for TEX/LATEX, as well as an interface for many computer algebra systems. The software is free and can be downloaded from http://www.texmacs.org. Although formulas only carried barely more semantics than LATEX in old versions of TEXMACS, we have recently started to integrate more and more semantic editing features. Let us briefly discuss some of the main ideas behind these developments; we refer to [11] for more details and historical references to related work.

First of all, we are only interested in what we like to call "syntactical semantics". In the formula $2 + 3$, this means that we wish to capture the fact that $+$ is an infix operator with arguments 2 and 3, but that we are uninterested in the fact that $+$ stands for addition on integers. Such syntactical semantics can be modeled adequately using a formal grammar. Several other mathematical formula editors are grammar-based [1–3, 6, 8, 9], and they make use of various kinds of formal grammars. In TEXMACS, we have opted for so-called packrat grammars [4, 5], which are particularly easy to implement and customize.

A second question concerns the precise grammar that we should use to parse formulas in scientific documents. Instead of using different grammars for various areas with different notations, we were surprised to empirically find out that a well-designed "universal" mathematical grammar is actually sufficient for most purposes; new notations can still be introduced using a suitable macro-mechanism.

The last main point concerns the interaction between the editor and the grammar. So far, we implemented a packrat parser for checking the correctness of a formula. While editing a formula, its correctness is indicated using colored boxes. It is also possible to detect and visualize the scopes of operators through the grammar. In addition to the parser, we implemented a series of tools which are able to detect and correct the most common syntactical mistakes and enhance existing documents with more semantics.

In the present paper, we wish to go one step further and enforce syntactic correctness throughout the editing process. Ideally speaking, the following requirements should be met:

- As far as user input is concerned, there should be no essential difference between editing formulas with or without the new mechanism for preserving syntactic correctness. For instance, we do not wish to force users to provide additional "annotations" for indicating semantics. It should also be possible to perform any editing action which makes sense from the purely visual point of view.

- The implementation should be as independent as possible from the actual grammar being used. In other words, we strive for a generic approach, not one for which specific editing routines are implemented for each individual grammar symbol.

The main technique that we will use for sticking as close as possible to the old, presentation-oriented editing behavior is to automatically insert "transient" markup for enforcing correctness during the editing process. For instance, when typing x +, TeXmacs will display

$$x + \square$$

The transient box is used to indicate a missing symbol or subexpression and will be removed as soon as the user enters the missing part.

The use of transient boxes for missing symbols or subexpressions is common in other editors [7]. The question which interests us here is how to automatically insert such markup when needed in a way that is essentially independent from specific grammars. In this paper, we work out the following approach which was suggested in [11]: before and after each editing operation, subject the formula to suitable "correction" procedures that are only allowed to add or remove transient markup. Correcting all errors in a general formula is a very difficult problem, but the power of our approach comes from the fact that the editing process is incremental: while typing, the user only introduces small errors—mostly incomplete formulas—which are highly localized; we may thus hope to deal with all possible problems using a small number of "kinds of corrections".

Obviously, the simplest kinds of corrections are adding or removing a transient box at the current cursor position. This is indeed sufficient when typing simple formulas such as $x + y + z$, but additional mechanisms are needed in other situations. For instance, in the formula $\alpha + |\beta$ (with the cursor between the "+" and the "β"), entering another + results in $\alpha + \square + \beta$ (instead of $\alpha + + \square \beta$ or $a + +b$). Hitting backspace in the same formula $\alpha + |\beta$ yields $\alpha + \beta$; in this case, the transient "+" should be parsed as an infix addition, and not as an ordinary symbol (as was the case for a transient box).

The appropriate corrections are not always so simple. For instance, consider the quantified expression $\forall x, \exists y, P(x, y)$. Just after we entered the existential quantifier "$\exists$", the formula will read $\forall x, \exists \square, \square$, i.e. it was necessary to add three transient symbols in order to make the expression syntactically correct. The fact that our approach should apply to general scientific documents with mathematical formulas raises several further problems. For instance, in the formula

$$a^2 + b^2 = c^2,$$

the trailing punctuation "," is incorrect from a mathematical point of view, but needed inside the surrounding English sentence. Similarly, more work remains to be done on the most convenient way to include English text inside formulas while maintaining syntactic correctness.

Yet another difficulty stems from the implementation: one needs to make sure that the necessary corrections take place after *any* kind of editing operation. However, for efficiency reasons, it is important to only run the correction procedures on small parts of the document. Inside an existing editor such as TeXmacs, these requirements turn out to be quite strong, so some trade-offs may be necessary.

In what follows, we report on our first implementation of these ideas inside TeXmacs. We describe and motivate the current design, discuss remaining problems, and outline directions for future improvements. Of course, more user feedback will be necessary in order to make the new mechanisms suitable for widespread use.

2 Survey of Formula Editing with TeXmacs

In this section, we briefly recall the main design philosophy behind the TeXmacs formula editor. We start with the description of the original, purely presentation-oriented mathematical editing mode. We pursue with the more recent grammar-based editing features, which are presented in more detail in [11].

2.1 Presentation-Oriented Editing

The original goal behind TeXmacs was to provide a user friendly editor for mathematical papers with a similar typesetting quality as TeX. The challenge was to design a real-time WYSIWYG editor for complex, structured documents. Some early inspiration came from the idea [1] that graphically oriented math editors achieve the highest level of user friendliness. For instance, when pressing the right arrow key, the cursor should move to the right if possible (instead of moving forward in some abstract document tree, as was the case in some other existing editors). Early versions of TeXmacs used algorithms for the cursor movement which achieved this in a systematic way [10], while still making sure that all possible cursor positions in the corresponding document tree could be reached.

Another aspect of user friendliness concerned the efficiency of mathematical input methods. We designed highly efficient (and easy to memorize) keyboard shortcuts for entering common mathematical symbols, such as `-` `>` for $\rightarrow$, `<` `=` for $\leqslant$, `<` `Tab` `/` for $\notin$, `R` `R` for $\mathbb{R}$, etc. TeXmacs also implements many "structured editing operations", so as to fully exploit the structure of documents. For instance, adding a row or column to a matrix can be done by pressing a single key or keyboard combination. Similarly, it is easy to change a matrix into a determinant or vice versa.

2.2 Grammar-Based Editing

The next challenge for TeXmacs is to ensure that we can only enter syntactically correct formulas, while keeping a presentation-oriented interface, which proved to be most user friendly. The first steps of this program were made in [11]. Now syntactic correctness is usually modeled as "parsability against a suitable grammar". Before anything else, one should decide on the grammar. In particular, does a single "universal grammar" suffice, or do we need many different grammars, depending on the preferred notations of authors?

For reasons that are explained in detail in [11], we opted for the development of a universal packrat grammar [4, 5] for parsing all our mathematical formulas. In order to conserve a sufficient degree of flexibility for the introduction of new notations, we rely on a combination of two techniques: on the one hand, TeXmacs comes with a powerful macro-language for introducing new markup elements. On the other hand, we introduced a special construct which allows a symbol or expression to be behave (i.e. be parsed) as an arbitrary other symbol or expression. This allows you for instance to annotate the symbol $\vee$ to behave as $+$, which implies that $a = b \vee c$ will be parsed as $a = (b \vee c)$ instead of $(a = b) \vee c$.

One of the major difficulties of semantic editing is a clean treatment of *homoglyphs*, i.e. symbols with the same graphical shape, but a different syntactical meaning. The most annoying homoglyph is the multiplication/function-application ambiguity mentioned in the introduction. Another good example concerns the wedge product $\mathrm{d}x \wedge \mathrm{d}y$ and logical conjunction $a = b \wedge x = y$, which admit different binding forces. Fortunately, there are not that many mathematical homoglyphs; for this reason, we advocate the introduction of separate symbols for them into the UNICODE standard.

3 Preservation of Correctness

In this section, we describe several strategies that can be used to preserve the syntactic correctness of formulas under editing operations. TeXmacs currently implements the "multiple correction schemes" strategy from Sects. 3.2 and 3.3. The reader may try this implementation by downloading version $\geqslant$1.99.3 or SVN revision $\geqslant$9718. The new editing mode is still experimental and can be enabled inside math mode by clicking on the 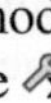 icon and checking Semantic correctness.

3.1 The Ideal Strategy for Preserving Correctness

Ideally speaking, maintaining the syntactic correctness of mathematical formulas throughout the editing process can be done by

1. Writing a "formula correction" procedure which takes any (correct or incorrect) formula on input and which inserts or removes transient markup in order to make it correct.
2. Run the correction procedure on all modified formulas in the document(s) after every editing operation.

This ideal strategy is simple and robust; it trivially guarantees the correctness of all formulas throughout the editing process. However, it does not take into account the specific nature of certain editing operations. In particular, it does not exploit the locality of many editing actions.

Example 1 Consider the strict application of the ideal strategy to the creation of a subscript in the formula $x + \square|$. Since $\square$ is a valid symbol, the main editing action would create an empty subscript for it. We next launch the correction procedure, which replaces the empty subscript by a transient box, yielding $x + \square_{\square|}$. However, the $\square$ being transient, the user would rather expect to endow the "+" operator with a subscript: this is indeed what happens in the old presentation-oriented editing mode when ignoring all transient markup. In other words, we rather expect to obtain $x +_{\square|} \square$.

The above example shows that an indiscriminate global correction procedure does not provide enough control. In fact, there are usually many ways to correct a formula by adding or removing transient markup. In order to determine the "best" solution, one typically needs to take into account the precise editing operation and the current cursor position.

Another constraint is that we would like the editor to behave as closely as possible as the old presentation-oriented editing mode when ignoring all transient markup. The above example shows that a global correction procedure does not necessarily respect this constraint. One theoretic solution to this problem is to remove all transient markup before performing the editing action and then put it back in when running the correction procedure. However, this approach may lead to non local changes in the document for every editing action, which is obviously not desirable.

Remark 1 For the above reasons, we have not implemented the correction strategy from this section yet. The idea nevertheless remains interesting for future research. Indeed, on the one hand side it raises the interesting theoretical question of correcting a string so as to make it parsable by a given (packrat) grammar. From the practical point of view, the ideal strategy has the important advantage of trivially guaranteeing syntactic correctness all along. In cases where this is hard to achieve using other means, it thereby remains a good fallback strategy.

3.2 Multiple Correction Schemes

Instead of implementing one global correction procedure, our current TeXmacs implementation relies on multiple "correction schemes". Each correction scheme is allowed to add or remove transient markup both before and after the actual editing operation. In other words, it really encapsulates the editing action into a semantically enhanced editing action. Furthermore, the correction scheme is allowed to fail (i.e. to produce an incorrect formula at the end). For this reason, we try multiple correction schemes in a row (the set of "eligible" schemes depends on the specific editing action), and stop as soon as we managed to obtain a correct formula.

In summary, we proceed as follows:

1. Depending on the editing action, determine a list of eligible correction schemes.
2. Try each eligible correction scheme in the list until we managed to obtain a correct formula.
3. If none of the correction schemes succeeded, then cancel the editing action.

For the actual implementation, it is clearly crucial to be able to undo editing actions whenever necessary, and in a way that is orthogonal to the usual undo/redo operations in $\mathrm{T_EX_{MACS}}$.

Example 2 When inserting a mathematical symbol, the first correction scheme we try is the following: first remove all transient markup around the cursor, then insert the symbol, and finally insert a transient box at the cursor position (if needed). For instance, typing `a` `+` `b` in an empty mathematical formula successively yields $|\square$, $a|$, $a+|\square$, and $a+b|$.

Example 3 The basic correction scheme from the previous example sometimes fails. For instance, assume that we are in the situation $a \wedge |\square$, and that we add a second "$\wedge$". When applying the basic correction scheme, we need to correct $a \wedge \wedge|$ through the insertion of a single transient box. However, the formula $a \wedge \wedge\square$ is still incorrect. For this particular case, we therefore use the following correction scheme: first add a transient box ($a \wedge \square|\square$), then perform the editing action ($a \wedge \square \wedge |\square$), and finally correct (nothing needs to be done at this step).

In Step 3, we simply canceled the editing action if all correction schemes failed. Several other fallback strategies can be considered. If we do not aim to maintain correctness at all costs, then we may apply the editing action without any corrections, and temporarily tolerate incorrect formulas. We might also implement an unconditionally successful fallback strategy as in Remark 1; by always adding such a strategy at the end of our list of eligible correction schemes, we will never reach Step 3. Yet another idea is to introduce a correction scheme which annotates subexpressions with exotic notations in such a way that they become correct.

3.3 *Quick Survey of Some of the Implemented Correction Schemes*

Our approach of using multiple correction schemes allows for fine-grained control, but also requires an increased amount of manual labor. Indeed, we both have to cover the complete set of editing actions, and for each editing action, we have to implement at least one correction scheme that will succeed in all possible situations.

Fortunately, the most common editing operations fall into four main categories: insertions and deletions that operate either on selections or not. Some other operations such as "search and replace" have not yet been adapted (see also the next section). Ultimately, the idea would be to provide manual support for the most common operations and to implement a suitable fallback strategy for the other ones.

Correction schemes for insertions Let us briefly list how we perform the most prominent correction schemes for insertions, in the absence of active selections. For each of the schemes, we show the successive states of the formula for a simple example.

- The basic scheme from Example 2.
- "Starting a prime or right script after a transient box" (e.g. inserting a new subscript in the formula $x + \square|$ from Example 1): first jump over the box with the cursor ($x + |\square$), then perform the action ($x +_{|} \square$), and finally add a transient box if necessary ($x +_{|\square} \square$).
- "Inserting a pure infix operator after a transient box" (e.g. inserting the infix operator "$\circ$" in $x + \square|$): perform the editing action ($x + \square \circ |$) and add a transient box if necessary ($x + \square \circ |\square$).
- The scheme from Example 3 for inserting two infix operators in a row.
- "Starting an extensible arrow with a script" (e.g. in the situation $E|$): remove all transient markup around the cursor ($E|$), perform the operation ($E \overset{|}{\rightarrow}$), add a transient box after the arrow ($E \overset{|}{\rightarrow} \square$), as well as a transient box at the cursor position ($E \overset{|\square}{\longrightarrow} \square$).
- "Insert content after an ordinary symbol" (e.g., entering ψ after $\varphi|$): remove all transient markup around the cursor ($\varphi|$), insert a transient "explicit space" ($\varphi\_|$), perform the editing action ($\varphi\_\psi|$), insert further transient boxes if needed ($\varphi\_\psi|$).
- "Insert content before an ordinary symbol" (e.g. entering ψ before $|\varphi$): remove all transient markup around the cursor ($|\varphi$), insert a transient "explicit space" after the cursor ($|\_\varphi$), perform the editing action ($\psi|\_\varphi$), insert further transient boxes if needed ($\psi|\_\varphi$).
- "Insert content in the middle of an operator" (e.g. starting a fraction in arc|sin): remove all transient markup around the cursor (arc|sin), insert transient "explicit spaces" before and after the cursor (arc_|_sin), perform the editing action ($\mathrm{arc}\_\frac{|}{}\_\mathrm{sin}$), insert further transient boxes if needed ($\mathrm{arc}\_\frac{|\square}{\square}\_\mathrm{sin}$).

The last three schemes also show that it is sometimes necessary to insert transient markup with different semantics as an ordinary symbol in order to make the formula correct.

Correction schemes for deletions For completion, we continue our list of examples with the most prominent correction schemes for deletions.

- "The basic deletion scheme if there is transient markup around the cursor" (e.g. hitting backspace in $a + \square|$ or in $-_{\square|}\square$): remove the transient markup around the cursor ($a + |$ resp. $-_{|}\square$), perform the editing action ($a|$ resp. $-|\square$), again remove all transient markup around the cursor if we deleted any composite tag ($a|$ resp. $-|$), add transient box if needed ($a|$ resp. $-|$).

- "The basic deletion scheme" (e.g. hitting backspace in $a + b|$): remove transient markup around the cursor ($a + b|$), perform the deletion ($a + |$), again remove all transient markup around the cursor if we deleted any composite tag ($a + |$), add transient box if needed ($a + |\square$).
- "Removal of actual infix operators" (e.g. hitting backspace in $a + |b$, but *not* in $-|a$): remove transient markup around the cursor ($a + |b$), perform the deletion ($a|b$), add a transient version of the deleted infix operator after the cursor ($a|+b$), add transient boxes around the cursor if needed ($a|+b$).
- "Need to jump over cursor before deletion" (e.g. hitting backspace in $\sum_{k=1}^{\infty} \square| \circ \varphi_k$): jump over the cursor ($\sum_{k=1}^{\infty} |\square \circ \varphi_k$), perform the "deletion" ($\sum_{k=1}^{\infty|} \square \circ \varphi_k$), add transient boxes around the cursor if needed ($\sum_{k=1}^{\infty|} \square \circ \varphi_k$).

These examples show that the correction schemes have to be implemented with quite a lot of care. This is due to the fact that it is convenient to design the schemes to apply with the right level of generality (e.g. not only to the deletion of symbols for the basic schemes, but also to the deletion of more complex structures, such as subscripts, fractions, etc.).

4 Problematic Cases and Challenges

Several problems arose during the implementation of the new semantic mathematical editing mode which preserves syntactic correctness. Some of them were more or less expected and have been solved; others require more work and further experimentation. So far, all problematic cases that we encountered fall into two categories

1. The incorrect treatment of special syntactic forms (and informal content in particular).
2. Complex editing operations (such as search and replace) that require special attention.

In this section, we will survey the most interesting issues that came up and highlight some of the remaining challenges.

4.1 Informal Content Inside Formulas

One difficulty with mathematical formulas in scientific papers with respect to formulas in, say, computer algebra systems, is that they may contain punctuation, decorations, typesetting directives, or explicative text. For instance, consider the following formula:

$$\begin{aligned} Z &= \{i \in I : f_i(x) = 0 \text{ and } g_i(x) = 0 \text{ almost everywhere}\} \\ &= \left\{i \in I : (f_i^2 + g_i^2)(x) = 0 \text{ almost everywhere}\right\}. \end{aligned}$$

This formula concentrates three difficulties:

- We used a trailing punctuation period "." to finish the formula.
- Since the formula does not fit on a single line, we used an "equation array" to manually break it into two rows. The cells of the underlying table should not be regarded as separate formulas (in which case the empty lower left cell would be incorrect), but rather be concatenated from left to right and from top to bottom.
- The formula involves English text "and" and "almost everywhere". The word "and" has the same semantics as the "$\wedge$" operator, whereas "almost everywhere" should be interpreted as a "postfix quantification".

The best approach to these problems is to introduce suitable annotation markup which describes the semantics of informal content of this kind. For instance, we might introduce a tag "punctuation" for annotating the trailing period, and which would be ignored by the parser. Alternatively, one might use a special symbol "punctuation period in math mode". In a similar spirit, AMS-LaTeX provides special environments (`split`, `align`, `gather`, etc.) for typesetting large formulas while preserving some of the intended semantics. TeXmacs also contains a general purpose tag "syntax", which may be used to parse an expression according to the rules of another specified expression. This allows us for instance to parse the word "and" in the same way as the infix operator "$\wedge$". However, we have no "postfix quantification" rule in our grammar yet. More generally, the design of a complete DTD for informal annotations is an interesting challenge.

Assuming suitable markup, the design of user-friendly ways to perform the necessary annotations is another matter. Trailing periods are so common that we actually would like to enter them simply by pressing [.]. There are two approaches to this problem. Our current solution is to adapt the grammar for displayed formulas so as to accept trailing punctuation (which also means that we do not need any special annotation semantics). A better solution would be to "requalify" symbols whenever needed. For instance, in the formula

$$x + y,$$

the trailing comma would be interpreted by default as a "punctuation symbol". However, as soon as we add a new character z to the line, we remove the annotation markup and requalify the comma to become a separator.

Of course, for arbitrarily complex informal text (such as the "almost everywhere" example), it will be hard to completely avoid user feedback on how to insert the necessary annotations. Nevertheless, some of the most common words ("and", "or", "iff", etc.) might be annotated automatically.

4.2 Special Syntactic Constructs

One obvious drawback of our strategy to manually design the necessary correction schemes is completeness: every additional mathematical notation potentially requires one or more new correction schemes. Fortunately, most mathematical notations are quite simple, so this disadvantage is not as bad as it might seem. General purpose scientific papers nevertheless involve far more special syntactic constructs than, say, computer algebra input. Let us illustrate some typical issues that occur on the hand of a few somewhat unorthodox constructs.

- The "universal grammar" from [11] contains special rules for decorated operators (as in $a +'_E b \hat{*} c$) and big operators (as in $\sum_{k=1}^{\infty} 1/k^2 = \pi^2/6$). The usual correction schemes are mostly sufficient for editing this kind of formulas. One example of a remaining problem is entering $a \hat{+} b$. In the old, presentation-oriented editing mode, we would type `a` `Alt-^` `+` `→` `b` (insert a, start an empty hat, enter $+$, move out of the hat, insert b). However, in the new semantic mode, this successively yields $a|$, $a\_|\widehat{\square}$, $a\_\widehat{\square + |\square}$, $a\_\widehat{\square + \square}\_b|$; a new correction scheme should be designed to treat this case more smoothly. Notice that an alternative way to enter $a \hat{+} b$ is to first type $a + b$, then select "+", and finally insert a hat.
- The "universal grammar" from [11] also contains a few rules that are uncommon in programming languages, but crucial for general purpose mathematical texts. For instance, the formula $a \lll b \leqslant c = d \neq e$ is interpreted as $a \lll b \wedge b \leqslant c \wedge c = d \wedge d \neq e$, and the formula $x_1, \ldots, x_n \in E$ as $x_1 \in E \wedge \cdots \wedge x_n \in E$. Less common is $n = 1, \ldots, 10$; what is the correct semantics? Fortunately, these special rules do not require any special correction schemes.
- Different authors use wildly varying notations for quantified expressions:

$$\forall x, \exists y, P(x, y)$$
$$\forall x \exists y : P(x, y)$$
$$(\forall x)(\exists y) P(x, y)$$
$$\vdots$$

  We already noticed in the introduction that it is "nice" to correct $\forall x, \exists$ into $\forall x, \exists \square, \square$. However, $(\forall x)$ might be corrected just as well as $(\forall x, \square)$ or as $(\forall x)\square$, depending on the author's preferred style. Our present solution to this kind of ambiguities is to further relax our grammar, by considering $(\forall x)$ to be a correct expression.
- One of the advantages of the new correctness-preserving editing mode is that missing expressions are clearly indicated to the user. When entering a 2×2 matrix $\begin{pmatrix} \square & \square \\ \square & \square \end{pmatrix}$ in a computer algebra system, this is indeed quite pleasant. But in the example

$$\begin{pmatrix} \lambda_1 & & \\ & \ddots & \\ & & \lambda_n \end{pmatrix}$$

of a diagonal matrix, this also forces users to manually fill out six of the cells with "invisible zeros". Our present solution is therefore to only require tables cells to be explicitly entered inside computer algebra sessions.

- The "universal grammar" from [11] also contains a few rules for "personal use". In particular, inside subscripts, we allow for notations such as $L_{\times\varphi,+\psi}$ and $f_{n;}$. Here $\times\varphi$ has the semantics of "post-multiplication" with φ. Given a power series $f = f_0 + f_1 z + f_2 z^2 + \cdots$, the notation $f_{n;}$ stands for $f_n z^n + f_{n+1} z^{n+1} + \cdots$. Now we are facing a dilemma: on the one hand, we are fond of these notations, which do not harm anyone. On the other hand, some users might *want* to be constrained to input something behind the ";" in $f_{n;\square}$. One solution would be to depart from the idea from [11] to promote using a "universal grammar". Instead, we might provide special style packages for specific notations. Another approach is to introduce suitable prefix and postfix homoglyphs of $\times$ and ;, together with simple keyboard shortcuts for entering them.

4.3 Special Editing Operations and Markup

Let us finally investigate to which extent existing editing operations have to be adapted to the new, more semantic editing mode. We will start with a few issues that are already dealt with and then turn our attention to the remaining challenges.

- TeXmacs provides a special "\-style" input method for people who already know LaTeX. For instance, one may enter α by typing `\` `a` `l` `p` `h` `a` `Enter`, or start a fraction by typing `\` `f` `r` `a` `c` `Enter`. The fact that a wide variety of editing actions can be triggered in this way required us to implement special correction schemes for this input method.
- The main TeXmacs input method for mathematical symbols is particularly powerful and intuitive. For instance, one may enter $\rightarrow$ and $\leqslant$ by typing `-` `>` resp. `<` `=`. However, this facility requires a lot of control over the undo-mechanism: when typing a shortcut `-` `>`, TeXmacs "forgets" the incomplete keystroke `-` and treats the shortcut `-` `>` as an atomic operation. In other words, typing `-` `>` and pressing "undo" will remove the entire arrow and not leave any $-$. Now remember from Sect. 3.2 that trying several correction schemes in succession also makes use of the undo-mechanism inside TeXmacs. Trying corrections while entering shortcuts such as `-` `>` necessitates the mechanism to work in a nested way. We had to further tweak our implementation so as to make that possible.
- Certain editing operations such as "save the current selection as an image" have side-effects that cannot be undone. Additional care is needed when implementing

correction schemes for such operations. Fortunately, such operations usually do not need to be corrected.

- One interesting editing action which is not necessarily local is "search and replace". Global editing actions of this kind are harder to support since the corresponding correction schemes need to track all modifications made throughout the document, and less indication is provided by the local context which corrections to choose in case of ambiguities. The "search and replace" operation also raises the question whether adapting the operation to a semantic context actually involves more than corrections *via* the addition or removal of transient markup: if we replace y by $a+b$ in $x \cdot y$, do we expect to obtain $x \cdot a + b$ or $x \cdot (a+b)$?
- For some editing operations, it is not always clear what their semantic counterparts should be. One good example concerns the facility to compute and inspect the structured differences between two versions of a document. When applied to the formulas $a + bc - d$ and $a + bcy$, the differences are indicated using red and green colors: $a + bc{-}dy$. How should we parse this formula? Both $a + bc - d$ and $a + bcy$ do make sense, but not $a + bc - dy$. It is not clear to us yet how the editor should behave in this situation.

References

1. Arsac, O., Dalmas, S., Gaëtano, M.: The design of a customizable component to display and edit formulas. In: ACM Proceedings of the 1999 International Symposium on Symbolic and Algebraic Computation, July 28–31, pp. 283–290 (1999)
2. Bertot, Y.: The CtCoq system: design and architecture. Formal Aspects Comput. **11**(3), 225–243 (1999)
3. Borras, P., Clement, D., Despeyroux, T., Incerpi, J., Kahn, G., Lang, B., Pascual, V.: Centaur: the system. SIGSOFT Softw. Eng. Notes **13**(5), 14–24 (1988)
4. Ford, B.: Packrat parsing: a practical linear-time algorithm with backtracking. Master's thesis, Massachusetts Institute of Technology, Sept (2002)
5. Ford, B.: Packrat parsing: simple, powerful, lazy, linear time. In: Proceedings of the Seventh ACM SIGPLAN International Conference on Functional Programming, ICFP '02, pp. 36–47. ACM Press, New York (2002)
6. Kajler, N.: Environnement graphique distribué pour le calcul formel. PhD thesis, Université de Nice-Sophia Antipolis (1993)
7. Padovani, L., Solmi, R.: An investigation on the dynamics of direct-manipulation editors for mathematics. In: Asperti, A., Bancerek, G., Trybulec, A. (eds.) Mathematical Knowledge Management, vol. 3119, Lecture Notes in Computer Science, pp. 302–316. Springer, Berlin (2004)
8. Soiffer, N.M.: The Design of a User Interface for Computer Algebra Systems. Ph.D. thesis, University of California at Berkeley (1991)
9. Théry, L., Bertot, Y., Kahn, G.: Real theorem provers deserve real user-interfaces. SIGSOFT Softw. Eng. Notes **17**(5), 120–129 (1992)
10. van der Hoeven, J.: GNU TeXmacs: a free, structured, wysiwyg and technical text editor. In: Filipo, D. (eds.) Le document au XXI-ième siècle, vol. 39–40, pp. 39–50. Metz, 14–17 mai 2001. Actes du congrès GUTenberg (2001)
11. van der Hoeven, J.: Towards semantic mathematical editing. J. Symb. Comput. **71**, 1–46 (2015)

About Balanced Application of CAS in Undergraduate Mathematics

Elena Varbanova

Abstract Educational goals and values of the teaching, learning and assessment of undergraduate mathematics are considered. A partial overview of the author's experience is represented and the necessity of balanced application of CAS in undergraduate mathematics education is discussed. The idea is to make what is important CAS supported, rather than what is CAS-supported important.

Keywords Undergraduate mathematics · CAS · Educational goals

1 Introduction

Aspire to Inspire ... before to Expire.

Inasmuch as the results of human activity depend on the environment, the introduction of new technological tools into the university mathematics education has become a must over the past 3 decades. The effectiveness, E, of a teaching–learning–assessment (TLA) process is given [7] by the following: E = f (Human abilities; Technology).

The universal language of Mathematics and its independence of cultural influences made possible the advent of the Computer Algebra Systems (CAS)—the most thought-provoking tool we have ever had for teaching and doing mathematics. However, the great power and potential of CAS cannot automatically transfer mathematics knowledge to learners and enhance their competency and appreciation of mathematics. Like any other technological tool, CAS requires adequate methodology for its integration in order to be converted into an effective instrument [1, 5, 10]. The main point is to find a balanced combination of educational values and successful tradition in the country and the power of this technology. This is a way of keeping the identity in mathematics education that can give birth to authentic, unique and adequate new practices in education. In this sense, technology could serve as a bridge over the ocean

E. Varbanova (✉)
Faculty of Applied Mathematics and Informatics, Technical University of Sofia, Sofia, Bulgaria
e-mail: elvar@tu-sofia.bg

I.S. Kotsireas and E. Martínez-Moro (eds.), *Applications of Computer Algebra*, Springer Proceedings in Mathematics & Statistics 198,
DOI 10.1007/978-3-319-56932-1_30

of a great variety of effective modern practices in education and as an international platform for exchange of educational know-how and of multitude of approaches. The latter is a part of the content and sense of the "new ethics in education": cooperation and collaboration instead of comparison, competition and confrontation.

2 Tradition Welcomes Technology

Challenge is the energy of life.

Technology exists and mathematics teachers need to be involved in reflecting on its effect. They have to show wisdom and should not ignore technology, because it is a strategic challenge for mathematics education. Our students will live in a highly automated future that will require highly organized, disciplined and creative minds: the latter can be developed through mastering basic mathematics knowledge and ideas in a digital environment [4].

It has to be noted that tradition guards knowledge; in its turn, knowledge is an emanation of tradition. Knowledge creates psychological comfort and forms certain templates in mind. At the same time, it changes and develops and, thus, enhances the tradition, and it also instils transformations in the latter.

Why could experienced teachers get the sense of danger or chaos when they come in contact with CAS? Because their main concern is the impact of this technology not to become "Deep impact". This means that the impact should by no means result in replacing an effective methodology, which is the core of a successful educational tradition, by a pseudo-methodology. A good tradition however enables its development through new technologies. Development implies disintegration of some old forms and integration of new ones without destroying some of the beneficial old elements.

3 Technology Salutes Tradition

As long as education can change,
the world can change.

Mathematics teaching and learning is a skill-and-habit-forming process [6] and it leaves its mark on the learners for life. Technology-meets-tradition activities have been designed and integrated in our mathematics courses [8–13] to facilitate and enhance this process. Our work aims at combining effectively the technology and the powerful tradition to make them go hand in hand in order to achieve a threefold educational goal:

- make students think better than they did before, e.g. to construct correct mental models, using CAS to help them see the consistency or inconsistency of their evolving mental models [5]

- develop improved student understanding and the right appreciation of mathematics and of its role in the everyday life
- help students acquire a life long habit of doing things not just anyhow including the habit of working smarter not harder.

Our strategy of integrating CAS is based on general methodological principles: systematization and consistency of subject content; accessibility; visualization; personalization; students' conscious involvement in learning and reflection on the activities.

4 (What + When + Where) · Why: Mathematics with Less/More CAS

In the methodology of mathematics teaching and learning the main question is "Why?". It is about the learning outcomes and educational goals as well as about educational values. The chain of questions "What–When–Where" are also associated with it. On one side, they are related to the curricula and courses. On the other side, they are mostly relevant to the teaching–learning–assessment (TLA) process. In the past 2–3 decades remarkable creative work with application of Computer Algebra Systems (CAS) has been done in undergraduate mathematics concerning these three questions. The care about the Why-question will never end because the world constantly changes. CAS are full of opportunity in this direction for adequate decisions in correspondence with the educational values and tradition in the country.

After many years of exploring CAS in education, we had to turn back to the main statement in the European Qualification Framework (EQF) of University Education: the quality of the education is to be evaluated by its results: Knowledge, Skills and Competency of the students. In this sense, the principle "Consider All Factors" introduced by De Bono [2] is to be followed by both teachers and students in their work with CAS. Otherwise CAS would not be used effectively and would not serve to the achievement of these results because "One can see as much as one knows".

There is no doubt that the students' basic knowledge on properties of elementary functions, on basic concepts such as sequences, series, functions, limits, integrals, matrices, simultaneous equations, curves, … is a crucial factor for good results of CAS-supported activities and for the enhancement of the students' achievements in mathematics. So, we face the old question "Knowledge for what?" or "What knowledge?"

Concerning application of CAS the educational tradition in the country is to be taken into account. We would like our students to be able to use CAS smarter, to make an intelligent extraction of information from any format it is represented, to distinguish structures/patterns, that is to be able to see the "global picture" and the details at same time, to solve some problems by observation only, to be emotionally involved doing mathematics, to verify the obtained solutions at least roughly or by different approaches (analytical, numerical, graphical, mixed) … .

According to our tradition we require students to develop analytical and critical thinking, to look for a concise solution, to acquire and connect knowledge from different topics and areas, to select appropriate prior and newly acquired knowledge in order to manage with the situation. Here are examples allowing to approach them with "less computer/CAS".

(1) Find the critical (stationary) point of the function $f(x, y) = 2x^2 + y^2 + 8x - 6y + 20$ and determine its nature.

Comment By means of the Second Partials Test, the student can easily answer the question with or without using technology.

Such examples however we give to students not for exercising this Test—as it is in most textbooks. On the contrary, we give it to provoke and develop their sense of simplicity and conciseness and to impress on them that they should not leave off the good old ideas as the one of "completing the square" $f(x, y) = 2(x + 2)^2 + (y - 3)^2 + 3$. Hence, $f(x, y) \geq f(-2, 3) = 3 = f_{min}$.

In this sense, both teachers and students should not become "captives" of technology: in cases like the above one its use would be against the role of mathematics for developing effectiveness of mind. Technology is not to act as a cloak of not good enough practices in education.

(2) The evaluation of a definite integral on a symmetric interval and of an odd function. Solving it with CAS is by no means the way students to appreciate the power and take advantage of CAS.

(3) Calculation of a limit of a function without being able to interpret the result. This would convince the student that technology is not a panacea for the lack of knowledge and thinking.

(4) The first derivative of a function is a sum, product or ratio, say, of perfect squares and exponential functions. Does the student need CAS to decide whether the function is increasing?

(5) The integrand is equal to the first derivative (obtained by the Chain Rule) of a function. Do students need CAS or Substitution method? ("Incorrect thoughts of people are due to their not enough developed ability to distinguish." Paramahansa Yogananda).

(6) Calculation of the determinant of a square triangular or diagonal matrix. With or without CAS?

5 Technology-Meets-Tradition Activities

> Small things make perfection
> but perfection is not a small thing.

We have been developing our approach having non-mathematics students in mind. We tried not to hide the difficulties and oversimplify the matters: it would be of no real help to the students in preparing them for their professional career.

To overcome the constraints of a technology-free teaching and learning environment we use CAS to:

- enhance learning of topics and solving problems where graphical illustrations facilitate the learning process
- introduce new concepts with multiple use in later sections of mathematics courses
- teach topics that are difficult to master, in which conceptual prototypes enhance the learning
- challenge or extend existing ideas, and encourage students to construct new cognitive models.

During the TLA process we observe "the student's trajectory of learning" [3], mainly the kind of material they are able to go into, the sort of components they pick up or do not pick up, their mathematical ideas and interpretation skills. We try to provide an appropriate support at the right moment to help students make progress. Along with acquiring knowledge and developing skills for its application, another educational goal is to create the habit of verification of the results.

In the next part of the paper, a number of topics and representative examples are considered to illustrate some aspects of our experience in integrating CAS into undergraduate mathematics education; one can easily understand why we have selected them.

5.1 Innovate Not Imitate ("More CAS")

Appropriately structured systems of questions and problems for each topic and for the entire course are at the heart of any mathematics course. Examples of the type below could be discussed in exercises on inverse functions, which precede exercises on differentiation. At the same time, it is perfect for technology-meets-tradition activity and it is here that CAS can come to the rescue [10].

Example 1. Find the first derivative of $f(x) = \arctan(\sqrt{\frac{1-\sin 2x}{1+\sin 2x}})$.

Solution. The graph of the function (Fig. 1) gives the teacher and the student the cue how to approach the problem: "First simplify!"—to the equivalent periodic piecewise linear function, and then differentiate:

$$f(x) = \begin{cases} -x + \frac{\pi}{4}, & x \in (-\frac{\pi}{4}, \frac{\pi}{4}] \\ x - \frac{\pi}{4}, & x \in (\frac{\pi}{4}, \frac{3\pi}{4}] \end{cases}$$

Such a function is definitely not suitable for exercising the Chain Rule, namely:

$$y' = \frac{1}{1+(\sqrt{\frac{1-\sin 2x}{1+\sin 2x}})^2} \cdot \frac{1}{2\sqrt{\frac{1-\sin 2x}{1+\sin 2x}}} \cdot (\frac{1-\sin 2x}{1+\sin 2x})' = \cdots$$

In the twenty-first century, this could be done just to show the digital generations how they are not supposed to approach such problems. It is tedious, time-wasting and could put students off mathematics for life.

The derivative of interest could be straightforward obtained with CAS but student's reflection on the result needs to approve it, for instance geometrically. The Russian proverb "Trust but Check Up" proved to be useful in doing mathematics with CAS.

Fig. 1 Example 1

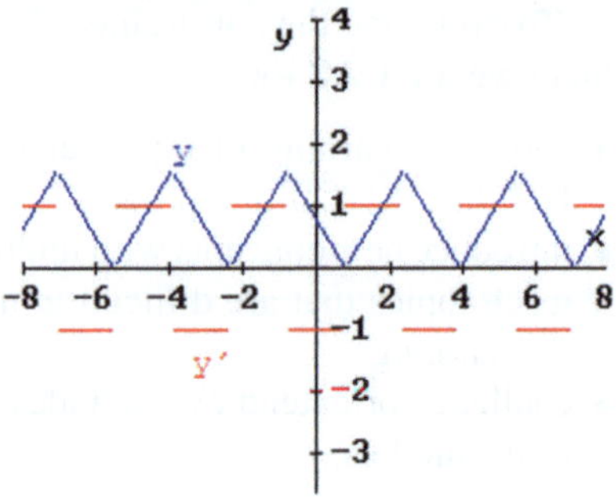

The above given approach would help student develop the habit

- to solve problems not just anyhow
- to verify the correctness of the result by different means
- to use technology for effective solution of a problem following the Nature Law "To do More with Less".

5.2 *Challenge Existing Ideas ("More CAS")*

The activities with application of CAS in the teaching and learning of the newly introduced concepts of Taylor/MacLaurin polynomial approximation, remainder of a series and error of approximation have been carried out to help students become aware about the necessity of rigorous knowledge and critical thinking [6]. The work aims at student's conceptual understanding of the new mathematical object, namely Taylor polynomial. The following traditional questions are posed by the teacher [1, 10, 11]:

1. Given the order of approximation and interval of interest, calculate the error of approximation.
2. Given the order of approximation and desired accuracy, find the appropriate interval for x.
3. Given the interval for x and the maximum error of approximation, find the minimal order of approximation.

CAS provide an environment for challenging the fundamental idea of approximation as they help students visualize the mathematical objects. Expansions of different orders superimposed on the graph of the approximated function can remove students misunderstanding of "+..." at the end of Taylor series and facilitate the construction of their mental models about it [5].

In the general case Question 3 cannot be solved analytically. The graphical opportunities of CAS make this additional activity possible. Following the rule "From Known to Unknown" the sine function has been used as a target function.

Example 2. For which values of n the error in the Taylor polynomial approximation to the sine function is no greater than 0.001 over the interval $[-\frac{\pi}{2}, \frac{\pi}{2}]$.

Solution. Figure 2 shows the operations performed to find the minimal order of approximation n. The inequality for n is based on the theoretical truncation error bound. This type of activities can be used to expose the well-known fact that "Knowledge is power, technology is powerful tool". Often we use another expression: "Knowledge is driver, technology is car".

$$\mathrm{SOLVE}\left(\frac{\left(\frac{\pi}{2}\right)^n}{n!} \le 5\cdot 10^{-3},\ n,\ \mathrm{Real}\right)$$

$$200\cdot\left(\frac{\pi}{2}\right)^n - n! \le 0 \le n! \vee n! \le 0 \le 200\cdot\left(\frac{\pi}{2}\right)^n - n!$$

Fig. 2 Example 2

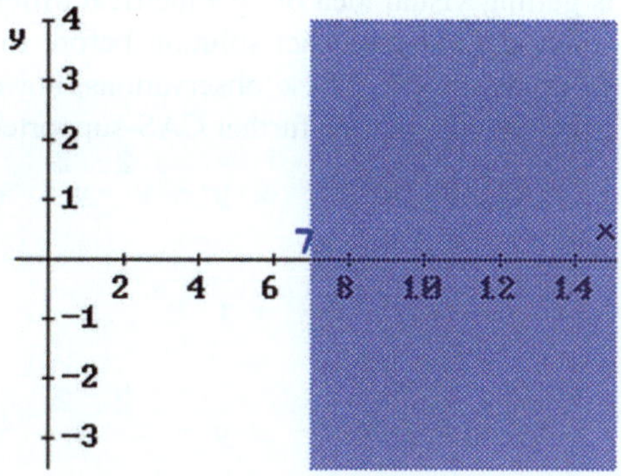

To demand student reflection on the work done by CAS a discussion with students about the difference between the actual and approximate values is carried out. The teacher points out the importance of knowing the properties of the sine function in order to make correct use of the constructed polynomial.

5.3 *Extend Existing Ideas ("Only CAS")*

The example below is given to students to illustrate that CAS can make a valuable contribution to any aspect of mathematics as a science.

Example 3. Use CAS to solve the initial-value problem $y' + xy = -x^2y^2$, $y(0)=1$.

Solution. First approach: The student can use a library function to attempt to find the required particular solution.

$$\text{BERNOULLI\_ODE}(x, -x^2, 2, x, y, 0, 1)$$

$$\frac{1}{y} = \hat{e}^{x^2/2} \cdot \left(\frac{\sqrt{2} \cdot \sqrt{\pi} \cdot \text{ERF}\left(\frac{\sqrt{2} \cdot x}{2}\right)}{2} + 1 \right) - x$$

The CAS response shows that it cannot proceed to the exact solution in closed form. Using the graphical capabilities of CAS the student can sketch this solution curve (Fig. 3).

Second approach: Any CAS has available a Taylor series method. An application of students' knowledge of Taylor's series can be thus successfully made for solving ODE. Approximate Taylor series solutions of ODE help the student easier understand the idea of approximate solution and the important concept of error in (or equivalently, accuracy of) the approximate solution.

For Example 3 Taylor series solutions of order 3 and 7 are obtained with CAS and compared to the exact solution. On the basis of their graphs in Fig. 3 the student is getting visual idea of "geometrical differences" between approximate analytical solutions and the exact solution before to start discussion on their accuracy and its improvement. These observations, however, are next thoroughly discussed and generalized and then further CAS-supported experiments are carried out.

$$\text{TAYLOR\_ODE1}(-x \cdot y - x^2 \cdot y^2, x, y, 0, 1, 3)$$

$$-\frac{x^3}{3} - \frac{x^2}{2} + 1$$

$$\text{TAYLOR\_ODE1}(-x \cdot y - x^2 \cdot y^2, x, y, 0, 1, 7)$$

$$-\frac{23 \cdot x^7}{210} + \frac{13 \cdot x^6}{144} + \frac{4 \cdot x^5}{15} + \frac{x^4}{8} - \frac{x^3}{3} - \frac{x^2}{2} + 1$$

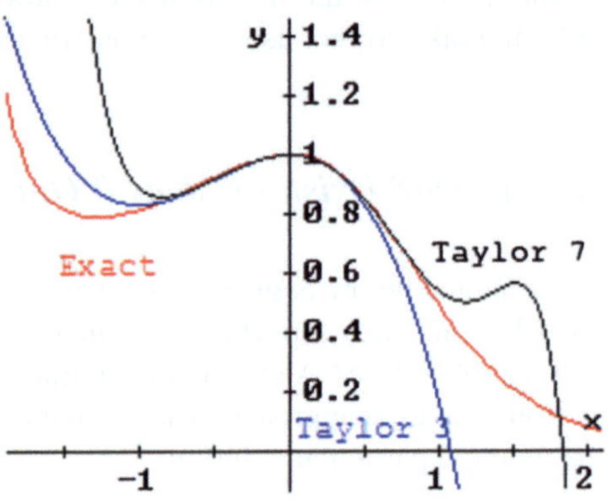

Fig. 3 Example 3

It is the teacher's concern for the student not to lose the sight of the main aim in constructing approximate solutions: to choose an order of approximation that ensures the "fit" of the approximate solution to the exact one over a given interval of the independent variable (x).

This kind of CAS-supported activity enhances student's comprehension of main concepts related to exact and approximate solutions and promotes links between the two types of mathematics knowledge: conceptual and procedural.

5.4 Work Smarter Not Harder ("Only CAS")

Action is Enemy of Thought.

One more example is provided here to illustrate one of the advantages of CAS in the development of students' resourcefulness and creativity in case they possess knowledge of geometrical interpretation of simple integrals.

Example 4. (a) Calculate the value of the integral: $\int_0^3 \sqrt{9 - x^2} dx$.

Solution: Actually, no calculation is needed in this particular case. Plotting the integrand and shading the area of interest (Fig. 4) the student can compute the value "by observation".

To actively and interactively involve the students in their own learning the teacher can ask them to experiment with similar types of integrals progressing in difficulty (according to a didactic principle: "From simple to complex, from easy to difficult, from known to unknown" [1–3]):

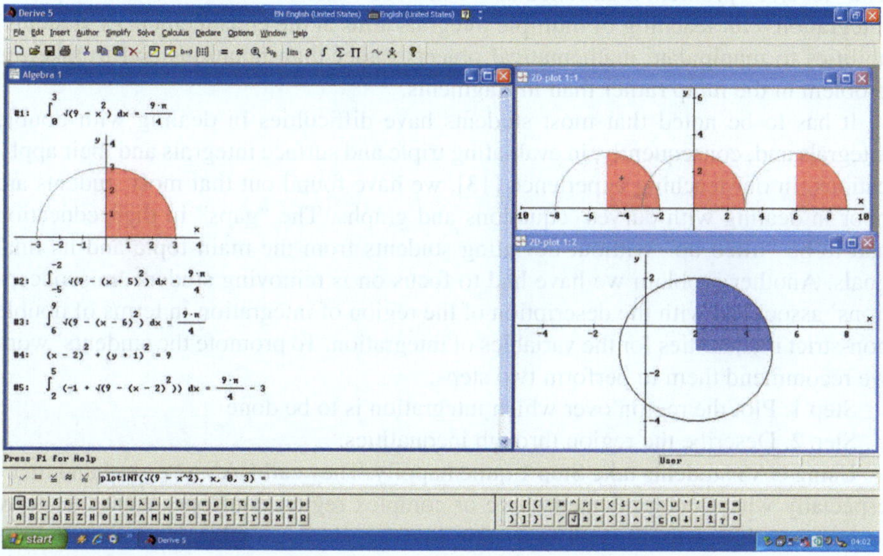

Fig. 4 Example 4

$\int_0^r \sqrt{r^2 - x^2}dx \Rightarrow \int_0^r \sqrt{2rx - x^2}dx \Rightarrow \int_a^{a+r} \sqrt{r^2 - (x-2)^2}dx$
$\Rightarrow \int_a^{a+r} (b + \sqrt{r^2 - (x-2)^2})dx.$

They help the students observe the phenomena of describing different objects by the same concept of definite integral. They become aware of the meaning and the importance of its "components": the integrand and the limits of integration (Fig. 4).

In mathematics textbooks either Substitution Method or Integration by Parts is usually applied for solving these types of integrals. Having CAS at disposal however, application of these methods simply recall the proverb "To Kill Sparrow by Cannon". There exists a great number of integrals suitable for their application and which cannot be solved using "conventional weapons": common sense, definitions, properties, geometric interpretations of concepts, etc.

(b) Using the approach in (a) calculate
$S = \int_0^1 \sqrt{1 - x^2}dx = \int_0^1 \sqrt{\frac{1}{4} - x^2}dx$ in your head. Have you got $\frac{3\pi}{16}$?

The "reverse" problem is also used to enhance students' feel of mathematics. They are asked to plot and shade the area first and then construct and calculate the corresponding integral. And CAS is irreplaceable for this purpose. Later, they can easily "toggle" on double integrals.

5.5 *Save Time and Facilitate Problem Solving ("More CAS")*

Imagination is more important than knowledge.
(A. Einstein)

The "reverse" problem mentioned above has been used as an introduction to double integration. Our teaching of multiple integrals aims at the enhancement of students' abilities to manipulate mathematical objects rather than notations and to take the problem in the lump rather than in fragments.

It has to be noted that most students have difficulties in dealing with double integrals and, consequently, in evaluating triple and surface integrals and their applications. In our teaching experience [13], we have found out that most students are poor in dealing with curves' equations and graphs. The "gaps" in their education had to be "filled up" without deviating students from the main topic and its final goals. Another problem we have had to focus on is removing student 'misconceptions' associated with the description of the region of integration in terms of double non-strict inequalities for the variables of integration. To promote the students' work we recommend them to perform two steps:

Step 1. Plot the region over which integration is to be done

Step 2. Describe the region through inequalities.

Using CAS students take Step 1 quite happily. They call CAS a "right-hand man" especially when a complicated curve or complex region confronts them. We have to mention that without such tools this first step is impossible for most students and, hence, much time has to be spent for teaching plane curves instead of double integrals.

To describe the region, the student has to operate with equations of the bounding curves. As some students find it difficult to solve equations for one of the variables, or they solve them wrongly, we recommend CAS as a helpful and reliable assistant.

How to evaluate the constructed double integral—this is a question which mathematics teachers are nowadays also confronted by. Only the educational goals and values can give the answer. Whatever they may be, CAS can considerably contribute to their achievement. The answer is no more unique as it was "before the age of CAS".

5.6 CAS-Supported Environment for Difficult to Master Topics ("More CAS")

Every picture tells a story.

The experience has shown that non-mathematics students have a preference for visual and practical illustrations of mathematical concepts, objects and results and for learning from concrete examples before moving on to abstract theory. The lack of a good visualization forms a gap in students' learning. The Window Shuttle Method makes CAS a powerful instrument for learning through "prototypes": numerical, analytical and graphical. The two examples below have been chosen to meet the students' preference.

Example 5. Apply the Second Derivative Test to locate and classify the critical points of $f(x, y) = x^3 + 3xy^2 - 15x = 12y$. Calculate the values of the function at these points. Plot the surface and use the option of Animation to observe the surface behaviour making difference between the different types of critical points.

CAS-supported activities have been developed to teach and discuss different types of limits of functions of two variables (FTV) and interrelations between them as well as to generate dynamic graphics to give geometrical meaning to definitions, properties, rules and theorems. The topic has become attractive thanks the opportunity for experimental observations on a variety of selected insightful examples. Through them many difficulties in the student comprehension of the concept of limit have been removed. Plotting the surfaces the student observes the existence or non-existence of gaps or holes in the graph of the function that hitherto were only possible in thought. The activities reveal the key aspects of the underpinning theory giving CAS techniques conceptual dimensions. They support the teaching–learning process in linking prior (simple limit, level curves) and newly acquired (iterated limit, double limit) knowledge. This helps teachers find out what mix of which types of learning environments is better suited for a particular topic or type of problems.

Example 6. (The iterated limits exist and are equal; the function approaches the same value along any straight line; the limit does not exist)

Show that the iterated limits of the function $H(x, y) = \frac{x^2 y}{x^4 + y^2}$ are equal to zero when $x \to 0$ and $y \to 0$ and the limit along any direction through the origin is also zero.

1. Does the limit $L = \lim_{(x,y)\to(0,0)} H(x, y)$ exist?
2. Find the limit of $H(x, y)$ as $(x, y) \to (0, 0)$ along the parabola $y = x^2$. Is your conclusion in (1) correct?

Our experience has proved CAS as a powerful and time saving instrument for students to be quick on the uptake of the diversity of FTV's behaviour.

6 About the National (Bulgarian) Student Olimpyad on Computer Mathematics

The participants are free to choose the technology for digital mathematics they prefer for each single problem. They are allowed to combine CAS and Excel as well as two or more CAS. Thirty problems from different mathematics subjects (Algebra, Analytic Geometry, Calculus, Differential Equations) have to be solved within 4 h.

Example 7. Determine the values of the real parameter M so that the polynomial $P_3(x) = (m-2)x^3 - 3mx^2 - 3mx + 2 - m$ has a double real root.

Solution. The solution can be determined at least by three different methods. Most participants have shown good theoretical background and higher order learning: they have preferred the application of Calculus in Algebra and quickly calculated the correct answer $m = \frac{1}{2}$.

7 Conclusion

It is very important to properly use an appropriate educational technology (ET) defined as [7]:

ET = Technology OF Education + Technology IN Education.

Tradition and Technology have to go hand in hand. Technology can "replace" hundreds of teachers but a powerful tradition and teachers can give thousands of technologies vitality.

Any CAS itself is not an ordinary tool: it is a thought-provoking tool that requires knowledge, provides information, and stimulates the acquisition of knowledge and development of skills and habits.

The proper utilization of technology in mathematics education requires a policy of support for research in the field of Educational Science, high quality software and teacher training. We need to go further: from

3T = Teachers Teaching with Technology

to

4C = Challenging Changes in Curricula and Courses

and

$$4L = \text{Life Long Love to Learning.}$$

In short, after 17-year teaching experience with application of CAS I can simply repeat the words said by the Spanish painter Francisco Goya: "I am still studying".

Acknowledgement to the Developers of CAS

We feel fortunate that we have the rare opportunity to experience through CAS the dramatic unity of knowledge, capability and tool. Such unity is a guaranty to achieve one of the main educational goals: Enhancement of Human Development Index.

References

1. Arens, T., Hettlich, F., Karpfinger, Ch., Kockelkorn, U., Lichtenegger, K., Stachel, H.: Mathematik. Spektrum, Heidelberg (2008)
2. De Bono, E.: Teaching Thinking. Penguin Books, London (1984)
3. Ganchev, G., Ninova, Yu., Nikova, V.: Methodology of Mathematics Education. SWU "Neofit Rilski", Blagoevgrad (2007) (in Bulgarian)
4. Gardner, H.: Five Minds for the Future. Harvard Business School Press, Boston, MA (2007)
5. Graham, T., Rowlands, S.: Using computer software in teaching mechanics. Int. J. Math. Educ. Sci. Technol. **31**(4), 479–493 (2000)
6. Grozdev, S.: For High Achievements in Mathematics. The Bulgarian Experience (Theory and Practice). ADE, Sofia (2007)
7. Kumar, K.L.: Educational Technology. New Age International Publishers, New Delhi (1996)
8. Todorov, M., Varbanova, E.: Application of DERIVE in the investigation of explicit functions of two variables. In: Proceedings of the Summer School "Applications of Mathematics in Engineering'24", Herron Press, Sofia (1999)
9. Varbanova, E.A.: CAS supported environment for learning and teaching calculus. CBMS—Issues in Mathematics Education: Enhancing University Mathematics, vol. 14, AMS & MAA (2007)
10. Varbanova, E.: Calculus—I. Lecture Notes. TU-Sofia, Sofia (2009) (in Bulgarian)
11. Varbanova, E.: Calculus—I. Exercises. TU-Sofia, Sofia (2011) (in Bulgarian)
12. Varbanova E., Stoynov, Y.: DERIVE-Approach to first order ordinary differential equations. In: Proceedings of the XIX International Summer School "Applications of Mathematics in Engineering and Economics", Sozopol (2003)
13. Varbanova, E.A., Patel, M.K., Marinova, D.: Tradition and innovation in teaching and learning double integral. In: Proceedings of ICTMT5, Klagenfurt (2001)

Some Remarks on Taylor's Polynomials Visualization Using Mathematica in Context of Function Approximation

Włodzimierz Wojas and Jan Krupa

Abstract In this paper the authors critically analyse popular way of graphic presentation Taylor's polynomials in context of function approximation. They discuss the difficulties of presentation the best local polynomial approximation of function by Taylor's polynomials. Proposed by the authors method of graphical presentation based on table of function and Taylor's polynomials values in neighbourhood of a chosen point. For graphical presentation ListPlot and Plot functions with logarithmic scale in Mathematica System are used.

Keywords Mathematical analysis · Taylor's theorem · Taylor's polynomials · Higher education · Local approximation · Application of CAS

Mathematics Subject Classification (2010): 97R20 · 97B40 · 97I30 · 97I40 · 41A10

1 Introduction

Taylor's theorem is one of the most classic results of university course in calculus or mathematical analysis. For the case of one variable function $y = f(x)$ and a point $x = x_0$, Taylor's polynomial of the nth order is defined as:

$$T_n(x) = f(x_0) + \frac{f'(x_0)}{1!}(x - x_0) + \frac{f''(x_0)}{2!}(x - x_0)^2 + \cdots + \frac{f^{(n)}(x_0)}{n!}(x - x_0)^n$$

where a function $f(x)$ have at a point x_0 finite derivatives up to the nth order inclusively. Many academic books e.g. [1, 3, 5, 7] contain graphs presented Taylor's polynomials for some elementary functions. For example for $f(x) = e^x$ or $f(x) = \sin x$

W. Wojas (✉) · J. Krupa
Department of Applied Mathematics, Warsaw University of Life Sciences (SGGW),
ul. Nowoursynowska 159, 02-776 Warsaw, Poland
e-mail: wlodzimierz_wojas@sggw.pl

J. Krupa
e-mail: jan_krupa@sggw.pl

I.S. Kotsireas and E. Martínez-Moro (eds.), *Applications of Computer Algebra*,
Springer Proceedings in Mathematics & Statistics 198,
DOI 10.1007/978-3-319-56932-1_31

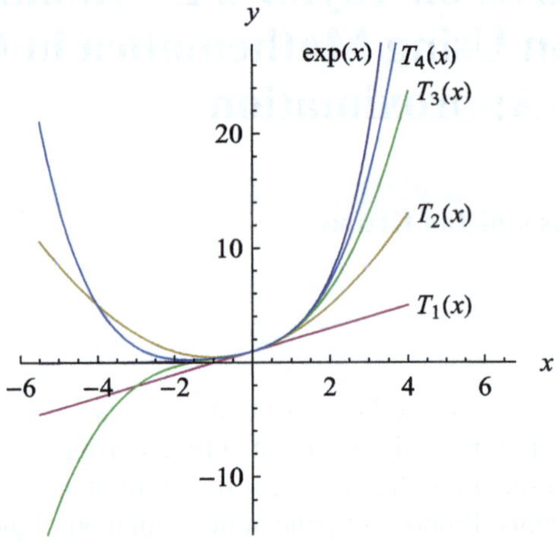

Fig. 1 Taylor's polynomials $T_1(x)$, $T_2(x)$, $T_3(x)$, $T_4(x)$ for function $f(x) = e^x$ at point $x_0 = 0$

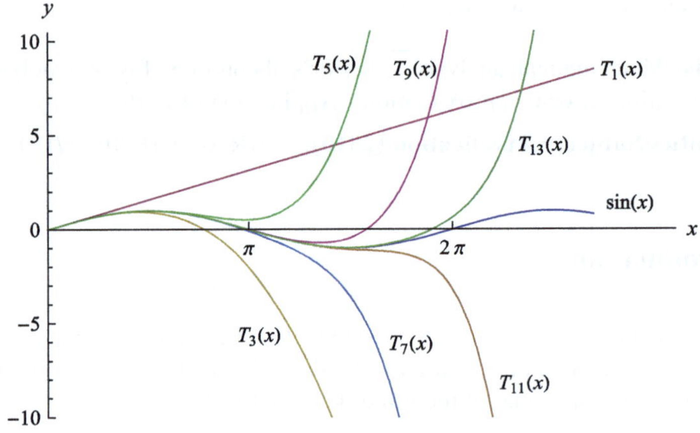

Fig. 2 Taylor's polynomials $T_1(x), T_3(x), \ldots, T_{13}(x)$ for function $f(x) = \sin x$ at point $x_0 = 0$

as is shown in Figs. 1 and 2. Often these graphs are presented with comments that it shows how well these polynomials approximate $y = f(x)$ near a point $x = x_0$ when n increases.

2 Visualization of Taylor's Polynomials in Context of Function Approximation

Visualization of Taylor's polynomials is easy and comfortable using CAS packages such as Mathematica, Maple, Derive or others. For one variable functions Math-

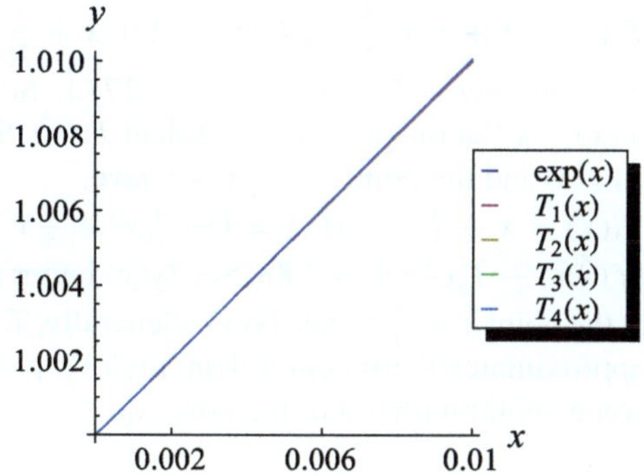

Fig. 3 Function $f(x) = e^x$ and its Taylor's polynomials $T_1(x)$, $T_2(x)$, $T_3(x)$, $T_4(x)$ in the reduced right neighbourhood (0, 0.01) of the point $x_0 = 0$

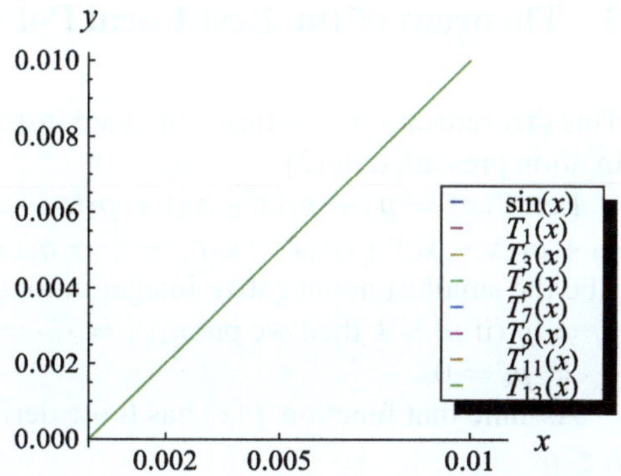

Fig. 4 Function $f(x) = \sin x$ and its Taylor's polynomials $T_1(x)$, $T_3(x), \ldots, T_{13}(x)$ in the reduced right neighbourhood (0, 0.01) of the point $x_0 = 0$

ematica package contains standard procedure Series[f, $\{x, x_0, n\}$] which generates Taylor's polynomial of the n-th order for the function $f(x)$ and point $x = x_0$. Using procedure Plot[$\{f_1, f_2, \ldots, f_k\}$, $\{x, x_{\min}, x_{\max}\}$] we can present graphs function $f(x)$ and some its Taylor's polynomials as is shown in Figs. 1 and 2. In academic books these graphs are often presented with comments that they show how well these polynomials approximate $y = f(x)$ near a point $x = x_0$ when n increases. The question may appear: in which sense these approximation is good? This kind of presentation can be misleading for students if we do not emphasize the fact of local character of this approximation. In Figs. 3 and 4 we see that graph of the function $f(x)$ and graphs of Taylor's polynomials seem to overlap close point $x = x_0$. On the base of these figures we cannot settle which Taylor's polynomial better approximates the function close to the point x_0.

In Figs. 1 and 2 we see that graph of the function $f(x)$ and graphs of Taylor's polynomials seem to overlap close the point $x_0 = 0$. Next, Taylor's polynomials separate from the graph of the $f(x)$. Closer to the point x_0 separates Taylor's polynomial of lower order, further from the point x_0 separates Taylor's polynomial of higher order. Figures 1 and 2 may suggest that overall Taylor's polynomial of higher order better approximates the function than Taylor's polynomial of lower order. But for example, for the function $f(x) = e^x$, $x_0 = 0$ and the point $x = -4$ it is easy to check that:

$T_2(x) = 1 + x + \frac{1}{2!}x^2$, $T_3(x) = 1 + x + \frac{1}{2!}x^2 + \frac{1}{3!}x^3$, $|f(-4) - T_2(-4)| = |e^{-4} - 5| < |f(-4) - T_3(-4)| = |e^{-4} + 17/3|$. So, $T_2(x)$ better approximates the function $f(x) = e^x$ at the point $x = -4$ than $T_3(x)$. Similarly, for the function $f(x) = \sin x$, $x_0 = 0$ and the point $x = \frac{5}{4}\pi$ we have:
$T_3(x) = x - \frac{1}{3!}x^3$, $T_5(x) = x - \frac{1}{3!}x^3 + \frac{1}{5!}x^5$, and $|f(\frac{5}{4}\pi) - T_3(\frac{5}{4}\pi)| \approx 5.46 < |f(\frac{5}{4}\pi) - T_5(\frac{5}{4}\pi)| \approx 7.88$. So, $T_3(x)$ better approximates the function $f(x) = \sin x$ at the point $x = \frac{5}{4}\pi$ than $T_5(x)$. Generally, Taylor's polynomial of higher order better approximates the function than Taylor's polynomial of lower order only locally in some neighbourhood of the point x_0.

3 Theorem of the Best Local Polynomial Approximation

This theorem and corollaries from it are inspired by theorem of the best local approximation presented in [2].

Let $P(x) = p_0 + p_1(x - x_0) + p_2(x - x_0)^2 + \cdots + p_m(x - x_0)^m$ and $Q(x) = q_0 + q_1(x - x_0) + q_2(x - x_0)^2 + \cdots + q_k(x - x_0)^k$ are different polynomials. Let r be the smallest nonnegative integer among numbers $i = 0, 1, 2, \ldots$ which satisfy $p_i \neq q_i$ (if $m > k$ then we put $q_{k+1} = \cdots = q_m = 0$, if $m < k$ then we put $p_{m+1} = \cdots = p_k = 0$).

Assume that function $f(x)$ has finite derivative of n order at point x_0 and assume $r \leq n$.

Theorem *If $p_i = \frac{f^{(i)}(x_0)}{i!}$ for all $i < r$ and $|\frac{f^{(r)}(x_0)}{r!} - p_r| < |\frac{f^{(r)}(x_0)}{r!} - q_r|$ then there exists such neighbourhood S of point x_0 that $\bigwedge_{x \in S \setminus \{x_0\}} |f(x) - P(x)| < |f(x) - Q(x)|$.*

Proof By Taylor's theorem we have: $f(x) - T_n(x) = (x - x_0)^n \omega(x)$, where $\omega(x)$ is a function continuous at x_0 and $\omega(x_0) = 0$. Thus:

$$
\begin{aligned}
&|f(x) - P(x)| \\
&= \left| \left(\frac{f^{(r)}(x_0)}{r!} - p_r \right)(x - x_0)^r + \frac{f^{(r+1)}(x_0)}{(r+1)!}(x - x_0)^{r+1} + \cdots + \frac{f^{(n)}(x_0)}{n!}(x - x_0)^n \right. \\
&\qquad \left. + (x - x_0)^n \omega(x) - p_{r+1}(x - x_0)^{r+1} - \cdots - p_m(x - x_0)^m \right|,
\end{aligned}
$$

$$
\begin{aligned}
&|f(x) - Q(x)| \\
&= \left| \left(\frac{f^{(r)}(x_0)}{r!} - q_r \right)(x - x_0)^r + \frac{f^{(r+1)}(x_0)}{(r+1)!}(x - x_0)^{r+1} + \cdots + \frac{f^{(n)}(x_0)}{n!}(x - x_0)^n \right. \\
&\qquad \left. + (x - x_0)^n \omega(x) - q_{r+1}(x - x_0)^{r+1} - \cdots - q_k(x - x_0)^k \right|.
\end{aligned}
$$

Taking the factor $(x - x_0)^r$ out we have:

$$|f(x) - P(x)|$$
$$= |(x - x_0)^r| \cdot \left| \left(\frac{f^{(r)}(x_0)}{r!} - p_r \right) + \frac{f^{(r+1)}(x_0)}{(r+1)!}(x - x_0) + \cdots + \frac{f^{(n)}(x_0)}{n!}(x - x_0)^{n-r} \right.$$
$$\left. + (x - x_0)^{n-r}\omega(x) - p_{r+1}(x - x_0) - \cdots - p_m(x - x_0)^{m-r} \right|.$$

$$|f(x) - Q(x)|$$
$$= |(x - x_0)^r| \cdot \left| \left(\frac{f^{(r)}(x_0)}{r!} - q_r \right) + \frac{f^{(r+1)}(x_0)}{(r+1)!}(x - x_0) + \cdots + \frac{f^{(n)}(x_0)}{n!}(x - x_0)^{n-r} \right.$$
$$\left. + (x - x_0)^{n-r}\omega(x) - q_{r+1}(x - x_0) - \cdots - q_k(x - x_0)^{k-r} \right|.$$

The above equalities are true if $m > r$ and $k > r$. If $m \leq r$, then defining $p_{m+1} = p_{m+2} = \cdots = 0$ we have:

$$|f(x) - P(x)|$$
$$= \left| \left(\frac{f^{(r)}(x_0)}{r!} - p_r \right)(x - x_0)^r + \frac{f^{(r+1)}(x_0)}{(r+1)!}(x - x_0)^{r+1} + \cdots + \frac{f^{(n)}(x_0)}{n!}(x - x_0)^n \right.$$
$$\left. + (x - x_0)^n\omega(x) \right|$$
$$= |(x - x_0)^r| \cdot \left| \left(\frac{f^{(r)}(x_0)}{r!} - p_r \right) + \frac{f^{(r+1)}(x_0)}{(r+1)!}(x - x_0) + \cdots + \frac{f^{(n)}(x_0)}{n!}(x - x_0)^{n-r} \right.$$
$$\left. + (x - x_0)^{n-r}\omega(x) \right|$$

and if $k \leq r$ then defining $q_{k+1} = q_{k+2} = \cdots = 0$ we have:

$$|f(x) - Q(x)|$$
$$= \left| \left(\frac{f^{(r)}(x_0)}{r!} - q_r \right)(x - x_0)^r + \frac{f^{(r+1)}(x_0)}{(r+1)!}(x - x_0)^{r+1} + \cdots + \frac{f^{(n)}(x_0)}{n!}(x - x_0)^n \right.$$
$$\left. + (x - x_0)^n\omega(x) \right|$$
$$= |(x - x_0)^r| \cdot \left| \left(\frac{f^{(r)}(x_0)}{r!} - q_r \right) + \frac{f^{(r+1)}(x_0)}{(r+1)!}(x - x_0) + \cdots + \frac{f^{(n)}(x_0)}{n!}(x - x_0)^{n-r} \right.$$
$$\left. + (x - x_0)^{n-r}\omega(x) \right|$$

(both numbers k and m cannot be at the same time less than r).

As x approaches to x_0 we obtain:

$$\lim_{x\to x_0}\left|\left(\frac{f^{(r)}(x_0)}{r!}-p_r\right)+\frac{f^{(r+1)}(x_0)}{(r+1)!}(x-x_0)+\cdots+\frac{f^{(n)}(x_0)}{n!}(x-x_0)^{n-r}+(x-x_0)^{n-r}\omega(x)-p_{r+1}(x-x_0)-\cdots-p_m(x-x_0)^{m-r}\right| = \left|\frac{f^{(r)}(x_0)}{r!}-p_r\right|,$$

$$\lim_{x\to x_0}\left|\left(\frac{f^{(r)}(x_0)}{r!}-q_r\right)+\frac{f^{(r+1)}(x_0)}{(r+1)!}(x-x_0)+\cdots+\frac{f^{(n)}(x_0)}{n!}(x-x_0)^{n-r}+(x-x_0)^{n-r}\omega(x)-q_{r+1}(x-x_0)-\cdots-q_k(x-x_0)^{k-r}\right| = \left|\frac{f^{(r)}(x_0)}{r!}-q_r\right|.$$

Because of our assumption $\left|\frac{f^{(r)}(x_0)}{r!}-p_r\right| < \left|\frac{f^{(r)}(x_0)}{r!}-q_r\right|$ and the last two limits we conclude that there exists such neighbourhood S of point x_0 that:

$$\bigwedge_{x\in S\setminus\{x_0\}}\left|\left(\frac{f^{(r)}(x_0)}{r!}-p_r\right)+\frac{f^{(r+1)}(x_0)}{(r+1)!}(x-x_0)+\cdots+\frac{f^{(n)}(x_0)}{n!}(x-x_0)^{n-r}+(x-x_0)^{n-r}\omega(x)-p_{r+1}(x-x_0)-\cdots-p_m(x-x_0)^{m-r}\right|$$
$$<\left|\left(\frac{f^{(r)}(x_0)}{r!}-q_r\right)+\frac{f^{(r+1)}(x_0)}{(r+1)!}(x-x_0)+\cdots+\frac{f^{(n)}(x_0)}{n!}(x-x_0)^{n-r}+(x-x_0)^{n-r}\omega(x)-q_{r+1}(x-x_0)-\cdots-q_k(x-x_0)^{k-r}\right|.$$

Multiplying both sides of the last inequality by $|(x-x_0)^r|$ we obtain that

$$\bigwedge_{x\in S\setminus\{x_0\}} |f(x)-P(x)| < |f(x)-Q(x)|.$$

In cases $m\le r$ or $k\le r$ the proof is analogous. □

Corollary 1 *Let $Q(x)$ be a polynomial which satisfies: there exists i $(i\le n)$ such that $q_i\neq\frac{f^{(i)}(x_0)}{i!}$ (if $m<n$ then we define $q_{m+1}=q_{m+2}=\cdots=q_n=0$). Then there exists such neighbourhood S of point x_0 that, $\bigwedge_{x\in S\setminus\{x_0\}} |f(x)-T_n(x)|<|f(x)-Q(x)|$. Particularly $Q(x)$ can be any polynomial of order not greater than n different than $T_n(x)$.*

Corollary 2 *There exists such neighbourhood S of point x_0 that,*

$$\bigwedge_{x\in S\setminus\{x_0\}} |f(x)-T_n(x)|\le|f(x)-T_{n-1}(x)|\le\cdots\le|f(x)-T_1(x)|,$$

where every inequality from the last sequence of inequalities becomes equality if and only if when the two consecutive Taylor's polynomial of $f(x)$ in both sides of the inequality are identical.

4 Visualization of the Best Locally Approximation by Taylor's Polynomials with Mathematica

Let us visualize Corollarys 1 and 2 for reduced right neighbourhood (0, 0.01) of the point $x_0 = 0$ using Wolfram Mathematica System [4, 6].

Example 1 For the Corollary 1 we define: $f(x) = e^x$ $x_0 = 0$, $T_2(x) = 1 + x + \frac{1}{2!}x^2$ and $P(x) = 1 + x - \frac{1}{2!}x^2$ for $x \in (0, 0.01)$. By Taylor's theorem we get: $e^x - T_2(x) = e^x - (1 + x + \frac{1}{2!}x^2) = \frac{1}{3!}(e^{\tilde{x}})x^3 > 0$ and $e^x - P(x) = 1 + x + \frac{1}{2!}x^2 + \frac{1}{3!}(e^{\tilde{x}})x^3 - (1 + x - \frac{1}{2}x^2) = x^2 + \frac{1}{3!}(e^{\tilde{x}})x^3 > 0$ for $\tilde{x} \in (0, x)$, $x \in (0, 0.01)$.

Hence, we have: $|f(x) - T_2(x)| - |f(x) - P(x)| = e^x - (1 + x + \frac{1}{2}x^2) - e^x + 1 + x - \frac{1}{2}x^2 = -x^2 < 0$ and finally $\bigwedge_{x \in (0,0.01)} |f(x) - T_2(x)| < |f(x) - P(x)|$.

Let us visualize this inequality by creating a table of numerical values for both sides of inequality with step 0.001.

We see in Table 1 that for all considered points inequality is true. Based on the Table 1 we can prepare Fig. 5 using logarithmic scale. Increasing WorkingPrecision and Accuracy in Mathematica Plot function we can get the continuous graphs presented in Fig. 6.

In Figs. 5 and 6 we see that the graphs of $|f(x) - T_2(x)|$ and $|f(x) - P(x)|$ are separated and that $|f(x) - T_2(x)| < |f(x) - P(x)|$ for $x \in (0, 0.01)$.

Example 2 For the Corollary 2 we define: $f(x) = \sin x$, $x_0 = 0$,
$T_3(x) = x - \frac{1}{3!}x^3$,
$T_7(x) = x - \frac{1}{3!}x^3 + \frac{1}{5!}x^5 - \frac{1}{7!}x^7$,
$T_{11}(x) = x - \frac{1}{3!}x^3 + \frac{1}{5!}x^5 - \frac{1}{7!}x^7 + \frac{1}{9!}x^9 - \frac{1}{11!}x^{11}$.
By Taylor's theorem, for all $x \in (0, 0.01)$ we have:
$f(x) - T_3(x) = (\frac{1}{4!}\sin\tilde{x})x^4 > 0$,
$f(x) - T_7(x) = (\frac{1}{8!}\sin\tilde{\tilde{x}})x^8 > 0$,
$f(x) - T_{11}(x) = (\frac{1}{12!}\sin\tilde{\tilde{\tilde{x}}})x^{12} > 0$,
where $\tilde{x}, \tilde{\tilde{x}}, \tilde{\tilde{\tilde{x}}} \in (0, x)$.

Hence, for all $x \in (0, 0.01)$ we get:

$$|f(x) - T_3(x)| - |f(x) - T_7(x)| = f(x) - T_3(x) - f(x) + T_7(x) = \frac{1}{5!}x^5 - \frac{1}{7!}x^7$$

$$= \frac{1}{7!}x^5(42 - x^2) = \frac{1}{7!}x^5(\sqrt{42} - x)(\sqrt{42} + x) > 0,$$

Table 1 The values of $f(x)$, $T_2(x)$, $P(x)$, $|f(x) - T_2(x)|$ and $|f(x) - P(x)|$ with step 0.001

x	$f(x)$	$T_2(x)$	$P(x)$	$\lvert f(x)-T_2(x)\rvert$	$\lvert f(x)-P(x)\rvert$
0.	1	1.	1.	0	0
0.001	1.001	1.001	1.001	$1.667083416680558\times10^{-10}$	$1.000166708341668\times10^{-6}$
0.002	1.002	1.002	1.002	$1.334000266755581\times10^{-9}$	$4.001334000266756\times10^{-6}$
0.003	1.003	1.003	1.003	$4.503377026012934\times10^{-9}$	$9.004503377026013\times10^{-6}$
0.004	1.00401	1.00401	1.00399	$1.067734187235881\times10^{-8}$	0.00001601067734187236
0.005	1.00501	1.00501	1.00499	$2.085940106338357\times10^{-8}$	0.00002502085940106338
0.006	1.00602	1.00602	1.00598	$3.605406486485558\times10^{-8}$	0.00003603605406486486
0.007	1.00702	1.00702	1.00698	$5.726684855523160\times10^{-8}$	0.00004905726684855523
0.008	1.00803	1.00803	1.00797	$8.550427343117207\times10^{-8}$	0.00006408550427343117
0.009	1.00904	1.00904	1.00896	$1.217738678140626\times10^{-7}$	0.00008112177386781406
0.01	1.01005	1.01005	1.00995	$1.670841680575422\times10^{-7}$	0.0001001670841680575

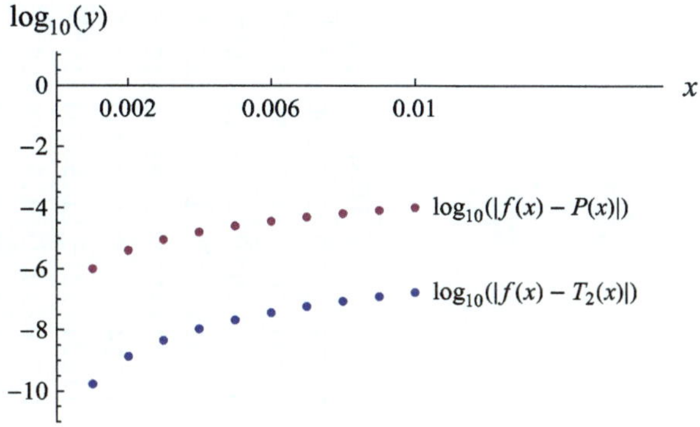

Fig. 5 Discrete graphs of $|f(x) - T_2(x)|$ and $|f(x) - P(x)|$ in reduced right neighbourhood $(0, 0.01)$ of the point $x_0 = 0$ with logarithmic scale using Mathematica Plot function

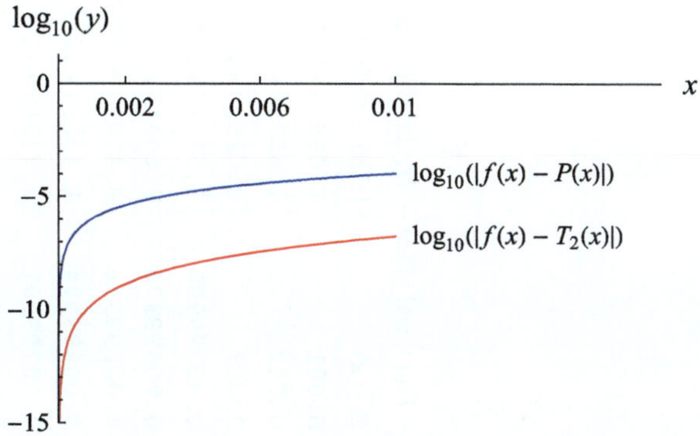

Fig. 6 Continuous graphs of $|f(x) - T_2(x)|$ and $|f(x) - P(x)|$ in reduced right neighbourhood $(0, 0.01)$ of the point $x_0 = 0$ with logarithmic scale using Mathematica Plot function

$$|f(x) - T_7(x)| - |f(x) - T_{11}(x)| = f(x) - T_7(x) - f(x) + T_{11}(x) = \frac{1}{9!}x^9 - \frac{1}{11!}x^{11}$$

$$= \frac{1}{11!}x^9(110 - x^2) = \frac{1}{11!}x^9(\sqrt{110} - x)(\sqrt{110} + x) > 0.$$

So, finally $\bigwedge_{x \in (0,0.01)} |f(x) - T_3(x)| > |f(x) - T_7(x)| > |f(x) - T_{11}(x)|.$

Let us visualize this double inequality by create a table of values for all sides of inequality with step 0.001.

We see in Table 2 that for all considered points double inequality is true. Based on the Table 2 we can prepare Fig. 7 using logarithmic scale.

Table 2 The values of $f(x)$, $T_3(x)$, $T_7(x)$, $T_{11}(x)$, $|f(x) - T_3(x)|$, $|f(x) - T_7(x)|$ and $|f(x) - T_{11}(x)|$ with step 0.001

x	$f(x)$	$T_3(x)$	$T_7(x)$	$T_{11}(x)$	$\|f(x) - T_3(x)\|$	$\|f(x) - T_7(x)\|$	$\|f(x) - T_{11}(X)\|$
0.001	0.001	0.001	0.001	0.001	8.45678×10^{-18}	2.75573×10^{-33}	1.60590×10^{-49}
0.002	0.002	0.002	0.002	0.002	2.66714×10^{-16}	1.41093×10^{-30}	1.31556×10^{-45}
0.003	0.003	0.003	0.003	0.003	2.02529×10^{-15}	5.42411×10^{-29}	2.56033×10^{-43}
0.004	0.00399999	0.00399999	0.00399999	0.00399999	8.53397×10^{-15}	7.22398×10^{-28}	1.07770×10^{-41}
0.005	0.00499998	0.00499998	0.00499998	0.00499998	2.60417×10^{-14}	5.38229×10^{-27}	1.96033×10^{-40}
0.006	0.00599996	0.00599996	0.00599996	0.00599996	6.47997×10^{-14}	2.77714×10^{-26}	2.09742×10^{-39}
0.007	0.00699994	0.00699994	0.00699994	0.00699994	1.40058×10^{-13}	1.11204×10^{-25}	1.55594×10^{-38}
0.008	0.00799991	0.00799991	0.00799991	0.00799991	2.73066×10^{-13}	3.69868×10^{-25}	8.82855×10^{-38}
0.009	0.00899988	0.00899988	0.00899988	0.00899988	4.92073×10^{-13}	1.06763×10^{-24}	4.08199×10^{-37}
0.01	0.00999983	0.00999983	0.00999983	0.00999983	8.33332×10^{-13}	2.75573×10^{-24}	1.60590×10^{-36}

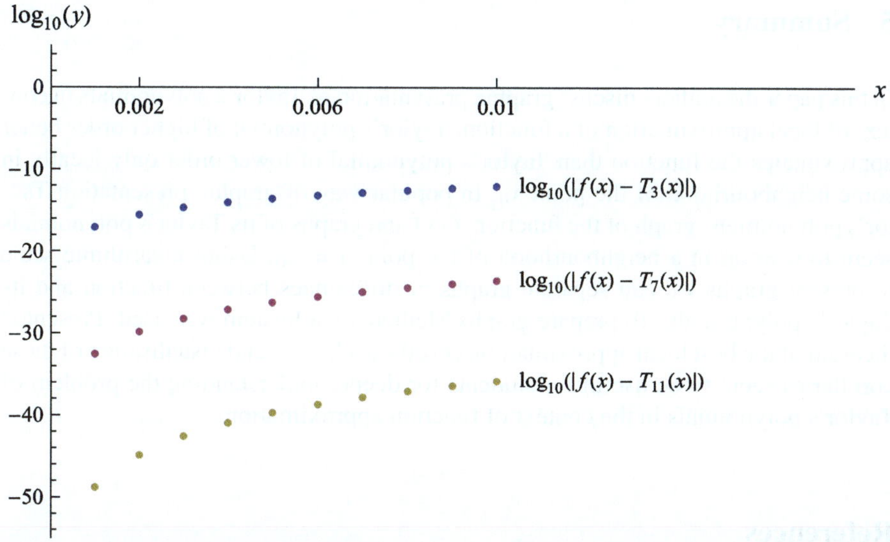

Fig. 7 Discrete graphs of $|f(x) - T_3(x)|$, $|f(x) - T_7(x)|$ and $|f(x) - T_{11}(x)|$ in reduced right neighbourhood (0, 0.01) of the point $x_0 = 0$ with logarithmic scale using Mathematica ListPlot function

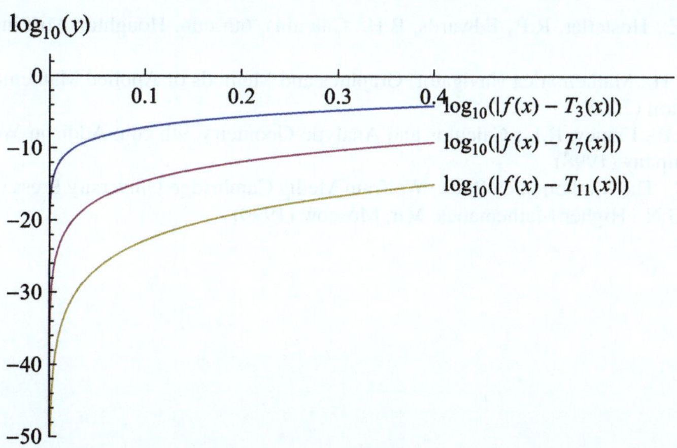

Fig. 8 Continous graphs of $|f(x) - T_3(x)|$, $|f(x) - T_7(x)|$ and $|f(x) - T_{11}(x)|$ in reduced right neighbourhood (0, 0.4) of the point $x_0 = 0$ with logarithmic scale using Mathematica Plot function

In Figs. 7 and 8 we see that graphs of $|f(x) - T_3(x)|$, $|f(x) - T_7(x)|$ and $|f(x) - T_{11}(x)|$ are separated.

5 Summary

In this paper the authors discuss graphic presentation of Taylor's polynomials in context of local approximation of a function. Taylor's polynomial of higher order better approximates the function than Taylor's polynomial of lower order only locally in some neighbourhood of the point x_0. In popular way of graphic presentation Taylor's polynomials, graph of the function $f(x)$ and graphs of its Taylor's polynomials seem to overlap in a neighbourhood of the point $x = x_0$. Using logarithmic scale to present graphs we can separate graphs of differences between function and its Taylor's polynomials. To prepare graphs Mathematica System was used. Presented theorem of the best local approximation, corollaries from it and visualisation of these corollaries seem to be useful for students for deeper understanding the problem of Taylor's polynomials in the context of function approximation.

References

1. Edwards, C.H., Penney, D.E.: Calculus, 6th edn. Prentice Hall College Div (1998)
2. Grebencia, M.K., Novosielov, C.I.: A Course of Mathematical Analysis. Moscow (1951) (in Russian)
3. Larson, R.E., Hostetler, R.P., Edwards, B.H.: Calculus, 6th edn. Houghton Mifflin Company (1998)
4. Ruskeepaa, H.: Mathematica Navigator: Graphics and Methods of Applied Mathematics. Academic, Boston (2005)
5. Thomas, G.B., Finney, R.L.: Calculus and Analytic Geometry, 9th edn. Addison-Wesley Publishing Company (1998)
6. Wolfram, S.: The Mathematica Book. Wolfram Media Cambridge University Press (1996)
7. Yakovlev, G.N.: Higher Mathematics. Mir, Moscow (1990)

Zooming Algorithms for Accurate Plotting of Functions of Two Real Variables

David G. Zeitoun and Thierry Dana-Picard

Abstract The study of a real function of two real variables can be supported by visualization using a Computer Algebra System (CAS). One type of constraints of the system is due to the implemented algorithms, yielding continuous approximations of the given function by interpolation. This masks often discontinuities of the given function and its curvature at small scales. It can also provide strange plots, rather inaccurate. In recent years, point based geometry associated with grid approximation has gained increasing attention as an alternative surface representation, both for efficient rendering and for flexible geometry processing of complex surfaces. In this paper we present different visualisation techniques used for 2D plots of a real function and propose two new zooming algorithms for accurate visualisation near discontinuities. First we show the limitations of the classical zooming procedure used in current software, then a mathematical analysis of the zooming process leads to two different treatment of the images. A first algorithm stores representations of the function at different scales, which enables different plots, depending of the screen scale. The second algorithm uses a unique high level grid with quadratic representation. The two algorithms are illustrated and a comparison is performed.

Keywords CAS · Zooming algorithms · Surface plots · Discontinuities · Linear interpolation · Meshing

1 General Frame of Study

With the introduction of the computer into the learning environment the way Mathematics is conveyed has changed. The traditional sequence Definition–Theorem–

D.G. Zeitoun
Orot College of Education, Rehovot, Israel
e-mail: ed.technologie@gmail.com

T. Dana-Picard (✉)
Jerusalem College of Technology, Havaad Haleumi Street, 21, 91160 Jerusalem, Israel
e-mail: ndp@jct.ac.il

I.S. Kotsireas and E. Martínez-Moro (eds.), *Applications of Computer Algebra*, Springer Proceedings in Mathematics & Statistics 198,
DOI 10.1007/978-3-319-56932-1_32

Proof has received a complement with examples where visualization plays an important role. This is true in numerous mathematical domains, such as Geometry, using the so-called Dynamical Geometry Packages, and also Analysis and Algebra with Computer Algebra Systems (CAS). Questions about the study of functions and curve discussion have been studied by [11, 12, 20] and others. In particular, various limitations of the usage of the computer have been discussed. Many educators have replaced the traditional sequence mentioned above by another one, beginning with computer assisted experimentation; see [14] as an example.

A well known case of CAS experimentation prior to theoretical study is curve discussion. It happens that a discontinuity of the function or the non-existence of a limit at some point are not shown by the display, despite the fact that a proof is easy to write. Such a cognitive conflict can appear, in a stronger form, when studying functions of two real variables. In order for the eye to catch the situation, most Computer Algebra Systems enable a dynamical point of view, using the mouse to turn the surface around. This can help to discover discontinuities or points where partial derivatives cannot exist, but how to be sure that the visual impression is correct? As we will see, discontinuities can be hidden.

Functions of two real variables are generally introduced in an Advanced Calculus course, where the students discover generalizations of notions learnt in their first Calculus course. The respective roles of the first and second derivatives are extended in the new frame to discover extrema and saddle points. When arriving at the visualization stage, drawings are harder to obtain by hand-work and computerized help is welcome.

Students learning towards a degree in a STEM related domain learn quite early a course in Advanced Calculus, i.e. a course where the main objects under study are multivariable functions. It happens that students cannot *see* how these objects behave. In particular, difficulties appear in classroom for functions having different limits at a point, according to the path approaching the point. Therefore dynamical visualization techniques are important.

The study of functions of two real variables can be supported by visualization using a Computer Algebra System (CAS). Contour plots were the first type of graphic representations. With the development of scientific computing, 3D plots were introduced and plotting the graph of a two-variable function has been made possible, including parametric plotting and implicit plotting.

Depending both on the hardware and on the software, constraints exist, making sometimes the plots non accurate. For example, the choice of the mesh (using a triangulation of the domain, or geodesics on the surface, etc.) has a great influence on the representation. Actual values of the function are computed on the border of the cells, then interpolations are performed; these interpolations may hide discontinuities. Other constraints come from the need to control the geometric transformations of the plotted surface: transformations such as zooming, rotation around a given axis, displacement along a given path are examples where the visualization device needs to understand the mathematical behavior of the function.

Let us consider the function f given by $f(x, y) = \frac{1}{1-(x^2+y^2)}$. The first plots displayed in Fig. 1 are examples of non-accurate plots, obtained with a brute force usage

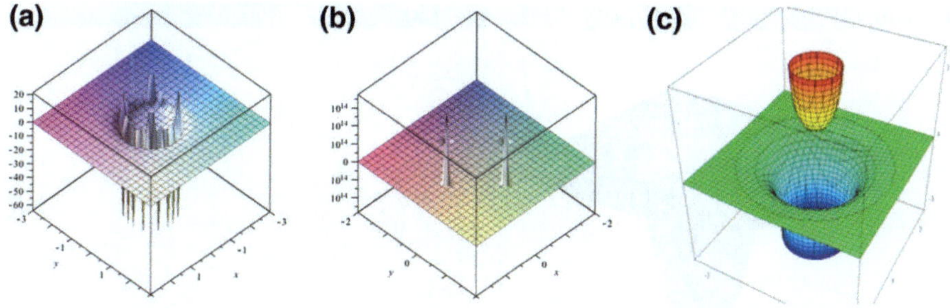

Fig. 1 Strange plots

of commands, without a suitable analysis of the function. The command differed only in the given x-intervals and y-intervals, $[-2, 2]$ in (a) and $[-3, 3]$ in (b). The right-most plot is quite accurate, it uses other techniques.

In the inaccurate plots, interpolations hide the actual discontinuities in different ways. Regular zooming cannot solve this problem, as it inflates the cells but does not recompute the needed numerical data.

Rotation around a given axis often masks discontinuities of the function and can also provide strange plots, not compatible with the hand-made analysis. In recent years, point based geometry has gained increasing attention as an alternative surface representation, both for efficient rendering and for flexible geometry processing of complex surfaces. More sophisticated representations that use lighting effects and virtual reality are available. We analyze the efficiency of the different representations with respect to the mathematical behavior of a function of two real variables.

An interface written in Matlab 14 has been developed for producing different representations of a given explicit function of two variables; see Fig. 2. After having received an analytic expression for the function, the software produces four types of representation, namely:

1. A color contour map of the function.
2. A three dimensional plot of function.
3. A relief terrain map, colors corresponding to the "height" of the function above the (x, y)-plane.
4. A virtual reality scene where the function is a terrain and the user is flying over it.

The first three representations are shown for $f(x, y) = (x^2 - y^2)/(x^2 + y^2)$ in Fig. 3. Note that this function has a discontinuity at $(0, 0)$. The virtual reality scene is shown for the same function in Figs. 4 and 5. Comparison between the different kinds of representation has to be done with respect to the accuracy of:

1. The global plotting of the function.
2. The appearance of the existing discontinuities and the non-appearance of non-existing discontinuities.
3. The visualization of directional derivatives.

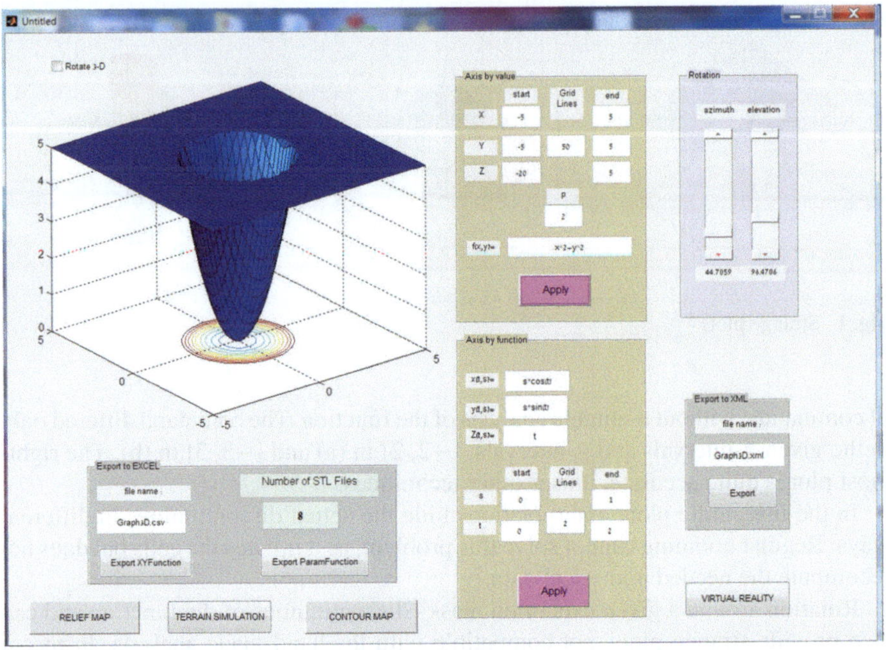

Fig. 2 Interface

(a) Contours (b) 3D plot (c) Terrain relief

Fig. 3 Color plottings

4. The different types of optima, i.e. extrema and saddle points.

Some of the issues that are to be discussed are as follows, all of them having an influence on the joint work of the students and the educator:

- The domain of the given function may be unbounded. Nevertheless, the plotting domain is always bounded; this is one of the constraints mentioned previously.
- Meshing techniques versus isoclines plotting.

This comparison and its outcome were an incitement to the usage of virtual reality in order to represent a function of two variables. In particular, the VR device under development is designed to generate paths on the surface under study and to replace classical zooming with advanced zooming. Zooming algorithms have been developed

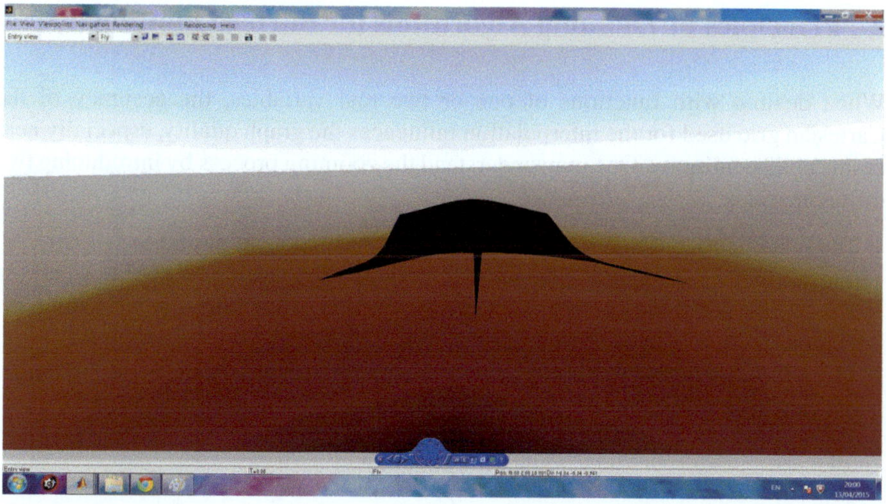

Fig. 4 Virtual scene

Fig. 5 Path along a discontinuity

as an image processing tool that allows to reveal small details by improving the resolution. The purpose of this work is different. Classical zoom inflates the existing cells of the mesh and no re-computation of the data are performed, therefore new details may not be revealed. Such a zooming does not improve the plot accuracy, especially near discontinuities and different types of extrema.

In this paper, we present new algorithms in order to overcome the above problems.

2 Principles of Zooming Algorithms for Surface Plotting

When dealing with functions of one or two real variables, the accuracy of the Cartesian grid used for the interpolation influences the graph quality, especially near points of discontinuity. One may understand the zooming process by introducing two different scales:

- The scales screen. These are a series of scales starting from S_1 and S_i; $i = 1 \ldots n$, that corresponds to length of the screen at a given view. During the zooming process $S_i = \alpha_i S_1$. When we enlarge the view $\alpha_i < 1$ and at the opposite we reduce the view $\alpha_i > 1$.
- The grid size, currently denoted by h; it corresponds to the size of the discretization patches used in the interpolation process.

2.1 *Zooming of a Cartesian Mesh*

When using a cartesian mesh, the basic size of the patch is h; however, the relative size of the mesh with respect to the scale of the screen S_i is $t_i = \frac{h}{S_i}$. Also, the size of the patch in a visualisation of the surface at a given screen size is $t_i = \frac{h}{S_i}$ (and not h). Then the general approximation of the view of function $f(x, y)$ at a screen scale S_i is denoted by $f_i(x, y)$. When using linear rectangular patches, the approximation of the function on the basic rectangular patch is linear and is given by the following formula:

$$f_i(x_0 + t_i, y_0 + t_i) = f(x_0, y_0) + t_i \frac{\partial f}{\partial x}|_{(x_0,y_0)} + t_i \frac{\partial f}{\partial y}|_{(x_0,y_0)} + O(t_i^2)$$

In this case, the error in the viewing of the plot at a screen scale S_i is at an order of magnitude $O(t_i^2)$. One can write that the error at a scale screen S_i is given by $\varepsilon_{iL} = K_1 t_i^2$ (index L for linear patches).

When using quadratic rectangular patches, the approximation of the function on the basic rectangular patch is quadratic and is given as follows:

$$\begin{aligned} f_i(x_0 + t_i, y_0 + t_i) = \; & f(x_0, y_0) + t_i \frac{\partial f}{\partial x}|_{(x_0,y_0)} + t_i \frac{\partial f}{\partial y}|_{(x_0,y_0)} \\ & + t_i^2/2 \left[\frac{\partial^2 f}{\partial^2 x}|_{(x_0,y_0)} + \frac{\partial^2 f}{\partial^2 y}|_{(x_0,y_0)} 2 \frac{\partial^2 f}{\partial x} \partial y|_{(x_0,y_0)} + O(t_i^3) \right] \end{aligned}$$

In this case, the error in the viewing of the plot at a screen scale S_i is in an order of magnitude $O(t_i^3)$. One can write that the error at a scale screen S_i is given by $\varepsilon_{iQ} = K_2 t_i^3$ (index Q for quadratic patches).

In a classical zooming algorithm the patch size h stay constant and therefore the relative size of the patch at a screen size S_i is given by $t_i = \frac{1}{\alpha_i} \frac{h}{S_1}$, where S_1 is the initial screen size. During a view enlargement, we have $\frac{1}{\alpha_i} > 1$, therefore the error for linear patches may be very large: $\varepsilon_{iL} = K_1 t_i^2 >> K_1(\frac{h}{S_1})^2$. When the graph is smooth, the function is at least continuous, then the view is approximately correct. However, when the function has a discontinuity the plot is not mathematically correct. Note also that for small view enlargement, the mathematical correctness of the view is not altered.

When using quadratic patches, the error at a scale screen S_i is given by $\varepsilon_{iQ} = K_2 t_i^3 = K_2(\frac{1}{\alpha_i})^3(\frac{h}{S_1})^3$. In this case, most of the enlargement may have an accurate representation. In order to understand this point, consider the case where $\alpha_i = \frac{1}{2^i}$ and the mesh is chosen such as $\frac{h}{S_1} = 2^{-p}$. Then for $i = p$, the screen view contains only a basic patch of the surface. Therefore,we will consider enlargement processes for $1 \le i \le p - 1$. Now suppose that we are working at a resolution of 2^{-q}, then the quadratic interpolation will be correct till $i = p - \frac{q}{3}$. However, when using linear interpolation, the interpolation is correct till $i = p - \frac{q}{2}$. For example, for $q = 6$, linear patches allow an accurate visualisation till $i = p - 3$, while for quadratic patches, the accuracy of the visualization is till $i = p - 2$ which is almost all the ranges of allowed view enlargement.

In most of the current scientific software, linear patches are used and therefore classical zooming does not permit an accurate visualisation at small scales.

We propose another strategy for a zooming algorithm: to maintain the ratio t_i constant. For example $t_i = \frac{h}{S_1}$. This requires that different grids are stored for different scale screens $h_i = \alpha_i\, h$.

3 Techniques of Visualization of Plots of Surfaces Using Mesh Generators

Triangular meshes are the most common surface representation in computer graphical applications. Because of their simplicity and flexibility, they replace traditional CAD surface representations, like NURBS surfaces (see [23]), in many areas where processing performance is an important issue. The reason for this is that triangle meshes are significantly more flexible, since surfaces of any shape and topology can be represented by a single mesh without the need to satisfy complicated interpatch smoothness conditions. The simplicity of the triangle primitive allows for easier and more efficient geometry generation and geometry processing algorithms.

Obviously, since the triangle primitive is mathematically much simpler compared to a NURBS patch, more of them have to be used to obtain the same approximation quality. However, if a smooth surface has to be represented by a triangle mesh (a piecewise linear surface), the approximation order is quadratic, i.e., halving the edge lengths reduces the error by a factor of 4 which means the number of triangles is inversely proportional to the approximation error. Hence, even with the weaker asymptotic behavior, a good approximation (for the typical precision requirements in graphics applications) can be achieved with a moderately fine mesh whose vertex density and distribution is adapted to the surface curvature, i.e., to the shape complexity.

Despite their being more flexible than NURBS, triangular meshes can also have restrictions and disadvantages in some special applications. Most algorithms working on triangular meshes require topologically consistent surfaces. As a consequence, manifold extraction or topology cleanup steps are necessary for mesh generation methods (see [2, 3]).

Representing sharp features, like edges or corners in technical data sets, is a well studied problem for triangle meshes. Because the surface is no longer differentiable, the approximation power breaks down to linear order. Moreover, alias artifacts are introduced by insufficient sampling, which cannot be removed by increasing the sampling density (see [23]). In order to remove these artifacts and reduce normal noise, the sampling has to be aligned to the principal curvature directions (see [24]). If surface splats are to represent sharp features, all splats that sample these features have to be clipped against one or two clipping lines (for edges and corners respectively) that are specified in their local tangent frames. Therefore, for 2D plots, an accurate representation may be achieved when building the mesh on the isoclines of the function.

Three-dimensional plot commands represent a real function of two real variables in a three dimensional view by approximating the function on a Cartesian grid. The two dimensional domain is discretized in a series of rectangular lagrangian elements, and on each element, the approximation is used. As an example, the two-dimensional Lagrangian interpolation arising out of linear bases defined on a rectangular element of size $h_x \times h_y$ are built as follows. Denote $\xi = \frac{x}{h_x}$; $\nu = \frac{y}{h_y}$, the four bases functions of the element are given by:

$$\phi_1(\xi, \nu) = \left[\frac{1}{2}(1-\xi)\right]\left[\frac{1}{2}(1-\nu)\right],$$

$$\phi_2(\xi, \nu) = \left[\frac{1}{2}(1+\xi)\right]\left[\frac{1}{2}(1-\nu)\right],$$

$$\phi_3(\xi, \nu) = \left[\frac{1}{2}(1+\xi)\right]\left[\frac{1}{2}(1+\nu)\right],$$

$$\phi_4(\xi, \nu) = \left[\frac{1}{2}(1-\xi)\right]\left[\frac{1}{2}(1+\nu)\right].$$

Referring to the previous section, the error at a scale screen S_i of such an element is given by $\varepsilon_{iL} = K_1 t_i^2$. For a quadratic bases defined on a rectangular lagrangian element, the nine bases functions of the element given by:

$$\text{Corner nodes } \psi_i(\xi, \nu) = \frac{1}{4}[\xi\ \xi_i(1 + \xi\ \xi_i)][\nu\ \nu_i(1 + \nu\ \nu_i)], \quad i = 1, 2, 3, 4$$

$$\text{Side nodes } \xi_j = 0\ \psi_j(\xi, \nu) = \left[\frac{1}{2}[\nu\ \nu_j(1 + \nu\ \nu_j)](1 - \xi^2)\right], \quad j = 5, 6$$

$$\text{Side nodes } \nu_k = 0$$

$$psi_k(\xi, \nu) = \left[\frac{1}{2}[\xi\ \xi_k(1 + \xi\ \xi_k)][(1 - \nu^2)]\right], \quad k = 7, 8$$

$$\text{Interior node } \psi_9(\xi, \nu) = \left[\frac{1}{4}(1 - \xi^2)\right][(1 - \nu^2)].$$

Referring to the previous section, the error at a scale screen S_i of such element is given by $\varepsilon_{iQ} = K_2 t_i^3$.

More sophisticated and accurate interpolations such as cubic, and Hermitian cubic 2-dimensional interpolation functions may be used (see [23], Section 2). These interpolations are often used in computational software. In order to build a graph of a function, the domain is decomposed into a finite number of elements based on a collection of $n + 1$ points $P_0(x_0, y_0), P_1(x_1, y_1), \dots, P_n(x_n, y_n)$ for each element $[x_k, x_{k+1}] \times [y_j, y_{j+1}]$. On this element the function is approximated by a function built in a way similar to what has been described in the previous subsection. For a given function f of the two real variables x and y, the interpolation used may be written as a double sum: a first summation over all the elements of the discretization of the domain, and a second one for the interpolation over the given element:

- For linear element:

$$F_{\text{app}}(x, y) = \sum_{\text{Cells}} \sum_{i=1}^{i=4} f(x_i, y_i)\phi_i\left(\frac{x}{h_{ix}}, \frac{y}{h_{iy}}\right). \tag{1}$$

- For quadratic element:

$$F_{\text{app}}(x, y) = \sum_{\text{Cells}} \sum_{i=1}^{i=9} f(x_i, y_i)\psi_i\left(\frac{x}{h_{ix}}, \frac{y}{h_{iy}}\right). \tag{2}$$

3.1 *Visualization Based on a Parametric Representation of the Curve*

In this type of plotting the Cartesian coordinates are functions of two real parameters $x = x(s, t);\ y = y(s, t)$.[1] Then the discretization of the domain is performed in the

---

[1]Polar coordinates $x = r\cos t;\ y = r\sin t$, where $s \geq 0; t \in \mathbb{R}$, are often used but tens of other kinds of coordinates are available in most of the CAS. We illustrate the case of polar coordinates in Sect. 3.4.

(s, t)-plane. The domain in this plane is separated into a finite number of elements based on a collection of $n + 1$ points

$$P_0(x(s_0, t_0), y(s_0, t_0)), P_1(x(s_1, t_1), y(s_1, t_1)), \dots, P_n(x(s_n, t_n), y(s_n, t_n)) \quad (3)$$

for each cell $[s_k, s_{k+1}] \times [t_j, t_{j+1}]$. On this cell the function is approximated by a function built as above. For a given function f, the interpolation is given as follows:

$$f_{\text{app}}(x, y) = \sum_{\text{Cells}} \sum_{i=1}^{i=4} f(x(s_i, t_i), y(s_i, t_i))\phi\left(\frac{s}{h_{ix}}, \frac{t}{h_{iy}}\right). \quad (4)$$

3.2 Discontinuities and Critical Points of a Function of Two Variables on a Cartesian Mesh

The plot command yields visualization as a surface plotting. The main difference between the two situations described above is in the mathematical treatment of the discontinuity[2] and of the critical points of the function. For a function f defined on a domain D in the (x, y)-plane, local maxima, local minima or saddle points can occur either at boundary points of R, or at interior points (x_0, y_0) of D where the first partial derivatives vanish, i.e. $f_x(x_0, y_0) = f_y(x_0, y_0) = 0$, or at points where f_x or f_y fail to exist.

Here is the core of the strange apparitions in Fig. 4 (the needles): depending on the type of interpolation, the condition

$$f_x(x_0, y_0) = f_y(x_0, y_0) = 0 \quad (5)$$

is different from the condition

$$f_{\text{app}x}(x_1, y_1) = f_{\text{app}y}(x_1, y_1) = 0. \quad (6)$$

Moreover, these extrema conditions may appear at different points $(x_0, y_0) \neq (x_1, y_1)$.

For polynomial quadratic interpolation functions (Hermite and cubic basis functions), these last conditions request the solution of a system of linear equations in order to determine the (possible) local extrema. Such critical points may exist on each discretization cell of the domain. This simple analysis permits to understand why we are viewing plots with local extrema without any connection with the known mathematical behavior of the function: the local extrema plotted by the software

[2]In certain situations, an option has to be added to the command in order to force it to consider the discontinuity.

correspond to extrema of the approximation function and not of the actually given function.

For linear interpolation functions of cubic type (most frequently used in mathematical software), it could be shown that the condition given by Eq. (6) is independent from the discretization steps h_{ix} and h_{iy} for a command based on a Cartesian grid (**plot3d**). In this case, the local extremum of the discretization is obtained on the corner of the elements.

To illustrate (and understand) the creation of needles in the plot of a function of two variables, consider the rational function defined by

$$f(x, y) = \frac{1}{x + y - 1}.$$

A MatLab 7 plot of this function for $-4 \le x \le 4$ and $-2 \le y \le 2$ is displayed in Fig. 6.

The needles appear near the discontinuity line whose equation is $x + y = 1$. In order to understand this effect consider a Cartesian discretization for the square given by $0 \le x \le 2$; $0 \le y \le 2$, using n^2 points. We have $h = h_x = h_y = \frac{2}{n}$. The approximated function for plotting is determined by the values of the points $x = ih$; $y = jh$ and an interpolation function

$$f_{\text{app}}(i, j, h) = \frac{1}{h(i + j) - 1}.$$

Using a linear interpolation, one may estimate the plotting value along a line close to the discontinuity line $x + y = 1$, let's say the line whose equation is $x + y = 1 + \varepsilon$, where ε is an arbitrary small real number. Along the line whose equation is

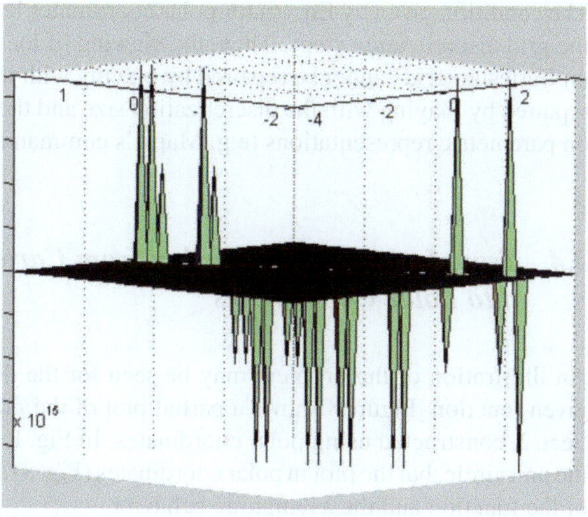

Fig. 6 MatLab plot of $f(x, y) = 1/(x + y - 1)$ with mesh discretization

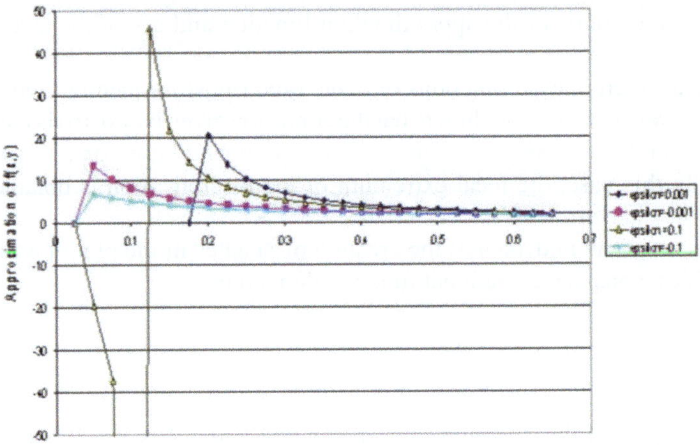

Fig. 7 Numerical solution of $f(x, y) = 1/(x + y - 1)$ along the line whose equation is $x + y - 1 = \varepsilon$

$x + y = 1 + \varepsilon$, the function should be constant and equal to $f_{(x+y=1+\varepsilon)} = \frac{1}{\varepsilon}$; $\varepsilon > 0$. However, because of the Cartesian grid discretization, the plotting software uses approximate values of the function at the relevant corner of the grid. Figure 7 shows plots of the approximated function along a line close to the discontinuity line for constant h and different values of ε.

3.3 *Discontinuities of a Function of Two Variables Using a Parametric Representation of the Surface*

The condition given by Eq. (6) for polar coordinates leads to a solution dependent on the grid discretization steps. Then, the viewing of local extrema on the graph based on Cartesian grid cannot be repaired by playing with the size of the mesh but may be repaired by playing with the discretization size and the domain range for plots based on parametric representations (e.g. Maple's command **parametricplot3d**).

3.4 *Transforming the Question from Cartesian Coordinates into Polar Coordinates*

An illustration of the problem may be seen for the different plots obtained for the given function. Figure 8 shows a partial plot of the graph of the function defined in Sect. 2, constructed using polar coordinates. In Fig. 1, no discontinuity appears near the unit circle, but the plot in polar coordinates (Fig. 8) shows clearly the discontinuity of the function and the asymptotic behavior.

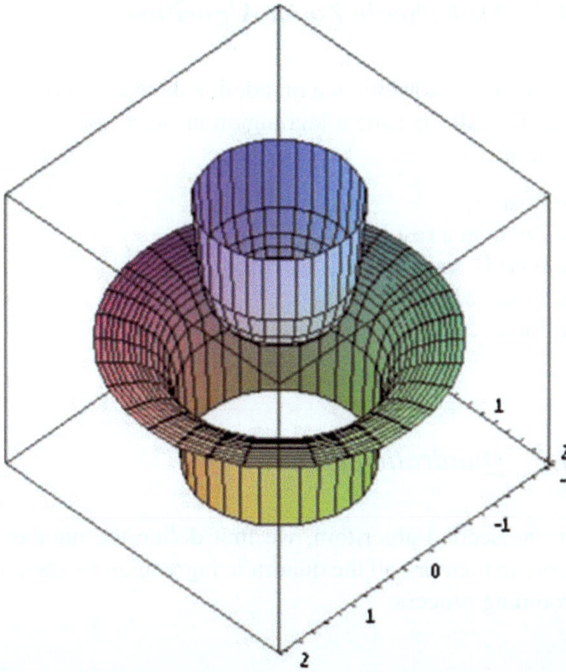

Fig. 8 A plot in polar coordinates

The use of isoclines of the function for the definition of the parametrization of the function is a very efficient solution for the reduction of artifacts. Unfortunately, the determination of isoclines for any function may be very difficult and require special algorithms for automatic meshing.

4 Zooming Algorithms

As mentioned earlier, the basic principle of the different algorithms proposed is to maintain the ratio $\frac{h}{S_1}$ constant during any enlargement process. This requires to work with a finite series of meshes (M_k); $k = 1, \ldots, p$. For the mesh (M_k), the size of the rectangular patch is $h_k = \alpha_k\, h$. The different zoom algorithms proposed here request to store in the memory of the graphic card the series of meshes (M_k); $k = 1, \ldots, p$. Also, the loading of each mesh has to be done interactively. This task is heavy and therefore, the number of stored meshes p has to be maintained as small as possible. In what follows,we propose two similar algorithms: the first one is based on linear lagrangian meshes, while the second one uses quadratic lagrangian mesh.

4.1 Multi Scale Zoom Algorithm

In the first algorithm, we first define the number of stored meshes p_1 and we store in memory all the Linear lagrangian meshes (M_k); $k = 1, \ldots, p_1$. Then the zooming process:

- Load S_1.
- Zoom at a resolution $S_i = \alpha_i \; S_1$.
- Find k_0 such that $\alpha_{k_0} \leq \alpha_i \leq \alpha_{k_0+1}$.
- Load (M_{k_0}).
- View.

4.2 Quadratic Patches

In the second algorithm, we first define the number of stored meshes p_2 and we store in memory all the quadratic lagrangian meshes (M_k); $k = 1, \ldots, p_2$. Then the zooming process:

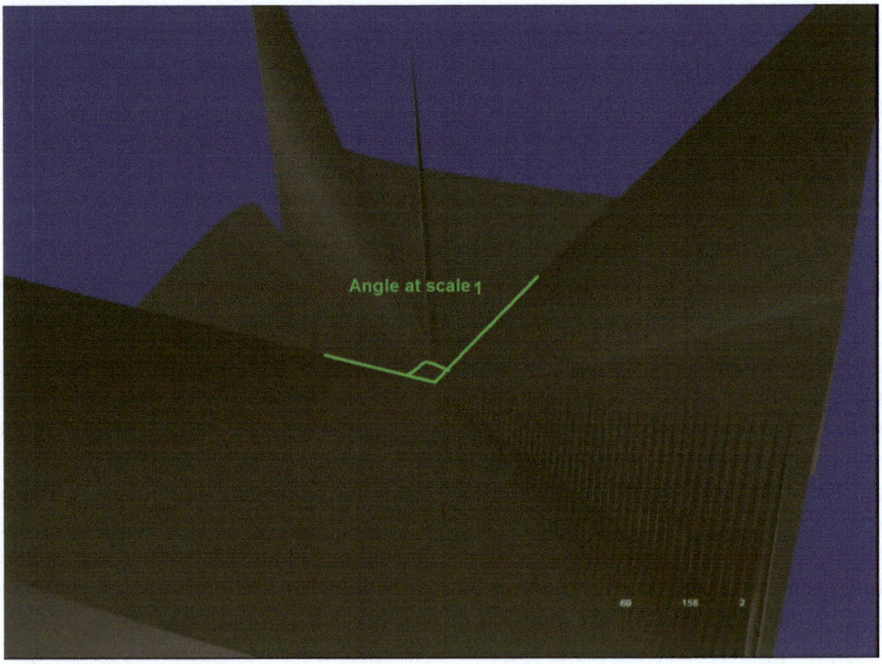

Fig. 9 View of discontinuity at a large scale

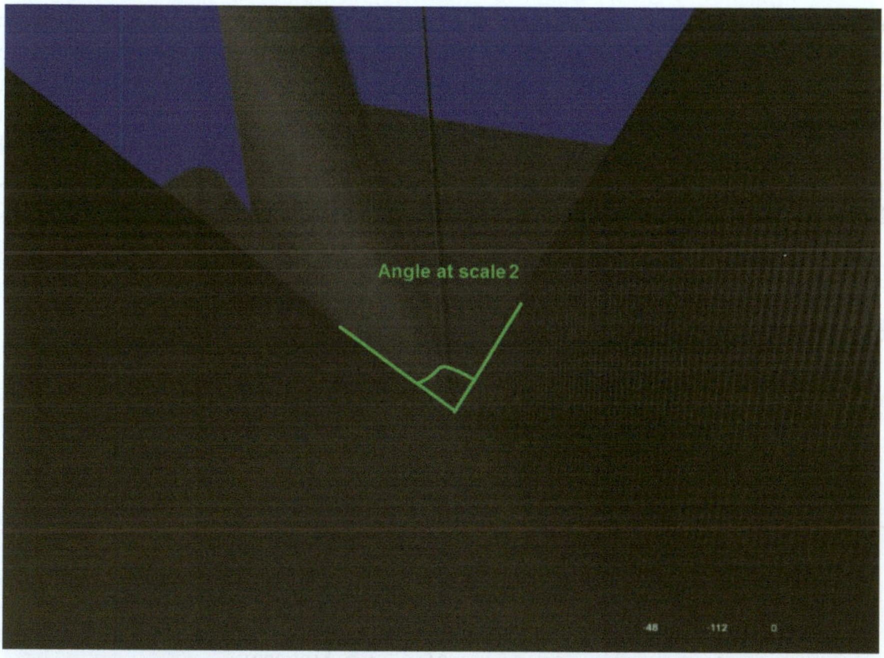

Fig. 10 View of discontinuity at a smaller scale

- Load S_1.
- Zoom at a resolution $S_i = \alpha_i \ S_1$.
- Find k_0 such that $\alpha_{k_0} \ \leq \alpha_i \ \leq \ \alpha_{k_0+1}$.
- Load (M_{k_0}).
- View.

The interest of this second algorithm is that $p_2 < p_1$.

4.3 Viewing

In Figs. 9 and 10 the multi scale zooming is shown around a point of discontinuity of the function. The angle of the view is changing with the scaling view.

5 Conclusion

Today, most of the 3D plot technologies use basic zooming procedure that are an enlargement of the view. This may lead to plots of mathematical surfaces that are not accurate. Moreover, strange plots have been shown when using continuous element at singular points of the function. Zooming algorithm have been developed in image processing for getting accurate edges at low scales. The algorithms used in image processing are based on pixel generation using neighboring rules. When dealing with zooming of a mathematical function the purpose of the zooming is to get an accurate plot of the function at any plots. In this contribution we present a general methodology based on relative scales and lagrangian linear and quadratic elements. The algorithms have already been implemented by the authors in concrete situations.

References

1. Amenta, N., Bern, M., Kamvysselis, M.: A new Voronoi-based surface reconstruction algorithm. In: Proceedings of ACM SIGGRAPH, vol. 98, pp. 415–421 (1998)
2. Avriel, M.: Nonlinear Programming Analysis and Methods. Prentice-Hall (1976)
3. Axelsson, O., Barker, V.A.: Finite Element Solution of Boundary Value Problems, Theory and Computation, Computer Science and Applied Mathematics. Academic Press (1984)
4. Bensabat, J., Zeitoun, D.G.: A least squares formulation for the solution of transport problems. Int. J. Num. Meth. Fluids **10**, 623–636 (1990)
5. Bertsekas, D.P.: Constrained Optimization and Lagrange Multiplier Methods. Academic Press (1982)
6. Botsch, M., Kobbelt, L.: Resampling feature and blend regions in polygonal meshes for surface anti-aliasing. In: Proceedings of Eurograph 01, pp. 402–410 (2001)
7. Bramble, J.H., Nitsche, J.: A generalized Ritz-least squares method for Dirichlet problems. S.I.A.M. J. Num. Anal. **10**(1), 81–93 (1973)
8. Bramble, J.H., Schatz, A.H.: Least squares methods for 2^{mth} order elliptic boundary value problems. Math. Comp. **25**(113), 1–32 (1971)
9. Bristeau, M.O., Pironneau, O., Glowinski, R., Periaux, J., Perrier, P.: On the numerical solution of nonlinear problems in fluid dynamics by least squares and finite element methods. Comp. Methods Appl. Mech. Eng. **17–18**, 619–65 (1979)
10. Chen, Ts.F.: On least squares approximations to compressible flow problems. Num. Meth. P.D.E. **2**, 207–228 (2001)
11. Dana-Picard, Th: Enhancing conceptual insight: plane curves in a computerized learning environment. Int. J. Techn. Math. Educ. **12**(1), 33–43 (2005)
12. Dana-Picard, Th, Kidron, I., Zeitoun, D.: To see or not to see II. Int. J. Techn. Math. Educ. **15**(4), 157–166 (2006)
13. Dana-Picard, Th., Badihi, Y., Zeitoun, D., Israeli, O.: Dynamical exploration of two-variable functions using virtual reality. In: Proceedings of CERME 6, Lyon (France) (2009)
14. Dana-Picard, Th., Zehavi, N.: Automated study of envelopes of 1-parameter families of surfaces (this volume)
15. Dos Santos, S.R., Brodlie, K.W.: Visualizing and investigating multidimensional functions. In: IEEE TCVG Symposium on Visualization, 1–10. Joint EUROGRAPHICS (2002)
16. Eason, E.D.: A review of least squares methods for solving partial differential equations. Int. J. Num. Meth. Eng. **10**, 1021–1046 (1976)
17. Fortin, M., Glowinski, R.: Augmented Lagrangian methods: applications to the numerical solution of boundary value problems. Stud. Math. Appl., **15**, North Holland (1983)
18. Golub, G.H., Van Loan, C.F.: Matrix Computations. The John Hopkins University Press, Baltimore (1984)
19. Jespersen, D.C.: A least squares decomposition method for solving elliptic equations. Math. Comput. **31**(140), 873–880 (1977)
20. Koepf, W.: Numeric versus symbolic computation. Plenary Lecture at the 2nd International Derive Conference, Bonn. http://www.zib.de/koepf/bonn.ps.Z (1995)

21. Kobbelt, L., Botsh, M.: A survey of point-based techniques in computer graphics. Comp. Graph. **28**, 801–814 (2004)
22. Kobbelt, L., Botsch, M., Schwanecke, U., Seidel, H.: Feature sensitive surface extraction from volume data. In: Proceedings of ACM SIGGRAPH, vol. 01, pp. 57–66 (2001)
23. Lapidus, L., Pinder, G.F.: Numerical Solution of Partial Differential Equations in Science and Engineering. Wiley (1982)
24. Levy, B.: Constrained texture mapping for polygonal meshes. In: ACM SIGGRAPH, pp. 12–17 (2001)
25. Sermer, P., Mathon, R.: Least squares methods for mixed-type equations. SIAM J. Num. Anal. **18**(4), 705–723 (1981)
26. Zeitoun, D.G., Dana Picard, Th.: Accurate visualization of graphs of functions of two real variables. Int. J. Comput. Math. Sci. **4**(1), 1–11 (2010)
27. Zeitoun, D.G., Laible, J.P., Pinder, G.F.: An iterative penalty method for the least squares solution of boundary value problems. Num. Methods P.D.E. **13**, 257–281 (1997)

MIX
Papier aus verantwortungsvollen Quellen
Paper from responsible sources
FSC® C105338

If you have any concerns about our products,
you can contact us on
ProductSafety@springernature.com

In case Publisher is established outside the EU,
the EU authorized representative is:
Springer Nature Customer Service Center GmbH
Europaplatz 3, 69115 Heidelberg, Germany

Printed by Libri Plureos GmbH
in Hamburg, Germany